Marine Geology

Marine Geology

Editor: Suzy Bullock

R CALLISTO REFERENCE

www.callistoreference.com

Callisto Reference,
118-35 Queens Blvd., Suite 400,
Forest Hills, NY 11375, USA

Visit us on the World Wide Web at:
www.callistoreference.com

This book contains information obtained from authentic and highly regarded sources. Copyright for all individual chapters remain with the respective authors as indicated. All chapters are published with permission under the Creative Commons Attribution License or equivalent. A wide variety of references are listed. Permission and sources are indicated; for detailed attributions, please refer to the permissions page and list of contributors. Reasonable efforts have been made to publish reliable data and information, but the authors, editors and publisher cannot assume any responsibility for the validity of all materials or the consequences of their use.

ISBN: 978-1-63239-849-9 (Hardback)

The publisher's policy is to use permanent paper from mills that operate a sustainable forestry policy. Furthermore, the publisher ensures that the text paper and cover boards used have met acceptable environmental accreditation standards.

Trademark Notice: Registered trademark of products or corporate names are used only for explanation and identification without intent to infringe.

Printed in the United States of America.

Cataloging-in-publication Data

Marine geology / edited by Suzy Bullock.
   p. cm.
Includes bibliographical references and index.
ISBN 978-1-63239-849-9
  1. Submarine geology. 2. Oceanography. 3. Marine biodiversity. 4. Marine ecology. I. Bullock, Suzy.
QE39 .M37 2017
551.46--dc23

Table of Contents

Preface

Marine geology is an emerging field of study which is highly beneficial in understanding the formation of sea floor and plate tectonics. This field is interdisciplinary in nature as it studies the structure and history of the ocean floor using different branches of study like geophysics, paleontology, etc. It is closely related to physical oceanography. This book is an important source of information as it provides its reader a comprehensive account of marine geology. A number of latest researches have been included to keep the readers up-to-date with the global concepts in this area of study. This book discusses the fundamentals as well as modern approaches of this field. Scientists and students actively engaged in the area of marine geology will find this book full of crucial and explored concepts.

This book aims to highlight the current researches and provides a platform to further the scope of innovations in this area. This book is a product of the combined efforts of many researchers and scientists from different parts of the world. The objective of this book is to provide the readers with the latest information in the field.

I would like to express my sincere thanks to the authors for their dedicated efforts in the completion of this book. I acknowledge the efforts of the publisher for providing constant support. Lastly, I would like to thank my family for their support in all academic endeavors.

Editor

Probability of Detecting Marine Predator-Prey and Species Interactions Using Novel Hybrid Acoustic Transmitter-Receiver Tags

Laurie L. Baker[1]*, **Ian D. Jonsen**[1], **Joanna E. Mills Flemming**[2], **Damian C. Lidgard**[1], **William D. Bowen**[3], **Sara J. Iverson**[1], **Dale M. Webber**[4]

1 Department of Biology, Dalhousie University, Halifax, Nova Scotia, Canada, **2** Department of Mathematics and Statistics, Dalhousie University, Halifax, Nova Scotia, Canada, **3** Population Ecology Division, Bedford Institute of Oceanography, Dartmouth, Nova Scotia, Canada, **4** VEMCO Ltd., Halifax, Nova Scotia, Canada

Abstract

Understanding the nature of inter-specific and conspecific interactions in the ocean is challenging because direct observation is usually impossible. The development of dual transmitter/receivers, Vemco Mobile Transceivers (VMT), and satellite-linked (e.g. GPS) tags provides a unique opportunity to better understand between and within species interactions in space and time. Quantifying the uncertainty associated with detecting a tagged animal, particularly under varying field conditions, is vital for making accurate biological inferences when using VMTs. We evaluated the detection efficiency of VMTs deployed on grey seals, *Halichoerus grypus*, off Sable Island (NS, Canada) in relation to environmental characteristics and seal behaviour using generalized linear models (GLM) to explore both post-processed detection data and summarized raw VMT data. When considering only post-processed detection data, only about half of expected detections were recorded at best even when two VMT-tagged seals were estimated to be within 50–200 m of one another. At a separation of 400 m, only about 15% of expected detections were recorded. In contrast, when incomplete transmissions from the summarized raw data were also considered, the ratio of complete transmission to complete and incomplete transmissions was about 70% for distances ranging from 50–1000 m, with a minimum of around 40% at 600 m and a maximum of about 85% at 50 m. Distance between seals, wind stress, and depth were the most important predictors of detection efficiency. Access to the raw VMT data allowed us to focus on the physical and environmental factors that limit a transceiver's ability to resolve a transmitter's identity.

Editor: João Miguel Dias, University of Aveiro, Portugal

Funding: This study and a graduate student stipend to L. Baker were supported by a research network grant (NETGP 375118 – 08) from the Natural Sciences and Engineering Research Council of Canada (NSERC) for the Ocean Tracking Network; L. Baker was additionally supported by a Dalhousie Graduate Fellowship. Additional support was provided by NSERC Discovery Grants to J. Mills Flemming, D. Bowen and S. Iverson, and by the Department of Fisheries and Oceans Canada. The funders had no role in study design, data collection and analysis, decision to publish, or preparation of the manuscript.

Competing Interests: Dale Webber is affiliated with the commercial company (VEMCO, Ltd) that develop Vemco Mobile Transceivers (VMT). VEMCO Ltd. did not finance the project, and was not directly involved in the the study design and analysis of the paper.

* E-mail: Laurie.Baker@dal.ca

Introduction

Electronic tracking and telemetry data have greatly improved our knowledge about the ecology of many marine species at the individual and population levels [1]. However, few studies have used these methods to investigate the nature of interactions between individual animals. Interactions among conspecifics and between species shape both social and ecosystem structures, and can affect population growth rates, distribution, diversity, and gene flow [2,3]. Studies of predator-prey, competitive and social interactions in marine species have largely been inferred from experiments [4], diet sampling [5], multi-species time series analyses [6,7], or direct observation [8]. These studies are often limited to accessible habitats (e.g. the intertidal, haul out sites) and may not provide insight at the individual level (e.g. time series analysis). Acoustic telemetry can overcome some of these shortcomings by providing information about interactions at the level of individuals from inaccessible marine environments, see

Barnet et al. [9] and Barnet and Semmens [10] who simultaneously tracked predator and prey.

The deployment of dual transmitting and receiving acoustic Vemco Mobile Transceivers (VMT, www.vemco.com) and satellite-linked GPS tags or geolocation tags [11] on large marine vertebrates provides an opportunity to understand species interactions in space and time. The VMT is a hybrid acoustic tag, housing a 69 kHz coded transmitter and a 69 kHz monitoring receiver (similar to the VR2W). Whereas arrays of stationary acoustic receivers are often necessarily confined to continental shelf areas (e.g. [12]), the deployment of VMTs on marine animals provides the ability to extend detection ranges of conspecific and other marine species to biologically interesting regions that may be missed by fixed arrays. The dual transmitter and receiver capabilities of the VMT create a mobile receiving station by which non-surfacing acoustic-tagged organisms, such as fish, can be detected. With these data we have the capacity to better understand the role of predators in ecosystems and to improve our understanding of their interactions with commercial fish stocks and fish species of conservation concern.

To interpret interactions between two organisms we must accurately describe the interaction locations, duration, and frequency. At the most basic level, this relies on knowing whether or not a tagged organism is present. Quantifying the probability of detecting a tag if it is near a given receiver, particularly under changing field conditions, is vital for making accurate biological inferences when using these VMTs (e.g. Argos, [13]; geolocation, [14]). In general, the probability of detecting a transmitter depends on the distance the transmitter is from the receiver, the properties of the medium and transmission (e.g. sound frequency), and the presence of physical obstructions and noise [15]. Sound intensity attenuates with the square of the range according to geometric spreading of the sound in water [15]. Therefore the distance a transmission travels in the ocean depends strongly on the sound frequency of the signal and characteristics of the propagation medium (i.e. sea water composition). Detection probability can also be affected if parts of the transmission are masked by background noise or distorted (e.g. changes in transmission frequency).

Changes in detection efficiency may occur in response to changes in oceanographic and environmental conditions: wind stress [16], [17]; water column stratification [18], [19]; water density [20], [18]; bottom topography [21]. Detection efficiencies have been quantified using a range of approaches: boat based, diver based, fixed sentinel tags, fixed tag with receiver at set distances, post-analysis, single tag at different distance, etc. [22]. While these studies provide valuable data on detection ranges, they cannot fully describe conditions experienced off-shore, and therefore cannot be expected to assess the performance of the VMT when deployed on a free-ranging marine animal. Our case study is distinct from standard acoustic studies, where only the tag is in motion; in our case both the tag and receiver are in motion. The importance of understanding how a tagged marine animal's behaviour affects tag performance is therefore increased. Differences between VMTs may arise because some individuals spend a greater proportion of their time in noisier locations or near complex geomorphology, which may lead to more obstructed transmissions [23] than in other locations. Understanding these behavioural patterns and how they differ seasonally, by age, sex, and physiological state is of the utmost importance.

Pinnipeds are well suited for testing the performance of VMTs. Their frequent return to the surface provides highly accurate GPS locations. Grey seals (*Halichoerus grypus*) fitted with VMTs are known to interact frequently with each other [24], and exhibit high site fidelity, making them easy to recapture to retrieve archived data. Evaluating VMTs when deployed on grey seals provides an opportunity to assess the efficiency of VMTs under realistic behavioural and environmental conditions. Here, we define detection efficiency as how well VMTs are able to detect another VMT transmitter (i.e. with what probability) within a defined range.

We conducted two analyses of detection efficiency of VMTs deployed on grey seals using post-processed detection data (complete transmissions) and summarized raw VMT data (complete and incomplete transmissions), to explore the effect of environmental factors: wind stress, distance between VMTs, and temperature and depth gradients. The raw VMT data consists of a record of all acoustic pings (the smallest sound unit) recorded by the VMT receiver, and differs from the post-processed detection data in that it contains records of incomplete transmissions in addition to complete transmissions (confirmed detections) as well as pings from environmental and anthropogenic sources. Vemco

provided us with summarized raw data for four VMTs consisting of acoustic pings classified by the time intervals between them and summed for each 10-minute period.

We evaluated the detection efficiency of VMTs, using calculated distances (based on GPS locations) between seals to generate a series of instances when detections are likely to have occurred. Access to the summarized raw VMT data allowed us to focus on the physical and environmental factors that limit a receiverability to resolve a transmitteridentity.

Materials and Methods

Ethics Statement

This research was conducted in accordance with guidelines for the use of animals in research [25] and of the Canadian Council on Animal Care. The research protocol for deployment of tags on grey seals was approved by the University Committee on Laboratory Animals, Dalhousie University's animal ethics committee (animal care protocol: 08–088) and the Department of Fisheries and Oceans, Canada (animal care permit: 10–65).

Study Site

The study was conducted between 8 September 2010 and 17 January 2011 on Sable Island, Nova Scotia, Canada (43°55, 60°00 and the Eastern Scotian Shelf in the northwest Atlantic Ocean (Figure 1). Sable Island is an important breeding site for grey seals [26] and the Eastern Scotian Shelf is an important foraging area [24,27].

Study Animals

Seventeen adult grey seals, *Halichoerus grypus* (Fabricius, 1791), selected from a pool of known-age adults were captured between 8 and 18 September 2010 on Sable Island and fitted with a VHF transmitter (164–165 MHz, www.atstrack.com), GPS satellite-linked tag (MK10-AF, www.wildlifecomputers.com) and a VMT according to the methods described in Lidgard et al. [24]. Briefly, the VHF and GPS tags were attached just below the neck to maximize the time the GPS tag spent above water where it could record the satellites in range. The VMT was attached to the lower back of the seal to increase the time the VMT spent in the water transmitting and receiving detections and to reduce electrical interference with the satellite tag. The GPS tag was programmed to collect light intensity, depth (m), and temperature (°C) every ten seconds and to record a GPS location every 15 minutes. GPS attempts were suspended when the unit was dry more than 20 minutes or when a location had been attained.

Peak sensitivities for hearing in phocids are between about 10 and 50 kHz with a high frequency limit of 100 kHz [28]. It is likely that seals could hear the 69 kHz VMT transmissions, given the power output of the transmitters (146–149 dB re 1 μPa SPL 1 m) [29]. However, we did not observe any differences in behaviour: seals in this study exhibited similar foraging and breeding patterns to seals previously tagged with satellite transmitters without an acoustic tag [27], [30], [31]. Ambient background noise, reflection and refraction of the signal, and habituation to the signal over time, make it unlikely that seals could localize other VMT tagged seals. Individuals were recaptured on Sable Island during the subsequent breeding season (December 2010 to January 2011) and their tags retrieved (median deployment period = 112 d range = 92–121 d).

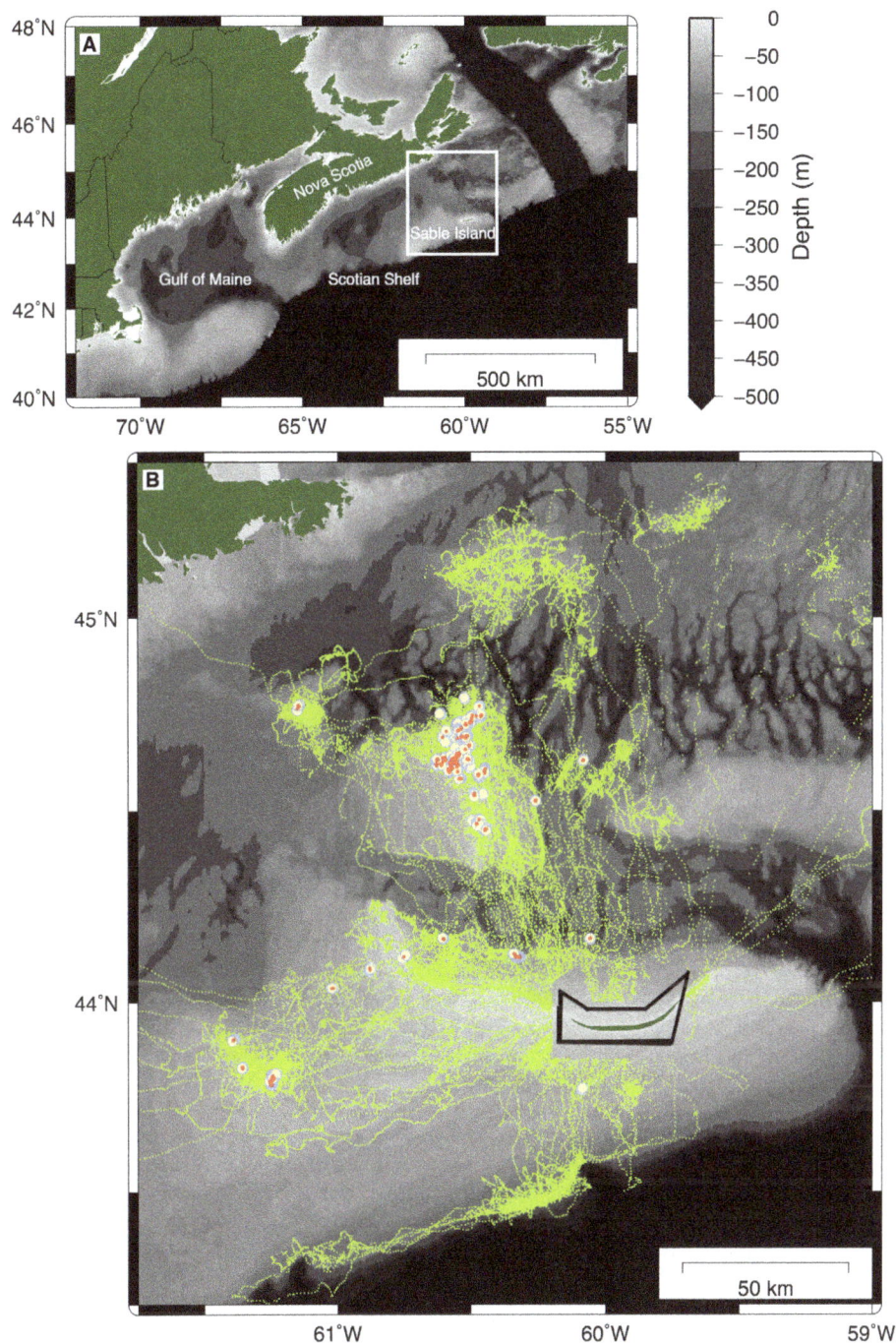

Figure 1. Study area and seal tracks. Nova Scotia and the Scotian Shelf (A) with the study area showing GPS tracks (green) and VMT expected (white) and observed (red) detections (B). The main shallow banks in the region are outlined with their 100 m isobaths (grey).

Post-processed Detection Data vs. Summarized Raw VMT Data

VMTs are coded transmitters, meaning they transmit a sequence of pings that form an acoustic code unique to each individual VMT. VMTs are programmed to transmit an acoustic code on an irregular schedule, every 60 to 180 seconds. During each code transmission the VMT turns off its receiver for approximately 3.5 s, to avoid receiving echos from its own transmission that could interfere with code validation, and records the date and time of the transmission. Each code transmission

comprises a sequence of eight acoustic pings (acoustic code). Each acoustic code begins with a synchronisation interval (sync)– the time between the first two acoustic pings– that identifies the transmission format. The series of acoustic pings that follows creates the unique identification code; the interval between each of the eight acoustic pings creates the unique identification code (Figure 2). A checksum is applied to the entire acoustic code to identify the legitimacy of the transmission. Hereafter, we use the terms transmission and acoustic code synonymously.

Complete Transmission

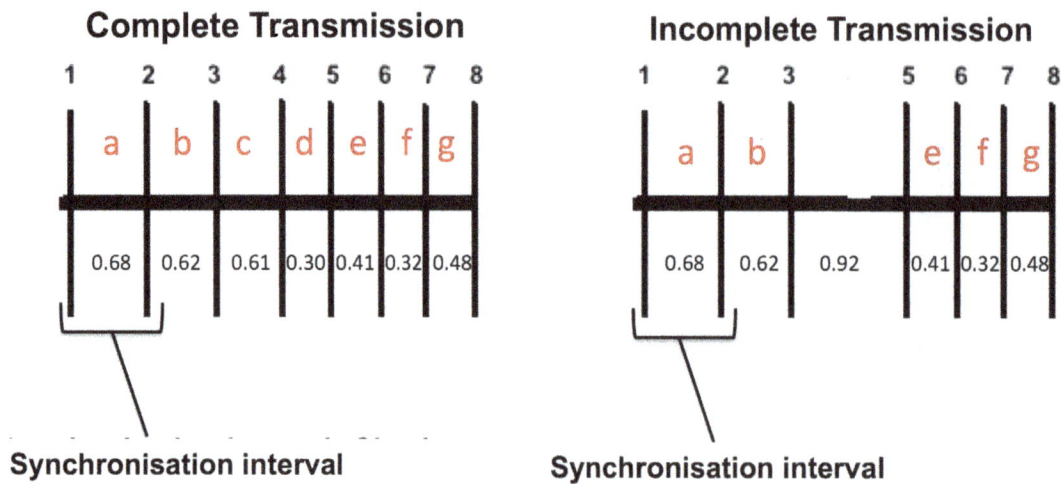

Incomplete Transmission

Synchronisation interval

Synchronisation interval

Figure 2. Complete vs. incomplete transmission. VMT transmissions comprise a series of 8 acoustic pings. Each oustic ping stringontains a synchronization interval (between the first two pings), used to identify acoustic-tag transmission format, followed by a series of pings unique to each individual tag. Intervals between 0.30–0.70 s correspond to consecutive pings. An interval between 0.70–1.50 s may indicate that one ping (of duration 0.01 s) is missing, e.g. time interval of 0.92 s in the incomplete transmission diagram. All 8 acoustic pings must be received for a detection to be recorded.

Post-processed detection data, available to all VMT users, comprises the complete received 69 kHz transmission– which may originate from a VMT or other 69 kHz transmitter– and a daily summary of the total number of acoustic pings, syncs, and rejected false detections. Received complete transmissions (detections), in VMT memory, comprise a date-time stamp and the identities of the transmitting and receiving acoustic tags. False detections are identified by VEMCO, using proprietary software, and removed from the dataset upon VMT retrieval. False detections may result from the collision of codes from other active transmitters that either generate a code that does not exist or an existing code that is known to be present elsewhere (e.g. tags deployed on freshwater species or on non-migratory species in other ocean basins).

The summarized raw VMT data is different from the post-processed detection data in that it includes all acoustic pings received by the transmitter, including those from incomplete transmissions. Acoustic pings may originate from a variety of sources such as other VMTs, acoustic transmitters and abiotic and biotic noise. Acoustics pings originating from VMTs and other VEMCO transmitters may be distinguished from background noise by the signature intervals between each ping in their acoustic codes (Table 1). VMTs are programmed such that consecutive acoustic pings in an acoustic code occur between 0.30 s and 0.70 s. Acoustic pings may also occur at intervals within 0.70 s and 1.50 s in cases where one or more acoustic pings in a code are missing (Figure 2). We therefore defined the range at which probable VMT pings occur as 0.30 s to 1.50 s. Acoustic pings occurring at intervals between 0.26 s and 0.30 s are thought to indicate possible echos, multipath transmissions, or transmission collisions. Acoustic pings occurring at intervals greater than 1.50 s are likely the result of environmental noise or are cases where VMTs are near their acoustic range limit.

Track Data and Expected vs. Observed Detections

We determined GPS locations by analyzing archival GPS data from each tag using software from the manufacturer. To be considered accurate, locations had to be acquired from >5 satellites with a residual error <30 m [32,33].

To link encounters between instrumented seals to locations interpolated at 3 min intervals from the seal tracks, clocks in the VMT and GPS tags were synchronized upon deployment and time corrected upon retrieval based on the respective clock drift calculated from GPS and VMT tags over the deployment time [24]. Distances between seals (m) were calculated from the 3-min interpolated locations.

Each seal's travel rate (m/s) was calculated using the original archival GPS location data. We matched these estimates to the respective transmitting and receiving VMTs using a date-time stamp. We assumed expected detections to occur every 180 s, based on tag specifications (every 60–180 s), when two VMTs encountered each other. We operationally defined an expected encounter as occurring when the VMTs were within 100–700 m of one another. We used 100 m as the lower limit of this range to avoid a decreased probability of detection, which may sometimes occur at close encounter ranges. We used 700 m as the upper limit of our range based on the manufacturers specifications and inspection of our detection data (Figure 3).

Despite being within range of VMTs that recorded data, two VMTs (66487, 66548) failed to record any detections, and one VMT (66494) was only recorded once by another VMT. Closer inspection of the seal tracks associated with these VMTs indicated they were spatially peripheral to the majority of the VMT-tagged seals, but still within range of certain known working VMTs. We

Table 1. Criteria used to determine ping origins.

Interval Length	Description
0.26–0.29 s	Possible echos or multipath transmissions
0.30–0.70 s	Interval range between **consecutive pings**
0.71–1.50 s	Interval range between **1 or more skipped pings**
>1.50 s	Spurious pings or 3 or more skipped pings

*Ping origins deduced from intervals between consecutive pings.

A   Frequency of Observed and Expected Detections

B   Ratio of Observed to Expected Detections

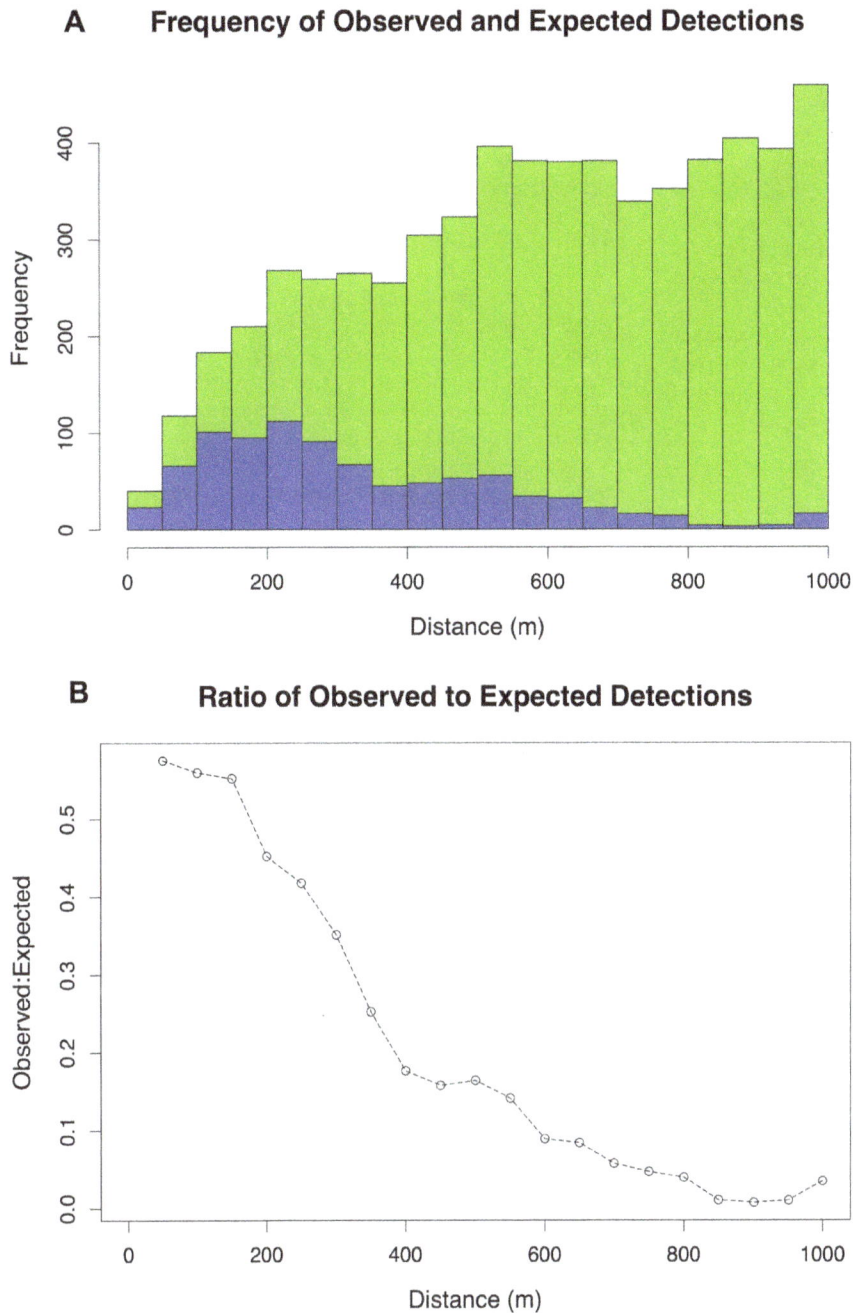

Figure 3. Density and ratio of detections. A. Density of observed (blue) and expected detections (green) with distance. B. Plot of the ratio of observed to expected detections.

excluded these non-functioning VMTs (66487, 66548, 66494). There were also confounding elements that could have affected the summarized raw VMT and post-processed detection data around the VMT deployment point, Sable Island. VMTs do not record signals out of the water; therefore it is important to exclude any periods the seal is out of the water from the analysis. Close to the island, it was difficult to determine if a VMT-tagged seal was out of the water if these durations were shorter than the wet-dry sensors on the GPS tag could detect. Furthermore, due to the shallow bathymetry and thus high noise disturbance around the island, we expected the capability of the VMT to record transmissions to be compromised. Thus, detection data around

Sable Island were removed prior to analyses (see polygon outlined in Figure 1B).

Conversion Efficiency

Vemco provided summarized raw VMT data for four of the VMTs (66556, 66504, 66555, 66541). From these data we calculated the VMT conversion efficiency. Conversion efficiency was defined as the ratio of acoustic pings translated into detections (complete VMT transmissions) to those received (complete and incomplete VMT transmissions, Figure 2).

Statistical Model and Environmental Variables

We used a generalized linear model (GLM) with a negative binomial distribution to model VMT detection and conversion efficiency, where the response variable was the number of observed detections from new encounters in a 12 h period. New encounters were identified as detections (expected or observed) occurring when there was at least a 30 min interval between consecutive detections for a defined pair of seals. The number of expected detections in each 12 h period was included in the model as an offset term to account for the time VMT-tagged seals spent near each other.

Conversion efficiency was evaluated by modelling the number of acoustic pings from complete VMT transmissions (observed detections×8 pings), including the total number of pings from VMTs received (pings occurring at intervals between 0.3–1.5 s) in 10 min intervals as an offset.

Environmental Variables

Environmental variables were selected according to their relevance to sound propagation on the Scotian Shelf and their availability (Table 2). To avoid temporal and spatial scale mismatches, most variables were limited to those that we could collect from the MK 10-AF tags which sampled every 10 seconds and at the seal's exact location. Temperature (°C) and depth gradients (m) between the transmitting and receiving seals were included in the model to test for the effect of water stratification and density changes. The directional (positive or negative) difference in depth and temperature was included because the direction of signal travel with respect to the temperature or depth gradient affects sound transmission differently. Horizontal distance (m) was included in the model to represent detection range.

Wind stress (N/m^2) was included in the model to test the effect of increased noise and changes in the air-sea interface through the introduction of air bubbles. Wind stress (N/m^2) was calculated from hourly estimates of wind speed on Sable Island (Department of Fisheries and Oceans, Canada) in MATLAB (MathWorks, Inc.), using the function stresslp.m (package: air and sea) following Large and Pond [34]. We hypothesized that the effect of noise and/or air bubbles generated by wind stress would be greatest at the surface; we therefore tested for a possible interaction between wind stress (N/m^2) and the depth of the shallowest seal (m) in the model. Seal identity was included as a factor to account for variation in VMT performance and differences in seal behaviour and movement patterns. Travel rate (m/s) was included to describe the seal's horizontal movement rates.

Model Selection

Terms in the model were added and subtracted using forward and backward selection [35]. Variable selection was based on hypothesis testing (p-values) and by comparing the pseudo adjusted R^2 calculated from the residual and null deviance of the model. Residual diagnostics were examined to determine goodness of fit. To explore how sensitive the results were to the subsample distance range, we explored the data subset by distance ranging from 100–250 m, 100–400 m, and 100–700 m. This was done to control for varying amounts of time spent by seals at different distances from one another.

Results

All 17 deployed VMT and GPS tags were recovered from seals upon their return to Sable Island during the breeding season. GPS locations were acquired with a median of 9 satellites (<15 m

residual error). A total of 1,168 detections were recorded, occurring at distances between 4 m and 1880 m (median = 320 m, mode = 250 m). Fewer detections occurred at both close range and and beyond 500 m. 60% of all detections occurred when the VMTs were within 500 m of one another (Figure 3A). We observed a decrease in the proportion of observed vs. expected detections with increased distance (Figure 3B). Only about half of the expected detections were recorded even when two VMT-tagged seals were estimated to be within 50–200 m. At a separation of 400 m, only about 15% of expected detections were recorded. The summarized raw VMT data provided a clearer picture of whether any part of a transmission was received with distance (Figure 4): the ratio of pings from complete transmission to pings from complete and incomplete transmissions fluctuated around 70%, with a minimum of around 40% at 600 m and a maximum of about 85% at 50 m (Figure 4).

Model 1: Expected and Observed Detections

The best model explained 35.7% of the variability in the detection efficiency. The probability of detection decreased with increasing distance between seals (−2.77, SE: 0.64), wind stress (−7.40, SE: 1.87), and depth of the shallowest seal (−0.03, SE: 0.01), (Figure 5).

Model 2: Conversion Efficiency

Wind stress (−1.59, SE: 0.35) and distance (−0.54, SE: 0.14) were both important predictors of conversion efficiency. Conversion efficiency decreased with increasing wind stress and increasing distance (Figure 6). Wind stress had the most significant effect on detection efficiency.

Sensitivity of Detection Efficiency to Distance Range

The results from each data subset were generally consistent with those of the main analyses. When encounters were defined at the 100–400 m range, results were consistent with the main analysis (100–700 m), but when encounters were defined at the 100–250 m range depth of the shallowest seal did not have a significant effect on detection efficiency. The signs and coefficients of model terms were conserved across distance ranges. The pseudo adjusted R^2 values were 19.5%, 28.1%, and 35.72% for the interval ranges: 100–250 m, 100–400 m, and 100–700 m respectively. These changes in explanatory power are likely the result of the increased influence of distance on decreases in detection efficiency.

Discussion

While it is relatively easy to ascertain if a tagged animal is present (true positive), it is more difficult to determine with certainty that it is absent (true negative) as it could be present but

Table 2. Environmental variables explored in VMT efficiency analyses.

Variable	Description
negtempdif	Directional temperature difference (±°C)
mindepth	Depth of the shallowest seal (m)
distance	Horizontal distance between seals (km)
negdepdif	Directional depth difference (±m)
travel rate	Travel rate of the receiving seal (m/s)

*Description of environmental variables tested in VMT efficiency analyses.

A **Received Pings and Pings from Complete Transmissions with Distance**

B **Ratio of Pings from Complete Transmissions to VMT Pings**

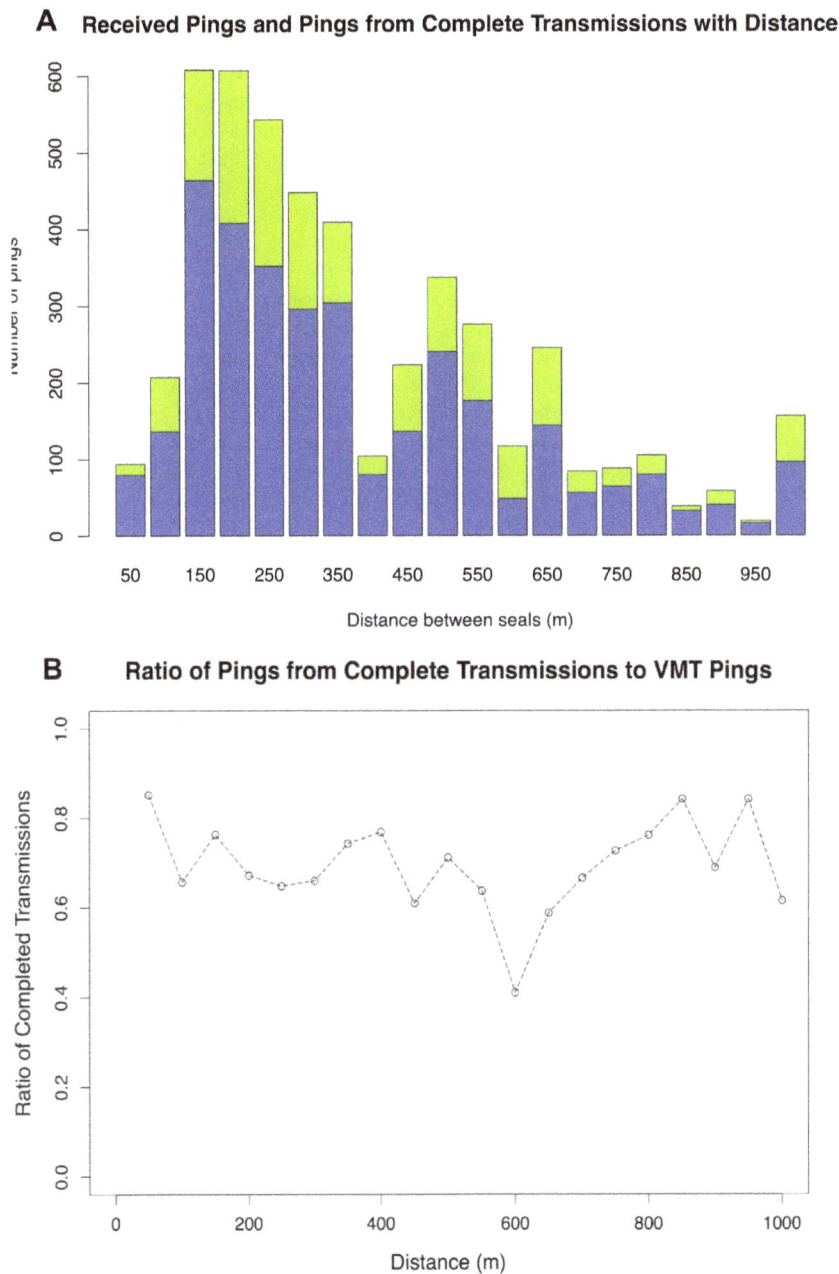

Figure 4. Density and ratio of VMT acoustic pings. A. Density of VMT acoustic pings received (green) and acoustic pings from VMT transmissions (blue) with distance. B. Plot of the ratio of pings from complete transmission to VMT pings received.

not detected (false negative). Quantifying the proportion of VMT transmissions that are not received and determining to what extent this is due to physical and environmental factors and the behaviour of the tagged animals, is vital to form accurate ecological conclusions from VMT data. Without an appreciation of these issues, these effects may lead to erroneous inferences.

We present one of the first studies to investigate the detection efficiency of acoustic VMT receivers deployed on marine animals and to analyze detection efficiency using summarized raw VMT data. Wind stress, depth of the shallowest seal, and distance between seals were significantly correlated with VMT performance. The summarized raw VMT data allowed us to determine the extent to which within-range VMTs are successfully detected

and provided a clearer picture of whether any part of a VMT transmission is received. The ratio of VMT pings from complete transmissions to VMT pings received fluctuated around 70% with a minimum of around 40% at 600 m and a maximum of about 85% at 50 m. This shows a vast improvement when compared with at best 50% of expected detections received between 50–200 m, dropping to 15% at 400 m when using only the post-processed detection data. Examining conversion efficiency (the ratio of complete transmissions to all transmissions received) provides additional insight into VMT detection efficiency by focusing on factors that limit a transceiver's ability to resolve a transmitter's identity.

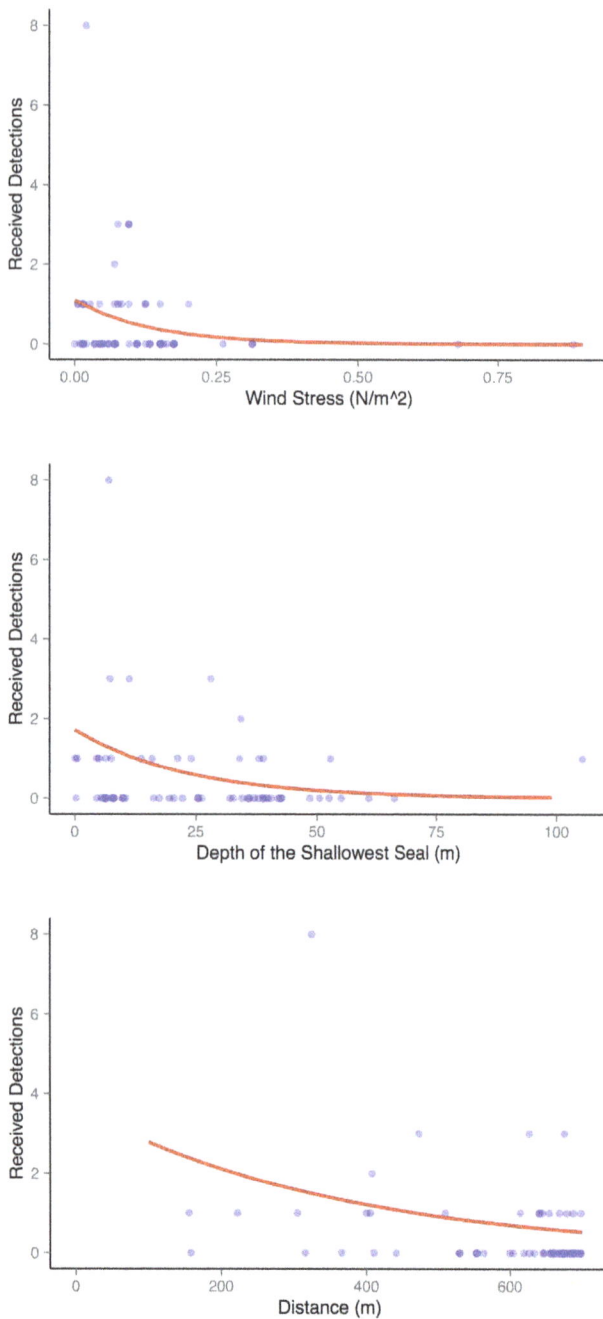

Figure 5. Factors affecting detection efficiency. The predicted effect on detection efficiency of the significant variables (red line): wind stress, minimum depth, and distance. Fitted values (observed detections offset by expected detections) as points. Points: dark blue indicates high intensity, light blue indicates low intensity.

To date, GPS tags provide the best location estimates for *in situ* studies of this nature. GPS locations were obtained with a small residual error (<15 m) [32], resulting in little uncertainty in the GPS locations and subsequently, little uncertainty in the actual detection distances observed. Therefore, although it is possible for the seals to be 60 m closer or further away than that reported, the chance of this occurring are low.

Environmental Factors Affecting VMT Performance

Distance between seals was a significant predictor of detection and conversion efficiency. In both cases, the probability of detection or conversion decreased with distance as expected. Detection range has long been identified as an important factor affecting the detection of acoustic tags [20]. Detection probability is hypothesized to decline proportionally to the decline in sound intensity, which is a combination of geometric and exponential decline due to sound spreading and attenuation resulting from water viscosity [15]. However, the exact shape of this relationship is unknown and modelling approaches vary. We were unable to resolve the shape of this relationship from our data due to its observational nature. However, results from our sensitivity analysis illustrate that the detection range, assumed *a priori*, did not affect the relationships observed.

We also observed a decrease in detection efficiency and conversion efficiency with increasing wind stress. Wind stress can introduce noise as well as air bubbles into the marine environment. Noise makes it difficult to distinguish the acoustic signal above the background noise and may result in failure to detect one or more of the pings. Air bubbles absorb a sound transmission because the acoustic signal has to pass between water and air. The absence of a significant interaction between wind stress and the depth of the shallowest seal suggests that the effect of wind stress on detection efficiency is not confined to surface waters. The observed decrease in detection efficiency with increasing depth may be indicative of sound attenuation occurring as a result of bathymetric effects [21].

Despite well established effects on sound transmission, we observed no effect of the propagation medium (temperature/depth gradients) on detection efficiency [15]. Sound propagation may be absorbed and deflected when traveling through density gradients (i.e., pycnocline). The coastal currents that transport source waters to the Scotian shelf exhibit strong seasonal cycles as well as significant interannual variability [36]. The Nova Scotia current reaches a peak velocity in winter, transporting low salinity and low temperature water originating in the Gulf of St. Lawrence [37] into the inshore waters. These forces generally result in a low salinity and low temperature signature inshore that is more pronounced during winter months [36]. Temperature and depth gradients are therefore more likely to affect detection efficiency after our deployment period (September–December) from January–March, than during our deployment period.

As animal-borne acoustic telemetry evolves beyond stationary receivers, it is unclear how factors such as the orientation of the VMT with respect to the animal or the size of the animal affect VMT performance. VMTs were placed on the lower back of the seal to maximize the time the VMT spent in the water receiving and transmitting signals. The sealbody might attenuate acoustic signals being transmitted to or received from a certain direction, regardless of VMT positioning. Although this effect has not been formally investigated, it would be extremely difficult to quantify *in situ*. A tri-axial accelerometer could be deployed to measure the sealspeed and VMT orientation, however, these devices also have limitations. Controlled experiments will be needed to investigate the influence of such factors on VMT performance. Other factors known to affect detection efficiency that were not included in our model are biotic and/or anthropogenic noise, e.g. [38], [39]. These, in addition to characteristics of the seals behaviour (e.g., the animalorientation during diving), may account for some of the unexplained variation in the model.

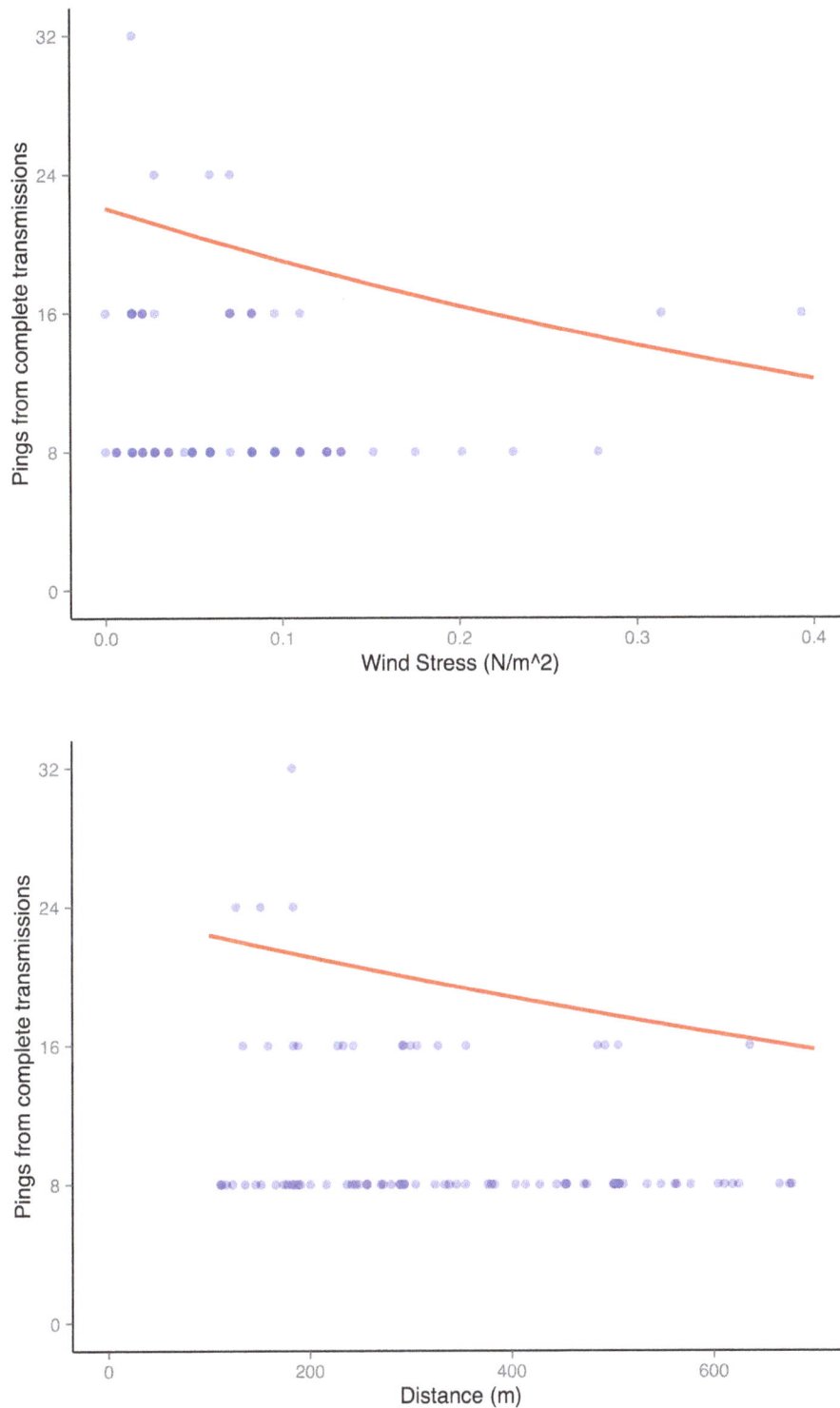

Figure 6. Factors affecting conversion efficiency. The predicted effect on conversion efficiency of the significant variables (red line): wind stress and distance. Fitted values (VMT acoustic pings from complete transmissions offset by total VMT acoustic pings received) as points. Points: dark blue indicates high intensity, light blue indicates low intensity.

VMT Engineering

To interpret interactions we must accurately define their location, duration, frequency, and confidently identify legitimate periods of silence (true negative), i.e. the absence of transmissions. For a detection to occur, the VMT receiver must be able to distinguish the acoustic signal from background noise. The background noise strength is dependent on weather and the fluid environment and other sources, including anthropogenic noise [15]. Distinguishing legitimate transmissions from background noise is an important component of measuring VMT performance. Simpfendorfer et al. [23] used syncs to estimate the volume of received incomplete and complete transmissions for a given period relative to the number of recorded transmissions; however, syncs are not precise. When tag transmissions collide, syncs can be created that are not from a tag transmission; consecutive pings from different tags may create a pseudo sync interval. The use of summarized raw VMT data addresses this shortcoming by utilizing aspects of the transmission that are less susceptible to false positives. With access to the summarized raw VMT data, users can examine the interval between consecutive pings to determine their origin and thus authenticity, i.e. whether the pings arose from echoes, multi-path collisions, environmental noise or are legitimate pings from a VMT.

Observational data in the ocean are often limited due to the technological, environmental, and physical challenges that accompany data collection. These constraints make it important to maximize what can be gleaned from such data. Currently, access to the summarized raw data is not routinely available. Wider access to data of this sort will provide users with an additional indicator of their tag's performance, and inform their analyses through the ability to identify false-negatives. In cases where the identity of the tagged individual is not pertinent, it may be sufficient to simply know that a seal was detected when part of a VMT transmission reached the VMT, even if we cannot account for the factors affecting the VMT transmission.

Without understanding the factors affecting detection efficiency, biological inferences regarding the prevalence and nature of species interactions via VMT/acoustic data will very likely be biased. For example, seasonal changes in environmental factors, that could reduce received transmissions, may be falsely attributed to seasonal changes in interaction rate. It is therefore vital that we account for changes in detection efficiency, as without this information, it is impossible to interpret what any given detection event represents.

Acknowledgments

We are very grateful to Suzanne Budge, Nell den Heyer, Susan Heaslip, Shelley Lang, Elizabeth Leadon, Jim McMillan, Sarah Wong, Rob Ronconi, and Sean Smith, for assistance in the field. S. Kessel, ML Bianchini, and one anonymous reviewer provided valuable feedback on the manuscript and improved its quality.

Author Contributions

Conceived and designed the experiments: DCL WDB IDJ SJI. Performed the experiments: DCL WDB. Analyzed the data: LLB IDJ JEMF. Contributed reagents/materials/analysis tools: DMW. Wrote the paper: LLB IDJ JEMF SJI WDB DCL DMW.

References

1. Cooke SJ, Hinch SG, Wikelski M, Andrews RD, Kuchel IJ, et al. (2004) Biotelemetry: a mechanistic approach to ecology. Trends in Ecology and Evolution 19: 334–343.
2. Whitehead H (2009) Socprog programs: analysing animal social structures. Behavioral Ecology and Sociobiology 63: 765–778.
3. Gorini L, Linnell JDC, May R, Panzacci M, Boitani L, et al. (2012) Habitat heterogeneity and mammalian predator-prey interactions. Mammal Review 42: 55–77.
4. Paine RT (1966) Food web complexity and species diversity. American Naturalist : 65–75.
5. Beck CA, Iverson SJ, Bowen WD, Blanchard W (2007) Sex differences in grey seal diet reect seasonal variation in foraging behaviour and reproductive expenditure: evidence from quantitative fatty acid signature analysis. Journal of Reviews in Ecology 76: 490–502.
6. Worm B, Myers RA (2003) Meta-analysis of cod-shrimp interactions reveals top-down control in oceanic food webs. Ecology 84: 162–173.
7. Lindegren M, Möllmann C, Nielsen A, Brander K, MacKenzie BR, et al. (2010) Ecological forecasting under climate change: the case of Baltic cod. Proceedings of the Royal Society B: Biological Sciences 277: 2121–2130.
8. Cantor M, Whitehead H (2013) The interplay between social networks and culture: theoretically and among whales and dolphins. Philosophical Transactions of the Royal Society B: Biological Sciences 368.
9. Barnett A, Abrantes KG, Stevens JD, Bruce BD, Semmens JM (2010) Fine-scale movements of the broadnose sevengill shark and its main prey, the gummy shark. PLOS ONE 5: e15464.
10. Barnett A, Semmens JM (2012) Sequential movement into coastal habitats and high spatial overlap of predator and prey suggest high predation pressure in protected areas. Oikos 121: 882–890.
11. O'Dor RK, Stokesbury M, Jackson GD (2007) Tracking marine species-taking the next steps. In: Australian Society for Fish Biology 2006 Workshop Proceedings. 6–12.
12. Cooke SJ (2012) Measuring the energetics and physiological status of wild fish using biotelemetry and biologging tools. In: AFS 142nd Annual Meeting. Afs.
13. Jonsen ID, Flemming JM, Myers RA (2005) Robust state-space modeling of animal movement data. Ecology 86: 2874–2880.
14. Winship AJ, Jorgensen SJ, Shaffer SA, Jonsen ID, Robinson PW, et al. (2012) State-space frame work for estimating measurement error from double-tagging telemetry experiments. Methods in Ecology and Evolution 3: 291–302.
15. Medwin H, Clay CS (1997) Fundamentals of acoustical oceanography. Academic Press.
16. How JR, de Lestang S (2012) Acoustic tracking: issues affecting design, analysis and interpretation of data from movement studies. Marine and Freshwater Research 63: 312–324.
17. Gjelland K, Hedger R (2013) Environmental influence on transmitter detection probability in biotelemetry: Developing a general model of acoustic transmission. Methods in Ecology and Evolution 3: 665–674.
18. Finstad B, Økland F, Thorstad EB, Bjørn PA, McKinley RS (2005) Migration of hatchery-reared Atlantic salmon and wild anadromous brown trout post-smolts in a Norwegian fjord system. Journal of Fish Biology 66: 86–96.
19. Singh L, Downey NJ, Roberts MJ, Webber DM, Smale MJ, et al. (2009) Design and calibration of an acoustic telemetry system subject to upwelling events. African Journal of Marine Science 31: 355–364.
20. Heupel MR, Semmens JM, Hobday AJ (2006) Automated acoustic tracking of aquatic animals: scales, design and deployment of listening station arrays. Marine and Freshwater Research 57: 1–13.
21. Kuperman WA, Lynch JF (2004) Shallow-water acoustics. Physics Today 57: 55–61.
22. Kessel ST, Cooke SJ, Heupel MR, Hussey NE, Simpfendorfer CA, et al. (2013) A review of detection range testing in aquatic passive acoustic telemetry studies. Reviews in Fish Biology and Fisheries: 1–20.
23. Simpfendorfer CA, Heupel MR, Collins AB (2008) Variation in the performance of 410 acoustic receivers and its implication for positioning algorithms in a riverine setting. Canadian Journal of Fisheries and Aquatic Sciences 65: 482–492.
24. Lidgard DC, Bowen WD, Jonsen ID, Iverson SJ (2012) Animal-borne acoustic transceivers reveal patterns of at-sea associations in an upper-trophic level predator. PLOS ONE 7: e48962.
25. Anon (2006) Guidelines for the treatment of animals in behavioural research and teaching. Animal Behaviour 71: 245–253.
26. Austin D, Bowen WD, McMillan JI, Iverson SJ (2006) Linking movement, diving, and habitat to foraging success in a large marine predator. Ecology 87: 3095–3108.
27. Breed GA, Bowen WD, McMillan JI, Leonard ML (2006) Sexual segregation of seasonal foraging habitats in a non-migratory marine mammal. Proceedings of the Royal Society B: Biological Sciences 273: 2319–2326.
28. Kastelein RA, Wensveen PJ, Hoek L, Verboom WC, Terhune JM (2009) Underwater detection of tonal signals between 0.125 and 100 kHz by harbor seals (Phoca vitulina). The Journal of the Acoustical Society of America 125: 1222–1229.
29. Bowles AE, Denes SL, Shane MA (2010) Acoustic characteristics of ultrasonic coded transmitters for fishery applications: could marine mammals hear them? The Journal of the Acoustical Society of America 128: 3223–3231.
30. Mellish JE, J IS, D BW (1999) Variation in milk production and lactation performance in grey seals and consequences for pup growth and weaning characteristics. Physiol Biochem Zool 72: 677–690.
31. Lidgard DC, Bowen WD, Jonsen ID, Iverson SJ (2005) State-dependent male mating tactics in the grey seal: the importance of body size. Behavioural Ecology 16: 542–549.
32. Bryant E (2007) 2D location accuracy statistics for Fastloc® cores running firmware versions 2.2 & 2.3. Wildtrack Telemetry Systems Ltd.
33. Hazel J (2009) Evaluation of fast-acquisition GPS in stationary tests and fine-scale tracking of green turtles. Journal of Experimental Marine Biology and Ecology 374: 58–68.

34. Large WG, Pond S (1981) Open ocean momentum flux measurements in moderate to strong winds. Journal of Physical Oceanography 11: 324–336.

35. Zuur AF, Ieno EN, Walker N, Saveliev AA, Smith GM (2009) Mixed effects models and extensions in ecology with R. Springer, 233–236.

36. Smith PC, Schwing FB (1991) Mean circulation and variability on the eastern Canadian continental shelf. Continental Shelf Research 11: 977–1012.

37. Sutcliffe Jr WH, Loucks RH, Drinkwater KF (1976) Coastal circulation and physical oceanography of the Scotian Shelf and the Gulf of Maine. Journal of the Fisheries Board of Canada 33: 98–115.

38. Voegeli FA, Pincock DG (1996) Overview of underwater acoustics as it applies to telemetry, volume 50. Taylor & Francis, 277–300.

39. Thorstad EB, Økland F, Finstad B (2000) Effects of telemetry transmitters on swimming performance of adult Atlantic salmon. Journal of Fish Biology 57: 531–535.

2

A Carapace-Like Bony 'Body Tube' in an Early Triassic Marine Reptile and the Onset of Marine Tetrapod Predation

Xiao-hong Chen[1], Ryosuke Motani[2]*, Long Cheng[1], Da-yong Jiang[3], Olivier Rieppel[4]

1 Wuhan Center of China Geological Survey, Wuhan, Hubei, P. R. China, **2** Department of Earth and Planetary Sciences, University of California Davis, Davis, California, United States of America, **3** Laboratory of Orogenic Belt and Crustal Evolution, Ministry of Education, Department of Geology and Geological Museum, Peking University, Beijing, P.R. China, **4** Center of Integrative Research, The Field Museum, Chicago, Illinois, United States of America

Abstract

Parahupehsuchus longus is a new species of marine reptile from the Lower Triassic of Yuan'an County, Hubei Province, China. It is unique among vertebrates for having a body wall that is completely surrounded by a bony tube, about 50 cm long and 6.5 cm deep, comprising overlapping ribs and gastralia. This tube and bony ossicles on the back are best interpreted as anti-predatory features, suggesting that there was predation pressure upon marine tetrapods in the Early Triassic. There is at least one sauropterygian that is sufficiently large to feed on *Parahupehsuchus* in the Nanzhang-Yuan'an fauna, together with six more species of potential prey marine reptiles with various degrees of body protection. Modern predators of marine tetrapods belong to the highest trophic levels in the marine ecosystem but such predators did not always exist through geologic time. The indication of marine-tetrapod feeding in the Nanzhang-Yuan'an fauna suggests that such a trophic level emerged for the first time in the Early Triassic. The recovery from the end-Permian extinction probably proceeded faster than traditionally thought for marine predators. *Parahupehsuchus* has superficially turtle-like features, namely expanded ribs without intercostal space, very short transverse processes, and a dorsal outgrowth from the neural spine. However, these features are structurally different from their turtle counterparts. Phylogeny suggests that they are convergent with the condition in turtles, which has a fundamentally different body plan that involves the folding of the body wall. Expanded ribs without intercostal space evolved at least twice and probably even more among reptiles.

Editor: Peter Dodson, University of Pennsylvania, United States of America

Funding: This work was supported by the China Geological Survey Project 1212010611603 to Xiao-hong Chen, the China Geological Survey Projects 1212011120148 to Long Cheng, and the National Natural Science Foundation of China Projects 40920124002 and 41372016 to Da-yong Jiang. The funders had no role in study design, data collection and analysis, decision to publish, or preparation of the manuscript.

Competing Interests: The authors have declared that no competing interests exist.

* E-mail: rmotani@ucdavis.edu

Introduction

The modern marine ecological web entails complex interactions among species of multiple trophic levels, from primary producers to apex predators. The relative trophic level of each individual is often measured by a nitrogen isotope fractionation value, $\delta^{15}N$ [1]. The heavier-than-normal isotope accumulates in the body of predators through predation, thus reaching the highest values in the apex predators. The value cannot be compared across a geographic range because the base concentration of ^{15}N depends on the local environment.

It has been observed in the modern marine ecosystem that those predators that feed on marine tetrapods reach higher trophic levels than fish or cephalopod feeders. For example, individuals feeding on tetrapods tend to have higher $\delta^{15}N$ values than fish or squid eaters in both killer whales [2] and great white sharks [3]. This suggest that marine tetrapods as prey are an essential element that supports the highest trophic level in the modern ocean. Then, it is evident that such a high trophic level did not always exist throughout the history of life because marine tetrapods have a limited stratigraphic range. This raises a question of when in

geologic time marine tetrapods as prey species became available, and apex marine predators to feed on them evolved.

Ribs are an essential structure that is common to all vertebrates. They display different morphologies depending on taxonomy, ontogeny, and position along the body axis. Unlike in cervical or sacral ribs, the main bodies of dorsal ribs are largely uniform across taxa, being curved rods with spaces, bridged by intercostal muscles [4,5]. At least some intercostal space persists even in reptiles with expanded dorsal ribs or body armors, such as *Sinosaurosphargis* [6], *Largocephalosaurus* [7,8], cyamodontid placodonts [9], ankylosaurs [10], and *Eunotosaurus* [11], although the spaces may be partly closed.

A notable exception is the turtle, whose costal plate grows from the rib and completely eliminates the intercostal space except in *Dermochelys* and *Odontochelys* [6]. We report here a different lineage of marine reptile that independently eliminated the dorsal intercostal spaces through rib expansion, forming a bony 'body tube' rather than a carapace.

Hupehsuchia [12] is an enigmatic group of marine reptiles that is endemic to the Lower Triassic of Hubei Province, China (ca. 248 million years ago [13,14]). Two monotypic genera are known, namely *Nanchangosaurus* [15] and *Hupehsuchus* [16]. A third genus was suggested in the literature but has not been formally named [12]. The group is known for a suite of unusual features, such as an edentulous and beak-like snout, double-layered neural spines, a heavily ossified skeletal construction, and polydactyly [12,17]. The bizarre body plan of *Hupehsuchus* (Fig. 1B) has led to a controversy about its paleoecology [12,18].

In 2011, Wuhan Centre of China Geological Survey (WGSC hereafter) undertook a field excavation in Yuan'an County, Hubei Province, China to find Early Triassic marine reptiles. The fieldwork resulted in more than ten specimens of marine reptiles, one of which is reported here as a key species to indicate the onset of marine tetrapod predation, as well as a new example of species bearing turtle-like expansion of ribs.

Materials and Methods

Specimens

The specimens observed for the present study are IVPP (Institute of Vertebrate Paleontology and Paleoanthropology, Beijing, China) V3232 (holotype of *Hupehsuchus nanchangensis*) and V4070, WGSC 26004, 26005, and 0940. IVPP V4070 is the specimen that [12] recognized as representing the third genus of Hupehsuchia without formally naming it because of the poor preservation. The WGSC specimens were excavated with proper permit from the Bureau of Land and Resources, China, and are accessioned in the fossil collection at the central facility of WGSC in Wuhan, Hubei Province, China.

Phylogeny

Phylogenetic analysis of Hupehsuchia has never been conducted before because only two named species had been known. We therefore built a new data matrix containing 25 discrete morphological characters for four ingroup and two outgroup taxa. See Text S1 (Supporting Information) for the matrix and character descriptions. The small matrix size allowed branch and bound searches that are guaranteed to find all most parsimonious trees. We used the computer software PAUP*4b10 and TNT 1.1 for tree searches. Bremer support and bootstrap values (n = 1000) were estimated using TNT 1.1.

Nomenclatural Acts

The electronic edition of this article conforms to the requirements of the amended International Code of Zoological Nomenclature, and hence the new names contained herein are available under that Code from the electronic edition of this article. This published work and the nomenclatural acts it contains have been registered in ZooBank, the online registration system for the ICZN. The ZooBank LSIDs (Life Science Identifiers) can be resolved and the associated information viewed through any standard web browser by appending the LSID to the prefix "http://zoobank.org/". The LSID for this publication is: urn:lsid:zoobank.org:pub:0F2EED52-F0A2-4125-B96A-8E39E9854DBE. The electronic edition of this work was published in a journal with an ISSN, and has been archived and is available from the following digital repositories: PubMed Central and LOCKSS.

Results

Phylogenetic Analysis

PAUP*4b10 and TNT 1.1 both found a single most parsimonious tree (Fig. 2), which is unsurprising given the small number of taxa contained in the data matrix. The tree has TL of 28, CI of 0.964, and RI of 0.957. Bremer support for the basal node of Hupehsuchia is 6, indicating that a large number of unique anatomical features are shared by its members. *Parahupehsuchus* forms a clade with IVPP V4070, from which it differs in many morphological characters as described below. This clade also has a robust Bremer support, with a value of 4.

The data matrix suggests that IVPP V4070 is diagnostic at least to the species level. However, we refrain from naming it, following

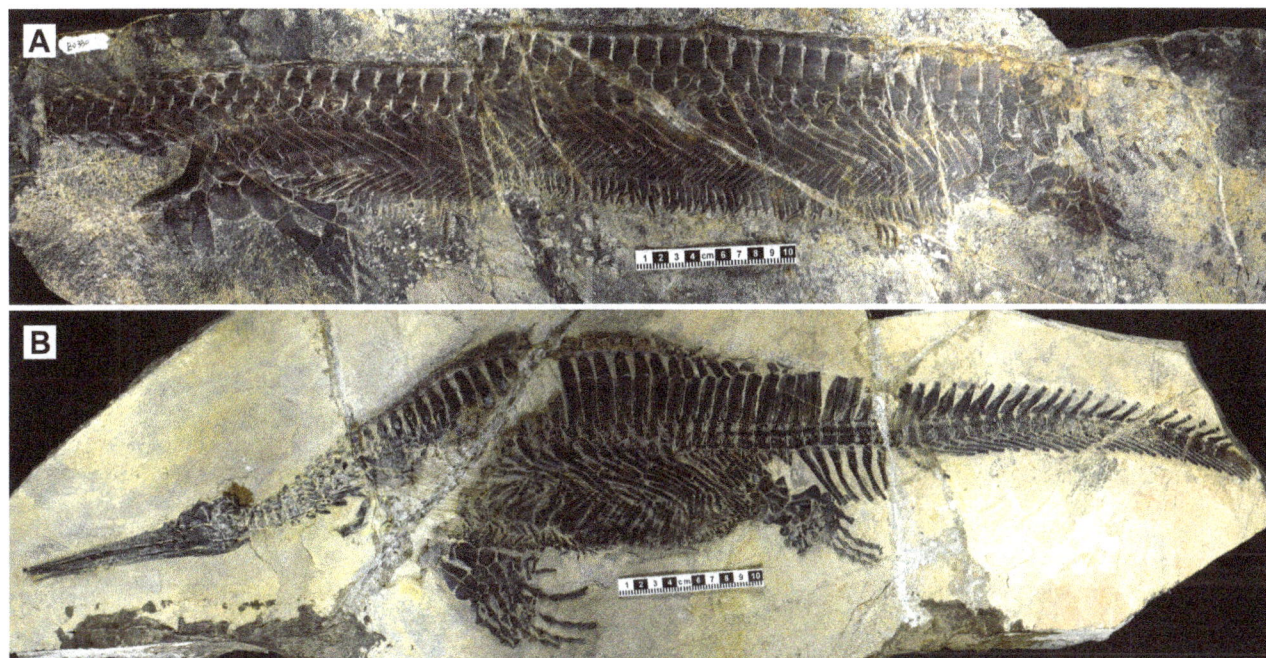

Figure 1. Holotype of *Parahupehsuchus longus* (WGSC 26005) and a specimen of *Hupehsuchus nanchangensis* (WGSC 26004). (A), whole view of WGSC 26005. (B), whole view of WGSC 26004. Scales are 10 cm long.

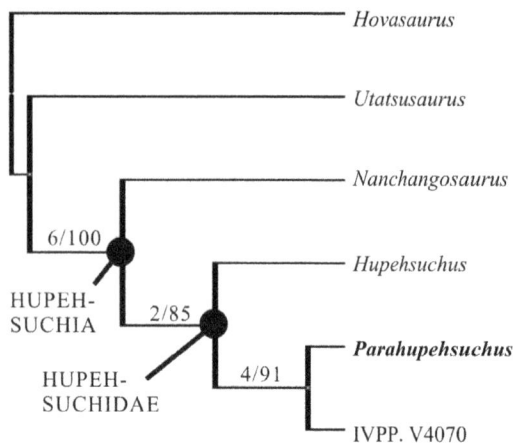

Figure 2. A phylogenetic hypothesis of hupehsuchian relationships. The tree is the single most parsimonious tree (TL = 28, CI = 0.964, RI = 0.957), given the small data matrix. Numbers are Bremer support/bootstrap (n = 1000) values. *Parahupehsuchus* is derived within a well-defined Hupehsuchia. See Text S1 for the data matrix.

the wisdom of [12]–the specimen is largely composed of natural molds of bone elements that are not always well defined. We will name the species in the future when describing an additional specimen that is probably conspecific with IVPP V4070.

Systematic Paleontology

Systematic hierarchy.

Reptilia Laurenti, 1768 [19].

Diapsida Osborn, 1903 [20].

Hupehsuchia Carroll and Dong, 1991 [12].

Revised diagnosis. Snout elongated, flat, and edentulous; humerus with anterior flange; radiale larger than other proximal carpals; presacral vertebral count exceeding 36; first segments of posterior dorsal neural spines without interspinal space; posterior flange of rib present at least proximally; lateral gastralia boomerang-shaped, pointing anteriorly, with short side directed medially; anterior flange of lateral gastralia overlapping adjacent gastralia.

Hupehsuchidae Young, 1972.

Revised Definition. The last common ancestor of *Hupehsuchus* and *Parahupehsuchus*, and all of its descendants.

Revised Diagnosis. Second neural spine segment in anterior dorsal region; third layer of dermal armor in dorsal region.

Type genus. *Hupehsuchus* Young, 1972.

Parahupehsuchus longus gen. et sp. nov.

urn:lsid:zoobank.org:act:0B2F1D4D-0435-496F-B1C0-6913F2 65240F.

Etymology. Generic name is a combination of παρά (Gr. near), hupeh (alternate spelling for Hubei), and ΣοῦΧος (Gr. name for the Egyptian crocodile deity Sobek). Specific name is from λονγοσ (Gr. long).

Holotype. WGSC 26005 (Figs. 1A, 3, 4A–B, 5A–B).

Diagnosis. Dorsal rib with extensive anterior and posterior flanges; dorsal intercostal space absent except near girdles; second rib facet on neural arch for anterior rib; trunk long, with about 38 dorsal vertebrae; ribcage with more or less unchanged dorsoventral depth; proximal carpal/tarsal row with extra element; extra anterior element in each of distal carpal, metacarpal, distal tarsal, and metatarsal rows.

Locality and Horizon. From the upper Spathian (Lower Triassic) Jialingjiang Formation, exposed in Yuan'an County, Hubei Province, China [14].

Description

General design. The preserved length of the skeleton is about 73 cm, of which the trunk makes up about 50 cm. The body is slender, being longer but narrower than in *Hupehsuchus* (Fig. 1). The depth of ribcage is about 65 mm throughout the trunk, giving rise to a 'parallel-sided' ribcage unlike the swollen one in *Hupehsuchus*. The difference in the degree of body elongation is reflected in vertebral count: there are 38 dorsal vertebrae in *Parahupehsuchus*, as opposed to about 28 in *Hupehsuchus* [12] and IVPP V4070. Five cervical, two sacral, and 11 caudal vertebrae are preserved but the cervical and caudal counts are incomplete, preventing comparisons with other hupehsuchians.

Rib. The dorsal ribs are the most peculiar of all bones in the specimen. It has anterior and posterior flanges that span the entire exposed length (Fig. 3). The extensive posterior flange overlies the posterior adjacent rib. The posterior flange also exists in some ribs of *Hupehsuchus* but only proximally. Given that only the external surface is exposed, it is unknown at this point if the medial side of the rib was also flat as the exterior, or whether it was T-shaped in cross-section as in turtles [6] and *Eunotosaurus* [11].

Each dorsal rib articulates with two adjacent vertebrae with a unique configuration (Fig. 3B). The most proximal few centimeters of the ribs are thick and lack both the anterior and posterior flanges. There is a single rib head, which is much broader than the corresponding diapophysis on the neural arch and bears two facets, one proximally and the other posteriorly. The posterior surface is only recognizable in a limited number of ribs because it is ventrally inclined and not obvious in dorsal view. The wide proximal rib facet articulates with the diapophysis (Fig. 3B, dia) that is narrower than itself but its anterior end seems to connect to the parapophysis on the centrum (Fig. 3, para and arf), which has an unusual shape; it has two articular facets, of which the postero-ventral one (Fig. 3B, para) seems to be homologous with the reptilian parapophyses and is almost confluent with the diapophysis–thus, this part of the parapophysis and diapophysis together form a synapophysis. There is an additional facet that stretches antero-dorso-medially from the main facet, forming a band of rough surface (Fig. 3B, arf). This additional facet articulates with the posterior facet of the rib that lies anteriorly. We will refer to this additional facet as the anterior rib facet hereafter. The anterior rib facet is elevated dorsally above the average dorsal margin of the centrum. Whereas the overlapping of ribs alone may have permitted some degree of sliding between ribs, this double articulation must have limited any mobility of these ribs. For the same reason, the longitudinal orientation of the ribs could not have been significantly different from what is preserved, at least proximally; the ribs are preserved perpendicular to the diapophyses, and parallel to the anterior rib facet. Such a double articulation is not known in *Hupehsuchus*. IVPP V4070 is too poorly preserved for the examination of the feature.

The synapophysis and anterior rib facet together form a shallow V-shaped articular surfaces for ribs in both dorsal and lateral views. The V is not tilted in lateral view because of the raised position of the anterior rib facet. Thanks to this configuration, the proximal parts of the ribs were not rotated around their respective axes, i.e., the parasagittal section of the rib flanges was nearly horizontal without pitching; note that this is not a cross-section perpendicular to the rib axis. This allowed the ribs to form a smooth tube in combination.

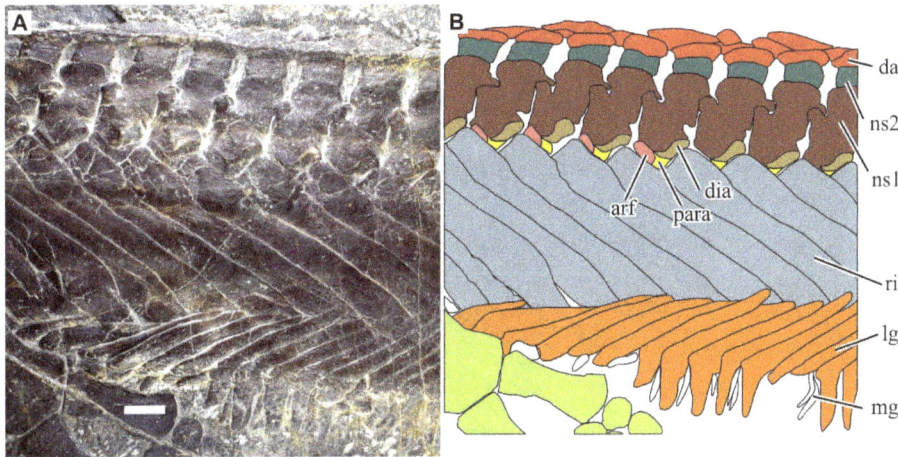

Figure 3. Anterior dorsal region of *Parahupehsuchus longus* **(WGSC 26005).** Bone identifications: arf, anterior rib facet extending from the parapophysis; da, dermal armor; dia, diapophysis of the neural arch; f, forelimb; lg, lateral gastralia; mg, median gastralia; ns1, first segment of neural spine; ns2, second segment of neural spine; para, the main facet of parapophysis; ri, rib. Scale is 1 cm. Note that ribs and gastralia overlap in a complex manner and the double rib articulation prevents rib motion.

Another important implication of the rib morphology is that there was no space for intercostal muscles, which must have been largely absent. Such an absence may explain the reason why the dermal ossicles are not found above the ribs when the animal clearly had a mechanism to form such ossifications. It is possible that thick dermis covered the ribs but there is no anatomical feature preserved to either reject or support such an hypothesis.

Gastralia. The ribs are extensively overlapped by the gastralia distally but it is unclear if the ribs and gastralia articulated with each other. If such an articulation is absent, then the degree of overlap may have been exaggerated through flattening of the body trunk during fossilization. The overlap between the lateral gastral elements and the distal parts of ribs is commonly seen in hupehsuchian specimens that are exposed in lateral view, i.e., those specimens that experienced compaction in bilateral direction during fossilization. However, the lateral gastral elements always lie external to the ribs. Such consistency in preservation posture across specimens is not expected unless at least the distal tip of the lateral gastral elements lay externally to the ribs in life, forming a bony tube.

There are three parts to the gastralia, namely a pair of lateral gastral elements that are flat and boomerang-shaped, and a single median gastral element that is much smaller and V-shaped. The median gastral element is round in cross-section, unlike its lateral

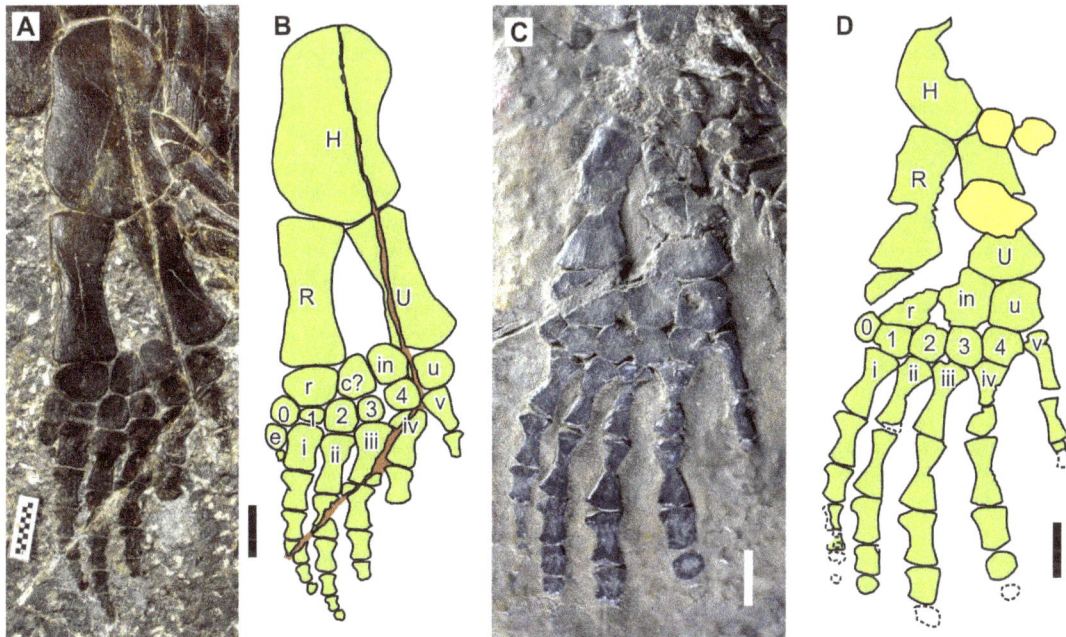

Figure 4. Forelimbs of the holotypes of *Parahupehsuchus longus* **(WGSC 26005) and** *Hupehsuchus nanchangensis* **(IVPP V3232).** (A), left forelimb of WGSC 26005. (B), map of A. (C), left forelimb of IVPP V3232. (D), map of C. Bone identifications: c?, bone identified as centralia by [12]; e, extra anterior metacarpal; H, humerus; in, intermedium; R, radius; r, radiale; U, ulna; u, ulnare; 0–4, distal carpals; i–v, metacarpals. Scales are 1 cm.

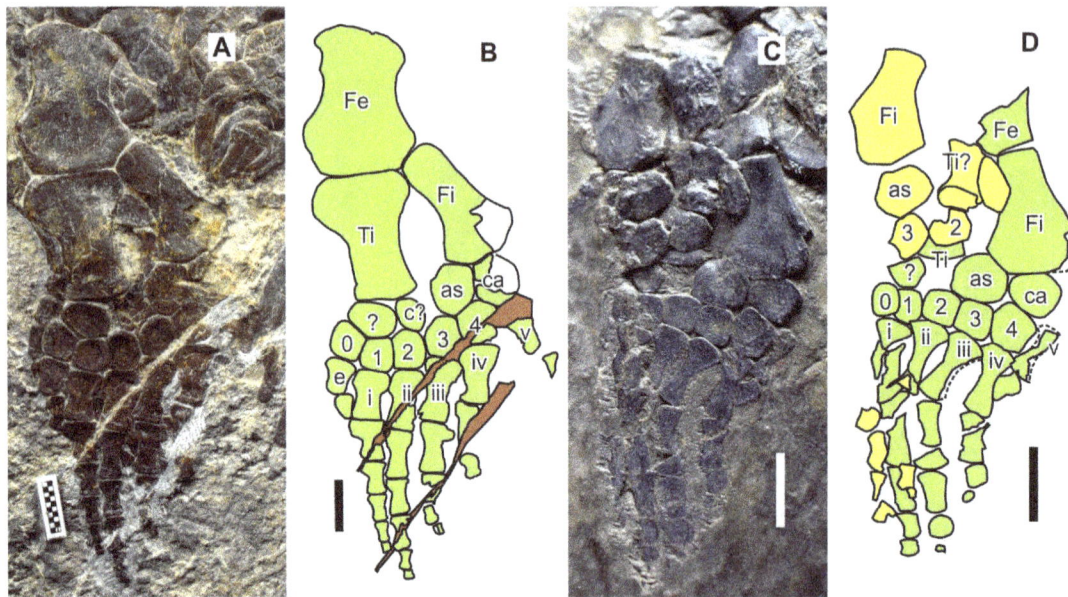

Figure 5. Hind limbs of the holotypes of *Parahupehsuchus longus* **(WGSC 26005) and** *Hupehsuchus nanchangensis* **(IVPP V3232).** (A), left hind limb of WGSC 26005. (B), map of A. (C), left hind limb of IVPP V3232. (D), map of C. Bone identifications: as, astragalus; c?, bone identified as centralia by [12]; ca, calcaneum; e, extra anterior metatarsal; Fe, femur; Fi, fibula; Ti, tibia; 0–4, distal tarsals; i–v, metatarsals, ? suspected neomorph. Scales are 1 cm.

counterpart. The bend of the boomerang of a lateral element is positioned anteriorly whereas the valley of v in a median element is pointing posteriorly. When articulated, the three together form a loose Σ shape in ventral view. *Hupehsuchus* also has a similar condition, with a pair of large lateral and a small median gastral elements. The lateral element was interpreted as the median one by [12].

Different rows of gastralia overlap with each other extensively, with the posterior element positioned externally to the anterior. The way they overlap is in the opposite direction to the pattern of rib overlap, where the posterior element is internal to the anterior one. This counter-overlapping pattern between ribs and gastralia must have further limited the flexibility of the trunk, even if the overlap between ribs and gastralia was less than what is preserved.

Neural spine. Despite the slender trunk, the dorsal neural spines of *Parahupehsuchus* are bipartite (Figs. 1A, 3), with a second segment above the original neural spine as in *Hupehsuchus* [12](Fig. 1B). The second segment is continuous with the first layer of dermal ossicles without a clear suture. Compared to the first segment, the height of the second segment is low, being less than half of the former. Also, the second segment is slightly narrower than the first segment–in *Hupehsuchus*, the base of the second segment is narrower than the top of the first segment only posteriorly in the trunk.

The second segment is already present in the most anterior cervical vertebra in the specimen. It is also present at least in the first six caudal vertebrae–the relevant parts are poorly preserved in more posterior vertebrae. In other words, every well-preserved neural spine in the specimen has a second segment. The second segment of *Hupehsuchus* is limited mostly to the dorsal region [12].

Dermal ossicles. There are up to three layers of dermal ossicles in the trunk (Fig. 3). The first layer extends immediately above the neural spine, occupying the entire width of the latter. These ossicles are somewhat triangular, pointing upward. The space between the first-layer ossicles is occupied by the second-layer ossicles, which are smaller and point downward to fit into the

triangular space between the first layer elements. The third layer ossicles lie above the first two layers. Each third-layer ossicle is larger than the ones below, and usually spans two to three vertebral segments.

As with the second segment of neural spine, dermal ossicles are present throughout the specimen when the relevant part is preserved. Thus, even the most anterior cervical vertebra is associated with the first layer of dermal ossicle, and so are at least the first six caudal vertebrae. The second layer elements are also present as long as there is a gap to fill between a pair of first layer elements. The third layer, however, has a more restricted distribution. The most cranial third layer element is above the eighth to tenth dorsal vertebrae, whereas the most caudad one is above the last dorsal and the two sacral vertebrae. This distribution pattern is very different from the more limited range in *Hupehsuchus*–dermal ossicles are present only between the last cervical and second caudal vertebrae, and the third layer is present between the 13th and last dorsal vertebrae in the genus.

Forelimb. The forelimb is flipper-shaped (Fig. 4A–B), unlike the paddle-shaped forelimb of IVPP V4070 or the polydactylous specimen of [17]. The phalangeal width clearly becomes narrower toward the tip of the manus. This is in contrast with the condition in *Hupehsuchus*, where the width reduction is almost absent (Fig. 4C–D). The manus is slightly longer than the zeugopodium. This again differs from *Hupehsuchus* whose manus is almost twice as long as the zeugopodium.

There is an additional anterior digit (digit '0') with a distal carpal, metacarpal and very small first phalanx. The first phalanx of manual digit 0 is so far unknown in *Hupehsuchus*. The preserved digital formula including this digit is (1)-5-5-3-1-1 but it is likely that distal phalanges are missing from digits 3 to 5 because the most distal elements are still large compared to those of digits 1 and 2. Further preparation of the relevant parts of the fossil using carbide needles, however, did not reveal any additional element. The type specimen of *Hupehsuchus nanchangensis* has a phalangeal formula of (0)-4-4-4-2, so the longest digits are longer in

Parahupehsuchus. The manus of the polydactylous specimen as figured by [17] seems to preserve a phalangeal formula of at least (3)-(4)-5-4-4-5-4, which is clearly different from the present formula in having another additional digit and increased numbers of phalanges posteriorly. The manual phalangeal formula of IVPP V4070 cannot be established with confidence because its forelimbs are incomplete distally.

The arrangement of the proximal carpals is puzzling. There is an extra element between the radiale and intermedium. A similar element in IVPP V4070 was identified as a centrale by [12] and lateral centrale by [17]. However, given that each of proximal and distal carpal rows has an extra element, it is also possible that this proximal element is a neomorph that was derived anteriorly from the intermedium. Also, *Parahupehsuchus* is more derived than *Hupehsuchus* (Fig. 2), which clearly lacks the suspected centrale. See below for further discussion in the section of hind limb.

Manual digits 1 and 2 of *Parahupehsuchus* are more tightly 'bundled' than the rest of the digits and converge distally. A similar bundling is seen in the type specimen of *Hupehsuchus nanchangensis* (Fig. 4C), so the condition is probably natural and not an artifact of preservation, unlike the interpretation presented in figures 5 and 11 of [12]. Such bundling is not obvious in IVPP V4070 or the polydactylous specimen described by [17].

Hind limb. The hind limb of *Parahupehsuchus* closely resembles its forelimb–it is flipper-shaped and its phalangeal width decreases rapidly toward the tip (Fig. 5A–B). Also, there is digit 0 with a distal tarsal, metatarsal, and the first phalanx. This phalanx, however, is larger than in the forelimb. The phalangeal formula is (1)-5-5-4-2-1 but distal phalanges are likely missing from digits 4 and 5. This is similar to the formula for IVPP V4070, which is (0)-4-5-5-?-1 but this latter hind limb is fan-shaped unlike the flipper-shaped hind limb of *Parahupehsuchus*. The phalangeal formula for the pes of *Hupehsuchus nanchangensis* is obscure; the distal end of the hind limb of the holotype (Fig. 5C, D) appears to be incompletely prepared. The preserved phalangeal formula in the polydactylous specimen of [17] is (4)-5-6-6-4-3, which is unique among hupehsuchians in having more than five phalanges in the longest digits (hyperphalangy).

The tibia is wider proximally than distally, as in most marine reptiles. The tibia of *Hupehsuchus* was previously reconstructed to be similar to the fibula [12], but an alternative interpretation may be that the element that was interpreted as the tibia is a laterally-flipped fibula, as in Fig. 5D. Notably, the hind limb of *Parahupehsuchus* is only slightly shorter than the forelimb–in *Hupehsuchus*, the forelimb is much larger than the hind limb (Figs. 1, 4 and 5).

Two additional proximal tarsals exist, rather than one as in the forelimb. Both are located distal to the tibia, with the posterior bone being smaller than the anterior element. IVPP V4070 also has two additional proximal tarsals but their relative size is the opposite of the condition in *Parahupehsuchus* because the posterior element is larger in that specimen [12]. The homology of these two bones is again debatable. The posterior element may be a centrale as suggested by [12] and [17] but that would still leave the anterior element as a neomorph, which is somehow more prominent than the suspected centrale in *Parahupehsuchus*. Given that the condition cannot be explained by involving at least one neomorph, the simplest interpretation may be to identify both of them as neomorphs. This, together with the appearance of the suspected centrale only in the derived member of Hupehsuchia, suggests that the bone may indeed be a neomorph. If so, the extra proximal carpal may also be a neomorph.

Discussion

The body tube of *Parahupehsuchus* provided the trunk with very limited flexibility despite its slender appearance. It undoubtedly restricted possible methods of locomotion. The limbs of *Parahupehsuchus* are too small relative to the body to be the main propulsive organs. The tail of *Parahupehsuchus* is unknown but, given that hupehsuchians generally have tails that are longer than the rest of the body, it is likely that *Parahupehsuchus* also relied on its tail for propulsion. Then, the swimming style of this genus likely resembled that of extant crocodylians, which have a stiff trunk and use the long tail for aquatic propulsion. The steering method, however, may have been different. The flipper shape of the limbs indicate their use as steering device as in many cetaceans, rather than drag-inducing maneuvering device as in the paddles of crocodiles and some aquatic turtles.

Another limitation imposed by the stiff trunk concerns the mechanics of respiration. The dorsal rib of *Parahupehsuchus* cannot rotate or move fore-and-aft because of the skeletal structures. Furthermore, there is no space for intercostal muscles that would move the rib. Therefore, it is impossible to change the volume of the body cavity to produce pressure differentiation for respiration through rib motion, unlike in many tetrapods [21]. Two other mechanisms used by *Crocodylus* to induce pressure differentiation in the chest is the translation of gastralia and pelvic rotation using abdominal muscles, and visceral movement by the diaphragmaticus muscle [22]. Of the two, the gastralia translation also seems impossible in *Parahupehsuchus* given the large overlap between the gastralia and ribs. This leaves the use of diaphragmaticus muscle and visceral movement, also known as hepatic piston [21], as the only alternative. This mechanism is not as important in *Crocodylus* as previously believed [22] but was a major mechanism among dinosaurs [23]. The holotype and only specimen of *Parahupehsuchus longus* does not preserve any positive or negative evidence regarding this interpretation.

The body tube of *Parahupehsuchus* is relevant to two ongoing debates, namely the speed of biotic recovery after the end-Permian mass extinction and the evolution of widely expanded ribs in reptiles, as in turtles. We will discuss them separately below.

Tetrapod Predation and Triassic Recovery

As mentioned in the Introduction, the modern trophic structure in the marine ecosystem did not always exist through geologic time, leaving the question of when it originated. Particularly interesting is the appearance of tetrapod eaters, which defines the highest trophic level in the modern marine ecosystem, together with their tetrapod prey. Tetrapods did not appear until the Carboniferous, and the only truly marine tetrapod in the Paleozoic were the mesosaurids of the Early Permian. This group was endemic to the Irati and White Hill Seas that were enclosed [24,25], and never invaded the open ocean. The appearance of open-ocean reptiles had to wait until the Triassic [26], in the new ecosystem that evolved after the devastating end-Permian mass extinction. The Nanzhang-Yuan'an fauna is one of the best preserved of such earliest marine tetrapod faunas of the Early Triassic.

The stiffened body trunk of *Parahupehsuchus* most likely had an anti-predatory function. The body tube is not a proper carapace because it does not form an outer shell of the body, exposing epaxial, pectoral, and pelvic muscles outside. However, the tube directly protects the internal organs from predators. Moreover, there were few or no intercostal muscles, so much of the trunk lacked exposed muscles that required protection. Furthermore, *Parahupehsuchus* has at least one row of three-layered dermal ossicles above the neural spines, where the external muscle mass is concentrated, suggesting that the dermal ossicles were protecting most of the exposed muscular mass–this marks a clear contrast with saurosphargids, which have a large mass of epaxial muscles overlying the long transverse processes to protect. Therefore, despite the limited extent of dermal ossicles, the body of *Parahupehsuchus* was well protected. The body plan of hupehsuchians in general is toward building a heavily ossified skeleton that would make ingestion and digestion by predators difficult. We interpret the condition in *Parahupehsuchus* as further development of this anti-predation structure. An alternative interpretation for the body tube may be an anti-pressure device for deep diving. However, it is unlikely that hupehsuchian were deep divers. Pachyostosis and increased bone density are common among marine invaders [27,28] and is expected to ballast the body against the movement of water, such as wave actions near the coastline. The added bone mass likely provides negative buoyancy even with air in the lungs. The skeleton of deep divers, in contrast, tends to have less bone mass [28,29] and such histological adaptations as spongy cortex bones [30]. Moreover, a solid body trunk is unnecessary for deep diving tetrapods–they experience various degrees of thoracic collapse during diving [31] except in sea turtles, and their internal organs are adapted to withstand the collapse.

Body protection in hupehsuchians suggests that there was a large predator that lived with these relatively small marine tetrapods. Among the new collection from Yuan'an County is a partial skeleton of a large unidentified eosauropterygian (WGSC 0940). The specimen is estimated to have been about 3–4 meters long. Such a sauropterygian predator would be sufficiently large to bite the trunk of *Parahupehsuchus*. Apart from *Parahupehsuchus*, three additional species of hupehsuchians [12], the ichthyopterygian *Chaohusaurus* [32], and two pachypleurosaurs *Hanosaurus* [33,34] and *Keichousaurus* [14,35], are known in the Nanzhang-Yuan'an fauna. All of them are about 1 m or less in total length–note that marine reptiles tend to have long tails so the trunk is much shorter and narrower in these reptiles than in the marine mammals of the same total length. Unlike heavy-built hupehsuchians, *Chaohusaurus* was lightly built and the pachypleurosaurs had moderately heavy skeletons. Then, there were at least seven species of potential prey marine reptiles with various degrees of body protection, together with at least one large predator. Notably, no fish fossil is known in the Nanzhang-Yuan'an fauna despite the abundance of marine reptiles, narrowing the prey choice for the large sauropterygian. Then, it is most likely that there was predation pressure upon these smaller marine reptiles.

The composition of the Nanzhang-Yuan'an fauna suggests that marine tetrapods potentially suitable as prey already existed in the Early Triassic, together with their predator. Then, a marine trophic structure similar to the modern one was already being established in the late Early Triassic, only about four million years after the end-Permian mass extinction. This timing is earlier than previously suggested [36]. Recovery after the end-Permian mass extinction was probably faster for marine predators than

previously thought [37] to allow the emergence of such a new trophic level that did not exist before the extinction.

Evolution of Rib Expansion

The skeleton of *Parahupehsuchus* shares three similarities with the turtle shell: expanded ribs without intercostal spaces, short transverse processes, and dorsal outgrowth of the neural spines. However, these features have different structures than those of turtles– for example the ribs of *Parahupehsuchus* overlap extensively and the neural spine outgrowth is fused with dermal armor, unlike in turtles. Also, hupehsuchians lack the folding of the body wall that limits the ribs to the axial domain of the body trunk in turtles [6,38], marking a fundamental difference in the body plan. Within Hupehsuchia, extensive rib expansion is known only in *Parahupehsuchus* and IVPP V4070, both of which are derived members (Fig. 2). Moreover, some intercostal spaces still remain in the stem-turtle *Odontochelys* [39]. Therefore, elimination of intercostal space is not homologous between *Parahupehsuchus* and turtles.

It is not impossible that the genetic foundation for rib expansion may be shared between the two without being expressed in intermediate taxa, as has been argued for the body wall folding in *Sinosaurosphargis*, cyamodontid placodonts, and turtles [6]. However, the fundamental difference in body plan due to the lack of body wall folding casts doubt on a close phylogenetic relationships between the two. Histological comparison is unfortunately impossible without damaging the holotype.

The present specimen suggests that rib expansion may not have been as rare among reptiles as previously believed. Overlapping ribs from extreme expansion evolved convergently at least twice and probably more times among reptiles. Most reptile groups with expanded ribs occurred in the marine Triassic of South China, between about 248.5 and 233.5 million years ago [13]. Regardless of whether there is a common genetic mechanism underlying this feature, it was at least expressed separately in each lineage. It is then possible that selection favored the feature because of common environmental factors. Candidates include chemical conditions, such as calcium availability, and biological factors, such as predation pressure. Future studies can test this hypothesis from multiple angles.

Acknowledgments

We thank N. Fraser, N. Kelley, H.-D. Sues, and X-c. Wu for discussions and suggestions. We also thank Ms. Fang Zheng for access to the holotype of *Hupehsuchus nanchangensis* at IVPP.

Author Contributions

Conceived and designed the experiments: XC RM LC DJ. Analyzed the data: RM. Wrote the paper: RM OR DJ. Fossil excavation: XC LC.

References

1. Newsome SD, Clementz MT, Koch PL (2010) Using stable isotope biogeochemistry to study marine mammal ecology. Marine Mammal Science 26: 509–572.

2. Newsome SD, Etnier MA, Monson DH, Fogel ML (2009) Retrospective characterization of ontogenetic shifts in killer whale diets via delta C-13 and delta N-15 analysis of teeth. Marine Ecology Progress Series 374: 229–242.

3. Kim SL, Tinker MT, Estes JA, Koch PL (2012) Ontogenetic and Among-Individual Variation in Foraging Strategies of Northeast Pacific White Sharks Based on Stable Isotope Analysis. PLoS One 7: Article No.: e45068.

4. Romer AS (1956) Osteology of the reptiles. Chicago,: University of Chicago Press. xxi, 772 p. p.

5. Hoffstetter R, Gasc J-P (1969) Vertebrae and ribs. In: Gans C, Bellairs AdA, Parsons TS, editors. Biology of the reptilia Volume I Morphology A. London: Academic Press. pp. 201–310.

6. Hirasawa T, Nagashima H, Kuratani S (2013) The endoskeletal origin of the turtle carapace. Nature Communications 4: Article No.: 2107.

7. Cheng L, Chen XH, Zeng XW, Cai YJ (2012) A new eosauropterygian (Diapsida: Sauropterygia) from the Middle Triassic of Luoping, Yunnan Province. Journal of Earth Science 23: 33–40.

8. Li C, Jiang D, Cheng L, Wu X, Rieppel O (2014) A new species of *Largocephalosaurus* (Diapsida: Saurosphargidae), with implications for the morphological diversity and phylogeny of the group. Geological Magazine 151: 100–120.

9. Scheyer T (2010) New interpretation of the postcranial skeleton and overall body shape of the placodont *Cyamodus hidegardis* Peyer, 1931 (Reptilia, Sauropterygia). Palaeontologia Electronica 13: 15A.

10. Weishampel DB, Dodson P, Osmólska H (2004) The dinosauria. Berkeley: University of California Press. xviii, 861 p. p.

11. Lyson TR, Bever GS, Scheyer TM, Hsiang AY, Gauthier JA (2013) Evolutionary Origin of the Turtle Shell. Current Biology 23: 1113–1119.

12. Carroll RL, Dong Z (1991) *Hupehsuchus*, an enigmatic aquatic reptile from the Triassic of China, and the problem of establishing relationships. Philosophical Transactions of the Royal Society of London Series B-Biological Sciences 331: 131–153.

13. Gradstein FM, Ogg JG, Schmitz MD, Ogg GM (2012) The Geologic Time Scale 2012. Oxford, UK: Elsevier. 1144 p.

14. Li J, Liu J, Li C, Huang Z (2002) The horizon and age of the marine reptiles from Hubei Province, China. Vertebrata Palasiatica 40: 241–244.

15. Wang K (1959) Ueber eine neue fossile Reptiliform von Provinz Hupeh, China. Acta Palaeontologica Sinica 7: 367–373.

16. Young CC (1972) *Hupehsuchus nanchangensis*. In: Young CC, Dong ZM, editors. Aquatic reptiles from the Triassic of China. Peking: Academia Sinica. 28–34.

17. Wu XC, Li Z, Zhou BC, Dong ZM (2003) A polydactylous amniote from the Triassic period. Nature 426: 516–516.

18. Collin R, Janis CM (1997) Morphological constraints on tetrapod feeding mechanisms: why were there no suspension-feeding marine reptiles? In: Callaway JM, Nicholls EL, editors. Ancient Marine Reptiles. New York: Academic Press. pp. 451–466.

19. Laurenti JN (1768) Specimen Medicum, Exhibens Synopsin Reptilium Emendatam cum Experimentis circa Venena. Viennae, Göttingen: Typ. Joan. Thom. Nob de Trattnern. 214 pp, erratum, 215 plaes p.

20. Osborn HF (1903) The reptilian subclasses Diapsida and Synapsida and the early history of the Diaptosauria. New York,: The Knickerbocker Press. p. 449–519 p.

21. Ruben JA, Jones TD, Geist NR, Hillenius WJ (1997) Lung structure and ventilation in theropod dinosaurs and early birds. Science 278: 1267–1270.

22. Munns SL, Owerkowicz T, Andrewartha SJ, Frappell PB (2012) The accessory role of the diaphragmaticus muscle in lung ventilation in the estuarine crocodile Crocodylus porosus. Journal of Experimental Biology 215: 845–852.

23. Schachner ER, Farmer CG, McDonald AT, Dodson P (2011) Evolution of the Dinosauriform Respiratory Apparatus: New Evidence from the Postcranial Axial Skeleton. Anatomical Record-Advances in Integrative Anatomy and Evolutionary Biology 294: 1532–1547.

24. Holz M (1999) Early Permian sequence stratigraphy and the palaeophysiographic evolution of the Parana Basin in southernmost Brazil. Journal of African Earth Sciences 29: 51–61.

25. Rossmann T (2002) Studies on mesosaurs (Amniota inc. sed., Mesosauridae): 3. New aspects on the anatomy, preservation and palaeoecology, based on the specimens from the Palaeontological Institute of the University of Zurich. Neues Jahrbuch Fur Geologie Und Palaontologie-Abhandlungen 224: 197–221.

26. Motani R (2009) The evolution of marine reptiles. Evolution: Education and Outreach 2: 224–235.

27. Houssaye A (2009) "Pachyostosis" in aquatic amniotes: a review. Integrative Zoology 4: 325–340.

28. Wall WP (1983) The Correlation between High Limb-Bone Density and Aquatic Habits in Recent Mammals. Journal of Paleontology 57: 197–207.

29. Debuffrenil V, Collet A, Pascal M (1985) Ontogenetic Development of Skeletal Weight in a Small Delphinid, Delphinus-Delphis (Cetacea, Odontoceti). Zoomorphology 105: 336–344.

30. Debuffrenil V, Mazin JM (1990) Bone-Histology of the Ichthyosaurs - Comparative Data and Functional Interpretation. Paleobiology 16: 435–447.

31. Piscitelli MA, Raverty SA, Lillie MA, Shadwick RE (2013) A Review of Cetacean Lung Morphology and Mechanics. Journal of Morphology 274: 1425–1440.

32. Chen XH, Sander PM, Cheng L, Wang XF (2013) A New Triassic Primitive Ichthyosaur from Yuanan, South China. Acta Geologica Sinica-English Edition 87: 672–677.

33. Young CC (1972) Marine lizard from Nanchang County, Hupeh Province. In: Young CC, Dong ZM, editors. Aquatic reptiles from the Triassic of China. Peking: Academia Sinica. pp. 17–27.

34. Rieppel O (1998) The systematic status of Hanosaurus hupehensis (Reptilia, Sauropterygia) from the Triassic of China. Journal of Vertebrate Paleontology 18: 545–557.

35. Young CC (1965) Onthe new nothosaurs from Hupeh and Kweichou, China. Vertebrata Palasiatica 9: 315–356.

36. Fröbisch NB, Fröbisch J, Sander PM, Schmitz L, Rieppel O (2013) Macropredatory ichthyosaur from the Middle Triassic and the origin of modern trophic networks. Proceedings of the National Academy of Sciences of the United States of America 110: 1393–1397.

37. Chen ZQ, Benton MJ (2012) The timing and pattern of biotic recovery following the end-Permian mass extinction. Nature Geoscience 5: 375–383.

38. Nagashima H, Sugahara F, Takechi M, Ericsson R, Kawashima-Ohya Y, et al. (2009) Evolution of the Turtle Body Plan by the Folding and Creation of New Muscle Connections. Science 325: 193–196.

39. Li C, Wu X, Rieppel O, Wang L, Zhao L (2008) An ancestral turtle from the Late Triassic of southwestern China. Nature 456: 497–501.

3

Edible Crabs "Go West": Migrations and Incubation Cycle of *Cancer pagurus* Revealed by Electronic Tags

Ewan Hunter*, Derek Eaton, Christie Stewart, Andrew Lawler, Michael T. Smith

Centre for Environment, Fisheries and Aquaculture Science, Lowestoft Laboratory, Lowestoft, Suffolk, United Kingdom

Abstract

Crustaceans are key components of marine ecosystems which, like other exploited marine taxa, show seasonable patterns of distribution and activity, with consequences for their availability to capture by targeted fisheries. Despite concerns over the sustainability of crab fisheries worldwide, difficulties in observing crabs' behaviour over their annual cycles, and the timings and durations of reproduction, remain poorly understood. From the release of 128 mature female edible crabs tagged with electronic data storage tags (DSTs), we demonstrate predominantly westward migration in the English Channel. Eastern Channel crabs migrated further than western Channel crabs, while crabs released outside the Channel showed little or no migration. Individual migrations were punctuated by a 7-month hiatus, when crabs remained stationary, coincident with the main period of crab spawning and egg incubation. Incubation commenced earlier in the west, from late October onwards, and brooding locations, determined using tidal geolocation, occurred throughout the species range. With an overall return rate of 34%, our results demonstrate that previous reluctance to tag crabs with relatively high-cost DSTs for fear of loss following moulting is unfounded, and that DSTs can generate precise information with regards life-history metrics that would be unachievable using other conventional means.

Editor: Maura Geraldine Chapman, University of Sydney, Australia

Funding: Department for Environment Food and Rural Affairs (Defra) contract M1103 "Spatial dynamics of edible crabs in the English Channel in relation to management," http://randd.defra.gov.uk/. The funders had no role in study design, data collection and analysis, decision to publish, or preparation of the manuscript.

Competing Interests: The authors have declared that no competing interests exist.

* E-mail: ewan.hunter@cefas.co.uk

Introduction

Routinely obtaining frequent, repeated and accurate estimates of the location of marine animals and an accompanying description of spatial and temporal behaviour patterns has for many years presented marine scientists with a significant technical challenge [1]. This challenge has been partially addressed in recent years for large to medium-sized marine vertebrates (e.g. [2,3]) with the rapid development of biologging technologies [4]. Progress with archival tagging in particular has significantly advanced understanding of the spatial structure and population dynamics of fishes, including smaller, sea-bed dwelling species including flatfish (e.g. [5,6]), gadoids (e.g. [7–9]) and elasmobranchs (e.g. [10,11]), often with application to fisheries management (e.g. [12–14]). Even some invertebrate species have been targeted (e.g. [15,16]). However progress has been slower for smaller, mobile species living at or near the seabed [1].

Obtaining long-term spatial information on Decapod crustaceans such as crabs, often key components in marine ecosystems (e.g. [17,18]) and subject to extensive and commercially valuable fisheries worldwide [19,20]), presents a further, unique set of challenges. Most crabs remain in contact with the sea-bed at all times rather than periodically rising into the water column, as may be the case for many demersal fish (e.g. [21]). Furthermore, crabs are adapted to a reptant ("creeping") lifestyle so it is important that tag attachment does not impede natural behaviours, such as burrowing in sediment and entry to rock crevices. Locally detailed information on habitat use can be gained from acoustic or ultrasonic tracking [22–24], but spatial coverage is limited and precision can be adversely affected by seabed features and can be lost altogether when the animal moves into a crevice or burrow. Electromagnetic telemetry also operates at a local scale over short time periods and studies are also limited by the short range of detection and the requirement for cables on the seabed [25–27].

There are however, aspects of crab morphometrics that offer some potential advantages for archival tagging. For example, tag weight may be less important than for finfish [28,29]. This may permit the use of larger tags with bigger, more powerful batteries and hence increased data storage capacity. Indeed, the utility of archival tagging has already been demonstrated in localised studies of the reef-wide movements of spider crabs [30], and in determining habitat choice behaviour in the Dungeness crab, *Cancer magister* [31,32] in an estuary. However, perhaps the main barrier to the routine use of archival tagging in crustacean studies historically, at least on a broad geographical scale, has been the unit cost of the tags relative to the perceived probability of tag loss due to moulting (i.e. periodic shedding of the exoskeleton as part of the growth process). However, the cost of archival tags has reduced substantially in recent years (from over £1000 per unit 15 years ago to under £300 per unit today, depending on the sensors). By targeting tagging at larger animals (which moult less frequently) early in their inter-moult period, the potential to retrieve large quantities of high quality behaviour data in large-scale tag releases now potentially outweighs the risks associated with tag loss, making the cost of large-scale tagging experiments scientifically and financially more viable.

In U.K. waters, edible crab (*Cancer pagurus*) is one of the most important commercial fisheries, yet there remain several important gaps in our understanding of their biology and ecology. Mark-recapture experiments carried out in the English Channel in the 1970s [33–35] indicated long distance movements, particularly by mature females, and predominantly along an east to west axis. However, in the 40 years since this work was completed [33], total landings in the English crab fishery have effectively quadrupled [19], and the average sea surface temperature in the English Channel has risen by approximately 1°C (www.mccip.org.uk).

In the present study, mature female edible crab were tagged with electronic data storage tags and released at selected locations in some of the most intensively fished crab fisheries in U.K. waters. As far as we are aware, this is the first bulk release of DST-tagged crabs over a wide geographical scale which has been solely reliant upon the commercial fishery for tag returns. With no cohesive current picture of stock identity, our aim was to describe adult crab movements, and to quantify the conditions, timing and duration of reproductive behaviour. Understanding how, where and when crabs undergo large scale migrations is the key to successful stock assessment and management and is important in identifying key life stages (e.g. egg incubation) and periods that may be vulnerable to local fishing or other human activities.

Materials and Methods

1. Release of Edible Crabs Tagged with Electronic Data Storage Tags

Between August 2008 and June 2009, 128 pot-trapped, female edible crabs (carapace width 138–228 mm, mean = 178.8 ± 19.0) were tagged with Cefas G5 long-life (2 MB memory capacity) electronic data storage tags (DSTs), with 2 MB memory capacity (CTL Ltd., Lowestoft, U.K.), configured with a 10 bar pressure sensor (reliable to ~100 m depth). Earlier aquarium trials identified that crevice burrowing by crabs between boulders could abrade unmodified DSTs through to the internal circuitry within periods as short as one month. To improve abrasion resistance, crab DSTs for wild deployment were therefore encased in secondary, lozenge-shaped perspex casings (bevelled along the outer edges) measuring $32 \times 12 \times 16$ mm. To maximise high resolution data collection, DSTs were programmed to record pressure at 30 s intervals and temperature at 5 min intervals for the first year at liberty, then both parameters at 5 min intervals thereafter.

Previous mark-recapture studies have suggested that migration by male edible crab is limited [33,23], therefore to maximise viable return rates, only female crab have been targeted in this study. To minimise the potential impact of moulting on tag loss, only recently moulted ("new shell") crabs with no obvious external damage were selected for tagging. Individual crabs were double-tagged. First, a uniquely numbered claw tag (coloured plastic cable-ties), was attached to the right-hand cheliped. The carapace was patted dry using absorbent paper and any fouling material on and around the area of attachment was removed. A small dab of superglue was applied centrally to the base of the DST, with fast-setting underwater epoxy resin applied around the basal perimeter. DSTs were then glued dorsally to the posterior carapace. The long axis of the tag was positioned parallel to, but not obstructing the epimeral line, with the tag label on the vertical plane (to avoid removal by abrasion on the top surface). The instant purchase obtained through the superglue counteracts the relatively slow curing time of the resin (approximately 2 h at room temperature). This allowed the resin to set fully with the tag firmly in place, obtaining a lasting bond between tag and crab. DST and tag

numbers were checked on release and the position and time recorded using a handheld GPS (Thales "Mobile Mapper", Thales Navigation Inc., France). The maximum time between capture, tagging and redeployment was approximately 30 min.

The tagged crabs were released at 4 locations in UK waters (Figure 1, Table 1): Eastern Channel ("EC"); Celtic Sea ("CS"); Western Channel Coastal ("WCC"); and Western Channel Offshore ("WCO"). No specific permissions were required for our tagging work, which was executed in international waters outside of the 12-mile UK territorial limit, and did not, therefore, require authorisation. Note that the edible crab, *Cancer pagurus*, is neither an endangered nor protected species. Although experimentation using live decapods crustaceans is not currently regulated in the UK or European Union, the highest standards of animal welfare were applied throughout our work. All tags were returned through the commercial fishery following a concerted publicity campaign, and the offer of financial rewards (£50) for each returned tag.

2. Data Download and Processing

Following return, individual DSTs were downloaded, and pressure data converted into depth. Plots were made of individual depth and temperature experience, and summarized by release on a monthly basis. Temperatures recorded by the tagged crabs were compared with temperatures monitored daily in the eastern Channel at 50.766N 0.300E ("Eastbourne", Cefas coastal monitoring) and approximately monthly in the western Channel (50.033 N, 4.367 W, "E1" CTD seabed temperature, Western Channel Observatory).

Where crabs appeared to be resting motionless on the sea-bed, we attempted to estimate their position using the tidal location method or "TLM" [36]. This technique estimates geographical location ("geolocation") based on the time of high water and tidal range, measured by the DST's depth (pressure) sensor when a tagged-individual remains motionless on the sea-bed over a full tidal cycle (or longer). Tidal ranges were extracted using a wave-fitting algorithm (see [36] for full details). Starting at each successive point in the DST pressure record, the algorithm searched for the best fitting sine-wave, applying a least-squares regression, using data from the following nine-hour period. The period of the model wave-form was constrained so that the half-tidal period could not fall below 4.5 hours or exceed 7.5 hours. The offset of the model was constrained so that the wave-form began with a maximum (or minimum) and continued beyond a minimum (or maximum). The daily best-fitting wave-form was used to calculate the times of high and low water, the tidal range, and to provide an indication of the quality of fit (sum of squares). Unlike recapture positions, which are dependent on recapture by fishers, the geolocations generated by the TLM are independent of the spatial distribution of the fishing fleet [37].

Where possible, incorrect geolocations were eliminated by comparing sea-bed depths (taken from British Admiralty Charts) at the derived geolocations with the actual depths recorded by the tag. Recorded temperatures were compared with averaged sea-surface temperatures (SST, taken from Bundesamt für Seeschiffahrt und Hydrographie, BSH). When depths for derived positions differed by more than ±10 m from the actual recorded sea-bed depth, or temperatures differed by more than ±1°C of the tag-recorded temperature, these positions were eliminated from the analyses. Where the method identified clusters of geolocations rather than individual points, the geographical midpoint was determined as the best fit.

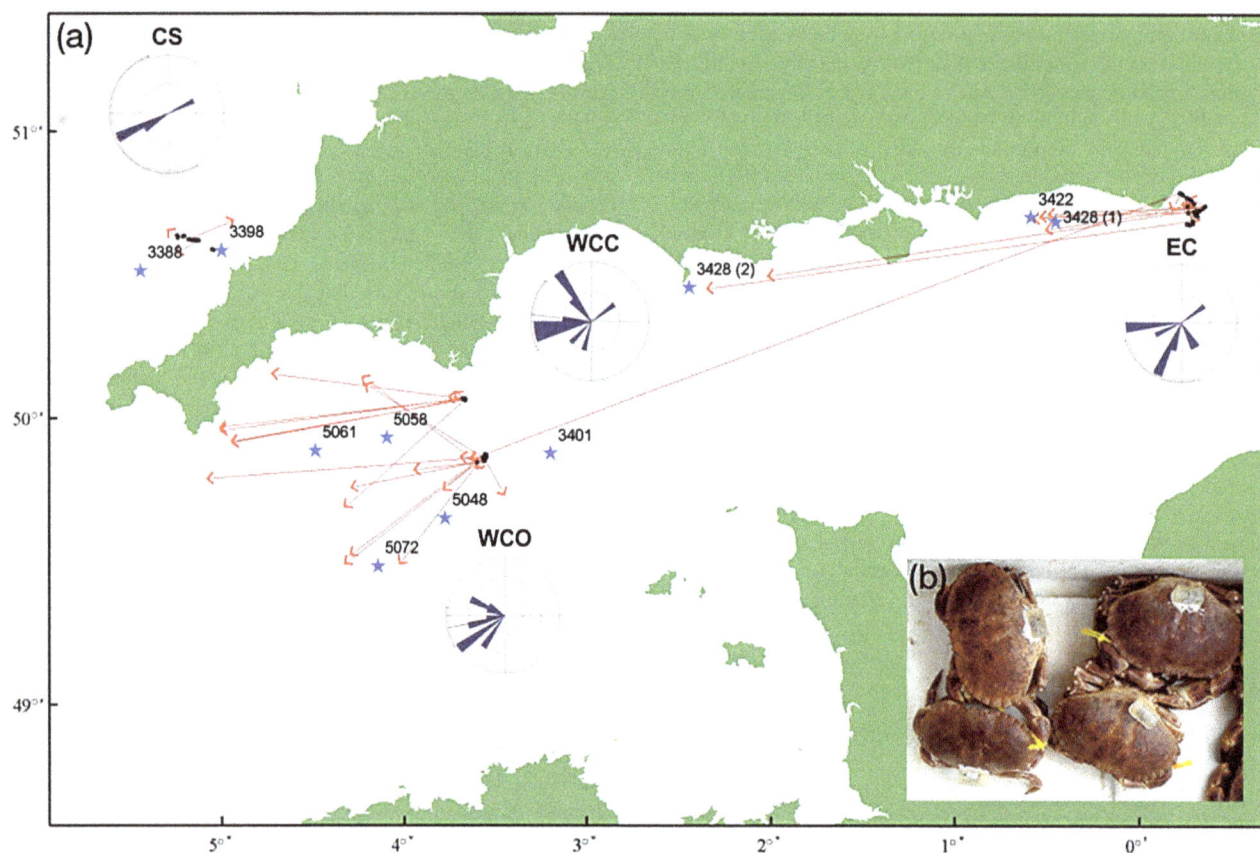

Figure 1. Migration routes and brooding locations of edible crabs *Cancer pagurus* **tagged with electronic data storage tags (DSTs).**
(a) Release and recapture locations, estimated brooding locations (see text), and distance and direction of movement by edible crabs tagged with electronic data storage tags and released in the English Channel and Celtic Sea from August 2008 to June 2009. Rose diagrams illustrate the mean axes (and 95% confidence limits) of migration. Key: Black dots, release locations; Red lines, direction of travel; red arrowheads, recapture locations; Stars, brooding locations of crabs (tag number) estimated using the tidal location method; EC, Eastern Channel; WCC, Western Channel Coastal; WCO, Western Channel Offshore; CS, Celtic Sea. (b) DST-tagged crabs prior to release.

Results

1. Recapture Rates of Tagged Crabs

To date (July 2012), 43 DSTs have been returned (34%, Table 1). Return rates between release sites varied between 17% (CS) and 40% (WCC). Only one individual was missing its DST on recapture (3413, EC). A second individual was recaptured, then immediately re-released once the tag details had been noted (5077,

WCO). Individual data records ranged between 8 and 575 days, and 4519 days of high resolution crab behaviour data were captured from 5540 days at liberty (Table 1). Fifty percent of DST recaptures were made during the first 40 days after release (Figure 2). The recapture rate thereafter did not follow a regular diminution pattern, but was seasonally distributed, with zero recaptures over the winter months.

Table 1. Mark-recapture results.

	Release date	Releases	Recaptures	Tag loss	Recorded days	Total days	Distance	Direction
Eastern Channel (EC)	Aug-08	32	12 (38)	1	1582 (8, 575)	1692	64.2±107.2	207.8° ±61.4°
Celtic Sea (CS)	Oct-08	29	5 (17)	0	514 (37, 254)	713	14.5±9.0	245.1° ±67.8°
Western Channel Coastal (WCC)	Jun-09	30	12 (40)	0	1025 (1, 268)	1171	48.6±43.9	276.6° ±54.9°
Western Channel Offshore (WCO)	Jun-09	37	14 (38)	0	1397 (9, 383)	1965	51.9±50.5	258.2° ±30.9°
Total		128	43 (34)	1	4519	5540		

Summary of release information by area and recapture data (recapture percentage in brackets) for edible crab tagged with electronic data storage tags. Recorded days (minimum and maximum data records in brackets) vs. Total days at liberty = the total number of days at liberty recorded by the electronic tags. Distances in kms ± s.d., directions in degrees ± circular s.d.

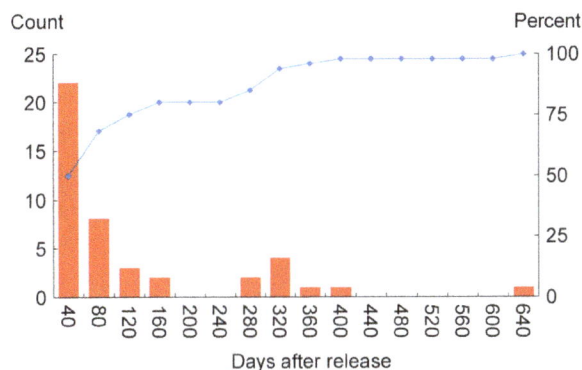

Figure 2. Return rates of tagged crabs. Return rate of edible crabs tagged with electronic data storage tags recaptured per 40 day time interval following release (histogram), and cumulative percentage recapture by time (line).

2. Distance and Direction Travelled

Tagged crabs were recaptured between 0.7 and 302.4 km from the point of release, from one to 679 days following release (Table 1), having recorded between 1 and 575 day long data records. The direction of movement in all releases followed a predominantly westward axis (Table 1). This was pronounced in the English Channel (Figure 1), where mean crab displacement was 64.2 km along an average vector of 207.8° in the EC (p = 0.05), and 48.6 km and 51.9 km along vectors of 276.6° (p = 0.01) and 258.2° (p>0.001) in WCC and WCO respectively (Table 1). CS crabs moved on average just 14.5 km from the point of release, and although this movement was also predominantly westwards (245.1°), this result was non-significant (p = 0.4). No long-distance migrations (≥15 km) followed a west-to-east axis (Figure 1).

3. Physical Data Recorded by the Tags

Individual variability in depth occupancy was low for CS and western Channel crabs (Figure 3A). By far the greatest levels of depth variation were associated with the EC crabs. Much of this variation was attributable to 3 individuals that recorded data for a year or more, and migrated between 173 and 302 km from the eastern Channel into the western Channel (see below). Individual depth records clearly demonstrate that crabs did not follow defined isobaths (depth contours) during migration (Figure 4).

The seasonal temperature cycle recorded by EC crabs was not fully mirrored by WCC and WCO crabs (Figure 3B). The western Channel crabs, located in deeper, colder, stratified water, experienced rising water temperatures at the start of autumn (i.e. Aug-Oct) 2009, when EC crabs in the previous year had experienced falling temperatures. This period also corresponds with the time when the crabs are most active in terms of migration (see below). However, a comparison with monitored temperatures in the eastern and western Channel suggests that temperatures recorded by the crabs reflect area-specific seasonal trends. The difference between the temperatures recorded by EC crabs and the coastal EC data was commensurate with westward movement in deeper water, further offshore (Figure 1, Figure 3B). Winter (October-January) temperature regimes were very similar for all areas (15°C declining to 9/10°C in January).

4. Annual Cycles and Egg Incubation

Six crabs recorded data over a full annual cycle: DST's 3401, 3422, 3428 (EC), 3398 (CS), 5048 and 5072 (WCO). A further 3

crabs recorded some of the egg-incubation period: DST's 3388 (CS), 5058 and 5061 (WCC). Individual plots of depth and temperature experience for these crabs are shown in Figure 4, and metrics on the timing and duration of egg incubation are given in Table 2.

For the three EC crabs that recorded data for a year or more, egg incubation was marked by a dramatic cessation of activity between 25/11/08 and 12/12/08 (Table 2). The crabs remained inactive for 175 to 188 days before active foraging recommenced (all crabs were recaptured from baited pots). A similar pattern was observed in western Channel crabs, although overall changes in depth occupancy were less pronounced (Figure 4). Only five DSTs were recaptured from CS. Both crabs that recorded data into the start of egg incubation (Nov-Dec) demonstrated a move into slightly shallower water prior to the onset of brooding. This was most pronounced for DST 3398, which moves up from 63 m to 48 m (Figure 4).

The average depth at which crabs incubated their eggs was 57±23 m, but ranged from 19 m in the shallower eastern Channel, to 84 m in the deeper Western Channel. Incubation lasted on average 177±24 days. The shortest incubation (or at least "incubation-like" behaviour) observed was 126 days during the second brooding season recorded by DST 3428. Temperature at brooding onset was 13±1.5°C, and 11±2°C when feeding recommenced. The lowest temperatures at the onset of incubation were recorded by CS crabs (10.7°C), and the highest in the deepest (WCO) grounds (15°C). By contrast, brooding appeared to stop earlier, and at lower temperatures in the western Channel (although due to sensor failure, we have no data for WCC). Again, the exception was DST 3428, which was the earliest to cease "brooding-like" behaviour (24/03/10), and at the lowest temperature (6.7°C), during her second year. As individual crabs were not returned, we cannot determine whether the crab was actually carrying eggs.

Occasional minor depth fluctuation (>0.5 m) during the incubation period by several females (Figure 4) suggests that the crabs were not always completely immobile during brooding.

5. Geolocation and Determination of Brooding Locations

The total number of TLM geolocations that could be generated from a single track was often limited, with >95% of interrogations during the active, migratory period failing to identify positions based on the tidal data recorded. Consequently, TLM estimates of position between release and recapture were effectively limited to those periods when the crabs were sedentary at the times of expected egg incubation.

Channel crabs were not restricted to a single, clearly defined brooding area. Incubation was effected at various locations throughout the Channel (Figure 1, Table 2). All brooding crabs were recaptured west of their brooding locations (some significantly so, e.g. 3422, Figure 1), suggesting that the same incubation sites are not occupied in successive years. DST 3428, the only individual to record data over two brooding seasons, settled down in two separate brooding locations separated by two degrees of longitude in the two successive years.

Discussion

Here we have described the results from the first large-scale geographical study of the migratory behaviour of crabs using electronic data storage tags. For the first time we have been able to chart the annual migrations of individual crabs, and describe the timing, physical conditions experienced and duration of egg incubation in *Cancer pagurus*.

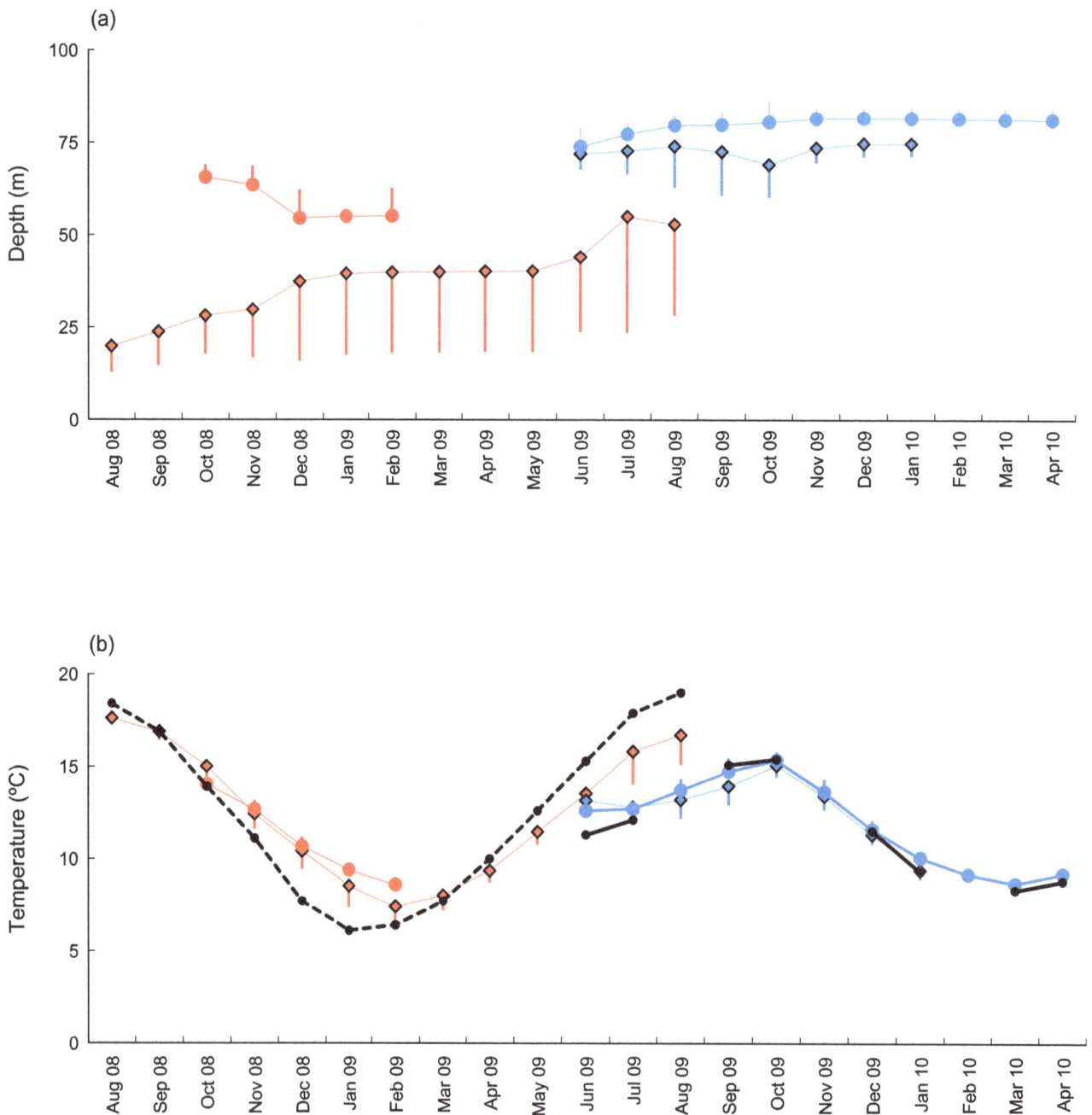

Figure 3. Average depth and temperature experience of edible crab. Average monthly (a) depths and (b) temperatures (± s.e) experienced by DST-tagged crabs released in the English Channel and Celtic Sea. Black lines (b only) indicate average monthly temperatures in the eastern Channel (broken line) and approximately monthly CTD measurements in the western Channel (intermittent solid line). Key: red diamonds, dashed line, Eastern Channel; red circles, solid line, Celtic Sea; blue circles, solid line, Western Channel Coastal; blue diamonds, dashed line, Western Channel Offshore.

1. Tag Performance and Geolocation

Tag return rates greatly surpassed expectations, with an overall return rate of 34% (compared with 17% overall in Bennett and Brown's [33] mark-recapture studies). At 17%, fewest DST's were returned from CS, which appears to reflect the levels of fishing effort experienced in the different areas (M.T. Smith, unpublished data). Recapture rates of 40% in the heavily fished South Devon fishery, are suggestive of intense levels of exploitation.

The interrupted pattern of tag recapture, was similar to that observed in other edible crab tagging programmes [33,23], and

was related to the reproductive activity of the crabs (see below). The tags proved robust to crab behaviour, with only one crab returned missing its DST, and no loss of information due to abrasion of the tag labels. Data download was not possible from two tags (DST's 5077 and 5098, both WCO) and sensor failure prior to recapture occurred for a further 6 tags. This negatively impacted on our findings on the annual cycle of behaviour for 5 of these individuals, all located in the western Channel. However an overall data capture rate of 82% compares well with analagous studies of fish behaviour [1]. Several tagged crabs were briefly

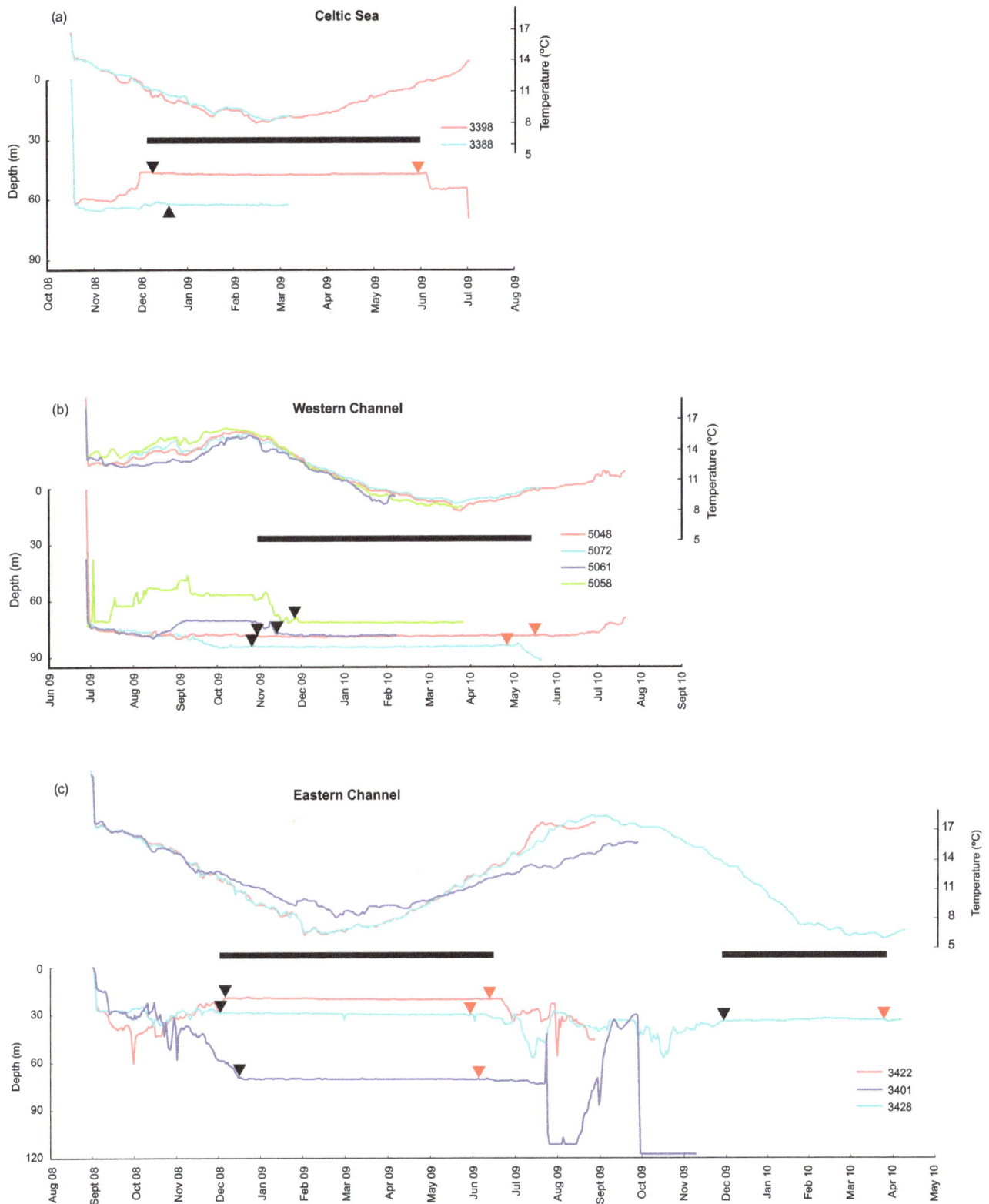

Figure 4. Seasonal depth and temperature experience of individual crabs. Individual temperature (upper plot) and depth (lower plot) records for edible crabs tagged with electronic data storage tags and released in the a) Celtic Sea, b) Western Channel and c) Eastern Channel. Black triangles indicate the onset of "incubation-like" behaviour, red triangles indicate the end of "incubation-like" behaviour. The approximate duration of the egg-incubation period in each area is indicated by a black bar.

recaptured soon after the initial release, and re-released. It was noted that the crab fishers widely believe that the true position can

accurately be determined from the tag, which is not in fact the case. DST 3401, one of the longest data records and furthest

Table 2. Metrics of egg incubation.

Location	Tag	Brooding period		Duration	Depth		Temperature		Brooding location	
		Start	Stop		Start	Stop	Start	Stop	Lat	Lon
EC	3401	12/12/2008	05/06/2009	175	70.29	71.14	11.45	12.12	49.92	−3.20
	3422	29/11/2008	11/06/2009	194	19.07	19.97	11.8	13.45	50.72	−0.48
	3428	25/11/2008	01/06/2009	188	29.03	30.34	12.1	12.69	50.70	−0.34
	3428	18/11/2009	24/03/2010	126	34.22	33.63	13.44	6.76	50.5	−2.42
CS	3388	16/12/2008	Tag fail		62.17		10.77		50.53	−5.54
	3398	07/12/2008	31/05/2009	175	46.77	46.97	10.74	12.2	50.61	−5.08
WC	5058	26/11/2009	Tag fail		71.15		12.63		49.97	−4.13
	5061	07/11/2009	Tag fail		77.11		13.74		49.92	−4.53
WCO	5048	30/10/2009	16/05/2010	198	78.64	78.23	14.95	10.05	49.69	−3.79
	5072	23/10/2009	25/04/2010	184	84.28	83.74	14.73	9.54	49.52	−4.17

Time period, depth and temperature at onset and completion of incubation, brooding location (estimated using tidal location method), and release and recapture locations of the eight electronic data storage-tagged edible crabs that recorded over all or part of the egg-brooding season. "Tag fail" indicates where data recording ceased before the end of the brooding period. EC, eastern Channel; WCC, Western Channel Coastal; WCO, Western Channel Offshore; CS, Celtic Sea.

migrating crabs, was also the only individual to move into water deeper than 100 m. Our tags were calibrated for use in depths down to 100 m. In this case the sensor functioned to 112 m (15/07/09), then recorded "112" continuously until the crab moved into marginally shallower water after 20 days. However, a second excursion below 112 m on 18/09/09 days resulted in a terminal failure of the sensor.

Unlike the North Sea, where submarine features such as Dogger Bank have proved informative in the reconstruction of DST-tagged fish migrations (e.g. [36]), the sea-bed relief of the English Channel tends to consist of relatively gentle gradients. An exception to this is the "Hurd Deep", a deep trench of >100 m formed by melting-ice during the previous ice age [38]. This trench effectively divides the northern and southern Channel midway, and in this case provides a useful "waymarker" in the migration route of DST 3401 (above), consistent with the recapture location. Hurd Deep is often cited as a physical barrier between the movements of fish between the northern and southern quadrants of the western English Channel.

The relatively poor performance of TLM was unexpected, and meant that geolocation-based migration-route reconstruction was not possible. Depth data recorded by the crabs often suggested superficially clean records of the tidal conditions necessary in the calculation of geoposition [36,37]. However subtle movements on the sea-bed, not clearly identifiable as depth changes, seem to have been sufficient to distort the tidal ranges and times of high water required by the data interrogation process. The tidal dynamics of the English Channel are also relatively complex: two tidal "solutions" (i.e. the same time of high water and tidal range) can often occur within a single degree of latitude (see [39]). Consequently, the spread of TLM "solutions" (not shown) were often parallel to, or concentric with, the area where the physical and recapture data hinted might be the more probable location of the crab, reducing the discriminative power of TLM. We were able, however, to identify the brooding locations of the crabs, when no significant horizontal movement occurred over periods of weeks and months.

A possible solution for geolocating crab tracks may be the application of the HMM model [40], where hidden Markov models are applied, incorporating both tidal location and bathymetry as well as using a Kalman filter to take account of previous and subsequent locations in the track record. This was originally developed to help reconstruct the ground-tracks of migrating demersal (sea-bed dwelling) fish which spend only intermittent periods resting on the seabed (principally cod, [40]). However such an exercise would require re-coding of the model to allow for the subtle movements exhibited by migrating crabs, an exercise outside the scope of the current project.

2. Reproductive Cycles and Westward Migration

The major breakthrough in the current study lay in describing the timing, conditions and locations of egg-incubation. Previous observations have been restricted primarily to indirect observations on the occurrence and distribution of egg-bearing crabs, either from landings, scuba surveys or aquarium experiments [41,42], and histological examination [43]. The onset of brooding, or at least the time at which the crabs ceased most activity, corresponded well with previous observations by Latrouite & Phillipe [42] suggesting a mid-November through to early January onset. Naylor et al. [44] observed that egg development in North Sea crabs ceased in late November, at which point the eggs entered a period of diapause. Development resumed in late March, with hatching in late June. The appearance of crab larvae in the plankton demonstrates a latitudinal gradient [45], with peak abundance occurring later at higher latitudes. This peak can occur from as early as March off the French Atlantic coast, to as late as August in the northern North Sea [45]. The earliest brooding onset observed in the current study was from late October (WCO). Indeed, westerly crabs appeared to start brooding slightly earlier than those in the eastern Channel, which tended not to commence brooding until mid- to late November.

Howard [41] observed empty guts sealed with a gelatinous "plug" in the majority of ovigerous females, in which the hepatopancreas condition was described as "poor" in 9/10 crabs. Our crabs became largely inactive throughout egg incubation, becoming re-animated towards the end of the brooding period, when foraging recommenced. Although some crab species may remain active during egg incubation [46], Brown and Bennett [47] suggest that ovigerous female *C. pagurus* are largely quiescent whilst incubating. Ungfors [43] collected ovigerous females in the Kattegat from baited nets deployed in April and May, while diver observations of brooding behaviour in a Norwegian fjord between

March and August identified some crabs leaving their brooding pits during May and June, and relocating to boulder refuges in the surrounding area [48]. In both latter studies, the activity described could be interpreted as marking the end of the incubation period. All of our crabs were recaptured in baited pots. It is generally accepted that the absence of egg-bearing females in commercial landings over winter is due to non-feeding during incubation. Certainly, some crabs did appear to show more activity within their brooding pits than others. DST 3428 recorded two successive "brooding" seasons. It is noted that she did not moult between broods (although it is thought that sperm retention by the females can facilitate multiple spawning from a single mating event [49,50]). "Brooding" commenced at similar times in the first and second years (table 2), but there was some evidence that the crab was more active during the second brooding season (daily activity rhythm), which lasted just 126, as opposed to 188 days. Unfortunately, the crab carcass was not generally recovered in these experiments, so we are unable to confirm that our crabs were carrying eggs throughout.

Brooding locations, estimated from TLM, show that brooding is not restricted to a single, clearly defined brooding ground, but may occur at various locations throughout the English Channel (although these are probably defined by substrate characteristics, [41], and that crabs did not show fidelity to the same brooding sites between years.

Both the results presented here and mark-recapture experiments executed in the 1970s [33–35] demonstrated long distance, predominantly westward movements by crabs in the English Channel, particularly mature females. In this, like the previous studies, we are unable to rule out the uneven distribution of fishing effort as having influenced these results. However, the westward migration has previously been interpreted as an example of counter-current spawning behaviour [51,52]. Pre-spawning migration to spawning grounds located in the western Channel have been thought to allow the hatching of planktonic larvae in prevailing tidal currents that will ultimately facilitate the return of settling larvae to their areas of maternal origin. None of our crabs exhibited west to east migration. It is noted however, that with the possible exception of the CS releases, all crabs that recorded the brooding period were migratory. However recent surveys of larval distribution, and interpretation using hydrodynamic modelling (D. Eaton, unpublished data) have suggested insufficient larval transport rates to return western Channel larvae to spawning areas located in the eastern Channel. Furthermore, genetic studies in the Kattegat-Skagerrak area [50] have provided some evidence of large-scale genetic mixing, but significant genetic variation at relatively local scales. The more wide-spread area from which larvae may originate demonstrated by the current study, may help explain at least some of these apparent discrepancies.

Conclusion

Results from this study provide a vivid demonstration of how the large-scale application of DST's in an intensively exploited crab fishery is a highly successful and effective means of gathering biological metrics for direct application in the management and conservation of shellfish fisheries. With significant ongoing expansion of coastal and offshore development (e.g. for gravel extraction, renewable energy installations, etc...), we anticipate that future archival tagging of crabs could not only provide a useful means of monitoring stocks, but may also be used to gauge the site-specific impacts of human activities impacting on crab and other stocks.

Acknowledgments

We are grateful to CTL Ltd. Lowestoft, for their help with the design of the crab tag, provision of "dummy" tags for testing, and download of returned data, notably Mike Challiss and Stephen Clarke. David Maxwell provided additional support for statistical analysis of the results. Mandy Roberts provided support for preparation of the final document and Karen Vanstaen and Roslyn McIntyre helped prepare final versions of the figures. We wish to thank everyone involved in the release of tagged crabs and return of the electronic tags.

Author Contributions

Conceived and designed the experiments: EH DE MTS. Performed the experiments: EH DE AL. Analyzed the data: EH CS. Contributed reagents/materials/analysis tools: CS AL. Wrote the paper: EH.

References

1. Metcalfe JD, Righton DA, Hunter E, Neville S, Mills DK (2008) New technologies for the advancement of fisheries science. In: Payne A, Cotter J, Potter ECE, editors. Advances in Fisheries Science: 50 years on from Beverton and Holt. Oxford: Blackwell Publishing. 255–279.

2. Block BA, Jonsen ID, Jorgensen SJ, Winship AJ, Shaffer SA, et al. (2011) Tracking apex marine predator movements in a dynamic ocean. Nature 475: 86–90.

3. Hammerschlag N, Gallagher AJ, Lazarre DM (2011) A review of shark satellite tagging studies. J Exp Mar Bio Ecol 398: 1–8.

4. Ropert-Coudert Y, Beaulieu M, Hanuise N, Kato A (2009) Diving into the world of biologging. Endang Species Res (2009): 21–27.

5. Metcalfe JD, Hunter E, Buckley AA (2006) Currents, clues and clocks: the migratory behaviour of North Sea plaice. Mar Freshw Behav Physiol 39: 25–36.

6. Yasuda T, Kawabe R, Takahashi T, Murata H, Kurita Y, et al. (2010) Habitat shifts in relation to the reproduction of Japanese flounder Paralichthys olivaceus revealed by a depth-temperature data logger. J Exp Mar Bio Ecol 385: 50–58.

7. Neat FC, Wright PJ, Zuur AF, Gibb IM, Gibb FM, et al. (2006) Residency and depth movements of a coastal group of Atlantic cod (Gadus morhua L.). Mar Biol 148: 643–654.

8. Svedang H, Righton D, Jonsson P (2007) Migratory behaviour of Atlantic cod Gadus morhua: natal homing is the prime stock-separating mechanism. Mar Ecol Prog Ser 345: 1–12.

9. Righton D, Quayle VA, Hetherington S, Burt G (2007) Movements and distribution of cod (Gadus morhua) in the southern North Sea and English Channel: results from conventional and electronic tagging experiments. J Mar Biol Assoc UK 87: 599–613.

10. Hunter E, Berry F, Buckley AA, Stewart C, Metcalfe JD (2006) Seasonal migration of thornback rays and implications for closure management. J Appl Ecol 43: 710–720.

11. Wearmouth VJ, Sims DW (2009) Movement and behaviour patterns of the critically endangered common skate Dipturus batis revealed by electronic tagging. J Exp Mar Bio Ecol 380: 77–87.

12. Kell LT, Scott R, Hunter E (2004) Implications for current management advice for North Sea plaice: Part I. Migration between the North Sea and English Channel. J Sea Res 52: 287–299.

13. Wiegand J, Hunter E, Dulvy N (2011) Evaluating management strategies for the thornback ray Raja clavata: Are marine protected areas better than traditional fisheries management? Mar Freshw Res 62: 722–733.

14. Loher T (2011) Analysis of match-mismatch between commercial fishing periods and spawning ecology of Pacific halibut (Hippoglossus stenolepis), based on winter surveys and behavioural data from electronic archival tags. ICES J Mar Sci 68: 2240–2251.

15. Hays GC, Doyle TK, Houghton JDR, Lilley MKS, Metcalfe JD, et al. (2008) Diving behaviour of jellyfish equipped with electronic tags. J Plankton Res 30: 325–331.

16. Lamare MD, Channona T, Cornelisen C, Clarke M (2009) Archival electronic tagging of a predatory sea star – Testing a new technique to study movement at the individual level. J Exp Mar Bio Ecol 373: 1–10.

17. Hawkins SJ, Sugden HE, Mieszkowska N, Moore PJ, Poloczanska E, et al. (2009) Consequences of climate-driven biodiversity changes for ecosystem functioning of North European rocky shores. Mar Ecol Progr Ser 396: 245–259.

18. Buhay JE (2011) Population Dynamics of Crustaceans. Integr Comp Biol 51: 577–579.

19. Bannister C (2009) On the Management of Brown Crab Fisheries. London: Shellfish Association of Great Britain.

20. Anonymous (2010) Fishery and Aquaculture Statistics 2008. Rome: Food and Agriculture Organisation of the United Nations.

21. Hunter E, Cotton RJ, Metcalfe JD, Reynolds JD (2009) Spatial and temporal variation in swimming activity and swimming patterns by plaice, *Pleuronectes platessa* L., in the North Sea. Mar Ecol Progr Ser 392: 167–178.

22. Skajaa K, Ferno A, Lokkeborg S, Haugland EK (1998) Basic movement pattern and chemo-oriented search towards baited pots in edible crab (*Cancer pagurus* L.). Hydrobiologia 371/372: 143–153.

23. Ungfors A, Hallbäck H, Nilsson PG (2007) Movement of adult edible crab (*Cancer pagurus* L.) at the Swedish West Coast by mark-recapture and acoustic tracking. Fish Res 84: 345–357.

24. Lynch BR, Rochette R (2007) Circatidal rhythm of free-roaming sub-tidal green crabs, *Carcinus maenas*, revealed by radio-acoustic positional telemetry. Crustaceana 80: 345–355.

25. Smith IP, Collins KJ, Jensen AC (1998) Movement and activity patterns of the European lobster (*Homarus gammarus*) revealed by electromagnetic telemetry. Mar Biol 132: 611–623.

26. Smith IP, Collins KJ, Jensen AC (2000) Digital electromagnetic telemetry system for studying behaviour of decapod crustaceans. J Exp Mar Bio Ecol 247: 209–222.

27. Smith IP, Jensen AC, Collins KJ, Mattey EL (2001) Movement of wild European lobster *Homarus gammarus* in natural habitat. Mar Ecol Progr Ser 222: 177–186.

28. Winter JD (1996) Advances in underwater biotelemetry. In: Murphy BR, Willis DW, editors. Fisheries Techniques, 2nd edition. Bethesda: American Fisheries Society. 555–590.

29. Jepsen N, Schreck C, Clements S, Thorstad EB (2005) A brief discussion on the 2% tag/bodymass rule of thumb. In: Spedicato MT, Lembo G, Marmulla G, editors. Aquatic telemetry: advances and applications. Proceedings of the Fifth Conference on Fish Telemetry held in Europe. Ustica, Italy, 9–13 June 2003. Rome: FAO/COISPA. 255–259.

30. Gonzales-Gurriaran E, Friere J, Bernardez C (2002) Migratory patterns of female spider crabs *Maja squinado* detected using electronic tags and telemetry. J Crustacean Biol 22: 91–97.

31. Curtis DL, McGaw IJ (2008) A year in the life of a Dungeness crab: methodology for determining microhabitat conditions experienced by large decapod crustaceans in estuaries. J Zool 274: 375–385.

32. Curtis DL, McGaw IJ (2012) Salinity and thermal preference of Dungeness crabs in the lab and in the field: Effects of food availability and starvation. J Exp Mar Bio Ecol 413: 113–120.

33. Bennett DB, Brown CG (1983) Crab (*Cancer pagurus*) migrations in the English Channel. J Mar Biol Assoc UK 63: 371–398.

34. Cuillandre J-P, Latrouite D, Le Foll A (1984) Le tourteau: biologie et exploitation. La Pêche Maritime 1278: 502–520.

35. Latrouite D, Le Foll D (1989) Données sur les migrations des crabes tourteau *Cancer pagurus* et araignées de mer *Maja squinado*. Océanis 15: 133–142.

36. Hunter E, Aldridge JN, Metcalfe JD, Arnold GP (2003) Geolocation of free-ranging fish on the European continental shelf as determined from environmental variables. I. Tidal location method. Mar Biol 142: 601–609.

37. Hunter E, Metcalfe JD, Holford BH, Arnold GP (2004) Geolocation of free-ranging fish on the European continental shelf as determined from environmental variables. II. Reconstruction of plaice ground tracks. Mar Biol 144: 787–798.

38. Gupta S, Collier JS, Palmer-Felgate A, Potter G (2007) Catastrophic flooding origin of shelf valley systems in the English Channel. Nature 448: 342–345.

39. Huntley DA (1980) Tides of the north-west European Continental Shelf. In: Banner FT, Collins MB, Massie KS, editors. The North-west European Shelf seas: the sea bed and the sea in motion. II. Physical and chemical oceanography and physical resources. Amsterdam: Elsevier. 301–351.

40. Pedersen MW, Righton D, Thygesen UH, Andersen KH, Madsen H (2008) Geolocation of North Sea cod (*Gadus morhua*) using hidden Markov models and behavioural switching. Can J Fish Aquat Sci 65: 2367–2377.

41. Howard AE (1982) The distribution and behaviour of ovigerous edible crabs (*Cancer pagurus*), and consequent sampling bias. J Conseil 40: 259–261.

42. Latrouite D, Philippe N (1993) Observations sur la maturité sexuelle et la ponte du torteau (*Cancer pagurus*) en Manche. ICES C.MC/K:23. Copenhagen: ICES Shellfish Committee.

43. Ungfors A (2007) Sexual maturity of the edible crab (*Cancer pagurus*) in the Skaggerak and Kattegat, based on reproductive and morphometric characters. ICES J Mar Sci 64: 318–327.

44. Naylor JK, Taylor EW, Bennett DB (1997) The oxygen uptake of ovigerous edible crabs (*Cancer pagurus*) (L.) and their eggs. Mar Freshw Behav Phy 30: 29–44.

45. Lindley JA (1987) Continuous plankton records: the geographical distribution and seasonal cycles of decapods crustacean larvae and pelagic post-larvae in the north-eastern Atlantic Ocean and North Sea, 1981–1983. J Mar Biol Assoc UK 67: 145–167.

46. Shields JD (1991) The reproductive ecology and fecundity of Cancer crabs. In: Wenner A, Kuris A, editors. Crustacean egg production. Rotterdam: A. A. Balkema. 193–213.

47. Brown CG, Bennett DB (1980) Population and catch structure of the edible crab (*Cancer pagurus*) in the English Channel. J Conseil 39: 88–100.

48. Woll AK (2003) In situ observations of ovigerous *Cancer pagurus* Linnaeus, 1758 in Norwegian waters (Brachyura, Cancridae). Crustaceana 76: 469–478.

49. McKeown NJ, Shaw PW (2008) Single paternity within broods of the brown crab *Cancer pagurus*: a highly fecund species with long-term sperm storage. Mar Ecol Progr Ser 368: 209–215.

50. Ungfors A, McKeown NJ, Shaw PW, Andre C (2009) Lack of spatial genetic variation in the edible crab (*Cancer pagurus*) in the Kattegat-Skagerrak area. ICES J Mar Sci 66: 462–469.

51. Pawson MG (1995) Biogeographical identification of English Channel fish and shellfish stocks. Fisheries Research Technical Report No. 99. Lowestoft: Ministry of Agriculture Fisheries and Food. 72 p.

52. Eaton DR, Brown J, Addison JT, Milligan SP, Fernand L (2003) Larvae surveys of edible crab (*Cancer pagurus*) off the east coast of England: implications for stock structure. Fish Res 65: 191–199.

4

Rhythms and Community Dynamics of a Hydrothermal Tubeworm Assemblage at Main Endeavour Field – A Multidisciplinary Deep-Sea Observatory Approach

Daphne Cuvelier[1]*, Pierre Legendre[2], Agathe Laes[3], Pierre-Marie Sarradin[1], Jozée Sarrazin[1]

1 Institut Carnot Ifremer EDROME, Centre de Bretagne, REM/EEP, Laboratoire Environnement Profond, Plouzané, France, 2 Département de Sciences Biologiques, Université de Montréal, succursale Centre-ville, Montréal, Québec, Canada, 3 Institut Carnot Ifremer EDROME, Centre de Bretagne, REM/RDT, Laboratoire Détection, Capteurs et Mesures, Plouzané, France

Abstract

The NEPTUNE cabled observatory network hosts an ecological module called TEMPO-mini that focuses on hydrothermal vent ecology and time series, granting us real-time access to data originating from the deep sea. In 2011–2012, during TEMPO-mini's first deployment on the NEPTUNE network, the module recorded high-resolution imagery, temperature, iron (Fe) and oxygen on a hydrothermal assemblage at 2186 m depth at Main Endeavour Field (North East Pacific). 23 days of continuous imagery were analysed with an hourly frequency. Community dynamics were analysed in detail for *Ridgeia piscesae* tubeworms, Polynoidae, Pycnogonida and Buccinidae, documenting faunal variations, natural change and biotic interactions in the filmed tubeworm assemblage as well as links with the local environment. Semi-diurnal and diurnal periods were identified both in fauna and environment, revealing the influence of tidal cycles. Species interactions were described and distribution patterns were indicative of possible microhabitat preference. The importance of high-resolution frequencies (<1 h) to fully comprehend rhythms in fauna and environment was emphasised, as well as the need for the development of automated or semi-automated imagery analysis tools.

Editor: Cristiano Bertolucci, University of Ferrara, Italy

Funding: This work was supported by the "Laboratoire d'Excellence" LabexMER (ANR-10-LABX-19), co-funded by a grant from the French government under the program "Investissements d'Avenir" and by a postdoctoral fellowship (D. Cuvelier) from Institut Carnot Ifremer EDROME. NEPTUNE Canada (Ocean Networks Canada) has provided us with access to their cabled network. The funders had no role in study design, data collection and analysis, decision to publish, or preparation of the manuscript.

Competing Interests: The authors have declared that no competing interests exist.

* E-mail: daphne.cuvelier@gmail.com

Introduction

The deep sea represents one of the least studied areas of our planet. To date, a mere 5–7% of the deep-sea floor has been explored. Even so, a myriad of unique species and ecosystems have been discovered in the deep with the chemosynthetic environments and their unique fauna as one of the most surprising discoveries of the last 40 years.

Currently, we are still striving to comprehend deep-sea ecosystem functioning, and to fulfil that objective time series are indispensable. To date, hydrothermal vent time series have been mostly based on annual or pluri-annual visits that were carried out with remotely operated vehicles (ROV's) or submersibles. Due to the peculiarities of the ecosystems and the use of both manned and unmanned submersibles equipped with video cameras, image analysis has always been an important tool to assess deep-sea community and ecosystem changes (see [1] for an overview of studies using imagery at hydrothermal vents). So far, temporal variation studies at hydrothermal vents have enlightened us on successional patterns as well as on communities' reactions to disturbances, natural or anthropogenic. In post-eruption studies, the nascence of hydrothermal vents is described, in which changes in community dynamics of pioneers and subsequent colonizers were linked with variations in temperature, sulphide supply, and

fluid composition [2,3,4,5]. First colonizers tend to tolerate higher temperatures and higher sulphide values while later arrivals live in lower temperature and lower sulphide concentrations. In studies under continuous venting conditions, the de- and reactivation of fluid exits, chimney collapse, and progressive mineralization of hydrothermal edifices contribute to fluid flow modification at small spatial scales, which, by rendering local habitats inhabitable and/ or unfavourable, affect faunal distribution and dynamics [6,7,8,9]. As mentioned before, these patterns are mostly deduced from observations separated by a year or more, not recording the cause and effect as it happens, which can only be observed through continuous monitoring. Next to these inter-annual visits, the use of time-lapse cameras at hydrothermal vents (a 26-day deployment at Axial seamount [10] and a 9 month deployment at TAG hydrothermal mound (Mid-Atlantic Ridge (MAR)) [11]) has already demonstrated that sub-annual processes, such as diurnal or semi-diurnal periods, also play a role in shaping hydrothermal vent communities and influence their dynamics and behaviour. High-frequency investigations, with year-round observations, are needed to inform us on natural changes, successional patterns, biotic interactions, reactions to disturbances and ecosystem resilience.

In this 21st century, technology has progressed in such a way that the difficulties of accessing and working in the deep sea are less of a threshold than before. Scientifically, this is illustrated by the development and implementation of deep-sea observatories, operational in a variety of settings [12]. In 2009, a regional-scale cabled observatory network, NEPTUNE (North-East Pacific Time-series Undersea Networked Experiments, an installation of Ocean Networks Canada www.oceannetworks.ca), came online. It consists of an 800 km electro-optic cable loop laid on the seabed over the northern Juan de Fuca tectonic plate, off the coast of British Columbia (Canada), granting real-time access to the data collected by the observatories. An ecological observatory module (TEMPO-mini) equipped with a video camera and environmental probes, is implemented at one instrument node of the NEPTUNE network and focuses on hydrothermal vent ecology. This observatory provides insights in the day-to-day activity of a hydrothermal faunal assemblage and the community dynamics occurring in a natural setting at Main Endeavour Field (MEF). In the present era of deep-sea exploitation, insights in natural change and community resilience are of utmost importance to evaluate the vulnerability and recovery of the hydrothermal vent ecosystems.

MEF was designated as a Marine Protected Area (MPA) in 2003, thus establishing one of the world's first deep-sea marine protected areas [13]. Its designation as a MPA prevents commercial exploitation, but scientific research is still permitted at specific sites. Nevertheless, sampling is restrained (sampling permits need to be applied for through 'Fisheries and Oceans Canada'), making imagery an adequate monitoring tool. The video imagery collected by the TEMPO-mini module monitors a *Ridgeia piscesae* tubeworm assemblage that has not been disturbed by sampling. A 23-day period was exhaustively investigated with an hourly frequency to analyse the changes over time documented in the faunal assemblage. Here we describe the (i) community dynamics for all taxa observed as well as their interactions and links with the environment and (ii) rhythms in faunal variation and environmental variables. This multidisciplinary observatory approach informs us on community dynamics and behavioural rhythms of the deep-sea hydrothermal vent fauna and its links with the environment.

Materials and Methods

1. Study Site and Observatory

NEPTUNE is a cabled observatory network located in the North-East Pacific off the coast of British Columbia (Canada), which contains various instrumented nodes (Fig. 1). TEMPO-mini is the ecological module [14] that focuses on hydrothermal vent ecology and is connected to the deep-sea Endeavour node (Fig. 1b). This module is the cabled counterpart of the autonomous Atlantic module TEMPO [15], and is used to record imagery and environmental variables in order to study temporal dynamics at deep-sea hydrothermal vents. TEMPO-mini was deployed at the Grotto hydrothermal edifice, at 2186 m depth, within the Main Endeavour Field (MEF, Fig. 1c). The Endeavour segment of the Juan de Fuca spreading ridge (Fig. 1a), and more specifically MEF, has a long history of hydrothermal vent research covering over 25 years [e.g. 7, 16, 17, 18, 19].

The main advantages of a cabled observatory are that energy, battery life and hard drive space are no longer main limiting factors, as opposed to the autonomous, wireless observatories. The instrumented nodes of the NEPTUNE observatory are powered by the on-shore facilities of Port Alberni on Vancouver Island. Thanks to live streaming, data collected is available in real-time on the internet (www.oceannetworks.ca) and is stored and managed at the University of Victoria in British Columbia (Fig. 1b).

2. Data Recordings and Observatory Set-up

TEMPO-mini's first deployment on the NEPTUNE network took place in September 2011 and it was recovered in June 2012; both were carried out with the Remotely Operated Vehicle (ROV) ROPOS. Since its deployment on September 29, 2011, TEMPO-mini transmitted data until the day of its recovery (June 19, 2012). The first week of the deployment was used to run tests in lighting, zoom of the camera, etc., while for the 23 subsequent days (October 7 (8 h UTC) – October 30, 2011 (14 h UTC)), imagery was recorded continuously. Lights were on constantly during the continuous imagery recording period. After these 23 days, the recording frequency changed to one half-hour (30 min) recording every 4 hours and corresponded with a substantial change in zoom (zoom-out, enlarging the filmed surface by a factor of 1.6). From this point onward, the imagery recorded was too zoomed-out to allow the quantitative assessment of macrofaunal densities. An additional problem arose with the oil in which the LED's were bathed causing a chemical reaction with the lenses of the projectors. This chemical reaction increased opacification of the lights' lenses which, consequently, emitted less light causing a darkening of the imagery recorded. Still, the module continued to record imagery until its recovery. Camera and light lenses as well as sampling inlets were protected against biofouling through a localised microchloration process [15].

The module was equipped with an Axis Q1755 camera featuring a 1/3″ Progressive Scan CMOS 2 Megapixel image sensor, which recorded videos with a resolution of 1440×1080 pixels and a frame rate of 24 fps. The faunal assemblage filmed was identified as 'community V low-flow' [7], characterised by slender *Ridgeia piscesae* tubeworms (Siboglinidae) along with abundant gastropod (Buccinidae, Lepetodrilidae, Provannidae) and pycnogonid fauna.

A 10 m long thermistor array with 10 temperature probes (separated by less than 1 m) was deployed in the neighbourhood of the filmed assemblage (Fig. 2). Only four temperature probes were positioned on the faunal assemblages at Grotto (T601 to T604); the fifth one (T605) was suspended in mid-water next to the TEMPO-mini module (Fig. 2). Probes T602 and T603 were positioned on assemblages identified as community IV [7], which was considered to be the most similar to the filmed assemblage as the latter (community V) is a subsequent stage of community IV in the succession model proposed by Sarrazin et al. [7]. The ecological framework of the observatory deployment is explained in more detail in the Discussion section. Temperature recordings had a resolution of 1 measurement every 30 seconds. Another tripod with environmental probes was deployed ca. 30–40 cm below the filmed assemblage. It comprised a CHEMINI Fe *in situ* analyser [20], recording three measurement cycles of dissolved iron (referred to as Fe from hereon) every 12 hours and an Anderaa optode, measuring oxygen concentrations (mL/L) and temperature every 30 seconds (Fig. 2).

3. Image Analyses

Main focus laid on the analysis of the imagery available for the continuous 23-day period, which featured the most constant zoom and few black-outs, adding up to ~550 h of video footage (Fig. 3). The surface of the tubeworm bush that was filmed totalled approximately ~0.0355 m^2 (ca. 20×18 cm). For this period (2011-10-07 8 h–2011-10-30 14 h), screen-stills were taken hourly, all times are noted in Coordinated Universal Time (UTC). A total of 23 video sequences were either unavailable or unusable (too dark or unfocused) for the predefined hourly frequencies. Overall, 536 hourly screen-stills were used as templates to map and count faunal abundances. During this continuous recording period, the

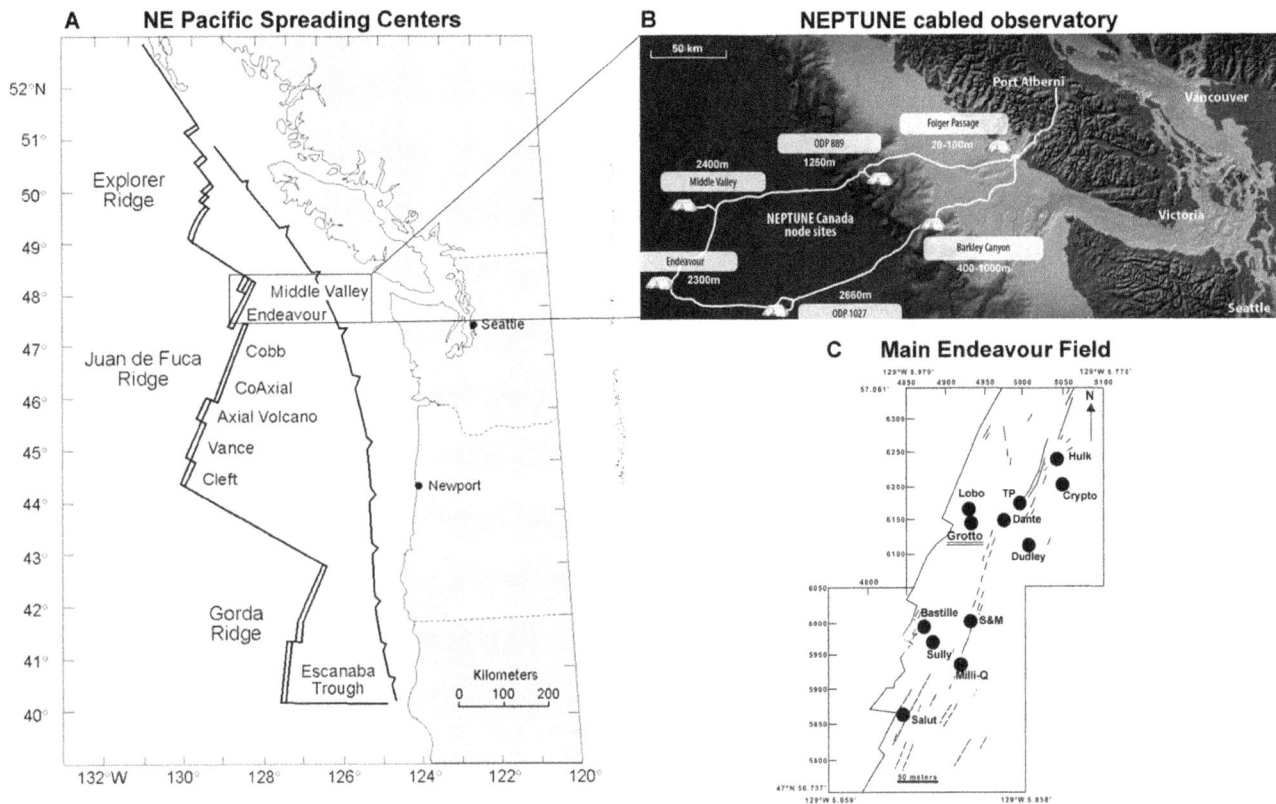

Figure 1. Location and lay-out of NEPTUNE observatory and TEMPO-mini study-site. Location (a) and lay-out (b) of the NEPTUNE cabled observatory on the northern Juan de Fuca plate. The NEPTUNE network contains multiple instrumented nodes (b). The Endeavour node connects the instrument platforms and the instruments at the Main Endeavour Field and contains the TEMPO-Mini module, which has been deployed at Grotto hydrothermal vent (c). MEF map based on [57] and V. Robigou (1995) (unpublished data).

zoom changed twice. The zoom changed only slightly, nevertheless we chose to work with the faunal densities to allow comparisons. Faunal densities were quantified for each hourly image, while for one image every 4 hours the microbial coverage was assessed. To pursue the latter, the microbial cover was marked in white and the rest of the image rendered in black. Using the "magic wand tool" of the ImageJ image analysis software [21], the surface covered by microbes was quantified and converted to percentages.

"Heat maps" were created for all mobile organisms. These maps are a graphical presentation of the numbers of observations using colour-codes. The colours thus reveal the most (dark red) and least (dark blue) frequented areas and distribution patterns for each taxon. For this purpose, the packages gplots [22] and hydroTSM [23], both in R 2.15.1 [24] were used. Several functions were adapted within the packages to meet our requirements and to fit our objectives. Pixel coordinates were attributed to individuals with the ImageJ "point picker" plugin on the hourly images for all images in the continuous imagery recording (n = 536). Subsequently, these coordinates were grouped into bins, each bin covering ~1.125 cm^2, and maps featuring spatial presence of fauna were produced. An index quantifying the measure of spatial segregation was calculated for each species and a pairwise segregation index was calculated between neighbouring species (Dixon package in R [25]).

4. Statistical Analyses

Various frequencies were investigated to study faunal variations, going from hourly, every 4 hours, every 6 hours to every 12 hours. Since the probes that recorded the environmental variables were not positioned on the filmed assemblage itself, timestamps were relative and could not be considered representative; hence, the values of the environmental variables used as explanatory variables for faunal variations over time were averaged per hour for each probe separately. Only the last two Fe measurement cycles were used to calculate Fe concentrations and these were averaged per recording time (be it twice or once a day). Gaps in the recorded variables data series were a restricting and determining factor in the selection of analyses.

Whittaker-Robinson (WR) periodograms, programmed in R by P. Legendre [26], were used to unravel periods in environmental variables and faunal densities. During periodogram analysis, data time series are folded into Buys-Ballot tables with periods of 2 to a maximum of $n/2$ observations. The number of columns in the table can be restricted to a range of periods assumed to be of interest for the study [27]. The underlying statistic, measuring the amplitude, used in this periodogram function is the standard deviation of the means of the columns of the Buys-Ballot table [27]; it is plotted in the resulting graph as a function of the periods. The WR periodogram can handle missing values in the datasets, when filled in with NA values ("Not Available"). Prior to periodogram analyses, datasets were tested for stationarity. In case of lack of stationarity, trends were removed by linear regression and residuals were used in the periodogram analysis.

Figure 2. TEMPO-mini module *in situ*. TEMPO-mini *in situ*, at Grotto hydrothermal vent at 2186 m depth. Insets show the filmed assemblage as well as the deployed environmental probes. T60x were the temperature probes and CHEMINI Fe and optode sampling inlets were mounted on a tripod deployed below the filmed assemblage.

Some degree of caution is needed during interpretation as this kind of periodogram also finds the harmonics of basic periods to be significant. Periodogram analyses were carried out for both temperature data and faunal densities using the hourly data for 23 days, with a maximum period of $n/2$ ($\sim$279.5 h).

Correlations between environmental variables and faunal densities (hourly resolution) were calculated using two-tailed correlations with permutations tests (n = 999), which is a correlation test that does not require normality of the data. P-values were subject to a Holm correction for multiple comparisons. Additional correlations using different frequencies (every 4 hours, every 6 hours and every 12 hours) were computed among faunal taxa.

Variation partitioning, by multiple regression and partial canonical analyses [28,29,30], was carried out on the faunal densities (excluding the visiting fish species), using two types (subsets) of explanatory variables: (1) significant temporal eigenfunctions (2) a selection of temperature probe measurements, retained in both cases by forward selection on the faunal density dataset. This analysis allowed a quantification of the proportion of variance in faunal densities explained by each subset of explanatory variables, while controlling for the effect of the other, as well as an estimation of the variation explained jointly by the two subsets.

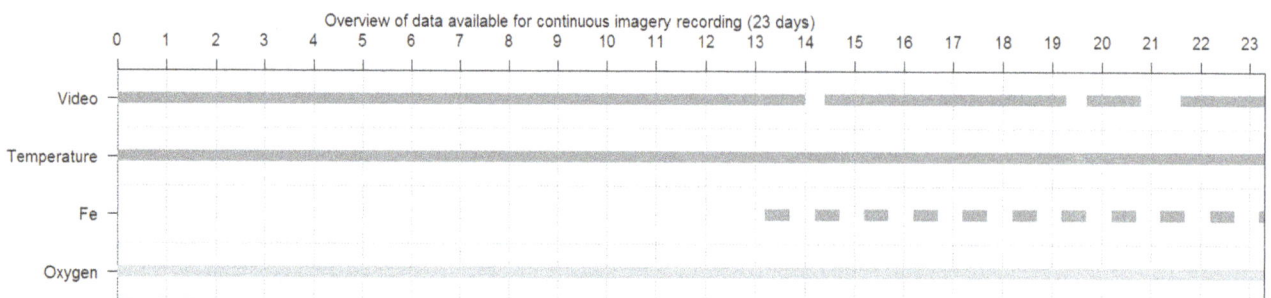

Figure 3. Data overview for continuous imagery duration. Data overview for continuous imagery duration, i.e. 559 h $\sim$23 days, during which video recording was programmed to be continuous. The gaps in the recordings were unusable video sequences (empty, black or unfocused). Temperature was recorded continuously (1 measurement every 30 seconds). For this period, Fe was measured once a day from October 20, 2011 onward. Oxygen was recorded continuously for the period, although it showed a steep decrease and other inexplicable patterns until day 83, rendering the data unusable (light-grey).

Results

1. Data Collected & Used

Total deployment time of the TEMPO-mini module amounted to ~9 months, from 2011-09-29 to 2012-06-20. All instruments, except for the CHEMINI Fe analyser (Fe measurements came to a complete stop on 2012-03-26), kept on recording until recovery. However, all time series contained gaps, due to black-outs of the NEPTUNE observatory (sometimes adding up to a couple of days for all instrument recordings) or localised failed recordings (instrument-dependent).

From 2011-10-07 at 8 h UTC to 2011-10-30 at 14 h UTC, ~550 hours of non-stop video were recorded and analysed with an hourly frequency, featuring 23 gaps (Fig. 3). For this period, temperature was recorded continuously (Fig. 4a), i.e. no gaps in the data. Fe measurements contained many gaps (Fig. 4b). The recording frequency of Fe measurements was changed on 2011-10-20 from three measuring cycles every twelve hours (twice a day) to once a day (at 6 h UTC), because of the drastic decrease observed in the reagents. For the oxygen measurements, a steep unexplainable decrease was noticed right after deployment; therefore oxygen measurements were omitted from the analyses of the continuous imagery recording (Fig. 3). The subsequent ~6.5 months of oxygen data showed normalised patterns, but were outside the scope of this study.

During the non-stop imagery recording period, changes in luminosity were evident. These were assessed using a HSL index on the images, combining hue, saturation, and light (Fig. S1). A first decrease was noticeable from hours 1 to 77 (corresponding to the period from 2011-10-07 to 2011-10-10), after which several adjustments on the camera's configuration were carried out (white balance and exposure settings) to compensate for this decrease in luminosity. Later on, similar smaller corrections were made during recordings (e.g. at about 160 h (2011-10-13 at 23 h UTC), Fig. S1). A continuous opacification of the LED lights' lenses caused the diminution of the amount of light emitted and consequently captured by the camera. This resulted in a blackening of the recorded imagery, which rendered the images unusable after December 2011.

2. Community Dynamics

2.1. Faunal variations and biological interactions. Six taxa were recognised on imagery and quantitatively assessed, i.e. *Ridgeia piscesae* tubeworms (Siboglinidae, Polychaeta), Polynoidae (Polychaeta), Pycnogonida (Arthropoda), Buccinidae (Mollusca), Zoarcidae (Chordata) and Majidae (Arthropoda). Their variations and observed interactions are presented in this section. For some taxa, identification was possible to the species level, for others, identification on imagery was not possible due to small sizes and/or the presence of more than one species in the area (more detailed information can be found in the Discussion section). Heat maps were created to show the distribution patterns of the mobile organisms, i.e. polynoids, pycnogonids and buccinids. Other gastropod species (limpets and snails) were visible on the imagery, though they were impossible to quantify accurately, and are only mentioned briefly. Results on the relationships between fauna and environmental variables are presented in 2.2.

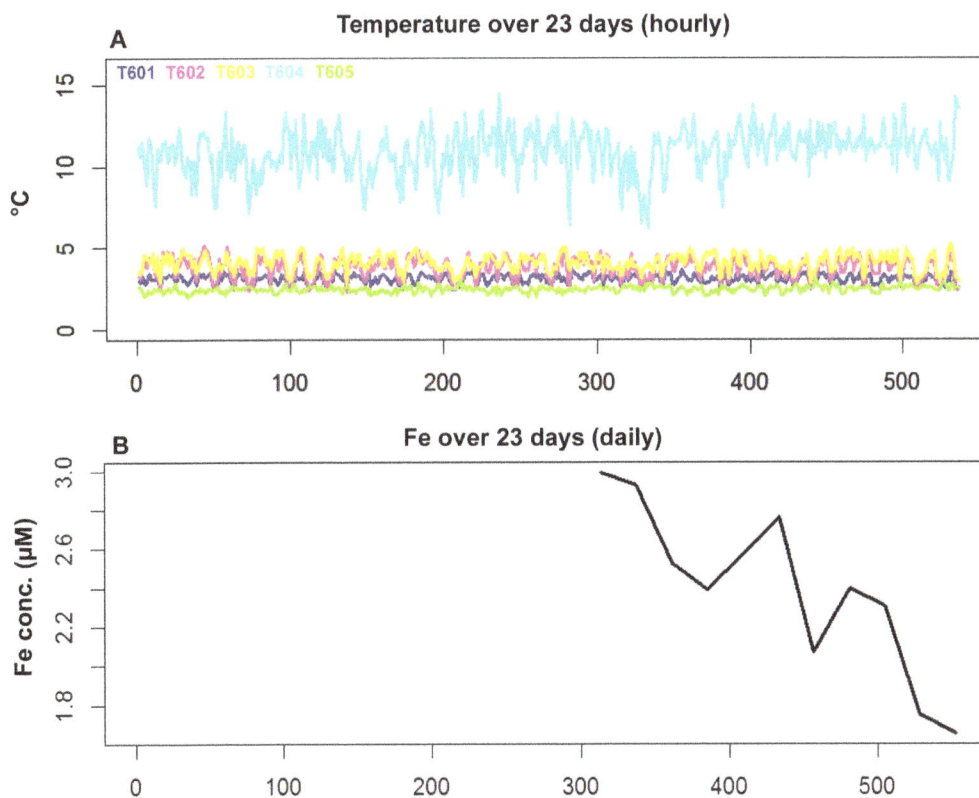

Figure 4. Overview of the environmental variables measured during the continous imagery recording. Overview of environmental variables measured in the vicinity of the filmed assemblage. Recording period represented here corresponds to the imagery analysis, i.e. 2011-10-07 at 8 h to 2011-20-30 at 14 h, for (A) Temperature (B) Fe concentrations. Gaps in the data during this period were due to instrument-dependent failed recordings.

Siboglinidae - *Ridgeia piscesae* siboglinid tubeworms were the prime constituent and most abundant species of the filmed assemblage. Tubeworm individuals were slender, narrow and more rusty-orange in colour compared to the white *Ridgeia piscesae* tubes observed in the nearby high-flow areas. The tubeworms showed elevated activity through extension-retraction movements of their branchial plumes extending outside the tube or being completely retracted sometimes until several centimetres inside the tube. Visible *R. piscesae* densities, i.e. individuals outside their tubes, ranged from 1038 to 8980 ind/m^2 (Fig. 5a) adding up to 8.5–78.6% of the total tubes that constituted the filmed tubeworm bush. 40.9% of the images analysed featured visible *Ridgeia piscesae* densities between 1500 and 3500 ind/m^2. The other 47.4% was almost equally divided among the 5500–7500 ind/m^2 (24.3%) and the 3500–5500 ind/m^2 (23.1%) categories, while the remaining 11.7% fell in the categories >7500 ind/m^2 (9.1%) and <1500 ind/m^2 (2.6%) (Fig. S2a). Some of the retraction movements observed were caused by other organisms, mainly by polynoid polychaetes. Polynoid movements or actions causing retraction in tubeworm plumes were threefold, with individuals either passing over and in-between the tubeworms, scanning the tube exit with their antennae (though in a number of cases the former two types of polynoid behaviour did not provoke any reaction in the tubeworms) or by attacking the exposed tubeworm plume with their proboscis, causing them to retract into their tubes (Video S1).

A Whittaker-Robinson periodogram computed for the siboglinid tubeworm densities revealed significant periods at 12 and 24 h as well as their harmonics at 36 and 48 h. The harmonics were neatly recognisable all along the periodogram (Fig. 6a).

Polynoidae. Polynoid polychaetes of different sizes were present on the tubeworm assemblage featuring differences in number of elytra and in colour. Several individuals had elytra missing and there were three sightings of a large polynoid with its elytra completely covered by white microbial filaments. Polynoids and pycnogonids did not interfere a lot with one another, nor did polynoids and buccinids. Short of two encounters between a zoarcid fish and a polynoid individual, in which the latter swam rapidly away after a possible contact with the fishes' head, no negative interactions of other species on polynoid presence were observed. As described in detail above, larger-sized polynoids, ca. 2–2.5 cm were occasionally observed to attack the *Ridgeia piscesae* tubeworm plumes that extended outside their tubes. Densities ranged from a minimum of 0 ind/m^2 to a maximum of 421 ind/m^2 (Fig. 5b), with a predominance of 70.2% of the images analysed featuring polynoid densities between 100–250 ind/m^2 (i.e. between 4–9 individuals in the field of view) (Fig. S2b).

A Whittaker-Robinson periodogram, for periods from 2 to $n/2$ (Fig. 6b), revealed significant periods at 4, 8, 12 and 24 h. Additional significant periods at 54 and 76 h were revealed, followed by a period corresponding to ~4.5 days (108–110 h, 111–116 h and 119–120 h), 152 h and ~9 days (207 h, 215–219 h, 224 h, 226–235 h and 250 h).

Pycnogonida - Pycnogonid densities were most likely to be an underestimation as individuals tended to cluster in crowded areas thus complicating counts especially as "stacking" was observed, i.e. various individuals were positioned on top of each other. Minimum and maximum densities observed amounted to 161 and 1075 ind/m^2 respectively (Fig. 5c), while 25.6% of the images showed densities between 600 and 700 ind/m^2, followed by 19.6% between 401–500 ind/m^2 and 16.4% between 501 and 600 ind/m^2 (Fig. S2c). No predatory behaviour was observed in the pycnogonids, nor were there any noteworthy negative interactions between the sea spiders and other taxa. A set of characteristic movements were observed of individuals dancing/bouncing up

and down, bending their legs, sometimes on top of another individual.

A Whittaker-Robinson periodogram showed no significant periods except for the 2 and 4 h periods, but none of their harmonics or other periods were significant (Fig. 6c).

Buccinidae - Buccinid whelks of the species *Buccinum thermophilum* were observed moving up and around the tubeworms bush. Often these individuals had multiple limpets attached to their shells. Minimum buccinid densities ranged from absent (0 ind/m^2) to a maximum of 188 ind/m^2, which corresponded to 0 to 7 individuals in the field of view (Fig. 5d). 60.5% of the images analysed featured densities between 51–100 ind/m^2 (or 2 to 4 individuals in the field of view), followed by 28.5% between 101–150 ind/m^2 (Fig. S2d). Active feeding of buccinids was not observed, neither on *Ridgeia piscesae* individuals nor on any other visible invertebrates. Instead, they were observed scanning the environment using their extended siphons. Sometimes these gastropods tended to bury themselves into the tubeworm bush and became almost invisible to the observer's eyes. Occasionally two individuals were positioned one on top of the other, moving jointly; sometimes the top one was "catapulted" away, rolling over and landing several cm's below. Paired alignment between two individuals which could allow copulation was observed, but no egg masses were found in the field of view. Even though buccinids were observed to move rather swiftly, the tubeworm bush did not appear to be an ideal surface for their locomotion, as they did not appear to travel efficiently and their locomotion was more like waggling.

A Whittaker-Robinson periodogram revealed significant periods corresponding to a ~4.5 day period (101–103, 105 and 108 h) and a large number of periods covering a ~9 day period (189–245 h) (Fig. 6d). Additional significant periods were found at 261, 263 and 266–268 h.

Other gastropods - Lepetodrilid limpets were observed moving about, forming strands of stacked individuals, sometimes sweeping over the assemblage. However due to the zoom level and their high abundances, no accurate quantification was possible. Provannid snails were also recognisable on some of the zoomed-in sequences (before 2011-10-07) but only seldom on the continuous recordings. Their small size and fairly dark colour made individuals hard to discern.

Zoarcidae - At most two individuals of zoarcid fish were sporadically encountered on the video imagery. Due to their low abundances and many zeros, they were not included in the statistical analyses, but their behaviour and activity was noted. Zoarcid individuals tended to position themselves most often on the tubeworm bush (20 observations or 3.73% of the hourly images analysed) often followed by hiding in-between the tubeworms or underneath the bush. Sometimes they were seen next to the bush (12 observations = 2.23% of images analysed). Two cases of possibly negative interactions were observed, involving a polynoid polychaete. The latter swam swiftly away after a possible contact with a fishes' head; this kind of fleeing behaviour appeared in both cases to be induced by the zoarcid fish. This fish species could have been impacted by the lights of the TEMPO-mini module, since they were always present in the field of view after a black-out.

Majidae - A majid spider crab visited the tubeworm bush several times. Eleven observations were made on the imagery, some of them only separated by a couple of minutes. A first visit took place on day 13, on 2011-10-20 at 7:59 h UTC, followed by several subsequent visits (n>5, last visit on 2011-10-27, 16 h UTC (day 20)). The nature of these visits consisted out of the spider crab "touring" the assemblage, often placing its legs straight into the

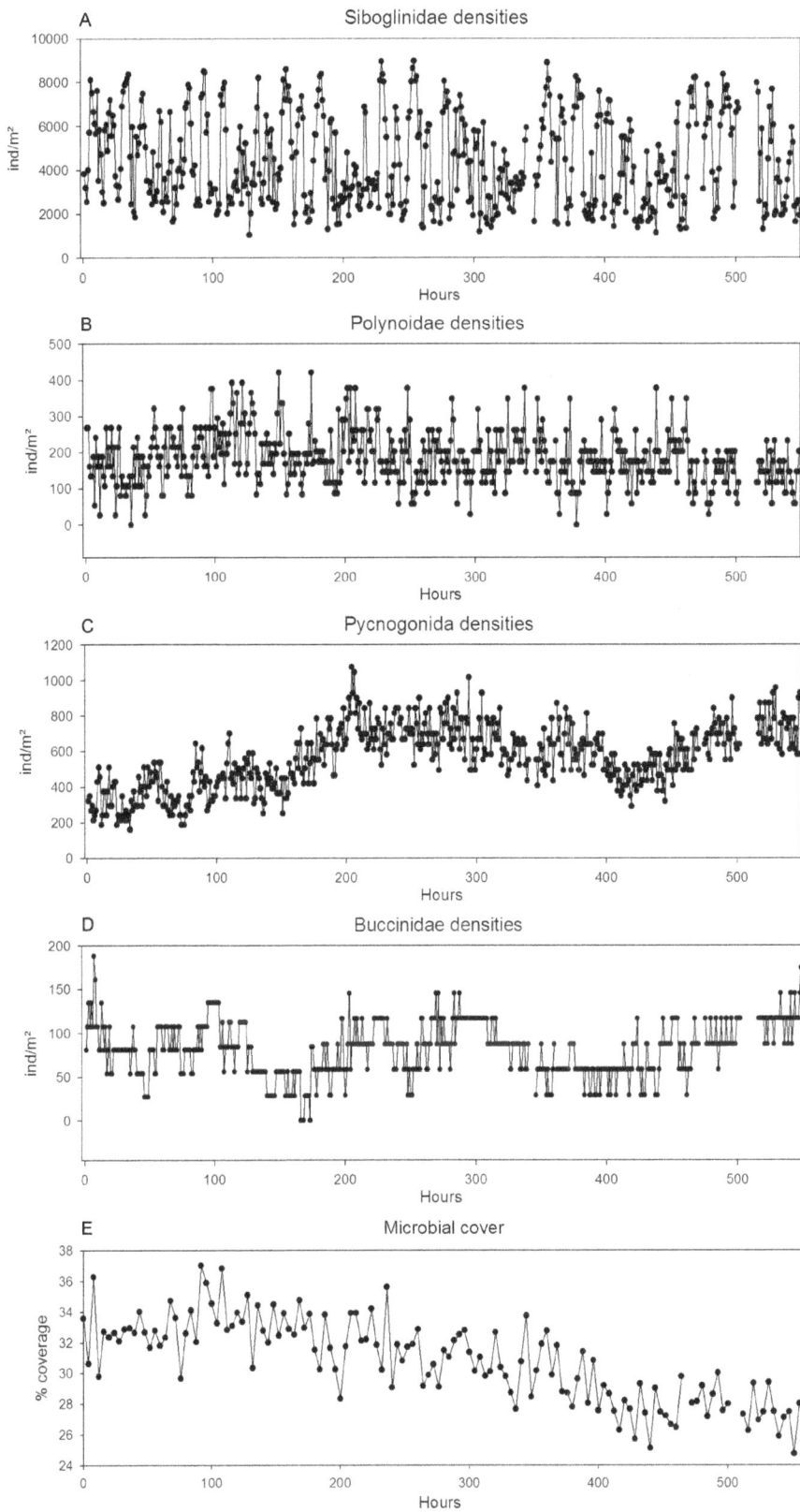

Figure 5. Hourly faunal densities for the continuous imagery period. Faunal densities (ind/m^2) as assessed through image analyses with an hourly frequency. The microbial coverage (%) was measured with a 4 hour frequency. All data span from 2011-10-07 at 8 h UTC to 2011-10-30 at 14 h UTC or 559 hourly observations with some missing values (n = 23).

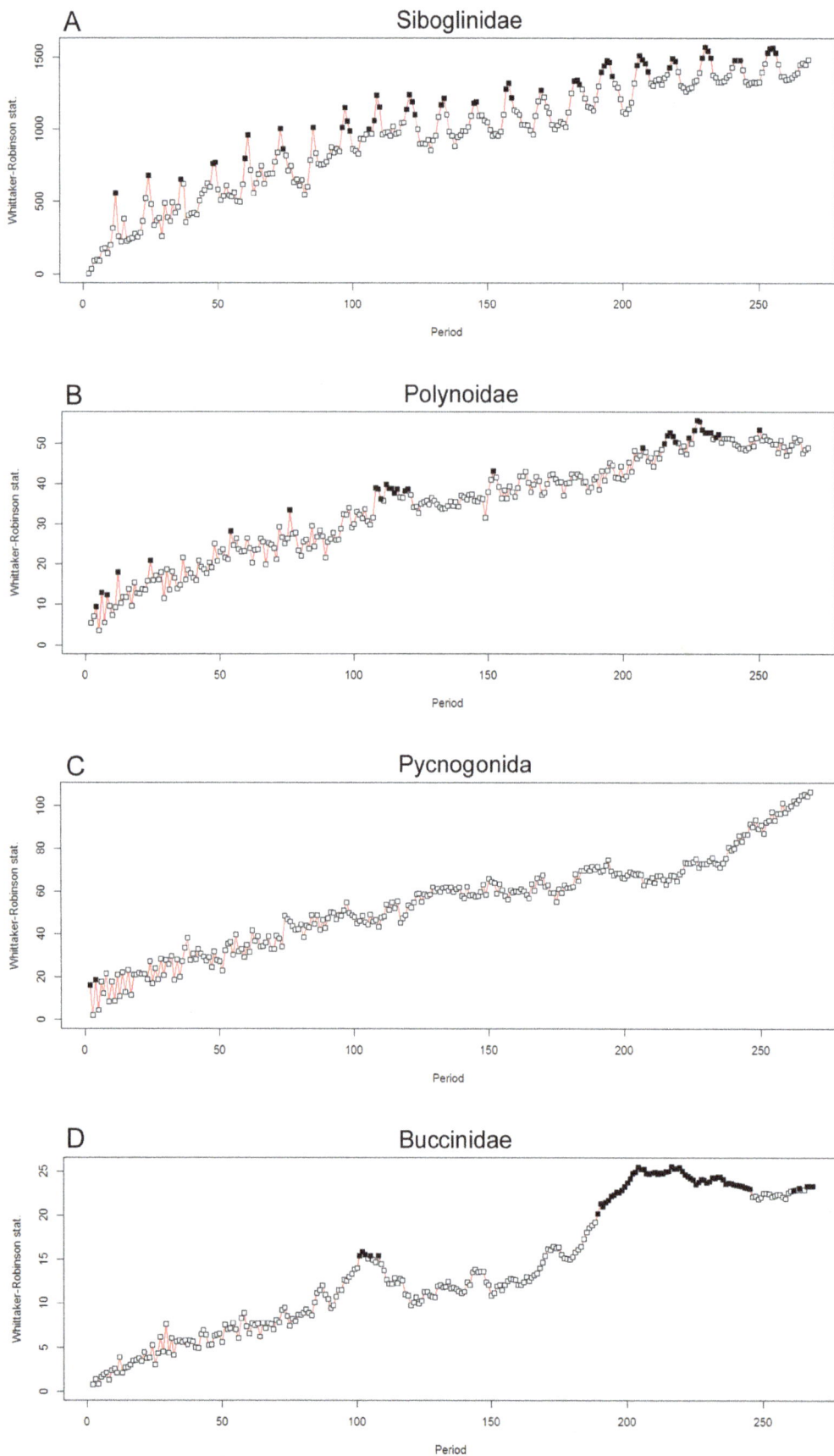

Figure 6. Whittacker-Robinson periodograms for faunal densities. Whittacker-Robinson periodograms computed for faunal densities, with periods spanning from 2 to *n*/2. Length of the series *n*: 559 hourly observations with some missing values (Fig. 5). Abscissa: periods in hours. The

Whittaker-Robinson statistic on the ordinate is the amplitude of variation in the Buys-Ballot table, measured as the standard deviation of the table column means of the table. Black squares are significant periods for p<0.05.

faunal assemblage. These actions did not prompt any reactions in the other taxa. Even when the majid crab placed one leg right next to a buccinid snail it did not trigger any reaction. During one of its other visits, the spider crab sat on the camera module with its legs dangling in front of the lens, thus putting the assemblage out of focus. Zooming in on the spider crabs' legs, two amphipods, of the Caprellidae family were recognised (Figure S3).

Microbial cover - The percentage (%) microbial cover, quantified every 4 hours, showed a clear decrease over time (Fig. 5e). A significant negative correlation was found with the pycnogonid densities ($r = 0.30$, $p = 0.016$, for 133 observations (n)).

Faunal relations - Following the polynoid attacks on the *Ridgeia piscesae* plumes, there was a significant negative correlation between the densities of tubeworms out of their tubes and polynoids ($r = -0.349$, $p = 0.012$, $n = 536$), and that for all frequencies studied. The relationship between pycnogonids and tubeworms was characterised by a negative correlation that was not significant for the hourly frequency ($r = -0.089$, $p = 0.28$, $n = 536$), nor for the other frequencies where it sometimes inverts (4 h: $r = -0.012$, $p = 1.00$, $n = 133$; 6 h: $r = 0.091$, $p = 1.00$, $n = 90$; 12 h: $r = 0.33$, $p = 0.42$, $n = 45$). There appeared to be little interference or interactions among polynoids, pycnogonids and buccinids. This was corroborated by heat maps, in which there was a clear distinction between the distributions of these three mobile taxa (Fig. 7). Spatial segregation tests were carried out on the point locations used in the heat maps for the three taxa by analysing the counts in a nearest neighbour contingency table. For all taxa, the observed counts of conspecific neighbours were larger than the expected counts, meaning that each taxon was associated with their conspecific neighbours (Table 1). For the polynoids, the measure of segregation (S [31]) also confirmed that their nearest neighbour was less likely to be a pycnogonid or a buccinid (negative S-values) and thus more likely to be another polynoid (positive S-values) (Table 1). The tendency to segregate was high in the polynoids (high S-values, Table 1), which was confirmed by looking at the heat maps where polynoids tended to move around a lot, covering the entire tubeworm bush (Fig. 7). The lack of distinct interactions between polynoids and pycnogonids was also reflected in the correlations, as for the hourly, 4 and 6-hourly frequencies no significant relations between the two taxa were revealed ($p > 0.4$), exception being the significant negative relation at the 12 h frequency ($r = -0.39$, $p = 0.036$, $n = 45$). Polynoids and buccinids, on the other hand, showed no significant correlations, regardless of the frequency studied ($p \sim 1.00$). The clustering behaviour in pycnogonids was illustrated by the heat maps (Fig. 7) and spatial segregation tests, showing the lowest tendency to segregate (Table 1). A significant positive relation was revealed on an hourly frequency for the pycnogonids and the buccinids ($r = 0.16$, $p = 0.02$, $n = 536$), but not for the other frequencies ($p \sim 1.00$). Of all taxa, buccinids had the highest tendency to segregate (Table 1), though the surface area covered was not as widespread as observed for polynoids.

2.2. Links with environment. Environmental variable time series started recording on 2011-09-29, though the results discussed here were selected to correspond to the continuous recording period, starting 2011-10-07 (Fig. 4). The correlations between fauna and environment presented here should be considered approximate, as the environmental variables were not recorded directly on the filmed assemblage itself (Fig. 2). More detailed information on this can be found in the Discussion section.

Temperature - Temperature values during the continuous imagery recordings for this study varied between 1.92°C and 14.47°C, with means ranging from 2.47 (T605) to 10.89°C (T604), though predominantly around 3.08°C to 4.06°C (T601, T602 and T603). For the 5 deployed temperature probes (Fig. 2), T604 showed the highest temperature values measured as well as the largest variations (Table 2), whereas the T605 probe recorded the lowest values, approaching the surrounding seawater temperature (~ 1.8°C). Significant correlations among the five temperature probes and the faunal variations were observed. Most significant were the positive correlations between T602 and T603 with the visible tubeworm densities ($r = 0.49$ and $r = 0.62$ respectively, $p = 0.02$, $n = 536$) and a significant negative one with T605 ($r = -0.30$, $p = 0.02$, $n = 536$). Significant negative correlations were observed between the same two temperature probes (T602 and T603) and polynoid densities ($r = -0.17$ and $r = -0.23$ respectively, $p = 0.02$, $n = 536$).

Fe - The minimum Fe concentrations ranged from 1.65 to a maximum of 2.99 µM during the continuous imagery recordings, while the mean was 2.39 ± 0.44 µM (Table 2). No significant relationships were revealed between the faunal variations and the Fe measurements available in the period October 20–30, 2011, featuring a resolution of one measurement a day (24 h frequency).

Variation partitioning - Variation partitioning allowed quantification of the variation explained by two subsets of variables with a Venn diagram representation (Fig. 8). The significant variables retained by forward selection against the faunal variations were the temperature measurements from probes T601, T602, T603 and T605, as well as a subset of 33 significant temporal eigenfunctions. Most of the variation in faunal densities was explained by temporal periods represented by the significant eigenfunctions, adding up to 36.5% (Y = [b+c], Fig. 8). X on the other hand represented the variation explained by the subset of temperature variables (X = [a+ b]); it explained 28.9% of the faunal variation. Of these percentages, 19.4% of the variation ([b]) is explained jointly by the temporal eigenfunctions and the temperature values, meaning that these variables showed collinearity (i.e. correlation). A small percentage ([a] = 9.5%) of the variation was explained solely by the temperature variables while other temporal periods found in the faunal densities were only modelled by the temporal eigenfunctions ([c] = 17.2%). All fractions were significant, with $p = 0.001$. The 54.1% residuals correspond to the faunal variation not explained by the temperature variables and eigenfunctions.

3. Temporal Variation in Environmental Conditions

3.1. Temperature. All 23 day temperature time series, from the different temperature probes, except T603, showed significant trends and thus did not comply with the stationarity requirement for periodic analysis. Hence, trends were removed and periodogram analysis was carried out on the residuals for periods of 2 to $n/2$ (279.5 h~11.5 days). Diurnal and semi-diurnal periods, and their harmonics, were the main significant frequencies discerned. No clear or distinct significant multiple day cycles were encountered. Significant periods at 25 h were revealed for all but one temperature time-series, the exception being probe T604. An additional significant period at 12 h was revealed for probes T601, T602 and T603, while for all temperature probes recurrent harmonics of both semi-diurnal (12 h) and diurnal (25 h)

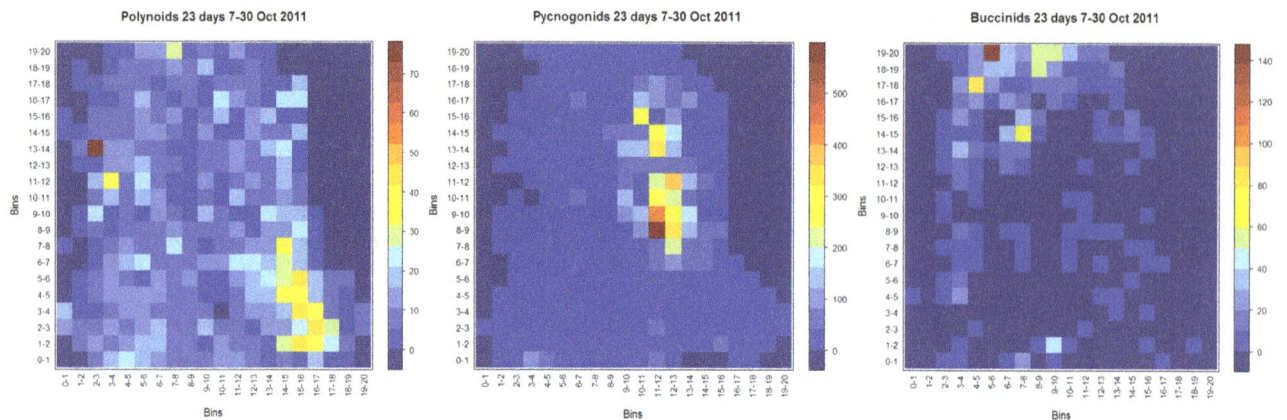

Figure 7. Heat maps of polynoids, pycnogonids and buccinids. Heat maps showing the distribution of the three mobile species on the filmed tubeworm bush, based on the number of observations during the 559 h or 23 day continuous imagery period. Coordinates were attributed to each observation and these were grouped and colour-coded per bin (1 bin = ~1.125 cm²). Colour-coded legends along with the corresponding number of observations can be found next to each plot.

frequencies were identifiable throughout the temperature time series, which agree well with the tidal cycle (12 h 25 min and 24 h 50 min) (Fig. 9). The periodograms of the first three probes (T601, T602 and T603) showed more similar patterns, while T604 had almost no significant periods and T605 appeared more intermediate between these two graph types.

3.2. Fe. Only a short period of ten days (2011-10-20 to 2011-10-30) with one measurement per day was available for Fe time-series analyses during the continuous imagery recordings, which proved too short to reveal any significant periods by the periodogram analysis.

Discussion

1. Ecological Framework of the Observatory Deployment

Fauna. The filmed assemblage was recognized as the low-flow variant of community V, described by [7], featuring narrow *Ridgeia piscesae* tubes and abundant gastropod fauna. On basalts, long skinny *R. piscesae* densities were shown to be able to reach densities up to 200000–300000 tubes/m² [18], while sampling of a similar

assemblage on a sulphide structure revealed densities up to 66903 ind/m² [17]; both are much higher than the visual counts made during this study (<9000 ind/m²). While gastropod species were very abundant in this type of assemblage [7,17], our image analysis study did not really reflect that, as only buccinids were large enough to be quantified on the imagery. High abundances of lepetodrilid individuals (*Lepetodrilus fucensis*) were detected and occasionally, a third gastropod species, *Provanna variabilis*, was visible (unpublished results, ground-truth sampling during 'Wiring the abyss cruise 2013' (June 2013)), but both were not quantifiable based on imagery. Community V low-flow was also characterised by high abundances of pycnogonids and polynoids, which were shown to be more numerous than in the other assemblages at Endeavour vents [17]. Pycnogonids were very abundant on the footage analysed here, with higher densities (ind/m²) than reported previously [17], even though our values still represent an underestimation due to stacking. The ground-truth sampling at Grotto - of an assemblage similar to the one filmed - also revealed the presence of *Paralvinella* cf. *palmiformis* (unpublished results, 'Wiring the abyss cruise 2013'), which were unaccounted for on

Table 1. Measures of spatial segregation [31] between three mobile taxa and neighbours based on all hourly observations during the 23 days continuous recording period (n = 559–23 missing).

From	To (neighbour)	Obs. Count	Exp. Count	S
Buccinidae	Buccinidae	**1024**	142.86	1.31
	Polynoidae	186	330.20	−0.30
	Pycnogonida	284	1020.94	−0.96
Polynoidae	Buccinidae	206	330.20	−0.22
	Polynoidae	**1750**	762.52	0.56
	Pycnogonida	1495	2358.28	−0.45
Pycnogonida	Buccinidae	323	1020.94	−0.53
	Polynoidae	1633	2358.28	−0.20
	Pycnogonida	**8714**	7290.78	0.31

The observed values (Obs. Count) larger than the corresponding expected values (Exp. Count) are in bold. Values of S calculated for a taxon and its conspecific neighbour larger than 0 indicate that the taxon is segregated; the larger the value of S, the more pronounced the segregation. These values of S (between conspecific neighbours) closer to 0 are consistent with random labelling of the neighbours of species. The S between two different taxa is less than 0 because the observed frequency of a taxon as a neighbour is smaller than expected under random labelling.

Table 2. Overview of the environmental data recorded in the vicinity of the TEMPO-mini ecological module at the Grotto hydrothermal vent.

	23 days continuous imagery				
	min	max	mean	stdev	var
Temperature (°C)					
T601	2.45	3.68	3.08	0.27	0.08
T602	2.28	5.14	3.76	0.65	0.42
T603	2.73	5.27	4.06	0.51	0.26
T604	**6.19**	**14.47**	**10.89**	**1.37**	**1.87**
T605	1.92	3.12	2.47	0.18	0.03
Fe (μM)	1.65	2.99	2.39	0.44	0.19

Highest values for temperature were in bold. Min: minimum, max: maximum, stdev: standard deviation, var: variance.

imagery but were present in the community V low-flow as described by [17].

Environmental variables. The positioning of the probes recording environmental variables was not optimal for the purposes and objectives of studying the community dynamics of the deep-sea hydrothermal vent fauna at MEF and linking it to its environment. The probes were deployed rather hastily because of a storm threat at the surface, cutting short the diving time. Nevertheless, following [7], T602 and T603 featured an assemblage corresponding to community IV. This specific community (IV), characterised by limpets, snails and small *R. piscesae* [7], was indicated as the one preceding the filmed assemblage (i.e. community V low-flow) in the succession model proposed by Sarrazin et al. [7], and as such the most similar to the filmed assemblage regarding the environmental conditions. For the other

probes, T601 appeared to be positioned on a similar assemblage as T602 and T603, although it was not really in touch with the assemblage in question. T604 featured the warmest temperatures on an assemblage recognisable by numerical abundant white limpets and *Paralvinella* palm worms, corresponding to the warmer community III [7]. Dense clouds of shimmering water bathed this community in warm fluids. Contrastingly, the T605 probe approached the ambient seawater temperature, though sometimes the influence of warmer fluids, through rising shimmering water plumes, was detectable. For the CHEMINI probe, there was some doubt that this sensor/probe was actually in contact with the fauna. It was deployed on a similar assemblage as the one filmed. The fact that the sampling inlet was not in touch with the faunal assemblage, along with the distance separating it from the field of view, might (at least partially) explain the lack of correlation/relationships with the faunal variations.

2. Temporal Community Dynamics

Temporal community dynamics between the taxonomic groups quantitatively assessed on imagery revealed insights in their daily interactions and mutual relationships. Depending on the frequency studied (hourly, 4 h, 6 h or 12 h), the correlations between faunal taxa might change, from statistically significant to not significant or sometimes even inverting the relationship.

Siboglinidae. Of all vestimeniferans, *Ridgeia piscesae* are known to exhibit greater tolerance to varying physicochemical conditions [32]. This was also confirmed by the different morphologies (ecotypes) of the species where a low-flow variant was characterised by skinny vestimentifera with poorly developed branchial filaments while high-flow vestimentifera had short wide white tubes and prominent, feathery branchial plumes [17,33]. Even with poorly developed branchial plumes, the filmed rusty-coloured skinny *Ridgeia* individuals showed quite some extension-retraction movements. While several plume retractions of *R. piscesae* were caused by polynoid attacks, in the majority of cases no visible external trigger could be identified. The extension/retraction movements of *R. piscesae* were following what appeared to be an individual rhythm. Nevertheless, this observatory approach allowed us to identify distinct semi-diurnal and diurnal rhythms in emergence/retraction of tubeworm plumes for the filmed assemblage as a whole, which were recognisable throughout the entire time series. Currents, fluid flow, temperature and oxygen have all been hypothesised to play a role in the occurrence of these rhythms in tubeworms. However, no consistent statistically significant patterns linking tubeworm (*R. piscesae*) extension/

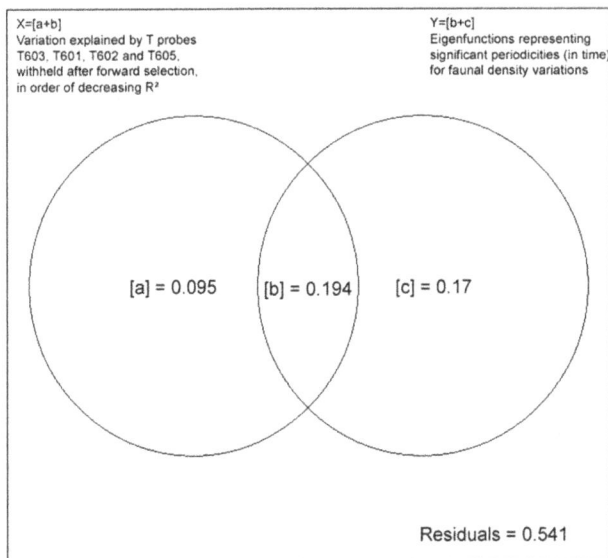

X=[a+b]
Variation explained by T probes T603, T601, T602 and T605, withheld after forward selection, in order of decreasing R²

Y=[b+c]
Eigenfunctions representing significant periodicities (in time) for faunal density variations

[a] = 0.095   [b] = 0.194   [c] = 0.17

Residuals = 0.541

Figure 8. Variation partitioning for the faunal density data. Variation partitioning for the faunal density data (excluding the zoarcid fish densities), using two types of explanatory variables: X = a selection of different temperature probes, and Y = significant temporal eigenfunctions, retained in both cases by forward selection on the faunal density dataset. [b] is the variation explained jointly by X and Y. The residuals represent the amount of variation not explained by a linear model of the two sets of explanatory variables.

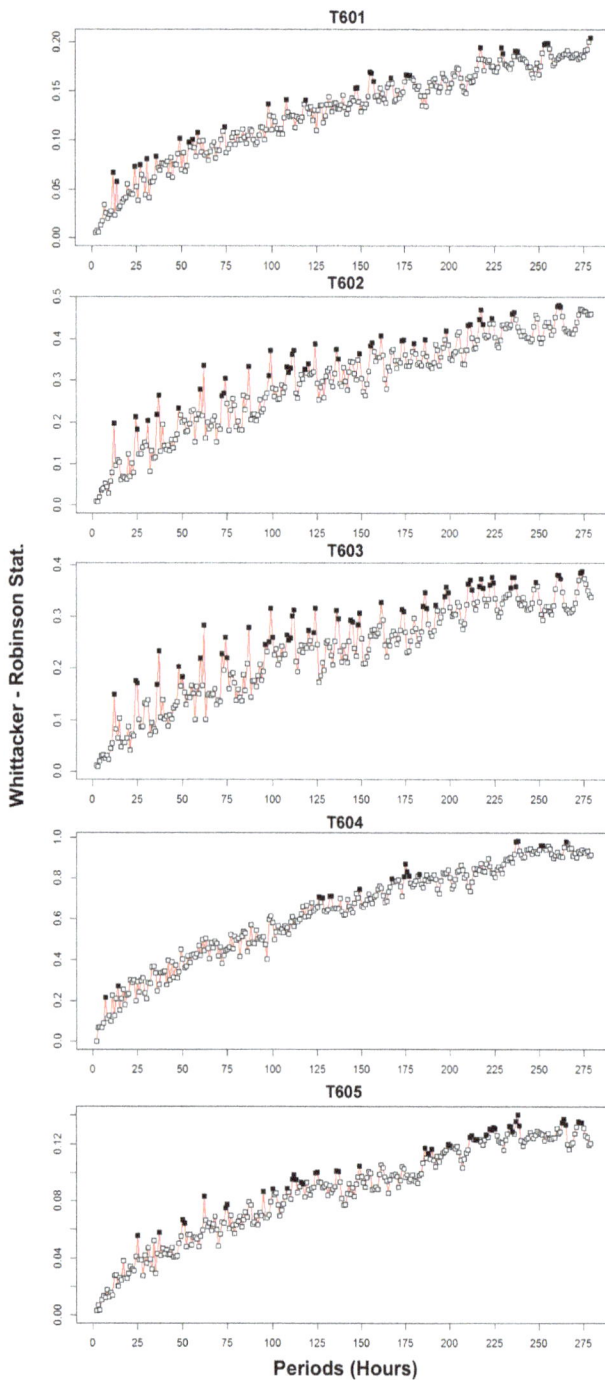

Figure 9. Whittaker-Robinson periodograms for 5 temperature probes. Whittaker-Robinson periodograms computed for the de-trended temperature values of all five probes (locations: see Fig. 2) during the 559 h or 23 day continuous imagery period. Abscissa: periods in hours. The Whittaker-Robinson statistic on the ordinate is the amplitude of variation in the Buys-Ballot table, measured as the standard deviation of the table column means. Black squares are significant periods at p<0.05.

retraction to high currents, changes in currents or turbidity were found [10]. In the present study, correlations with temperature were revealed, although these should be treated with some caution due to the sub-optimal probe deployment. Emergence/retraction movements of siboglinid tubeworms were also proposed to be a

thermoregulatory behaviour or suggested to be governed by oxygen requirements [10,34]. Urcuyo et al. [18] indicated that most of the time individuals in a long-skinny tubeworm assemblage were exposed to extremely low levels of vent fluid and sulphide while their posterior sections were consistently immersed in sulphide concentrations 1000 times higher, which led these authors to suggest that this long-skinny ecotype might have the capacity/ability to acquire sulphide through its posterior end (root-like structure). If this is taken into account in the interpretation of our results, the extension/retraction movements could indeed be a consequence of changing sulphide and oxygen saturation levels/requirements, along with an associated regulatory behaviour, i.e. through increasing or decreasing the distance between the source of the higher temperature and sulphide concentrations.

No reproductive 'actions' were observed during this study. While free spawning events have been observed for *Riftia pachyptila* [35], *Ridgeia piscesae* tubeworms transfer sperm packages (tiny white thread-like objects) from male to female using the branchial filaments of the branchial plumes [36]. The tiny size of the sperm packages, the poorly developed branchial plumes [36] of the studied assemblage as well as the relative short time span of the images analysed could explain the lack of such observations for the studied assemblage.

Polynoidae. Different species of polynoids were observed on the imagery, characterised by differences in sizes, colours and number of elytra. Polynoids known to inhabit MEF vents are *Lepidonotopodium piscesae* and *Branchinotogluma* sp. [19]. Four *Branchinotogluma* species were initially recognised on the Juan de Fuca Ridge: *B. hessleri*, *B. grasslei*, *B. sandersi* and *Opisthotrochopodus tunnicliffeae*, but the last three have recently been treated as one single species, i.e. *Branchinotogluma tunnicliffae* [37,38]. Even though it cannot be used as a distinguishing species characteristic, *L. piscesae* has been observed with bacterial-coating [7,19,37]. Several polynoids had scales missing, which could be evidence of predation or intraspecific fights [39], though no such behaviour was observed on the TEMPO-mini video footage. Additionally, their movements in between the tubeworms might prevent the elytra from regenerating (the loss of elytra in scale worms was hypothesised as being an adaptation to an interstitial environment [40]). Based on the heat maps and individual recognisability, certain polynoids were observed to appear and dwell in the same region, which raised the question of territoriality or homing patterns. Differences in local physico-chemical conditions on the tubeworm bush and microhabitat preference could contribute to this pattern.

Pycnogonida. At least one pycnogonid species, *Ammothea verenae*, is known from the neighbouring vents at MEF (known from Bastille edifice [41] and from scenescent/inactive samples at Endeavour [42]). Hence, it is likely that this species also inhabits the Grotto hydrothermal vent. It was reassigned to the genus of *Sericosura* by Bamber [43]. One of the noteworthy behaviours observed were the pycnogonids bouncing up and down on top of each other. This kind of 'dancing' or 'pumping' behaviour in females is recognised as (an initiation to) mating behaviour [44]. No negative interactions were observed between pycnogonids and tubeworms on the video imagery, more precisely, pycnogonids did not invariably cause retraction of tubeworm plumes, nor were they observed to attack or nibble on tubeworm individuals. Therefore their presence and abundance appeared more likely linked to different environmental conditions in which they tended to prevail. This was confirmed by pycnogonids showing opposite correlations with environmental variables when compared to tubeworms. Moreover, the pycnogonids were also gregarious in a very localised region of the field of view. Territoriality and suitable microhabitat could thus both explain this observed distribution pattern. There was also a significant negative correlation between microbial cover and pycnogonid densities. As *Sericosura (Ammothea) verenae* was shown to be a part of the bacterivore feeding guild [45] it could, through increasing densities and feeding (grazing), decrease the percentage microbial cover. No recurrent rhythms in pycnogonid abundance or temporal variation could be identified. Its quantification was most prone to sampling bias and possible observer effects/experience due to stacking of individuals; therefore densities presented here should be considered more indicative than quantitative.

Buccinidae. The buccinid gastropod species *Buccinum thermophilum* was also identified from other edifices of the MEF neighbouring Grotto [46]. Analysis of stomach contents of this species in a previous study included basalt chips, vestimentiferan tube parts, pycnogonid legs, polychaete setae and unidentified flesh [46]. We were unable to confirm that these whelks indeed fed on *Ridgeia piscesae* or on pycnogonids or polychaetes, as no such feeding actions were observed. The snails were seen burying themselves in the tubeworm bush, where they could possibly be feeding on smaller fauna (e.g. small crustaceans) living in interstitial spaces. This was also put forward by Martell et al. [46], who stated that, based on the lack of substratum preference, *B. thermophilum* were more likely to feed on small mobile prey rather than on vestimentiferans. The upward extended siphons in this species, observed in the present study, differed from its littoral conspecific, and could represent a response against rising sulphide-rich plumes [46]. This species featured the highest spatial segregation, not due to the large area covered, but rather because of their low densities, i.e. not much possibility to have or find conspecific neighbours. Significant periods were revealed at 4.5 days and ~9 days, though tidal-related (semi-diurnal) signals were not significant. For comparison, during a 62-day continuous video recording in a Sagami Bay cold seep at 1100 m depth, no infradian rhythms were found in visual *Buccinum* snail counts, though a significant semi-diurnal tidal component (12 h) was discernable [47].

Zoarcidae. Observed eelpout fishes of the Zoarcidae family could belong to the genus *Pachycara*, which has been reported from Endeavour. These eelpouts exhibited swift movements, which supposedly could represent feeding actions; however no specific impact was made out on the hydrothermal fauna in the field of view. There were at most 2 individuals present in the field of view and, based on their presence/absence patterns, an effect of lighting was assumed. The effect of lights was suggested to be stronger for eelpouts when compared to invertebrates [47].

Majidae. A majid spider crab, most likely *Macroregonia macrochira*, known from submersible and towed camera photographs from the Juan de Fuca and Explorer ridges and in high concentrations from around hydrothermal vents [48], visited on several accounts the filmed assemblage. Despite being known as a major predator of hydrothermal vent animals [2,18] and more specifically of *Buccinum thermophilum* snails [46], it did not cause fleeing reactions in the mobile fauna or retractions in tubeworms during its visits, nor were there any predation activities observed. Caprellid amphipods, as seen attached on the spider crab's legs, were previously identified as a new species attached to the *M. macrochira* spider crab's mouthparts and constituted the first record of this family close to hydrothermal vents (*Caprella bathytatos* from Endeavour vents [49]).

Effects of lighting. Lights were continuously powered on during the experimentation presented here. Whether this impacted the studied fauna is hard to discern, except for the apparent impact on fish presence. We observed no distinct decline in faunal invertebrate abundances linked with the duration of the experiment and consequently the exposure to continuous light.

Links with environmental variables. The distinct differences in spatial distribution for the three mobile taxa (polynoids, pycnogonids and buccinids) observed in this study could reflect different microhabitats, which could be corroborated by the different correlations between the faunal densities and temperature measurements from the different probes. Overall links between the community dynamics and the environmental variables were visualised with the variation partitioning. Most variation explained was shared between temporal periods and temperature, followed by the variation explained solely by the temporal periods. Additionally, only a very small fraction was explained exclusively by the temperature measured. This thus implies that there were temporal periods explaining variation in faunal densities that were not recognised in the environmental variables measured as such. The explanation for this can be three-fold, with the explanations not being mutually exclusive and more likely to be entwined. For one, this pattern could at least partially be attributed to the positioning of the thermistor array, which was deployed in proximity of the filmed assemblage instead of on it. Hydrothermal vents are characterised by steep gradients in environmental variables and a high local variability on a scale of centimetres [50,51], which could account for some of the discrepancies observed. The elevated spatial variation of abiotic conditions at vents was also illustrated by the differences observed between the different deployed temperature probes. Even though the probes were only separated by several decimetres at most, the same significant periods were not always as easily recognised amongst them. Local hydrodynamic conditions and bottom surface currents could influence temperature measurements on a scale of tens of centimetres [52]. Secondly, it is more than likely that there are other (non-measured) environmental factors at play, causing variations in the faunal abundances [50]. This was emphasised by the prominent amount of variation that remained unexplained. Thirdly, the organisms within the assemblage might have different individual rhythms based on energy requirement and saturation, hence revealing additional significant temporal periods. This could also be linked with the size of the individuals. Alternatively there could be a delay in reaction time between the environmental variables measured and the response of the fauna.

3. Temporal Variations in Environmental Variables

Temperature. The temperature measured in the vicinity of the filmed assemblage ranged within the limits of what is typically referred to as low-temperature fluid (5–75°C) or 'diffuse' hydrothermal flow [52]. In the temperature series, diurnal periods at ~25 h were discerned. Significant semi-diurnal periods were also found, although sometimes they could only be identified based on their harmonics. These semi-diurnal and diurnal periods and harmonics were particularly clear in T601, T602 and T603, but less in the data originating from the warmer T604. Explanation for this pattern could be due to the waning influence of the tides as the heat flux strength increases [52]. The influence of currents and local hydrodynamic conditions was visible in the case of T605 as well, occasionally bathing it in warmer fluids which might have possibly masked some of the tidal periods in the surrounding seawater.

Tidal patterns in environmental variables could be attributed to a modulation of the fluid flow or the fluid flow rate. In the Barkley canyon, another instrumented node of the NEPTUNE network at 870 m depth and closer to the shore, periods of enhanced bottom currents associated with diurnal shelf waves, internal semidiurnal tides, and also wind-generated near-inertial motions were shown to modulate methane seepage [53]. Measured tides, based on pressure data recorded by a camera platform CTD in the same canyon, were mixed semidiurnal/diurnal [54]. However, temperature variability at hydrothermal vents was shown to greatly diminish when current directions did not shift in direction with the tides [55]. Hence, it was suggested that temperature variability correlates with the variability of the current speed and direction, not with the ocean tidal pressure [55]. The modulation of temperature by tides is thus only indirect, through the modulation of horizontal bottom currents by tides. A current meter on a future deployment of the TEMPO-mini module would allow to test this hypothesis and to quantify a possible impact of the currents on the fauna in more detail.

Fe (iron). Fe is used as a proxy for vent fluid composition. However, no significant periods were found in the Fe measurements for the duration of the deployment. Nor were there any significant links established with the faunal variations in any of the analysed taxa. Semi-diurnal and diurnal periods could not be determined because the shortest period that can be resolved is twice the interval between the observations of the time-series (in this case 2×24 h after 2011-10-20) [27]. The 'lower' sampling resolution, in addition to the many gaps in the time series, might explain why no relationship was found between the faunal taxa and Fe concentrations.

On Rhythms and Recording Frequencies and Other Conclusions

Overall, tidal rhythms were shown to be at play at 2200 m depth in the hydrothermal vent ecosystem at MEF, both in fauna and environment. For the fauna, the identified rhythms may either be controlled by an internal biological clock or constitute a response of the organism to changing environmental conditions [47]. For the neat semi-diurnal and diurnal cycles in the tubeworms' appearance, the question thus remains whether these rhythms were induced by temperature, currents or other environmental variables (or a combination of them) or if they were driven by an endogenous individual/programmed trigger. For the other taxa (polynoids, pycnogonids and buccinids), tidal cycles could be blurred by biological forcing (e.g. predation and competition) or by higher-frequency changes in environmental conditions [56]. In our study, the environmental variables (a selection of several temperature probes) only explained about one-third of the variation encountered in the densities of the faunal assemblage over time. For future deployments, a more optimised environmental probe deployment, possibly including additional instruments (e.g. a current meter), should be conceived, taking into account the challenges of the deployment site at Grotto hydrothermal vent. This will allow for a more detailed assessment of entrainment between the rhythms in fauna and local environment.

The necessary importance should also be attributed to the programming of recording frequencies, taking into account the instrument's limitations, as these frequencies have an impact on the resolutions of the rhythms revealed (i.e. the shortest period that can be resolved is twice the interval between the observations of the time-series) and on their use as explanatory variables. Additional caution is needed because, depending on the frequencies investigated, the type of relationships (significance, positive or negative) between the faunal taxa might be subject to change.

In summary, there were clear differences in distribution for polynoids, buccinids and pycnogonids (with the latter showing a well-defined clustering behaviour) on the tubeworm bush, which could be indicative of different microhabitats. Negative (predatory) interactions with polynoids attacking branchial tubeworm plumes were observed, while visits of large predators (Zoarcidae and Majidae) did not cause quantifiable or apparent reactions in the observed taxa of the filmed community/assemblage. A general decrease in microbial cover coincided with increasing numbers of bacterivorous pycnogonids. Alleged mating behaviour was observed in pycnogonids, though for other taxa such observations were less obvious or unobserved. Longer time series, alternated with regulated zoomed-in recordings, will grant us more thorough and additional insights.

Even though our ecological analyses were, for now, limited to an hourly resolution, we need to emphasise the importance of the higher resolution (i.e. <1 h) observation frequencies. Additionally, longer time series of faunal densities will enlighten us on the 'representativeness' of the multiple-day periods as found in this study and possible other infradian rhythms. The time-consuming aspect of manual imagery analysis, however, is a definite limiting factor for the feasibility of such a study. In order to achieve such a fine-scale resolution and/or longer time-series analysis, the development of automated and/or semi-automated imagery analysis tools is indispensable.

Supporting Information

Figure S1 HSL (hue, saturation and light) index of imagery recorded. HSL index, combining hue, saturation and light, was calculated to assess the quality of the hourly imagery recorded during the continuous recording period (2011-10-07 to 2011-10-30). The HSL index, whose maximum value is 255 (white) and minimum is 0 (black), shows the general darkening of the recorded imagery with time.

Figure S2 Frequency of faunal density observations per taxon. Percentage of the images observed with faunal densities in defined categories. $N = 559$-23 gaps, i.e. 536 images analysed.

Figure S3 Spider crab on top of TEMPO-mini module. A majid spider crab, probably *Macroregonia macrochira*, sitting on top of the TEMPO-mini module. Two caprellid individuals attached to its legs can be recognised (possibly *Caprella bathytatos*, Caprellidae, Amphipoda).

Video S1 A polynoid polychaete attacking a tubeworm plume. A polynoid polychaete attacking, with its proboscis, an extended branchial plume of a Ridgeia piscesae tubeworm, causing it to retract into its tube. Footage recorded at 2186 m depth at the Grotto hydrothermal vent (Main Endeavour Field).

Acknowledgments

The authors would like to explicitly thank the TEMPO-mini engineers and technicians who developed and maintained the TEMPO-mini module: Julien Legrand, Yves Auffret, Stéphane Barbot, Jean-Yves Coail, Laurent Delauney, Anthony Ferrant and Gerard Guyader. We also would like to thank the NEPTUNE scientists and engineers, whose engagement and professionalism made this study possible. Extended thanks go to the captain and crews of the R/V Thomas G. Thompson and ROV ROPOS during the deployment and recovery cruises in 2011 and 2012. Thanks to IMAR/ DOP-UAç for support to DC during the final writing phase of this manuscript and to InterRidge for the cruise bursary (attributed to DC) for participation in the 'Wiring the abyss cruise 2013'.

Author Contributions

Conceived and designed the experiments: DC AL PMS JS. Performed the experiments: DC AL PMS JS. Analyzed the data: DC. Contributed reagents/materials/analysis tools: DC PL AL PMS JS. Wrote the paper: DC PL JS.

References

1. Cuvelier D, de Busserolles F, Lavaud R, Floc'h E, Fabri M-C et al (2012) Biological data extraction from imagery - How far can we go? A case study from the Mid-Atlantic Ridge. Mar Env Res 82: 15–27.

2. Tunnicliffe V, Embley RW, Holden J, Butterfield DA, Massoth G et al (1997) Biological colonization of new hydrothermal vents following an eruption on Juan de Fuca Ridge. Deep Sea Res Pt I 44(9–10): 1627–1644.

3. Shank TM, Fornari DJ, Von Damm KL, Lilley MD, Haymon RM et al (1998) Temporal and spatial patterns of biological community development at nascent deep-sea hydrothermal vents (9° 50N, East Pacific Rise). Deep Sea Res Pt II 45: 465–515.

4. Tsurumi M, Tunnicliffe V (2001) Characteristics of a hydrothermal vent assemblage on a volcanically active segment of Juan de Fuca Ridge, northeast Pacific. Can J Fish Aquat Sci 58(3): 530–542.

5. Marcus J, Tunnicliffe V, Butterfield DA (2009) Post-eruption succession of macrofaunal communities at diffuse flow hydrothermal vents on Axial Volcano, Juan de Fuca Ridge, Northeast Pacific. Deep Sea Res Pt II 56(19–20): 1586–1598.

6. Tunnicliffe V, Juniper SK (1990) Dynamic character of the hydrothermal vent habitat and the nature of sulphide chimney fauna. Prog Oceanogr 24(1–4): 1–13.

7. Sarrazin J, Robigou V, Juniper S, Delaney J (1997) Biological and geological dynamics over four years on a high-temperature sulfide structure at the Juan de Fuca Ridge hydrothermal observatory. Mar Ecol Prog Ser, 153: 5–24.

8. Desbruyères D (1998) Temporal variations in the vent communities on the East Pacific Rise and Galápagos Spreading Centre: a review of present knowledge. Cah Biol Mar 39: 241–244.

9. Cuvelier D, Sarrazin J, Colaço A, Copley JT, Glover AG et al (2011) Community dynamics over 14 years at the Eiffel Tower hydrothermal edifice on the Mid-Atlantic Ridge. Limn Oceanogr: 56(5), 1624–1640.

10. Tunnicliffe V, Garrett J, Johnson H (1990) Physical and biological factors affecting the behaviour and mortality of hydrothermal vent tubeworms (vestimentiferans). Deep Sea Res 37(1): 103–125.

11. Copley JTP, Tyler PA, Van Dover CL, Schultz A, Dickson P et al (1999) Subannual Temporal Variation in Faunal Distributions at the TAG Hydrothermal Mound (26° N, Mid-Atlantic Ridge). Mar Ecol 20(3–4): 291–306.

12. Puillat I, Lanteri N, Drogou JF, Blandin J, Géli E et al (2012) Open-Sea Observatories: A New Technology to Bring the Pulse of the Sea to Human Awareness. In: Marcelli M editor. Oceanography, ISBN: 978-953-51-0301-1, InTech 3–40.

13. Devey CW, Fisher CR, Scott S (2007) Responisble science at hydrothermal vents. Oceanography, 20(1): 162–171.

14. Auffret Y, Sarrazin J, Coail JY, Delauney L, Legrand J et al (2009) TEMPO-Mini: a custom-designed instrument for real-time monitoring of hydrothermal vent ecosystems. Martech 2009 conference proceedings.

15. Sarrazin J, Blandin J, Delauney L, Dentrecolas S, Dorval P et al (2007) TEMPO: A new ecological module for studying deep-sea community dynamics at hydrothermal vents. OCEANS '07 IEEE, Aberdeen, June 2007. Proceedings no. 061215-042.

16. Tivey MK, Delaney JR (1986) Growth of large sulfide structures on the Endeavour Segment of the Juan de Fuca Ridge. Earth Planet Sci Lett 77: 303–317.

17. Sarrazin J, Juniper S, (1999) Biological characteristics of a hydrothermal edifice mosaic community. Mar Ecol Progr Ser 185: 1–19.

18. Urcuyo IA, Massoth GJ, Julian D, Fisher CR (2003) Habitat, growth and physiological ecology of a basaltic community of Ridgeia piscesae from the Juan de Fuca Ridge. Deep Sea Res Pt I 50(6): 763–780.

19. Robert K, Onthank KL, Juniper SK, Lee RW (2012) Small-scale thermal responses of hydrothermal vent polynoid polychaetes: Preliminary in situ experiments and methodological development. J Exp Mar Biol Ecol 420–421: 69–76.

20. Vuillemin R, Le Roux D, Dorval P, Bucas K, Sudreau JP et al (2009). CHEMINI: A new in situ CHEmical MINIaturized analyzer. Deep Sea Res Pt I 56(8): 1391–1399.

21. Rasband WS (2012) ImageJ, U.S. National Institutes of Health, Bethesda, Maryland, USA, http://imagej.nih.gov/ij/, 1997–2012.

22. Warnes GR (2012) Includes R source code and/or documentation contributed by: B Bolker, L Bonebakker, R Gentleman, W Huber A Liaw, T Lumley, M Maechler, A Magnusson, S Moeller, M Schwartz and B Venables. gplots: Various R programming tools for plotting data. R package version 2.11.0. http://CRAN.R-project.org/package = gplotsRasband, W.S., 2012. ImageJ, U. S. National Institutes of Health, Bethesda, Maryland, USA, http://imagej.nih. gov/ij/, 1997–2012.

23. Zambrano-Bigiarini M (2012) hydroTSM: Time series management, analysis and interpolation for hydrological modelling. R package version 0.3–6. Available: http://CRAN.R-project.org/package = hydroTSM.

24. R Core Team (2012) R: A language and environment for statistical computing. R Foundation for Statistical Computing, Vienna, Austria. ISBN 3-900051-07-0, URL http://www.R-project.org/.

25. De la Cruz M (2008) Metodos para analizar datos puntuales. In: Introduccion al Analisis Espacial de Datos en Ecologia y Ciencias Ambientales: Metodos y Aplicaciones (eds. Maestre FT, Escudero A, Bonet A 76–127. Asociacion Espanola de Ecologia Terrestre, Universidad Rey Juan Carlos y Caja de Ahorros del Mediterraneo, Madrid. ISBN: 978-84-9849-308-5.

26. Legendre P (2012) Whittaker-Robinson periodogram. R program and documentation available: www.numericalecology.com.

27. Legendre P, Legendre L (2012) Numerical ecology. Third English Edition. Elsevier Ed. p. 306.

28. Borcard D, Legendre P, Drapeau P (1992) Partialling out the spatial component of ecological variation. Ecology 73: 1045–1055.

29. Peres-neto PR, Legendre P, Dray S, Borcard D (2006) Variation partitioning of species data matrices: Estimation and comparison of fractions. Ecology 87(10): 2614–2625.

30. Borcard D, Gillet F, Legendre P (2011) Numerical ecology with R. Use R! series, Springer Science, New York. p. 306.

31. Dixon PM (2002) Nearest-neighbor contingency table analysis of spatial segregation for several species. Ecoscience 9 (2): 142–151.

32. Bright M, Lallier FH (2010) The biology of Vestimentiferan tubeworms. Oceanogr Mar Biol, Annu Rev 48(2010): 213–266.

33. Andersen AC, Flores JF, Hourdez S (2006) Comparative branchial plume biometry between two extreme ecotypes of the hydrothermal vent tubeworm Ridgeia piscesae. Can J Zool 1822 : 1810–1822.

34. Chevaldonné P, Desbruyères D, Haitre M (1991) Time-series of temperature from three deep-sea hydrothermal vent sites. Deep Sea Res 38(11): 1417–1430.

35. Van Dover CL (1994) In situ spawning of hydrothermal vent tubeworms (Riftia pachyptila). Biol Bull 186: 134–135.

36. MacDonald IR, Tunniclife V, Southward EC (2002) Detection of sperm transfer and synchronous fertilization in Ridgeia piscesae at the Endeavour Segment, Juan de Fuca Ridge. Cah Biol Mar 43: 395–398.

37. Desbruyeres D, Segonzac M, Bright M (Eds.) (2006) Handbook of Deep-sea Hydrothermal Vent Fauna. Second Completely Revised Edition. Denisia, 18. Biologiezentrum der Oberösterreichischen Landesmuseen, Linz, Austria. 544p.

38. Levesque C, Juniper SK, Limen H (2006) Spatial organization of food webs along habitat gradients at deep-sea hydrothermal vents on Axial Volcano, Northeast Pacific. Deep Sea Res Pt I 53(4): 726–739.

39. Britayev TA, Doignon G, Eeckhaut I (1999) Symbiotic polychaetes from Papua New Guinea associated with echinoderms, with descriptions of three new species. Cah Biol Mar 40: 359–374.

40. Struck TH, Purschke G, Halanych KM (2005) A scaleless scale worm: Molecular evidence for the phylogenetic placement of Pisione remota (Pisionidae, Annelida) Mar Biol Res 1(4): 243–253.

41. Govenar BW, Bergquist DC, Urcuyo IA, Eckner JT, Fisher CR (2002) Three Ridgeia piscesae assemblages from a single Juan de Fuca Ridge sulphide edifice. Cah Biol Mar 43: 247–252.

42. Tsurumi M, Tunnicliffe V (2003) Tubeworm-associated communities at hydrothermal vents on the Juan de Fuca Ridge, northeast Pacific. Deep Sea Res Pt I 50(5): 611–629.

43. Bamber RN (2009) Two new species of Sericosura Fry & Hedgpeth, 1969 (Arthropoda: Pycnogonida: Ammotheidae), and a reassessment of the genus. Zootaxa 68: 56–68.

44. Bain AB, Govedich FG (2004) Courtship and mating behaviour in the Pycnogonida (Chelicerata: Class Pycnogonida): a summary. Invert Reprod Dev 43: 63–79.

45. Bergquist DC, Eckner JT, Urcuyo IA, Cordes EE, Hourdez S et al (2007) Using stable isotopes and quantitative community characteristics to determine a local hydrothermal vent food web. Mar Ecol Prog Ser 330(1): 49–65.

46. Martell KA, Tunnicliffe V, Macdonald IR (2002) Biological features of a buccinid whelk (Gastropoda, Neogastropoda) at the Endeavour vent fields of juan de Fuca Ridge, Northeast Pacific. J Molluscan Stud 68: 45–53.

47. Aguzzi J, Costa C, Furushima Y, Chiesa J, Company J et al (2010) Behavioral rhythms of hydrocarbon seep fauna in relation to internal tides. Mar Ecol Prog Ser 418: 47–56.

48. Tunnicliffe V, Jensen RG (1987) Distribution and behaviour of the spider crab Macroregonia macrochira Sakai (Brachyura) around the hydrothermal vents of the northeast Pacific. Can J Zool 65: 2442–2449.

49. Martin JW, Pettit G (1998) Caprella bathytatos New Species (Crustacea, Amphipoda, Capresslidae), from the Mouthparts of the Crab Macroregonia macrochira Sakai (Brachyura, Majidae) in the Vicinity of Deep-sea Hydrothermal vents off British Columbia. B Mar Sci 63: 189–198.

50. Sarrazin J, Juniper SK, Massoth G, Legendre P (1999) Physical and chemical factors influencing species distributions on hydrothermal sulfide edifices of the Juan de Fuca Ridge, northeast Pacific. Mar Ecol Prog Ser 190: 89–112.

51. Le Bris N, Govenar BW, Legall C, Fisher C (2006) Variability of physico-chemical conditions in 9°50′N EPR diffuse flow vent habitats. Mar Chem 98(2–4): 167–182.

52. Hautala S, Johnson HP, Pruis M, García-Berdeal I, Bjorklund T (2012) Low-temperature hydrothermal plumes in the near-bottom boundary layer at Endeavour Segment, Juan de Fuca Ridge. Oceanography 25: 192–195.

53. Thomsen L, Barnes C, Best M, Chapman R, Pirenne B et al (2012) Ocean circulation promotes methane release from gas hydrate outcrops at the NEPTUNE Canada Barkley Canyon node. Geophys Ress Lett 39: L16605.

54. Juniper SK, Matabos M, Mihály S, Ajayamohan RS, Gervais F et al (2013) A year in Barkley Canyon: A time-series observatory study of mid-slope benthos and habitat dynamics using the NEPTUNE Canada network. Deep Sea Res Pt II 92: 114–123.

55. Tivey MK, Bradley AM, Joyce TM, Kadko D (2002) Insights into tide-related variability at seafloor hydrothermal vents from time-series temperature measurements. Earth Planet Sci Lett 202: 693–707.

56. Aguzzi J, Company JB, Sardà F, Abelló P (2003) Circadian oxygen consumption patterns in continental slope *Nephrops norvegicus* (Decapoda: Nephropidae) in the western Mediterranean. J. Crustacean Biol. 23: 749–757.

57. Delaney JR, Kelley DS, Lilley MD, Butterfield DA, McDuff RE et al (1997) The Endeavour Hydrothermal System I: Cellular circulation above an active cracking front yields large sulfide structures, "fresh" vent water, and hyperthermophilic archae, RIDGE Events: 11–19.

5

Methane-Carbon Flow into the Benthic Food Web at Cold Seeps – A Case Study from the Costa Rica Subduction Zone

Helge Niemann[1,2*]**, Peter Linke**[3,4]**, Katrin Knittel**[2]**, Enrique MacPherson**[5]**, Antje Boetius**[2,6]**,
Warner Brückmann**[3,4]**, Gaute Larvik**[2]**, Klaus Wallmann**[3,4]**, Ulrike Schacht**[3¤]**, Enoma Omoregie**[2,7]**,
David Hilton**[8]**, Kevin Brown**[8]**, Gregor Rehder**[3,9]

1 Department of Environmental Sciences, University of Basel, Basel, Switzerland, **2** Max Planck Institute for Marine Microbiology, Bremen, Germany, **3** Sonderforschungsbereich 574, University of Kiel, Kiel, Germany, **4** Helmholtz Centre for Ocean Research Kiel, GEOMAR, Kiel, Germany, **5** Centro de Estudios Avanzados de Blanes (CEAB-CSIC), Blanes, Spain, **6** Alfred Wegener Institute for Marine and Polar Research, Bremerhaven, Germany, **7** Centro de Astrobiología (CSIC/INTA), Instituto Nacional de Técnica Aeroespacial Torrejón de Ardoz, Madrid, Spain, **8** Scripps Institution of Oceanography, University of California, San Diego, United States of America, **9** Leibniz Institute for Baltic Sea Research Warnemünde (IOW), Rostock, Germany

Abstract

Cold seep ecosystems can support enormous biomasses of free-living and symbiotic chemoautotrophic organisms that get their energy from the oxidation of methane or sulfide. Most of this biomass derives from animals that are associated with bacterial symbionts, which are able to metabolize the chemical resources provided by the seeping fluids. Often these systems also harbor dense accumulations of non-symbiotic megafauna, which can be relevant in exporting chemosynthetically fixed carbon from seeps to the surrounding deep sea. Here we investigated the carbon sources of lithodid crabs (*Paralomis* sp.) feeding on thiotrophic bacterial mats at an active mud volcano at the Costa Rica subduction zone. To evaluate the dietary carbon source of the crabs, we compared the microbial community in stomach contents with surface sediments covered by microbial mats. The stomach content analyses revealed a dominance of epsilonproteo-bacterial 16S rRNA gene sequences related to the free-living and epibiotic sulfur oxidiser *Sulfurovum* sp. We also found *Sulfurovum* sp. as well as members of the genera *Arcobacter* and *Sulfurimonas* in mat-covered surface sediments where *Epsilonproteobacteria* were highly abundant constituting 10% of total cells. Furthermore, we detected substantial amounts of bacterial fatty acids such as i-C15:0 and C17:1ω6c with stable carbon isotope compositions as low as −53‰ in the stomach and muscle tissue. These results indicate that the white microbial mats at Mound 12 are comprised of *Epsilonproteobacteria* and that microbial mat-derived carbon provides an important contribution to the crab's nutrition. In addition, our lipid analyses also suggest that the crabs feed on other ^{13}C-depleted organic matter sources, possibly symbiotic megafauna as well as on photosynthetic carbon sources such as sedimentary detritus.

Editor: Hauke Smidt, Wageningen University, The Netherlands

Funding: Financial support for the M66-Subflux cruise came through the Collaborative Research Center (SFB) 574 ("Volatiles and Fluids in Subduction Zones") at Kiel University funded by the DFG. This work was further supported by NSF (OCE-0242034 and OCE-0242091), the Max Planck Society, the Helmholtz Association and the University of Basel. The funders had no role in study design, data collection and analysis, decision to publish, or preparation of the manuscript.

Competing Interests: The authors have declared that no competing interests exist.

* E-mail: helge.niemann@unibas.ch

¤ Current address: CO2CRC, Australian School of Petroleum, The University of Adelaide, Adelaide, Australia

Introduction

Most deep-sea ecosystems on Earth are considered to be energy limited, because they depend on a small fraction of photosynthetically produced organic carbon (C), which sinks from the productive ocean surface to the seafloor [1,2]. They are contrasted by chemosynthetic ecosystems such as hydrothermal vents and seeps, which are fueled by chemical energy transported with subsurface fluids. Especially cold seeps, which form around mud, gas and oil escape structures and which are characterized by high methane effluxes [3,4], support high biomasses of deep-sea life, comprising chemosynthetic microbial mats and megafauna as well as associated heterotrophic animals [5–9]. The key biogeochemical process at cold seeps is the anaerobic oxidation of methane

with sulfate (AOM), which is a net conversion of methane and sulfate to carbon dioxide and sulfide [10–12]. The seeping sulfide fuels aerobic thiotrophic communities comprising free-living and symbiotic bacteria. The free-living thiotrophs often form dense microbial mats above gassy sediments [13–15]. Symbiotic megafauna such as bathymodiolin bivalves and siboglinid tubeworms host thiotrophic bacteria in specialized cells and tissues [7]. Oxidised cold seep surface sediments may also support free-living aerobic methanotrophs [16–19]. These are not known to form dense mats at cold seeps, but they also occur as endosymbiotic associations with megafauna, such as bivalves and tubeworms [20–23]. Furthermore, some highly adapted, hydrothermal vent or cold seep endemic annelids [24], gastropods [25] and crustaceans

[26,27] farm chemosynthetic, microbial epibionts on their skin and shells, which they graze upon.

An important question in the ecology of vent and seep ecosystems remains as to how chemosynthetically fixed carbon is transferred to the deep-sea food web [5,9,28–30]. Current knowledge is mostly based on measurements of the stable C isotope ratio of faunal bulk tissue [5,9,31,32]. At cold seeps, both methane and its oxidation product CO_2, are strongly depleted in ^{13}C [33]. Consequently, methanotrophic and thiotrophic bacteria, which incorporate ^{13}C-depleted methane and/or $^{13}CO_2$ in their biomass are characterized by $\delta^{13}C$-values much lower than -15 to $-30‰$, which is the range typical for photosynthetically fixed C [33,34]. Consumer species feeding on free living chemosynthetic bacteria or symbiotic fauna hosting these microbes in their tissue will also incorporate the ^{13}C-depleted C in their biomass [5,35]. A valuable addition to the measurement of bulk tissue is the analysis of compound-specific $\delta^{13}C$-values, for example of fatty acids (FAs), which are contained in cellular membranes [36,37]. These lipids are incorporated from the food sometimes without significant alteration into the consumer biomass; e.g. essential fatty acids [38]. Furthermore, some lipids are diagnostic biomarkers because they are synthesized by specific source organisms. The analysis of their presence and specific C-isotope composition help to identify multiple dietary C-sources utilized by a consumer. However, the isotopic composition of biomass typically integrates over significant parts of an organism's lifetime. In order to investigate food sources that a consumer ingested only recently, the analyses of stomach content, including DNA, are frequently used in food web studies [39–41].

The aim of this study was to assess the importance of CH_4-derived carbon for a dominant consumer – lithodid crabs – of the benthic food web at an active cold seep of the Costa Rica subduction zone. We combined bulk- and compound-specific stable C isotope analyses of muscle tissue and stomach contents as well as fluorescence in situ hybridization and screening for microbial 16S rRNA gene sequences to investigate the relevance of chemosynthetically-derived carbon for the crab's nutrition.

Materials and Methods

Site description

Mound 12 (Md. 12) is an active mud volcano located at the Central America convergent margin off the coast of Costa Rica at 1020 m water depth (8° 55.85′ N, 84° 18.75′ W; [42]). It belongs to a series of cold seeps along the Costa Rican Pacific margin, which are related to the subduction of the Cocos plate and erosion of continental material, subsequent dehydration of subducted clay minerals as well as production of thermogenic CH_4 [43–45]. At Md. 12, CH_4, geofluids and mud ascend to the seafloor along faults, which cut deeply through the basement and upper plate sediments [46]. Diapirism and mudflows have formed a roundish (~800 m diameter) cone-shaped relief (<30 m) with an irregular pinnacle in the NE and a lower profile ridge in the SW [42,47]. The mudflows are intercalated with slope sediments, indicating that Md. 12 is frequently active, alternated by low-activity phases. At present, the mound seems to be most active at its pinnacle and the SW flank, which is characterized by dense microbial mats and other chemosynthetic organisms (mytilid mussels and Lamellibrachia tube worms) [47–49]. At a microbial mat site, we previously measured a total CH_4 flux of ~ 10 mol m^{-2} yr^{-1} of which only half was oxidized with SO_4^{2-} [47]. Indeed, bottom waters above Md. 12 were enriched in CH_4 with 1–2 orders of magnitude higher concentrations compared to background values,

indicating that a significant fraction of the seeping CH_4 can escape into the water column [49].

Sea floor observations and sampling

We visited Md. 12 during two consecutive cruises with R/V Atlantis (AT11-28) and R/V Meteor (M66-2) in June and September 2005, respectively. Direct and/or video observations were carried out in June with DSV Alvin (Woods Hole Oceanographic Institute, USA) and in September with ROV Quest (Marum, Germany). In addition, we also photographed the sea floor during cruise M66-2 over a time period of 408 hours with a frequency of 2 pictures per hour. For this approach, a downward-facing digital still camera (Ocean Imaging Systems, North Falmouth, USA, 6.1 Mpix) was mounted on a lander frame (Deep-sea Observation System – DOS [48]) resulting in a field of vision of 0.4 m^2. The lander was deployed on top of a microbial mat (8° 55.69′ N, 84° 18.78′ W), which covered ~60% of the cameras field of vision.

A specimen of the abundantly observed lithodid crab (see results and discussion section for a taxonomic assessment), which was apparently feeding on microbial mats, was sampled using DSV Alvin's manipulators (8° 55.72′ N, 84° 18.83′ W). The crab was stored in a basket until surfacing of the submersible and directly thereafter photographed and dissected. A tissue sample from a leg muscle and the stomach were removed and frozen at $-20°C$ until further analyses in the home laboratory. A ~6 m wide sediment strip (8° 55.69′ N, 84° 18.82′ W) covered by the whitish, thiotrophic microbial mats as well as bare sediments 1–2 m adjacent to the microbial mat were sampled by push coring with ROV Quest.

Taxonomic identification of lithodid crabs

The lithodid crabs were taxonomically identified from photographs that we recorded in situ (i.e., with the deep-sea camera of the DOS lander; e.g. Fig. 1b), and on board from the specimen recovered with Alvin (e.g. Fig. 1c, d). Identification was based on morphological features such as spines, spinules and granules according to our previous work [50].

Lipid analyses and determination of C and N contents

Extraction of lipids, separation and derivatization was carried out as described previously [51,52]. Briefly, total lipid extracts (TLEs) were obtained from subsamples of the muscle tissue (~500 mg wet weight – ww.) and stomach (including its contents; ~400 mg ww.) by ultrasonication with organic solvent mixtures (methanol and dichloromethane) of decreasing polarity. The TLEs were then saponified and subsequently separated into fractions containing (i) fatty acids (FAs), (ii) hydrocarbons, (iii) ketons and (iv) alcohols (including glycerol ethers). FAs and alcohols were methylated prior to extraction using BF$_3$ in methanol and bis(trimethylsilyl)trifluoracetamide (BSTFA) to form fatty acid methyl esters FAMES and trimethylsilyl (TMS) ethers, respectively. Separation of single lipid compounds, their identification, quantification and the determination of their stable carbon isotope composition was achieved by gas chromatography (GC) coupled to flame ionization detection (GC-FID), quadrupole mass spectrometry with electron ionization (GC-MS) and isotope ratio mass spectrometry (GC-IRMS), respectively [53]. Bulk stable carbon isotope composition was measured from CO_2, released after flash combustion of ~100 mg (ww.) of muscle tissue in an automated elemental analyzer (Thermo Flash EA, 1112 Series) coupled to an isotope ratio mass spectrometer (Finnigan Deltaplus XP, Thermo Scientific).

Figure 1. A lithodid crab (*Paralomis diomedeae* relative) commonly encountered at Md. 12. (a) Bird's eye view from a lander mounted still camera (ca. 40×50 cm), (b) close up with visible feeding tracks, (c) dorsal and (d) ventral view of the captured specimen. The scale bars represent 6 cm.

Determination of bulk C and N contents was carried out according to standard methods (www.geomar.de/en/research/fb2/fb2-mg/benthic-biogeochemistry/mg-analytik/determination-of-cns/). Briefly, all inorganic and organic C and N compounds in sediment samples were flash combusted in a CNS analyzer (Carlo Erba Instruments, LTD) and the resulting combustion gases were analyzed with a thermal conductivity detector yielding total C and N contents. Organic C was determined in a similar fashion subsequently to the removal of carbonate-bound C with HCl. C:N-ratios are reported as the molar ratio of organic C versus total N.

DNA extraction and clone library construction

Total DNA of the microbial community in the crab's stomach was extracted from ~350 mg (ww.) of stomach material using the FastDNA spin kit for soil (Q-Biogene, USA) as described elsewhere [53]. PCR amplification of 16S rRNA genes, cloning, and sequencing was conducted according to [16]. For the construction of the epsilonproteobacterial clone library, a subsample of 50 μl of formaline-fixed sediment sample (the same sample as used for CARD-FISH, see next section) was centrifuged and the pellet was washed with 1× PBS and finally resuspended in 50 μl H_2O. Subsequently, we sonicated the sample (2×30 sec, 35 kHz) in a water bath sonicator. 1 μl of a 100-fold dilution was used as template for specific amplification of epsilonproteobacterial 16S rRNA gene sequences using primers Epsi682F (5′ TGTGTAGGGGTAAAATCCG 3′)/GM4. The PCR conditions were as follows: 32 cycles, annealing temperature 44°C. Ten parallel PCRs of each sample were pooled, purified using the QIAquick gel extraction kit (Qiagen, Hilden, Germany) and eluted in 30 μl H_2O. Cloning reactions were performed with the TOPO TA Cloning Kit (Invitrogen, San Diego, CA, USA) and inserts sequenced using the BigDye Terminator v3.1 Cycle Sequencing Kit (Applied Biosystems, Carlsbad, CA, USA) on an ABI PRISM 3130xl Genetic Analyzer. Sequences were checked for chimeras using the program UCHIME [54] and phylogenetically analyzed with the ARB software package using database SSURef_-SILVA_111 (July 2012, 739,633 sequences) downloaded from ARB SILVA resources [55]. The sequence data from the stomach sample will be published in the EMBL, GenBank and DDBJ nucleotide sequence databases under the accession numbers HE974888 to HE974904 as well as HF559372 and HF559373. Sequences from the epsilonproteobacterial clone library will be published under the accession numbers HG321355-HG321366.

Cell enumeration and catalyzed reporter deposition fluorescence in situ hybridization (CARD–FISH)

Sediment samples for CARD-FISH were fixed in formaldehyde solution, washed in PBS and stored at $-20°C$ as described previously [16]. CARD-FISH was carried out on two parallel surface sediment samples (0–2 cm) from the microbial mat habitat and on one sediment sample from the adjacent, non-covered sediment as described previously [56] with the following modifications: Samples were sonicated before filtration (20 s an amplitude of 42 μm <10 W; MS73 probe, Sonopuls HD70, Bandelin, Germany) and endogenous peroxidases were inactivated by incubation in 0.5% H_2O_2 in methanol for 30 min at room temperature. Cell walls were permeabilized with 10 mg ml^{-1} lysozyme in 1×TE-buffer for 45 min at 37°C [57]. For the specific detection of *Epsilonproteobacteria*, we used the HRP-labeled probe Epsi682 (5′-CGGATTTTACCCCTACACM- 3′; biomers.net) [58] applied at 20% formamide. Cells were stained with DAPI, embedded in mounting medium and counted under an epifluorescence microscope in 20–100 independent microscopic fields.

Methane oxidation- and sulfate reduction rate measurements

Microbial turnover of CH_4 and SO_4^{2-} in sediments of Md. 12 was measured with radiotracer assays according to previously published works [51,59,60]. Briefly, CH_4 oxidation and sulfate reduction (SR) rates were determined from 6 push cores distributed over the ~6 m wide sediment strip covered with bacterial mats and from 3 push cores recovered 1–2 m away from this mat.

Results and Discussion

Sea floor observations and biogeochemical environment

Sea floor habitat. We visited Md. 12 in 2005 and investigated the seafloor with DSV Alvin and ROV Quest. Visually, we could identify several habitats: reduced sediments covered by whitish microbial mats (e.g. Fig. 1a, b) and adjacent bare sediments without microbial mats (movie S1 in the supplements), colonies of bathymodiolin mussels (*Bathymodiolus* sp.) or siboglinid tubeworms (*Lamellibrachia* sp.) and CH_4-derived carbonate pavements. As reported previously [47–49], these habitats were distributed in a patchy fashion, interspersed by olive-green sediments. The size of the microbial mat patches varied from decimeters to several meters in diameter. The whitish color of the mats suggested that they consisted of thiotrophic bacteria, but the morphology of the mats differed in thickness and structure from those present at most known cold seep systems formed by large sulfur bacteria such as *Beggiatoa*, *Thiomargarita* or *Thioploca* [6,15,53,61]. They resembled more the *Arcobacter* type mats known from mud volcanoes, such as of the Eastern Mediterranean [14,62]. The sediments below the mats strongly smelled of sulfide, and previous measurements found ∼15 mM sulfide in porewaters from this habitat [47]. All cores recovered from the microbial mats were also rich in CH_4 as indicated by their degassing during recovery, and the oversaturated CH_4 concentrations of ≥1.4 mM in recovered sediments (data not shown). *Ex situ* rate measurements of AOM and SR showed peak values of up to 225 and 327 nmol cm^{-3} d^{-1}, and integrated rates of 7.4 and 6.5 mol m^{-2} yr^{-1} for AOM and SR, respectively (Tab. 1). The high sulfide concentrations are thus explained by AOM–dependent SR. We found a high variability in rate measurements when comparing replicates (Tab. 1), possibly related to heterogeneous fluid flow regimes [47]. Just ∼1 m outside the microbial mat habitat, sediments barely smelled of sulfide, and AOM and SR rates were <5 nmol cm^{-3} d^{-1}, equivalent to areal rates of <0.3 mol m^{-2} yr^{-1} (Tab. 1).

Lithodid crabs grazing on microbial mats. We frequently observed one type of lithodid crab, which dwelled and apparently fed on the microbial mats of Md. 12 (Fig. 1 a–d, movie S1). Based on the shape of the carapace, rostrum and abdomen documented by high-resolution photography, we identified this species as *Paralomis* sp. [50]. Its morphology is similar to *P. diomedeae*, known to populate continental margins from Costa Rica to Peru, but it differs by the granules on the dorsal carapace surface and the

armature of the chelipeds and walking legs. This suggests that the *Paralomis* type of Md. 12 could be a new *Paralomis* species, closely related to *P. diomedeae*. A conclusive determination of the crab's taxonomic status requires collection of new material and in-depth morphological and genetical investigations. We did not conduct off-site surveys during our sampling campaigns so that we can only speculate about the biogeography of the *P. diomedeae* related crabs and potential adaptations for the consumption of chemosynthtic biomass. Little is known about the ecology of *P. diomedeae* but the mouth parts (mandibles, maxillae, maxillulae and maxillipeds) of the previously examined specimens from the eastern Pacific Ocean off Costa Rica [50] and the ones of Md. 12 indicate that both are adapted to an omnivorous diet including detritus. Indeed, during submersible and ROV dives, we observed that the *Paralomis* sp. of Md. 12 grazed on the microbial mats (or on surface sediments including the mats) leaving clearly distinguishable feeding tracks of bare sediments behind (Fig. 1a, b). Other members of the genus *Paralomis*, possibly opportunistic scavengers or predators, have also been observed at other cold seeps and hydrothermal vents constituting a potential link for the export of seep carbon to the surrounding deep sea [28,63,64]. However, to our knowledge, only one other publication has reported similar, direct observations from a cold seep setting, i.e. hermit crabs feeding on *Beggiatoa* mats at the Gullfaks seeps, North Sea [65]. The longer-term recordings of the lander-mounted still camera provided further evidence that the microbial mats apparently attracted *Paralomis* sp. (movie S1). During the 408 hours of observation with the lander-mounted camera, we counted 184 sightings of this crab species on a microbial mat patch while only 6 sightings were recorded from surrounding sediments (Tab. 1). Our observations, furthermore, indicate a pattern where intensive grazing was followed by a time period between 8 and 33.5 h of little or no grazing during which the mat regrew until it was grazed of again. We could also record the occurrence of a larger food fall, i.e. a Pyrosome (tunicate colony), which was also consumed by the *Paralomis* sp. (movie S1), confirming that they are opportunistic predators/scavengers. In addition to the *P. diomedeae* relative, we also noticed a second but rather rarely occurring *Paralomis* species (Fig. 2), which we tentatively identified as *P. papillata* or a relative of this species [50]. However, the few available photo documents did not allow for a more reliable identification. One specimen of the so-called Yeti Crab (*Kiwa puravida*) could be seen once on the photo material of the lander mounted camera. We could not observe the *P.*

Table 1. Habitat characteristics of Md. 12 sediments covered- and devoid of microbial mats.

	microbial mat	adjacent sediments
tot. No. of crabs observed	184	6
oxidation state	strongly reduced	oxic/anoxic/slightly reduced
organic C (wt%)	2.6 (±0.1)	2.5 (±0.1)
C:N-ratio	9.9 (±0.5)	10.0 (±0.4)
sediment depth of AOM max. (cm)	3 cm	-
AOM max. (nmol cm^{-3} d^{-1})	225 (±60)	5 (±1)
areal AOM (mol m^{-2} yr^{-1})	7.5 (± 1.8)	0.2 (±0.04)
SR max. (nmol cm^{-3} d^{-1})	328 (±107)	8 (±5)
areal SR (mol m^{-2} yr^{-1})	6.5 (±1.8)	0.3 (±0.18)

Total number of crabs was counted from still camera images (2 pictures h^{-1}) during an observation period of 408 h. Note that we did not account for feeding tracks without a photo record of the originator and that single specimens could have been counted repeatedly. Org. C contents and C:N-ratios were averaged over the first 10 cm- and AOM and SR rates were integrated over the first 16 cm of surface sediment. Errors are presented as standard error.

Figure 2. A second type of lithodid crab that we observed rarely at Md. 12 (tentatively identified as *Paralomis papillata* relative).

papillata relative or the Yeti Crab feeding on the mats but we noticed snails, which seemed to feed on the mats (movie S1).

Dietary carbon sources for the Paralomis diomedeae relative

Sediment C and N contents. To investigate whether the *P. diomedeae* relative preferentially feeds on microbial mats compared to regular sediments as suggested by our observations, we compared the bulk chemical composition of surface sediments. Both habitat types were characterized by high contents of organic C (~2.5 weight%) and low C:N-ratios (~10, Tab. 1) throughout the upper 10 cm of surface sediments. These values are comparable to seafloor sediments from the highly productive upwelling regions of Peru [66] or Chile [67] at ~1000 m water depth and are indicative for a high fraction of fresh organic matter. This may be explained by the high pelagic primary production in the region of the Costa Rica Dome [68]. The organic deposits in sediments surrounding the seeps of Md. 12 could thus also serve as a relevant carbon source for the *Paralomis sp*. Nevertheless, the nutritional value of the microbial mat is probably higher than that of sediment detritus because of the low C:N-ratio of bacteria (typically 4–5), caused by a relatively high cellular protein content (~50%) [33]. As the bacterial mat was very thin, the rather coarse sediment sampling of 2 cm sections may thus have masked this signal. Besides the microbial mats and the sedimentary detritus, also the symbiotic megafauna at Md. 12 (i.e., *Bathymodiolus* sp. and *Lamellibrachia* sp.) could be an attractive food source for the crabs. Bivalves and annelids typically contain very high protein contents, which may comprise >70% of their organic matter [34]. However, we did not observe the crabs to feed on these potential food sources.

To further investigate the dietary importance of chemosynthetic vs. photosynthetic carbon for the *Paralomis* sp. we analyzed the molecular signatures of stomach contents, muscle tissue and surface sediments covered by microbial mats (see next 2 sections).

16S rRNA gene libraries and FISH. CARD-FISH analyses of two parallel samples of the microbial mat and underlying sediments with the *Epsilonproteobacteria*-specific probe EPSI682 indicated that *Epsilonproteobacteria* constituted 9.5 and 11.1% of single cells. In contrast, in the surface layer of the adjacent, bare sea floor, we could only detect <2% *Epsilonproteobacteria*. With respect to the morphological appearance of the mat, this confirmed dominance by *Epsilonproteobacteria* rather than by large gammaproteobacterial thiotrophs (*Beggiatoa*, *Thiomargarita* or *Thio-*

ploca). We used probe EPSI682 as a specific forward primer together with the general bacterial primer GM4 in a PCR to resolve the diversity of CARD-FISH-detected *Epsilonproteobacteria* in the microbial mat habitat. Of the epsilonproteobacterial 16S rRNA genes (53 epsilonproteobacterial sequences from 72 clones analyzed in total, Tab. 2), six sequences grouped within the genus *Sulfurovum*. Other epsilonproteobacterial sequences belonged to the genera *Arcobacter* (25 sequences), *Sulfurimonas* (17 sequences), and *Campylobacter* (5 sequences).

From the stomach contents, we could amplify bacterial 16S rRNA gene sequences successfully but repeated attempts to amplify archaeal rRNA genes failed. This likely indicates a very low abundance of archaea in the stomach contents, which is in accordance with our biomarker analyses (see next section). From the amplified bacterial 16S rRNA genes, we analyzed a total of 79 clones. We identified *Epsilonproteobacteria* of the genus *Sulfurovum* as the dominant bacterial group in the stomach of the *Paralomis* sp. (Tab. 3). Two groups (8 and 17 sequences, respectively) with a high intragroup sequence similarity of 98–99% and 94–95% between the two groups were detected. Sequences of cluster 1 were 99.8% similar to sequences from Eel River Basin methane seeps ([69] e.g. acc.no.FJ264599) and those of the second cluster were 97.8% similar to a sequence obtained from particulate detritus from grabs of the vestimentiferan tubeworm *Ridgeia piscesae* (Forget & Jupiter, database release, acc.no. JN662293). Furthermore, sediment *Sulfurovum* sp. was highly similar to the *Sulfurovum* sp. cluster 1 found in the crab's stomach (96.8–99.8% sequence similarity). Also gut and sediment *Campylobacter* spp. showed a high degree of similarity (up to 98.7%). Although *Arcobacter*- and *Sulfurimonas*-related sequences were not retrieved from the crab's stomach, these results provide evidence that *Epsilonproteobacteria* in the stomach originate from the thiotrophic mats, which the crab was observed to feed upon. Together with our observations of crabs feeding specifically on microbial mats, this strongly suggests that these mats are an important nutrition source for the *P. diomedeae* relative recovered from Md. 12.

Epsilonproteobacteria are known from a variety of hydrothermal vents [70] but have also been found at cold seeps [19,62,69,71] including brines [72,73]. Members of the genus *Sulfurovum* have been found as free-living bacteria [74,75], episymbionts associated with a hydrothermal vent shrimp [70,76] and with the cold seep associated Yeti Crab (*Kiwa puravida*), the latter of which was also found at Md. 12 [26]. Members of the *Sulfurovum* clade were also found in the gut system of the Yeti Crab and a hydrothermal vent shrimp [26,77]. However, these *Sulfurovum* types shared only ~95% similarity with our sequences. The biogeochemical functioning of the *Sulfurovum* relatives constituting the microbial mats at Md. 12 is not clear. Known members of the genus *Sulfurovum* use elemental sulfur or thiosulfate as an electron donor, and nitrate or oxygen as electron acceptors [74,78,79]. Whole genome sequencing of a *Sulfurovum* strain (NBC37-1) revealed the presence of *sox* genes (coding for enzymes involved in sulfide oxidation) and the strain also had cytoplasmic and periplasmic sulfide-quinone oxidoreductases that oxidize sulfide to elemental sulfur [80].

The stomach contents also contained sequences of other, seep-related chemosynthetic microbes including aerobic organisms thriving in the upmost, oxic surface sediment layer as well as anaerobic strains from deeper sediment layers. We detected one sequence of a relative of *Hypomicrobium* and *Acinetobacter*, which were previously found to grow aerobically on chloro- or dichloromethane [81] and long-chain alkanes [82], respectively. Among the anaerobic strains, we detected two deltaproteobacterial sequences belonging to relatives of the *Desulfobulbus*/Seep-SRB3

Table 2. Epsilonproteobacterial 16S rRNA gene library obtained from surface sediments (0–2 cm) covered with whitish microbial mats.

Order	Family	Genus	No. of clones	Clone representative	Acc. No.
Campylobacterales	Helicobacteraceae	Sulfurovum, cluster 1	6	CRsed_Md12_64_17A3	HG321355
		Sulfurimonas	17	CRsed_Md12_64_82B11	HG321360
	Campylobacteraceae	Campylobacter	5	CRsed_Md12_64_45E6	HG321356
		Arcobacter	25	CRsed_Md12_64_66B9	HG321357

cluster, one sequence of the SEEP-SRB2 cluster, which comprise SRB associated to ANME archaea [10,83], and three sequences related to *Desulfocapsa*, which is a typical SRB in marine sediments, including cold seeps [16]. Furthermore, we also found other bacteria of unknown biogeochemical function that have regularly been found in anoxic cold seep sediments, i.e. relatives of the Candidate Division OD1 and *Propionibacterium* (of which we found six and two sequences, respectively) [11,84,85]. However, the relatively low abundance of sequences of anaerobic cold seep microbes indicates that the crab specimen analyzed here mostly fed on oxic surface- and ingested rather little amounts of reduced sediments containing AOM biomass, at least during its last feeding activities. The relatively deep position of the AOM horizon (~3 cm, Tab. 1) could make archaeal biomass rather inaccessible

for the *P. diomedeae* relative or the expectedly high sulfide contents of the AOM horizon [47] could be too toxic.

25 out of 78 sequences were affiliated with *Candidatus Lumbricincola* and *Candidatus Bacilloplasma*, relatives that most likely belong to the gut flora of the *Paralomis* sp. *Candidatus Lumbricincola* has yet only been found in the gut systems of annelids [86]. *Candidatus Bacilloplasma* relatives, on the other hand, were found in the guts of decapod crustaceans (*Scylla* sp.; Sun & Li, database release acc.no. AY360354 and *Nephrops norvegicus* [87]) isopods [88] and chordates (Wu & Wang, database release ac.no. GU293173). Members of the class *Mollicutes* are often pathogenic or parasitic, but also commensal and beneficial associations with their hosts have been found [41,89].

Stable carbon isotope and lipid analyses. The bulk stable carbon isotope composition of the muscle tissue was −46‰ (Fig. 3)

Table 3. Bacterial 16S rRNA gene library obtained from the stomach sample of a lithodid crab (*Paralomis diomedeae* relative), which was observed feeding on surface sediments covered with whitish microbial mats.

Phylum	Class	Order	Family	Genus	# Clones	Clone Repres.	Acc. No.
Proteobacteria	Alphaproteobacteria	Rhizobiales	Hyphomicrobiaceae	uncultured	1	ATLA_Crab_Bac_E11	HE974888
	Gammaproteobacteria	Enterobacteriales	Enterobacteriaceae	Enterobacter	1	ATLA_Crab_Bac_F06	HE974889
		Pseudomonadales	Moraxellaceae	Acinetobacter	1	ATLA_Crab_Bac_B03	HE974890
	Deltaproteobacteria	Desulfobacterales	Desulfobulbaceae	Desulfocapsa	3	ATLA_Crab_Bac_C05	HE974891
				Desulfobulbus /Seep-SRB3	2	ATLA_Crab_Bac_H03	HE974892
		Desulfobacterales		Seep-SRB2	1	ATLA_Crab_Bac_A02	HF559372
	Epsilonproteobacteria	Campylobacterales	Helicobacteraceae	Sulfurovum, cluster 1	8	ATLA_Crab_Bac_H05	HE974893
				Sulfurovum, cluster 2	17	ATLA_Crab_Bac_E05	HF559373
			Campylobacteraceae	Campylobacter	1	ATLA_Crab_Bac_B12	HE974894
Bacteroidetes	Flavobacteria	Flavobacteriales	Flavobacteriaceae	Cloacibacterium	1	ATLA_Crab_Bac_E03	HE974895
Planctomycetes	Planctomycetacia	Planctomycetales	Planctomycetaceae	Rhodopirellula	1	ATLA_Crab_Bac_E08	HE974896
Acidobacteria	Acidobacteria	Acidobacteriales	Acidobacteriaceae	uncultured	1	ATLA_Crab_Bac_A11	HE974897
Firmicutes	Clostridia	Clostridiales	Lachnospiraceae	Cellulosilyticum	1	ATLA_Crab_Bac_D05	HE974898
	Bacilli	Lactobacillales			1	ATLA_Crab_Bac_G01	HE974899
Tenericutes	Mollicutes	Mycoplasmatales	Mycoplasmataceae	Lumbricincola and Bacilloplasma relatives	25	ATLA_Crab_Bac_D11	HE974900
Actinobacteria	Actinobacteria	Propionibacteriales	Propionibacteriaceae	Propionibacterium	2	ATLA_Crab_Bac_C06	HE974901
Chloroflexi					1	ATLA_Crab_Bac_G12	HE974902
Candidate Division OD1					6	ATLA_Crab_Bac_A12	HE974903
Cyanobacteria (chloroplast)					4	ATLA_Crab_Bac_C07	HE974904

and thus extremely ^{13}C-depleted in comparison to organic matter in regular, recent marine sediments (−10 to −35‰), which are usually of photosynthetic origin (Calvin Benson Cycle) [33]. In eukaryotes, such negative carbon isotope signatures are typically attributed to a methanotrophic food chain [30,36,90–92]. However, also sulfate reducing bacteria and thiotrophs may show similar signatures by incorporating isotopically depleted CO_2 derived from methane oxidation and by further fractionation in autotrophic assimilation pathways [93,94]. Together with our observations of the *Paralomis* sp. feeding habits and the presence of *Sulfurovum* sequences in the crab's stomach, the low $\delta^{13}C$-value of the muscle sample thus strongly indicates that the *Paralomis* sp. derives a substantial fraction of organic carbon from the thiotrophic microbial mats, apparently over significant parts of the crab's lifetime. However, the bulk stable isotope composition may also comprise contribution from other chemosynthetic- and/or phototrophic sources.

To investigate the potential dietary carbon sources in more detail, we analyzed lipids from stomach contents (including the stomach epithelium) and from muscle tissue of a walking leg. Only trace amounts of the isoprenoidal glycerol ethers archaeol and *sn2*-hydroxyarchaeol, which are typical for AOM-mediating ANME archaea [95] were found in the stomach sample (data not shown). This directly implies that the stomach of the *Paralomis* sp. contained comparably little archaeal biomass, which is consistent with our 16s rRNA analyses (see above).

Contrary to the archaeal compounds, we detected substantial amounts of FAs in both, the stomach and the muscle sample (Fig. 3, Tab. 4). These lipids are of bacterial and/or eukaryotic origin. In the muscle sample, the FAs may originate from *de novo* synthesis, direct incorporation of food-derived compounds or a mix of both and can thus be used to trace chemosynthetic biomass

in heterotrophs [37]. In the stomach sample, these lipids probably originate to a substantial degree from the crabs food source (however, note that the stomach sample contained not only stomach contents but also the stomach epithelium so that it comprises a mixed lipid signature of food and crab). The essential FAs C20:5ω3, C20:4ω6 constituted a major fraction of the analyzed FAs, in both samples (Fig. 3). These lipids cannot be synthesized by the crab *de novo* [38] and are thus derived from the crab's food source. With respect to the depleted isotopic signatures of about −40 (C20:5ω3) and −37‰ (C20:4ω6), it is very likely that these compounds substantially originate from chemosynthetic bacterial biomass corroborating the molecular, and bulk stable isotope data. Moreover, the higher fractional abundance of C20:5ω3 and C20:4ω6 in the stomach- compared to the muscle sample indicates that these FAs were enriched in the stomach contents and thus originate from a recently ingested food source, possibly the microbial mats. Further evidence for the dietary importance of chemosynthetic biomass for the crab is provided by the presence of unusual, ^{13}C-depleted FAs in the stomach and the muscle sample (Fig. 3), which contained substantial amounts of the iso- and anteiso-branched C15–C17 FAs, the moneonic FAs C16:1ω5 and C17:1ω6 as well as the cyclopropylic FA cyC17:0ω5,6. Generally, these lipids are not found in crustaceans, but are representative of AOM-associated SRB and/or thiotrophic communities [17,52,95,96]. Just as for the essential FAs C20:5ω3, C20:4ω6, the depleted stable carbon isotope signature of these compounds with values as low as −50.5 and −52.6‰ (C17:1ω6) in the stomach and muscle sample, respectively, point to CH_4-derived carbon as a dominant carbon source.

In addition to microbial mat biomass, our lipid data provide evidence that the crabs utilize detrital material as well. A second essential FA, C22:6ω3, had a much higher fractional abundance in

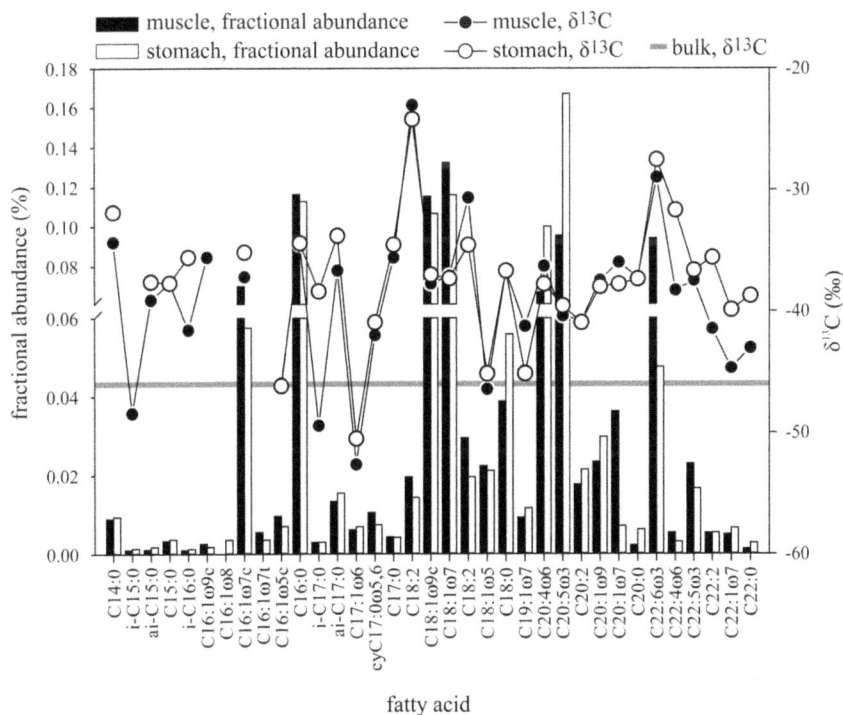

Figure 3. Fractional abundance and stable carbon isotope composition of fatty acids in a muscle- and a stomach sample of the
Paralomis diomedeae **relative.** Note that the stomach sample contained stomach contents and stomach epithelium. The bulk stable carbon isotope composition of the muscle is indicated (grey horizontal line).

Table 4. Concentrations (μg g^{-1} dry weight) and stable carbon isotope compositions of fatty acids, cholesterol and desmosterol.

copound	muscle		stomach	
	conc.	δ^{13}C	conc.	δ^{13}C
∑FA	38.6	−36.3	34.8	−36.5
cholesterol	5.5	−40.8	3.1	−37.5
desmosterol	1.9	−43	0.9	−39.8

The sum of fatty acids comprises all analyzed fatty acids with chain length between 12–22 carbon atoms. The fatty acid stable carbon isotope compositions were calculated as abundance-weighted averages.

the muscle tissue compared to the stomach sample (Fig. 3), which suggests that this compound originates from food sources not present in the stomach at the time of sampling. The high δC^{13}C-value of C22:6ω3 (about −28‰) indicates a photosynthetic origin of this FA. Most likely, the crab had consumed non-seep carbon during past feeding activities, for instance sedimented detrital organic matter, or food falls such as the Pyrosome colony (see movie S1).

In comparison to the bulk stable carbon isotope composition of the muscle tissue (−46‰), the abundance-weighted, average FA δ^{13}C-value was considerably less depleted (−36‰, Tab. 4). Therefore, the crab specimen must have consumed additional ^{13}C-depleted compounds other than FAs. One such compound class are steroids, of which we found ^{13}C-depleted cholesterol (cholest-5-ene-3β-ol) and its probable precursor desmosterol (cholest-5,24-diene-3β-ol) (Tab. 4). Just as the essential FAs, decapod crustaceans appear to lack the ability to synthesize steroids de novo [97,98] indicating a dietary origin of these compounds. Similar to the essential FA C22:6ω3, we found a much higher fractional abundance of steroids in the muscle tissue compared to the stomach sample. One source of steroids could be infauna organisms such as polychaetes and nematodes, which, at other cold seeps, were found feeding on organic carbon from deeper sediment layers including the AOM horizon [5]. A second source of steroids could be symbiotic megafauna such as *Bathymodiolus* sp. and *Lamellibrachia* sp, which are also a potential food source for heterotrophic megafauna [36,99]. We did not measure δ^{13}C-values of these organisms at Md. 12, but it is reasonable to assume that the bathymodiolin biomass is strongly ^{13}C-depleted just as has been found at other cold seeps [22,90,92], so that *Bathymodiolus* sp. could be a source of the crab's ^{13}C-depleted steroid pool. *Lamellibrachia* sp., on the other hand, is often not ^{13}C-depleted [92,100–102]. Nevertheless, a dietary mixture comprising symbiotic microbial mats, pelagic detritus and megafauna and/or infauna, probably accounts for

the difference between bulk- and (abundance weighted) FA stable carbon isotope composition.

Conclusions

Our sea floor observations together with the analyses of ribosomal RNA genes, lipid biomarkers and stable carbon isotope composition provides evidence that at Md. 12, the lithodid crabs closely related to *Paralomis diomedeae* feed on chemosynthetic biomass. This includes the *Epsilonproteobacteria* (*Sulfurovum* related spp., *Arcobacter* spp. *and Sulfurimonas* spp.), which form the thiotrophic microbial mats at Md. 12. Additionally, our analyses showed that other hydrocarbon degrading- and sulfate-reducing microbes as well as seep macro- and/or megafauna contribute to the nutrition of the crab. The stable carbon isotope- and lipid composition of the crab tissue confirmed that it is an opportunistic scavenger, using both, chemosynthetically as well as photosynthetically derived carbon in its diet. This agrees well with the shape of the crab's feeding appendages, which are functionally similar to other lithodid deep-sea crabs with an omnivorous diet (including detritus) and an opportunistic and vagrant life style. The results of this study suggest that cold seeps may have an important ecological role not only for seep-endemic but also for opportunistic, mobile megafauna.

Supporting Information

Movie S1 Time-lapse movie of sea floor observation recorded from a stationary, downward facing camera (2 pictures per hour, field of vision ≈0.4 m^2). Lithodid crabs (*Paralomis diomedeae* relative), which were apparently grazing on a thiotrophic, microbial mat were the most common observable fauna type (184 sightings during 408 hours total observation time).

Acknowledgments

We thank captain and crew of R/V Atlantis (cruise AT11-28) and R/V Meteor (cruise M66-2) for their excellent support with work at sea. We are particularly grateful to the teams of DSV Alvin and ROV Quest for their excellent help with sampling and observation. We thank Bernhard Bannert and Wolfgang Queisser for technical support during the lander deployments on R/V Meteor. We also thank Karen Stange for on-board methane and shore-based stable carbon isotope analysis, Viola Beier and Nicole Rödiger for excellent technical assistance with laboratory analyses, Maike Nicolai for video editing, Marcus Elvert for lipid identification and Lea Steinle and Moritz Lehmann for helpful comments on this manuscript.

Author Contributions

Conceived and designed the experiments: HN PL. Performed the experiments: HN PL GL KW EO. Analyzed the data: HN PL KK EM. Wrote the paper: HN PL KK EM AB WB GL KW US EO DH KB GR.

References

1. Suess E (1980) Particulate organic carbon flux in the oceans-surface productivity and oxygen utilization. Nature 288: 260–263.
2. Jahnke RA (1996) The global ocean flux of particulate organic carbon: Areal distribution and magnitude. Global Biogeochem Cy 10: 71–88.
3. Niemann H, Boetius A (2010) Mud Volcanoes. In: Timmis KN, editors. Handbook of Hydrocarbon and Lipid Microbiology. Berlin: Springer. 205–214.
4. Suess H (2010) Marine Cold Seeps. In: Timmis KN, editors. Handbook of Hydrocarbon and Lipid Microbiology. Berlin: Springer. 187–204.
5. Levin LA (2005) Ecology of cold seep sediments: Interactions of fauna with flow, chemistry and microbes. Oceanogr Mar Biol 43: 1–46.
6. Jørgensen BB, Boetius A (2007) Feast and famine – microbial life in the deep-sea bed. Nat Rev Microbiol 5: 770–781.

7. Dubilier N, Bergin C, Lott C (2008) Symbiotic diversity in marine animals: the art of harnessing chemosynthesis. Nat Rev Microbiol 6: 725–740.
8. Vanreusel A, Andersen AC, Boetius A, Connelly D, Cunha MR, et al (2009) Biodiversity of Cold Seep Ecosystems Along the European Margins. Oceanography 22: 110–127.
9. Bernardino AF, Levin LA, Thurber AR, Smith C (2012) Comparative Composition, Diversity and Trophic Ecology of Sediment Macrofauna at Vents, Seeps and Organic Falls. PLoS ONE 7: e33515.
10. Knittel K, Boetius A (2009) Anaerobic Oxidation of Methane: Progress with an Unknown Process. Annu Rev Microbiol 63: 311–334.
11. Holler T, Wegener G, Niemann H, Deusner C, Ferdelman TG, et al (2011) Carbon and sulfur back flux during anaerobic microbial oxidation of methane and coupled sulfate reduction. P Natl Aacad Sci USA 108: E1484–E1490.

12. Milucka J, Ferdelman TG, Polerecky L, Franzke D, Wegener G, et al (2012) Zero-valent sulphur is a key intermediate in marine methane oxidation. Nature 491: 541–546.

13. de Beer D, Sauter E, Niemann H, Kaul N, Foucher JP, et al (2006) In situ fluxes and zonation of microbial activity in surface sediments of the Håkon Mosby Mud Volcano. Limnol Oceanogr 51: 1315–1331.

14. Omoregie EO, Mastalerz V, de Lange G, Straub KL, Kappler A, et al (2008) Biogeochemistry and community composition of iron- and sulfur-precipitating microbial mats in the Chefren mud volcano (Nile Deep Sea fan, Eastern Mediterranean). Appl Environ Microb 74: 3198–3215.

15. Grünke S, Lichtschlag A, de Beer D, Felden J, Salman V, et al (2012) Mats of psychrophilic thiotrophic bacteria associated with cold seeps of the Barents Sea. Biogeosciences 9: 2947–2960.

16. Knittel K, Boetius A, Lemke A, Eilers H, Lochte K, et al (2003) Activity, distribution, and diversity of sulfate reducers and other bacteria in sediments above gas hydrate (Cascadia margin, Oregon). Geomicrobiol J 20: 269–294.

17. Niemann H, Lösekann T, de Beer D, Elvert M, Nadalig T, et al (2006) Novel microbial communities of the Haakon Mosby mud volcano and their role as a methane sink. Nature 443: 854–858.

18. Omoregie EO, Niemann H, Mastalerz V, de Lange G, Stadnitskaia A, et al (2009) Microbial methane oxidation and sulfate reduction at cold seeps of the deep Eastern Mediterranean Sea. Mar Geol 261: 114–127.

19. Niemann H, Fischer D, Graffe D, Knittel K, Montiel A, et al (2009) Biogeochemistry of a low-activity cold seep in the Larsen B area, western Weddell Sea, Antarctica. Biogeosciences 6: 2383–2395.

20. Childress JJ, Fisher CR, Brooks JM, Kennicutt II MC, Bidigare R, et al (1986) A methanotrophic marine molluscan (Bivalve, Mytilidae) symbiosis: mussels fueled by gas. Science 233: 1306–1308.

21. Fisher CR (1990) Chemoautotrophic and Methanotrophic Symbioses in Marine-Invertebrates. Rev Aquat Sci 2: 399–436.

22. Duperron S, Sibuet M, MacGregor BJ, Kuypers MMM, Fisher CR, et al (2007) Diversity, relative abundance and metabolic potential of bacterial endosymbionts in three Bathymodiolus mussel species from cold seeps in the Gulf of Mexico. Environ Microbiol 9: 1423–1438.

23. Petersen JM, Dubilier N (2009) Methanotrophic symbioses in marine invertebrates. Environ Microbiol Rep 1: 319–335.

24. Cary SC, Cottrell MT, Stein JL, Camacho F, Desbruyeres D (1997) Molecular identification and localization of filamentous symbiotic bacteria associated with the hydrothermal vent annelid Alvinella pompejana. Appl Environ Microb 63: 1124–1130.

25. Goffredi SK, Waren A, Orphan VJ, Van Dover CL, Vrijenhoek RC (2004) Novel forms of structural integration between microbes and a hydrothermal vent gastropod from the Indian Ocean. Appl Environ Microb 70: 3082–3090.

26. Thurber AR, Jones WJ, Schnabel K (2011) Dancing for Food in the Deep Sea: Bacterial Farming by a New Species of Yeti Crab. PLoS ONE 6: e26243.

27. Tsuchida S, Suzuki Y, Fujiwara Y, Kawato M, Uematsu K, et al (2011) Epibiotic association between filamentous bacteria and the vent-associated galatheid crab, Shinkaia crosnieri (Decapoda: Anomura). J Mar Biol Assoc UK 91: 23–32.

28. Chevaldonné P, Olu K (1996) Occurrence of anomuran crabs (Crustacea: Decapoda) in hydrothermal vent and cold-seep communities: A review. P Biol Soc Wash 109: 286–298.

29. Sahling H, Galkin SV, Salyuk A, Greinert J, Foerstel H, et al (2003) Depth-related structure and ecological significance of cold-seep communities - a case study from the Sea of Okhotsk. Deep-Sea Res Pt I 50: 1391–1409.

30. Sommer S, Linke P, Pfannkuche O, Niemann H, Treude T (2010) Benthic respiration in a seep habitat dominated by dense beds of ampharetid polychaetes at the Hikurangi Margin (New Zealand). Mar Geol 272: 223–232.

31. Cordes EE, Becker EL, Fisher CR (2010) Temporal shift in nutrient input to cold-seep food webs revealed by stable-isotope signatures of associated communities. Limnol Oceanogr 55: 2537–2548.

32. Decker C, Olu K (2012) Habitat heterogeneity influences cold-seep macrofaunal communities within and among seeps along the Norwegian margin – Part 2: contribution of chemosynthesis and nutritional patterns. Mar Ecol 33: 231–245.

33. Madigan MT, Martinko JM, Stahl DA, Clark DP (2012) Brock Biology of Microorganisms. Upper Saddle River: Pearson Prentice Hall. 1152 p.

34. Canfield DE, Kristensen E, Thamdrup B (2005) Aquatic Geomicrobiology. Oxford: Elsevier. 656 p.

35. Levin LA, Michener RH (2002) Isotopic evidence for chemosynthesis-based nutrition of macrobenthos: The lightness of being at Pacific methane seeps. Limnol Oceanogr 47: 1336–1345.

36. MacAvoy SE, Carney RS, Fisher CR, Macko SA (2002) Use of chemosynthetic biomass by large, mobile, benthic predators in the Gulf of Mexico. Mar Ecol-Prog Ser 225: 65–78.

37. MacAvoy SE, Macko SA, Carney RS (2003) Links between chemosynthetic production and mobile predators on the Louisiana continental slope: stable carbon isotopes of specific fatty acids. Chem Geol 201: 229–237.

38. Bergé J-P, Barnathan G (2005) Fatty Acids from Lipids of Marine Organisms: Molecular Biodiversity, Roles as Biomarkers, Biologically Active Compounds, and Economical Aspects. In: Le Gal Y, Ulber R, editors. Marine Biotechnology I. Springer Berlin/Heidelberg. 49–125.

39. Blankenship LE, Yayanos AA (2005) Universal primers and PCR of gut contents to study marine invertebrate diets. Mol Ecol 14: 891–905.

40. Meziti A, Kormas K, Pancucci-Papadopoulou M, Thessalou-Legaki M (2007) Bacterial phylotypes associated with the digestive tract of the sea urchin Paracentrotus lividus and the ascidian Microcosmus sp. Russ J Mar Biol 33: 84–91.

41. Wang W, Gu W, Gasparich GE, Bi KR, Ou JT, et al (2011) Spiroplasma eriocheiris sp. nov., associated with mortality in the Chinese mitten crab, Eriocheir sinensis. Int J Syst Evol Micr 61: 703–708.

42. Mörz T, Fekete N, Kopf A, Brückmann W, Kreiter S, et al (2005) Styles and productivity of mud diapirism along the middle American margin – Part II: Mound culebra and mounds 11 and 12. In: Martinelli G, Panahi B, editors. Mud Volcanoes, Geodynamics and Seismicity. Dordrecht, the Netherlands: Springer. 49–76.

43. Kopf A, Deyhle A, Zuleger E (2000) Evidence for deep fluid circulation and gas hydrate dissociation using boron and boron isotopes of pore fluids in forearc sediments from Costa Rica (ODP Leg 170). Mar Geol 167: 1–28.

44. Ranero CR, von Huene R (2000) Subduction erosion along the Middle America convergent margin. Nature 404: 748–752.

45. Sahling H, Masson DG, Ranero CR, Huhnerbach V, Weinrebe W, et al (2008) Fluid seepage at the continental margin offshore Costa Rica and southern Nicaragua. Geochem Geophy Geosy 9: 10.1029/2008GC001978mör.

46. Hensen C, Wallmann K, Schmidt M, Ranero CR, Suess E (2004) Fluid expulsion related to mud extrusion off Costa Rica – A window to the subducting slab. Geology 32: 201–204.

47. Linke P, Wallmann K, Suess E, Hensen C, Rehder G (2005) In situ benthic fluxes from an intermittently active mud volcano at the Costa Rica convergent margin. Earth Planet Sc Lett 235: 79–95.

48. Brückmann W, Bialas J, Kopf A, Rhein M, Rheder G (2009) SUBFLUX, Cruise No. 66, August 12 – December 22, 2005. METEOR-Berichte 09-2. University of Hamburg. 158 p.

49. Mau S, Sahling H, Rehder G, Suess E, Linke P, et al (2006) Estimates of methane output from mud extrusions at the erosive convergent margin off Costa Rica. Mar Geol 225: 129–144.

50. Macpherson E, Wehrtmann IS (2010) Occurrence of lithodid crabs (Decapoda, Lithodidae) on the Pacific coast of Costa Rica, Central America. Crustaceana 83: 143–151.

51. Niemann H, Elvert M, Hovland M, Orcutt B, Judd AG, et al (2005) Methane emission and consumption at a North Sea gas seep (Tommeliten area). Biogeosciences 2: 335–351.

52. Elvert M, Boetius A, Knittel K, Jørgensen BB (2003) Characterization of specific membrane fatty acids as chemotaxonomic markers for sulfate-reducing bacteria involved in anaerobic oxidation of methane. Geomicrobiol J 20: 403–419.

53. Niemann H, Duarte J, Hensen C, Omoregie E, Magalhães VH, et al (2006) Microbial methane turnover at mud volcanoes of the Gulf of Cadiz. Geochim Cosmochim Acta 70: 5336–5355.

54. Edgar RC, Haas BJ, Clemente JC, Quince C, Knight R (2011) UCHIME improves sensitivity and speed of chimera detection. Bioinformatics 27: 2194–2200.

55. Pruesse E, Peplies J, Glöckner FO (2012) SINA: Accurate high-throughput multiple sequence alignment of ribosomal RNA genes. Bioinformatics 28: 1823–1829.

56. Pernthaler A, Pernthaler J, Amann R (2002) Fluorescence In Situ Hybridization and Catalyzed Reporter Deposition for the Identification of Marine Bacteria. Appl Environ Microb 68: 3094–3101.

57. Ishii K, Mußmann M, MacGregor BJ, Amann R (2004) An improved fluorescence in situ hybridization protocol for the identification of bacteria and archaea in marine sediments. FEMS Microbiol Ecol 50: 203.

58. Moussard H, Corre E, Cambon-Bonavita M-A, Fouquet Y, Jeanthon C (2006) Novel uncultured Epsilonproteobacteria dominate a filamentous sulphur mat from the 13°N hydrothermal vent field, East Pacific Rise. FEMS Microbiol Ecol 58: 449–463.

59. Treude T, Boetius A, Knittel K, Wallmann K, Jørgensen BB (2003) Anaerobic oxidation of methane above gas hydrates at Hydrate Ridge, NE Pacific Ocean. Mar Ecol-Prog Ser 264: 1–14.

60. Jørgensen BB (1978) A comparison of methods for the quantification of bacterial sulfate reduction in coastal marine sediments – I. Measurement with radiotracer techniques. Geomicrobiol J 1: 11–27.

61. Girnth A-C, Grünke S, Lichtschlag A, Felden J, Knittel K, et al (2011) A novel, mat-forming Thiomargarita population associated with a sulfidic fluid flow from a deep-sea mud volcano. Environ Microbiol 13: 495–505.

62. Grünke S, Felden J, Lichtschlag A, Girnth A-C, De Beer D, et al (2011) Niche differentiation among mat-forming, sulfide-oxidizing bacteria at cold seeps of the Nile Deep Sea Fan (Eastern Mediterranean Sea). Geobiology 9: 330–348.

63. Sibuet M, Olu K (1998) Biogeography, biodiversity and fluid dependence of deep-sea cold-seep communities at active and passive margins. Deep-Sea Res Pt II 45: 517–567.

64. Martin JW, Haney TA (2005) Decapod crustaceans from hydrothermal vents and cold seeps: a review through 2005. Zool J Linn Soc-Lond 145: 445–522.

65. Hovland M (2007) Discovery of prolific natural methane seeps at Gullfaks, northern North Sea. Geo-Mar Lett 27: 197–201.

66. Niggemann J, Ferdelman TG, Lomstein BA, Kallmeyer J, Schubert CJ (2007) How depositional conditions control input, composition, and degradation of organic matter in sediments from the Chilean coastal upwelling region. Geochim Cosmochim Acta 71: 1513–1527.

67. Thamdrup B, Canfield DE (1996) Pathways of carbon oxidation in continental margin sediments off central Chile. Limnol Oceanogr 41: 1629–1650.

68. Fiedler PC (2002) The annual cycle and biological effects of the Costa Rica Dome. Deep-Sea Res Pt I 49: 321–338.

69. Beal EJ, House CH, Orphan VJ (2009) Manganese- and Iron-Dependent Marine Methane Oxidation. Science 325: 184–187.

70. Campbell BJ, Engel AS, Porter ML, Takai K (2006) The versatile epsilon-proteobacteria: key players in sulphidic habitats. Nat Rev Microbiol 4: 458–468.

71. Omoregie EO, Mastalerz V, de Lange G, Straub KL, Kappler A, et al (2008) Biogeochemistry and community composition of iron- and sulfur-precipitating microbial mats at the Chefren mud volcano (Nile Deep Sea fan, Eastern Mediterranean). Appl Environ Microb 74: 3198–3215.

72. Joye SB, Samarkin VA, Orcutt BN, MacDonald IR, Hinrichs KU, et al (2009) Metabolic variability in seafloor brines revealed by carbon and sulphur dynamics. Nat Geosci 2: 349–354.

73. Borin S, Brusetti L, Mapelli F, D'Auria G, Brusa T, et al (2009) Sulfur cycling and methanogenesis primarily drive microbial colonization of the highly sulfidic Urania deep hypersaline basin. P Natl Aacad Sci USA 106: 9151–9156.

74. Inagaki F, Takai K, Nealson KH, Horikoshi K (2004) Sulfurovum lithotrophicum gen. nov., sp. nov., a novel sulfur-oxidizing chemolithoautotroph within the ε-Proteobacteria isolated from Okinawa Trough hydrothermal sediments. Int J Syst Evol Micr 54: 1477–1482.

75. Schauer R, Roy H, Augustin N, Gennerich HH, Peters M, et al (2011) Bacterial sulfur cycling shapes microbial communities in surface sediments of an ultramafic hydrothermal vent field. Environ Microbiol 13: 2633–2648.

76. Tokuda G, Yamada A, Nakano K, Arita NO, Yamasaki H (2008) Colonization of Sulfurovum sp. on the gill surfaces of Alvinocaris longirostris, a deep-sea hydrothermal vent shrimp. Mar Ecol 29: 106–114.

77. Durand L, Zbinden M, Cueff-Gauchard V, Duperron S, Roussel EG, et al (2010) Microbial diversity associated with the hydrothermal shrimp Rimicaris exoculata gut and occurrence of a resident microbial community. FEMS Microbiol Ecol 71: 291–303.

78. Takai K, Campbell BJ, Cary SC, Suzuki M, Oida H, et al (2005) Enzymatic and genetic characterization of carbon and energy metabolisms by deep-sea hydrothermal chemolithoautotrophic isolates of Epsilonproteobacteria. Appl Environ Microb 71: 7310–7320.

79. Yamamoto M, Nakagawa S, Shimamura S, Takai K, Horikoshi K (2010) Molecular characterization of inorganic sulfur-compound metabolism in the deep-sea epsilonproteobacterium Sulfurovum sp. NBC37-1. Environ Microbiol 12: 1144–1153.

80. Nakagawa S, Takaki Y, Shimamura S, Reysenbach AL, Takai K, et al (2007) Deep-sea vent epsilon-proteobacterial genomes provide insights into emergence of pathogens. P. Natl Acad Sci USA 104: 12146–12150.

81. Fetzner S (2010) Aerobic Degradation of Halogenated Aliphatics. In: Timmis KN, editors. Handbook of Hydrocarbon and Lipid Microbiology. Springer. 866–885.

82. Rojo S (2010) Enzymes for Aerobic Degradation of Alkanes. In: Timmis KN, editors. Handbook of Hydrocarbon and Lipid Microbiology. Springer. 782–797.

83. Kleindienst S, Ramette A, Amann R, Knittel K (2012) Distribution and in situ abundance of sulfate-reducing bacteria in diverse marine hydrocarbon seep sediments. Environ Microbiol 14: 2689–2710.

84. Lanoil BD, Sassen R, La Duc MT, Sweet ST, Nealson KN (2001) Bacteria and Archaea physically associated with Gulf of Mexico gas hydrates. Appl Environ Microb 67: 5143–5153.

85. Bowles MW, Samarkin VA, Bowles KM, Joye SB (2011) Weak coupling between sulfate reduction and the anaerobic oxidation of methane in methane-rich seafloor sediments during ex situ incubation. Geochim Cosmochim Acta 75: 500–519.

86. Nechitaylo TY, Timmis KN, Golyshin PN (2009) 'Candidatus Lumbricincola', a novel lineage of uncultured Mollicutes from earthworms of family Lumbricidae. Environ Microbiol 11: 1016–1026.

87. Meziti A, Ramette A, Mente E, Kormas KA (2010) Temporal shifts of the Norway lobster (Nephrops norvegicus) gut bacterial communities. FEMS Microbiol Ecol 74: 472–484.

88. Kostanjšek R, Strus J, Avgustin G (2007) "andidatus Bacilloplasma,"a novel lineage of Mollicutes associated with the hindgut wall of the terrestrial isopod Porcellio scaber (Crustacea : Isopoda). Appl Environ Microb 73: 5566–5573.

89. Whitcomb RF (1981) The Biology of Spiroplasmas. Annu Rev Entomol 26: 397–425.

90. Paull CK, Jull AJT, Toolin LJ, Linick T (1985) Stable isotope evidence for chemosynthesis in an abyssal seep community. Nature 317: 709–711.

91. Van Dover CL (2007) Stable Isotope Studies in Marine Chemoautotrophically Based Ecosystems: An Update. In: Michener R, Lajtha K, editors. Stable Isotopes in Ecology and Environmental Science. Blackwell Publishing Ltd. 202–237.

92. Thurber AR, Kröger K, Neira C, Wiklund H, Levin LA (2010) Stable isotope signatures and methane use by New Zealand cold seep benthos. Mar Geol 272: 260–269.

93. Boetius A, Ravenschlag K, Schubert C, Rickert D, Widdel F (2000) A marine microbial consortium apparently mediating anaerobic oxidation of methane. Nature 407: 623.

94. Lösekann T, Robador A, Niemann H, Knittel K, Boetius A, et al (2008) Endosymbioses between bacteria and deep-sea siboglinid tubeworms from an Arctic Cold Seep (Haakon Mosby Mud Volcano, Barents Sea). Environ Microbiol 10: 3237–3254.

95. Niemann H, Elvert M (2008) Diagnostic lipid biomarker and stable carbon isotope signatures of microbial communities mediating the anaerobic oxidation of methane with sulphate. Org Geochem 39: 1668–1677.

96. Bergé J-P, Barnathan G (2005) Fatty Acids from Lipids of Marine Organisms: Molecular Biodiversity, Roles as Biomarkers, Biologically Active Compounds, and Economical Aspects. In: Le Gal Y, Ulber R, editors. Marine Biotechnology I. Springer Berlin/Heidelberg. 49–125.

97. Blumenberg M, Seifert R, Reitner J, Pape T, Michaelis W (2004) Membrane lipid patterns typify distinct anaerobic methanotrophic consortia. P Natl Aacad Sci USA 101: 11111–11116.

98. Van den Oord A (1964) The absence of cholesterol synthesis in the crab, Cancer Pagurus L. Comp Biochem Physiol 13: 461–467.

99. MacAvoy SE, Fisher CR, Carney RS, Macko SA (2005) Nutritional associations among fauna at hydrocarbon seep communities in the Gulf of Mexico. Mar Ecol-Prog Ser 292: 51–60.

100. Kennicutt MC, Brooks JM, Bidigare RR, Fay RR, Wade TL, et al (1985) Vent-type taxa in a hydrocarbon seep region on the Louisiana slope. Nature 317: 351–353.

101. MacAvoy SE, Macko SA, Joye SB (2002) Fatty acid carbon isotope signatures in chemosynthetic mussels and tube worms from Gulf of Mexico hydrocarbon seep communities. Chem Geol 185: 1–8.

102. MacDonald IR, Boland GS, Baker JS, Brooks JM, Kennicutt MC, et al (1989) Gulf of Mexico hydrocarbon seep communities. Mar Biol 101: 235–247.

6

Comparative Population Structure of Two Deep-Sea Hydrothermal-Vent-Associated Decapods (*Chorocaris* sp. 2 and *Munidopsis lauensis*) from Southwestern Pacific Back-Arc Basins

Andrew David Thaler[1]*, **Sophie Plouviez**[1], **William Saleu**[2], **Freddie Alei**[3], **Alixandra Jacobson**[1], **Emily A. Boyle**[1], **Thomas F. Schultz**[1], **Jens Carlsson**[4], **Cindy Lee Van Dover**[1]

1 Marine Laboratory, Nicholas School of the Environment, Duke University, Beaufort, North Carolina, United States of America, 2 Nautilus Minerals, Port Moresby, NCD, Papua New Guinea, 3 Environmental Science and Geography Division, School of Natural and Physical Sciences, University of Papua New Guinea, Port Moresby, Papua New Guinea, 4 School of Biology & Environmental Science, University College Dublin, Dublin, Ireland

Abstract

Studies of genetic connectivity and population structure in deep-sea chemosynthetic ecosystems often focus on endosymbiont-hosting species that are directly dependent on chemical energy extracted from vent effluent for survival. Relatively little attention has been paid to vent-associated species that are not exclusively dependent on chemosynthetic ecosystems. Here we assess connectivity and population structure of two vent-associated invertebrates—the shrimp *Chorocaris* sp. 2 and the squat lobster *Munidopsis lauensis*—that are common at deep-sea hydrothermal vents in the western Pacific. While *Chorocaris* sp. 2 has only been observed at hydrothermal vent sites, *M. lauensis* can be found throughout the deep sea but occurs in higher abundance around the periphery of active vents We sequenced mitochondrial *COI* genes and deployed nuclear microsatellite markers for both species at three sites in Manus Basin and either North Fiji Basin (*Chorocaris* sp. 2) or Lau Basin (*Munidopsis lauensis*). We assessed genetic differentiation across a range of spatial scales, from approximately 2.5 km to more than 3000 km. Population structure for *Chorocaris* sp. 2 was comparable to that of the vent-associated snail *Ifremeria nautilei*, with a single seemingly well-mixed population within Manus Basin that is genetically differentiated from conspecifics in North Fiji Basin. Population structure for *Munidopsis lauensis* was more complex, with two genetically differentiated populations in Manus Basin and a third well-differentiated population in Lau Basin. The unexpectedly high level of genetic differentiation between *M. lauensis* populations in Manus Basin deserves further study since it has implications for conservation and management of diversity in deep-sea hydrothermal vent ecosystems.

Editor: Erik Sotka, College of Charleston, United States of America

Funding: Funding was provided by a research contract from Nautilus Minerals Niugini Ltd. (CLVD, JC, TFS), Duke University (ADT, CLVD), InterRidge (ADT and FA), the International Seabed Authority (WS) and the National Science Foundation (OCE-1031050 to CLVD). Samples from North Fiji and Lau Basin were provided by Dr. Robert Vrijenhoek (NSF Grant: OCE-0241613). The funders had no role in study design, data collection and analysis, decision to publish, or preparation of the manuscript.

Competing Interests: The authors recognize funding from Nautilus Minerals Ltd., a deep-sea mining company operating in Papua New Guinea, as a potential competing interest. While WS is currently in the employ of Nautilus Minerals, his contribution to the project was completed prior to that employment. This manuscript represents the original work of the authors and the funder had no role in study design, analysis, decision to publish, or preparation of the manuscript.

* Email: andrew.david.thaler@gmail.com

Introduction

In the deep sea, studies of gene flow and population structure have disproportionately focused on numerically dominant taxa from chemosynthetic ecosystems, where it is relatively easy to collect sufficient individuals for population genetic analyses [1]. Within deep-sea chemosynthetic ecosystems, particularly hydrothermal vents, many population genetic studies target holobiont taxa (invertebrate host species with chemoautotrophic symbionts [2]), such as giant tubeworms (*Riftia pachyptila*: [3,4]) and mussels (*Bathymodiolus thermophilus*: [5–7]) on the East Pacific Rise, swarming shrimp (*Rimicaris exoculata* [8,9]) on the Mid-Atlantic Ridge, or provannid gastropods (*Ifremeria nautilei*:[10,11] and *Alviniconcha* spp.: [12]) in the western Pacific. Population genetic studies are also reported for holobiont taxa from deep-sea chemosynthetic communities at methane seeps in the Gulf of Mexico and eastern Atlantic [13,14]. From these studies, we are beginning to understand the extent of genetic connectivity among deep-sea populations associated with chemosynthetic ecosystems and establish natural management units.

Among the first-order questions that arise regarding gene flow and population structure in deep-sea chemosynthetic ecosystems is whether there is greater community and population differentiation among back-arc basins than along mid-ocean ridges, due to the isolated nature of the basins [15,16]. Vents along mid-ocean ridges tend to be linearly distributed, which may create dispersal corridors [17] where genetic differentiation emerges from geomorphological and hydrographic features rather than distance. In contrast, vents at back-arc basins are non-linearly distributed, forming a complex patchwork of habitat for vent endemic taxa. In the western Pacific, mitochondrial and nuclear microsatellite marker-based population studies in several holobiont taxa suggest genetic homogeneity within back-arc basins (i.e., *Bathymodiolus* spp. [18,19], *Alviniconcha* spp. [12,20], *Ifremeria nautilei* [11]). But at larger scales—among back-arc basins—there is evidence for limited connectivity among populations of the provannid gastropods *Alviniconcha* sp. 2 [10] and *Ifremeria nautilei* [10,11], with populations in Manus Basin isolated from well-connected populations throughout North Fiji and Lau Basins.

Dispersal filters are also observed for vent-associated taxa on the Eastern Pacific Rise (EPR) and often correspond to geomorphological features [3,7,21]. These filters tend to be species-dependent, with inconsistent patterns of isolation among taxa. For example, bathymodiolin mussels (*Bathymodiolus thermophilus*) on the East Pacific Rise are divided into northern and southern populations across the Easter Microplate (based on mitochondrial and allozyme markers [7]). *Alvinella pompejana* and *Branchipolynoe symmytilida*, two polychaete species with a similar distributions and one of which (*B. symmytilida*) is commensal in *Bathymodiolus thermophilus*, are undifferentiated across the same boundary (based on a mitochondrial marker [3]). In the northeast Pacific, populations of *Lepetodrilus* limpets diverge across the Blanco Transform Fault, a 450-km long ridge offset that separates the Juan de Fuca and Gorda Ridges (based on mitochondrial and allozyme markers [22]), while unidirectional gene flow was detected in a vent tubeworm, *Ridgeia piscesae*, across the same boundary (based on a mitochondrial maker [23]). Along the Mid-Atlantic Ridge, populations of the mussel *Bathymodiolus puteoserpentis* are connected across approximately 9 degrees of latitude (based on a mitochondrial marker [24]) and hybridize with *Bathymodiolus azoricus* at the Broken Spur vent field (based on mitochondrial and nuclear markers [25]); north of Broken Spur, *B. puteoserpentis* is replaced by *B. azoricus*.

In this study, we explore population structure in two western Pacific back-arc basin vent species—the alvinocarid shrimp *Chorocaris* sp. 2 and the galatheid squat lobster *Munidopsis lauensis*—to determine the extent of population differentiation within and among back-arc basins for vent-associated species. Shrimp in the genus *Chorocaris* are so far only reported at or near hydrothermal vents [26,27] and *Chorocaris* sp. 2 has, to date, only been observed in Manus and North Fiji Basin (T. Komai, personal communication). Squat lobsters in the genus *Munidopsis* occupy numerous deep-sea habitats, including cold seeps [28], seamounts [29], and wood- and whale-falls [30], and are also generally distributed along the continental slope and bathyal depths [31,32]. *M. lauensis* is frequently observed in large aggregations near active hydrothermal venting [26,33,34], but also occurs in lesser numbers at inactive sulfide edifices on the vent periphery [35] and in still lower abundance as one moves away from the vent field (i.e., > 500 m; Thaler, personal observation). Both species are relatively abundant at hydrothermal vents in Manus Basin [34]. Neither species possesses chemoautotrophic endosymbionts, which may make them less dependent on hydrothermal vent fluid than holobiont species, and possibly allows them to use other habitats as stepping-stones for connectivity, though, for both species, larval life history is poorly characterized. We infer that *M. lauensis* is an opportunistic species that is broadly distributed in the deep-sea but is attracted to the rich food source at hydrothermal vents, whereas *Chorocaris* sp. 2 is more patchily distributed and restricted to regions around active hydrothermal venting. We expected that opportunistic vent taxa, like *Munidopsis lauensis*, would show the least

population structure, while the vent-associated shrimp, *Chorocaris* sp. 2, would exhibit greater population structure in comparison.

Materials and Methods

Sample collection and DNA extraction

Samples were collected with permission of the governments of Papua New Guinea, Fiji, and Tonga. These field studies do not involve any endangered or protected species.

Chorocaris spp. and *Munidopsis lauensis* were collected from three hydrothermal-vent sites in Manus Basin—Solwara 8, Solwara 1, and South Su (Figure 1)—during the *M/V Nor Sky* research campaign in June-July 2008 (Chief Scientist: S. Smith). Samples were collected using an ST200 work class ROV modified for biological sampling. To allow comparisons of population structure at multiple spatial sales, two to four discrete aggregations of *Chorocaris* sp. 2 or *Munidopsis lauensis* were sampled from each site within Manus Basin (Table 1). Additional samples from other southwestern Pacific back-arc basins (Figure 1) were provided by collaborators: *Chorocaris* sp. 2 was collected from Ivory Tower in North Fiji Basin and *M. lauensis* was collected from Hine Hina and Tu'i Malila in Lau Basin in May-June 2005 using the *ROV Jason II* supported by the *R/V Melville* (Chief Scientist: R. Vrijenhoek).

Tissues were preserved in 95% ethanol prior to DNA extraction. Genomic DNA was isolated using a standard Chelex-Proteinase-K extraction (10–30 mg digested with 120 µg Proteinase K (Bioline: Taunton, MA) in 600 µl 10% Chelex-100 resin (Bio-Rad: Hercules, CA) overnight at 60°C, heated to 100°C for 15 min, and centrifuged at 10,000 rpm for 5 minutes; [36]) or Wizard SVG tissue extraction kit (Promega Corp: Madison, WI) following manufacturer's protocols. Extracted DNA was stored at 4°C until amplification and archived at −20°C.

COI sequencing and analysis

Chorocaris sp 2. mitochondrial *COI* fragments were amplified using the following reaction conditions: 10 to 100 ng of DNA template was combined with 2 µL 10× PCR buffer (200 mM Tris, pH 8.8; 500 mM KCl; 0.1% Triton X-100; 0.2 mg/ml BSA), 2 mM $MgCl_2$, 0.6 mM dNTPs, 0.5 µM LCOI1490 and 0.5 µM HCOI2198 primers [37], and 1 unit of Taq polymerase in a 20 µL reaction with the following PCR protocol: initial melting temperature of 94°C for 120 seconds; 25 cycles of 94°C for 35 seconds, 48°C for 35 seconds, 72°C for 80 seconds; and a final extension of 72°C for 600 seconds. *Munidopsis lauensis* mitochondrial *COI* fragments were amplified using the following reaction conditions: 10 to 100 ng of DNA template was combined with 2.5 µL PCR buffer, 2 mM $MgCl^+$, 0.6 mM dNTPs, 0.05 µM GALA COIR primer (5′-GAA YAG GRT CTC CTC CTC CTA C -3′) and 0.05 µM GALA COIF primer (5′- CAT CAC TWA GWT TRA TYA TTC CAG CAG AA-3′), and 1 unit of Taq polymerase in a 25 µL reaction with the following PCR protocol: initial melting temperature of 94°C for 240 seconds; 35 cycles of 94°C for 60 seconds, 55°C for 120 seconds, 72°C for 210 seconds; and a final extension of 72°C for 600 seconds. Reactions were stored at 4°C until purification.

To remove unincorporated nucleotides, 14 µl of PCR product was incubated with 0.2 µl 10× ExoAP buffer (500 mM Bis-Tris, 10 mM MgCl2, 1 mM ZnSO4), 0.05 µl Antarctic Phosphatase (New England Biolabs: Ipswich, MA), 0.05 µl Exonuclease I (New England Biolabs: Ipswich, MA) at 37°C for 60 min followed by 85°C for 15 min to inactivate the enzymes. Sequencing reactions (both directions) were performed using Big Dye Terminator v3 reactions (Applied Biosystems: Foster City, CA). Dye Terminator removal was performed using AMPure magnetic beads (Agen-

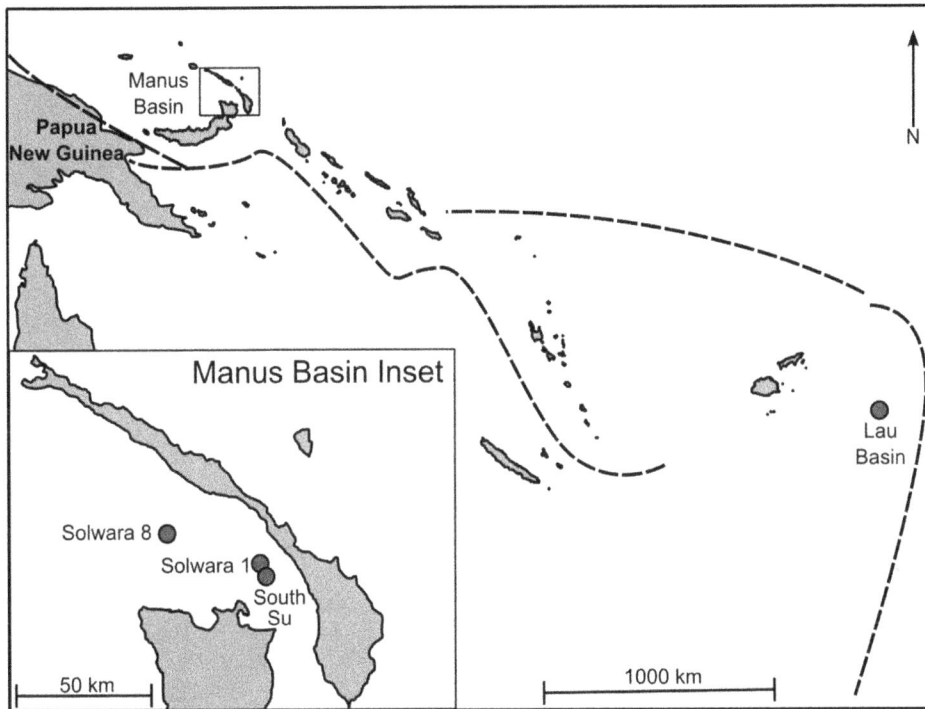

Figure 1. Sampling locations in Manus, North Fiji, and Lau Basins. Dashed lines: subduction zones. Figure originally published in Thaler *et al.* [11].

court: Morrisville, NC), sequencing products were analyzed on an ABI 3730xl DNA Analyzer (Applied Biosystems International), and sequence chromatograms were edited with CodonCode Aligner (version 3.7.1; CodonCode Corporation: Dedham, MA). Consensus sequences were compared against the NCBI GenBank database to confirm species identity when available [38] and

Table 1. *Chorocaris* spp. and *Munidopsis lauensis* sampling locations in Manus, North Fiji, and Lau Basin.

	Basin	Site	Mound/Aggregation	Latitude	Longitude	Depth (m)
Chorocaris spp.	Manus	Solwara 8	Aggregation 1	3° 43.751'S	151° 40.410'E	1720
			Aggregation 2	3° 43.826'S	151° 40.457'E	1720
			Aggregation 3	3° 43.669'S	151° 40.873'E	1710
		Solwara 1	Aggregation 4	3° 47.453'S	152° 5.485'E	1530
			Aggregation 5	3° 47.367'S	152° 5.781'E	1490
			Aggregation 6	3° 47.372'S	152° 5.619''E	1480
		South Su	Aggregation 7	3° 48.537'S	152° 6.284'E	1300
			Aggregation 8	3° 48.497'S	152° 6.298'E	1330
			Aggregation 9	3° 48.572'S	152° 6.312'E	1320
			Aggregation 10	3° 48.572'S	152° 6.317'E	1320
	North Fiji	Ivory Tower		16° 59.300'S	173° 54.900'E	1970
Munidopsis lauensis	Manus Basin	Solwara 8	Aggregation 1	3° 43.824'S	151° 40.458'E	1710
			Aggregation 2	3° 43.740'S	151° 40.404'E	1720
		Solwara 1	Aggregation 3	3° 47.370'S	152° 05.778'E	1490
			Aggregation 4	3° 47.370'S	152° 05.616'E	1480
		South Su	Aggregation 5	3° 48.564'S	152° 06.144'E	1300
			Aggregation 6	3° 48.492'S	152°0 6.186'E	1350
	Lau Basin	Hine Hina		22° 31.80'S	176° 41.82'W	1900
		Tu'i Malila		20° 59.350'S	176° 34.100'W	1880

sequence alignments were constructed using the MUSCLE alignment algorithm [39] implemented in CodonCode Aligner. Representative sequences of dominant haplotypes were deposited in GenBank (Accession # KF498731 - KF498847). Full *COI* sequences for each individual are provided as FASTA files (File S1 and S2).

Maximum-parsimony phylograms of aligned mitochondrial sequences were assembled in MEGA version 5 (10,000 replicates; Tamura 3-parameter substitution model determined by Mega 5: Find Best-Fit Substitution Model for both shrimp and squat lobsters; [40]). Potential cryptic species were determined by comparing the degree of divergence between two putative species to the extent of divergence among established species within the same genus [41].

Statistical-parsimony networks were assembled in TCS version 1.21 (default settings; [42]). For *M. lauensis*, 4 *COI* sequences were obtained from NCBI GenBank—EF157850 from the Desmos Caldera in Manus Basin, EF157851 from Mariner vent field in Lau Basin, EF157852 from Hine Hina vent field in Lau Basin, and EF157853 from Brothers Seamount, New Zealand [29]. These sequences were used only for statistical-parsimony network analyses. Number of haplotypes (*H*), haplotype diversity (*Hd*), nucleotide diversity (π), and Fu's F_S were calculated using DnaSP version 5.10.01 [43]. DnaSP was also used to construct mismatch distribution curves for expected values under constant population size and population growth/decline models. Arlequin version 3.5.1.2 [44] was used to estimate pairwise φ_{ST} and permutation tests were used to identify significant departure from genetic homogeneity and Sequential Bonferroni was used to correct for multiple tests [45].

Microsatellite genotyping and statistical analyses

Six microsatellite markers (*Cho30, Cho36, Cho63, Cho76, Cho91, Cho99*) were amplified from *Chorocaris* sp. 2 in Manus Basin following methods reported in Zelnio *et al.* [46]. Microsatellite loci were not amplified from *Chorocaris* sp. 2 in North Fiji Basin due to low sample size (n = 9). Nine microsatellite markers (*Mp8, Mp12, Mp14, Mp15, Mp16, Mp21, Mp24, Mp27, Mp29*) were amplified from *Munidopsis lauensis* in Manus and Lau Basin following methods reported in Boyle *et al.* [47]. Full microsatellite genotypes for each individual are provided as GENPOP files (Files S3 and S4).

Divergence from expected Hardy-Weinberg Equilibrium (HWE; GENEPOP; default settings; version 4.0; [48]) and allelic richness (Microsatellite Analyzer; version 4.05; [49]) were assessed. Permutation tests were used to determine significant variation in allelic richness (F-stat; default settings; version 2.9.3.2; [50]). MicroChecker (version 2.2.3; 1000 randomizations; [51]) was used to detect the potential presence of null alleles, stutter, and large allele dropout. To test for the potential influence of selection, loci were screened using LOSITAN (25,000 simulations; IA and SMM; [52,53]).

Analysis of molecular variance (AMOVA) was used to analyze hierarchal population structure in Arlequin. Pairwise genetic differentiation (F_{ST}) between aggregations, sites, and basins was analyzed using Microsatellite Analyzer. Alpha levels were adjusted via Sequential Bonferroni to correct for multiple tests [45]. Structure version 2.3.3 (admixture model, sampling locations as prior distributions; [54]) was used to visualize potential population structure. Analyses were conducted with a 1,000,000 step burn-in, 10,000,000 repetitions, and 3 replicates per level from K = 1 to 7. Effective population size was estimated based on microsatellite linkage-disequilibrium using LDNe (default parameters; [55]) and corroborated using ONeSAMP (version 1.2, default parameters

[56]). Pairwise relatedness to test for potential kinship effects was estimated with KINGROUP (Version 2_090501; [57]. BOTTLENECK (version 1.2.02.; [58]) was used to test for genetic bottlenecks among site for both *Chorocaris* sp. 2 and *Munidopsis lauensis*. Equilibrium heterozygosity (H_{eq}) was estimated under the TPM model allowing for 4% multi-step mutations (1000 iterations).

Results

Chorocaris sp. 2: Population structure

Samples of shrimp in the genus *Chorocaris* from Manus (191 individuals; Table 2) and North Fiji Basins (9 individuals; Table 2) comprised two putative species based on *COI* genetic divergence (5.4% divergence between species; Figure S1): *Chorocaris* sp. 1 (12 of 41 individuals collected from South Su in Manus Basin), and *Chorocaris* sp. 2 (179 individuals from all three sites in Manus Basin and 9 individuals from North Fiji Basin; Table 2). *Chorocaris* sp. 2, the numerically dominant shrimp species at Manus Basin vents, was further analyzed for population structure using *COI* and microsatellite markers.

For *Chorocaris* sp. 2, 106 *COI* haplotypes (454 bp) were identified from 179 individuals from Manus Basin. Haplotype diversity among aggregations in Manus Basin was high, ranging from 0.70 to 1.00 (Table 2). Fu's F_S values were all significantly negative (Table 2). Mismatch distribution curves followed unimodal distributions, consistent with a population growth/decline model (Figure S2).

The statistical parsimony network for *Chorocaris* sp. 2 has a web-like topology, with many singletons connected through multiple nodes, indicating high genetic variability (Figure 2). A small North Fiji clade branches off the larger Manus clade (Figure 2). One dominant haplotype (n = 20) is shared between Manus and North Fiji Basins and a second North Fiji singleton haplotype is found within the Manus Basin haplotype group (Figure 2). Several additional lineages within Manus Basin are divergent from the main Manus haplotype group by up to 6 mutational steps; this divergence exceeds the *COI*-based genetic divergence found between the main Manus haplotype group and North Fiji Basin haplotype groups (Figure 2).

Six microsatellite loci were amplified from *Chorocaris* sp. 2 within Manus Basin (64 to 92 individuals per site; Table 3). Total alleles per locus ranged from 3 to 11 (mean = 7). In permutation tests, allelic richness (Rs) did not vary significantly among patches, mounds, or sites (10,000 permutations, $P>0.05$; Table 3). Neither directional nor balancing selection was detected among microsatellites at any spatial scale (LOSITAN, $P>0.05$). Three loci deviated significantly from Hardy-Weinberg expectations and showed evidence for heterozygote deficiency at two sites (*Cho63, Cho76, Cho91*; Table 3). MicroChecker suggested that null alleles were present at all three loci and were responsible for heterozygote deficiencies. As these three markers fell within Hardy-Weinberg expectations when samples from all three sites in Manus Basin are pooled and the presence of null alleles has been shown not to severely bias assignment tests [59], these three markers were used for subsequent analyses.

Analysis of Molecular Variance (AMOVA) for *Chorocaris* sp. 2 and pairwise tests (F_{ST} and φ_{ST}) for population differentiation in Manus Basin indicated no significant differentiation at any spatial scale (Table 4). Assignment tests placed all *Chorocaris* sp. 2 from Manus Basin into a single population (Structure, K = 1, data not shown). AMOVA performed on microsatellites across basins indicated that almost 35% of the observed genetic variability was accounted for by differentiation at the basin level, i.e., between

Table 2. Chorocaris spp. and Munidopsis lauensis.

Species	Location		N	H	Hd	F_S
Chorocaris sp. 1	Manus Basin					
	South Su		12	8	0.92	**−3.96**
		Aggregation 7	1	1	n/a	n/a
		Aggregation 9	5	4	0.90	n/a
		Aggregation 10	6	4	0.87	n/a
Chorocaris sp. 2	Manus Basin		179	106	0.98	**−162.12**
	Solwara 8		88	60	0.98	**−69.19**
		Aggregation 1	46	37	0.99	**−35.36**
		Aggregation 2	23	20	0.99	**−14.81**
		Aggregation 3	19	15	0.97	**−7.05**
	Solwara 1		62	47	0.98	**−58.13**
		Aggregation 4	14	15	1.00	**−12.00**
		Aggregation 5	5	5	1.00	n/a
		Aggregation 6	43	32	0.98	**−32.95**
	South Su		29	24	0.98	**−17.19**
		Aggregation 7	5	3	0.70	n/a
		Aggregation 8	4	4	1.00	n/a
		Aggregation 9	8	8	1.00	n/a
		Aggregation 10	12	11	0.99	**−4.89**
	Fiji Basin		9	7	0.92	**−1.74**
Munidopsis lauensis	Manus Basin		81	4	0.07	n/a
	Solwara 8		43	3	0.09	n/a
	Solwara 1		10	1	0.09	n/a
	South Su		28	2	0.07	n/a
	Lau Basin		30	2	0.07	n/a

Summary statistics for *COI* sequences (*Chorocaris* sp. 2–616 bp; *M. lauensis* –454 bp) from Manus and North Fiji Basins. *N*: number of individuals, *H*: number of haplotypes, *Hd*: haplotype diversity, F_S: Fu's F_S. Significant Fu's F_S indicated in bold. Lau Basin samples are pooled from two sites (Hine Hina and Tu'i Malila) for population comparisons. n/a: index not estimated.

Manus and North Fiji Basin ($p<0.05$) and significant pairwise population differentiation was detected between Manus Basin and North Fiji Basin ($\varphi_{ST} = 0.334$ to 0.372; $p<0.05$; Table 4). Effective population size was estimated to be functionally infinite based on microsatellite linkage disequilibrium within Manus Basin samples (based on LDNe and ONeSamp) and KINGROUP indicated no significant pairwise relatedness among individuals within a site. In tests for population bottlenecks, *Chorocaris* sp. 2 departed from mutation-drift equilibrium at Solwara 1 and Solwara 8 (two-tailed $p<0.05$), but not South Su. (two-tailed $p>0.05$) suggesting a recent bottleneck.

Munidopsis lauensis: Population structure

A total of 111 *COI* haplotypes (454 bp) were amplified from *Munidopsis lauensis* (81 from Manus Basin, 30 from Lau Basin; Table 2). Three additional individuals from Lau Basin were identified as *Munidopsis antonii*, a closely related species [60]. Haplotype and nucleotide diversity was low in *M. lauensis* ($Hd<$ 0.09, $\pi<0.0002$; Table 2): a single *COI* haplotype was present in 107 of the individuals examined. Four singleton *COI* haplotypes were separated by only a single nucleotide mutation from the dominant haplotype (Figure 3). Four sequences obtained from GenBank from two additional sites in Lau Basin, one additional site in Manus Basin, and one from the Brothers Seamount (New

Zealand) were identical to the dominant haplotype from our study sites in Manus and Lau Basins (Figure 3).

No *COI*-based genetic differentiation was detected among *Munidopsis lauensis* from Manus or Lau Basin at any spatial scale (φ_{ST}; Table 4). The low genetic variability of these samples precludes an interpretation of Fu's F_S. AMOVA analyses indicated no hierarchical population structure. Mismatch distribution curves were consistent with a model of stable population size (Figure S2).

Within Manus Basin, all nine microsatellite loci amplified in 17 to 31 individuals of *Munidopsis lauensis* from each site (Table 5). Two loci deviated significantly from Hardy-Weinberg expectations—*Mp24* has an excess of heterozygotes at Solwara 1 and *Mp14* had an excess of homozygotes at South Su (Table 5). No evidence for selection, null alleles, stutter, or large allele dropout was detected. Seven microsatellite loci amplified in both Manus and Lau Basins (*Mp8*, *Mp14*, *Mp15*, *Mp16*, *Mp24*, *Mp27*, *Mp29*), all of which adhered to Hardy-Weinberg expectations (Table 4) with no evidence for selection, null alleles, stutter, or large allele dropout detected. Private alleles were present at all sampling locations, with the highest number of private alleles at Solwara 8 and Solwara 1 (14 each; Table 4) and the lowest number of private alleles at South Su and within Lau Basin (7 each; Table 5)

Pairwise tests for genetic differentiation based on seven microsatellite markers that amplified in samples from all sites in

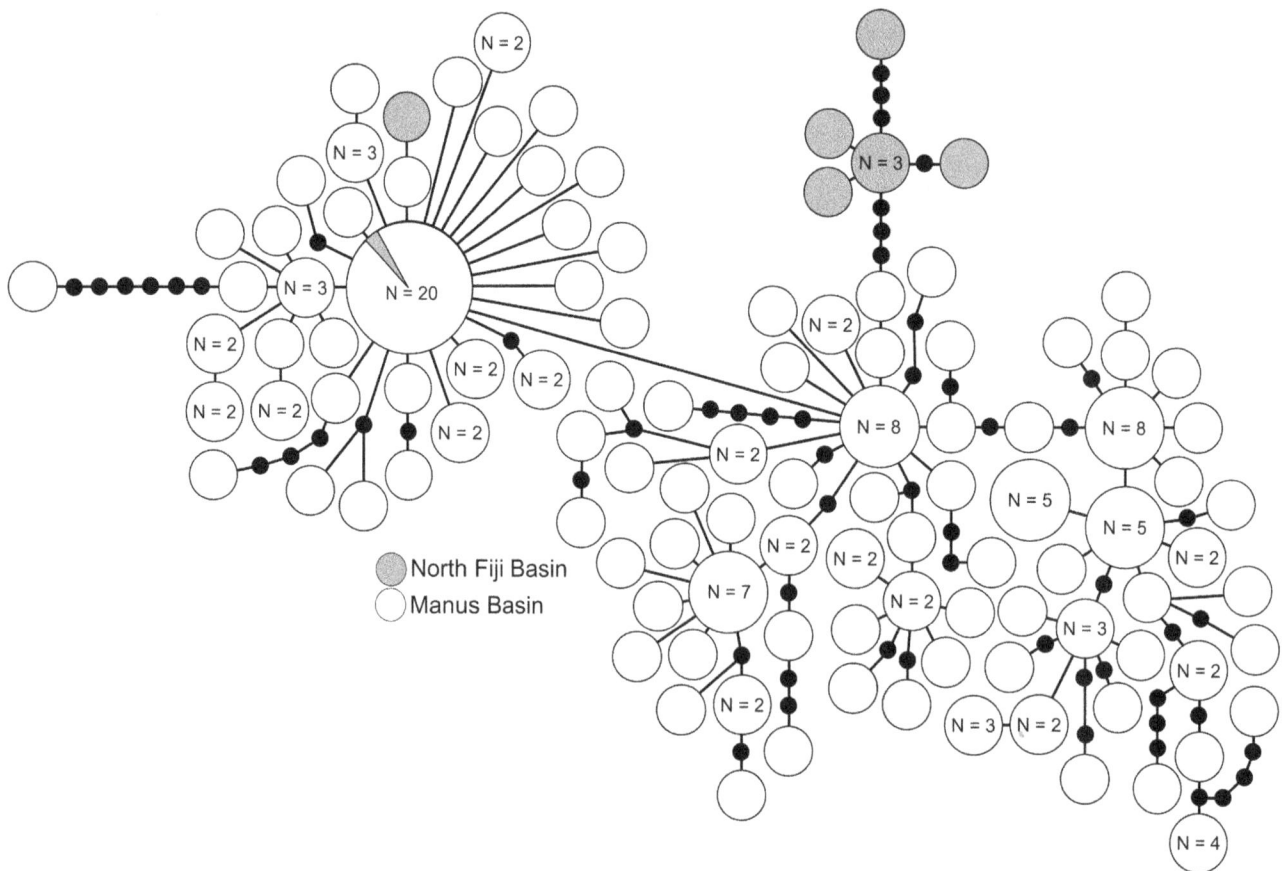

Figure 2. *Chorocaris* **sp. 2.**Statistical parsimony network for haplotypes from samples collected in Manus and North Fiji Basin. Large circles represent a single individual unless noted on the figure. Small black circles represent inferred haplotypes not observed in this data set. Each node represents 1 pb difference.

Manus as well as Lau Basin revealed significant genetic differentiation between Solwara 1 and the other two sites (South Su and Solwara 8) in Manus Basin ($F_{ST} = 0.07$; $p < 0.05$; Table 4) and significant genetic differentiation between populations of *Munidopsis lauensis* in Lau Basin and those in Manus Basin ($F_{ST} \geq 0.11$; $p < 0.05$; Table 4). AMOVA analyses indicated that between-basin effects accounted for nearly 90% of the hierarchal population structure; within-Manus effects accounted for ~10% of the structure. Assignment tests (Structure) suggested that the most likely number of populations is 3 (K = 3, average ln P(D) = −1489.6), with one population in Lau Basin, a second population at Solwara 1 in Manus Basin, and a third population shared at Solwara 8 and South Su in Manus Basin (Figure 4). Effective population sizes for both populations in Manus Basin as well as the Lau Basin population were estimated to be functionally infinite based on microsatellite linkage disequilibrium (based on both LDNe and ONeSamp). KINGROUP indicated no significant pairwise relatedness among individuals within sites in Manus Basin or within Lau Basin. No significant bottleneck effects were detected for *M. lauensis* as all 'sites' were found to be in mutation-drift equilibrium.

Discussion

Overview

Genetic differentiation of species endemic to discrete habitats tends to be positively correlated with the degree of patchiness of those habitats, especially in species with limited dispersal potential [61], though lack of genetic differentiation over large spatial scales has been observed for vent taxa and inferred to be a consequence of the ephemeral nature of vent patches [16]. Genetic differentiation may also be reduced in species associated with patchy habitats that can also exploit alternative habitats, albeit in lower densities (*e.g.*, coral-reef fish display increased gene flow when populations are connected via intermediate, non-coral-reef 'stepping stones', [62]). While we expected that the opportunistic *Munidopsis lauensis* would have less population structure when compared with *Chorocaris* sp. 2, we discovered that *M. lauensis* exhibited strong signals of genetic differentiation at relatively small spatial scales, whereas the population structure of *Chorocaris* sp. 2 was similar to other vent-associated species from the western Pacific (i.e., *Ifremeria nautilei* [11]).

Cryptic Species

Cryptic or miss-identified species were discovered in samples of shrimp and squat lobsters from western Pacific back-arc basins. *Chorocaris* sp. 1, a shrimp closely related to *Chorocaris* sp. 2 [46], was identified from *COI* sequences of shrimp from South Su (Manus Basin). *Chorocaris* sp. 1 is not known from other sites in the western Pacific (T. Komai, personal communication). Three individuals of *Munidopsis antonii*, a squat lobster closely related to *M. lauensis* [31] were found in Lau Basin samples. *M. antonii* is broadly distributed [31], but this is the first report of the species in Lau Basin. Cryptic species have the potential to confound population genetics studies

Table 3. *Chorocaris* sp. 2, Manus Basin.

		Cho30	Cho36	Cho63	Cho76	Cho91	Cho99
Solwara 8	n	91	91	84	78	89	88
	a	3	7	10	9	6	5
	Rs	2.66	6.88	9.93	8.90	5.89	4.65
	as	185–189	157–189	155–179	232–266	201–213	222–234
	H_O	0.11	0.53	**0.57**	**0.31**	0.21	0.47
	H_E	0.15	0.52	0.65	0.49	0.30	0.46
Solwara 1	n	92	81	86	67	89	82
	a	4	9	10	10	9	4
	Rs	3.76	9.00	9.90	9.90	8.23	4.00
	as	183–189	149–189	159–179	248–276	199–215	222–232
	H_O	0.12	0.40	**0.48**	**0.21**	**0.29**	0.41
	H_E	0.11	0.44	0.64	0.45	0.47	0.43
South Su	n	90	88	82	64	81	87
	a	5	11	8	10	7	6
	Rs	5.00	9.59	8.00	10.00	7.00	6.00
	as	183–191	145–189	157–173	244–268	201–213	212–234
	H_O	0.21	0.42	**0.37**	**0.20**	**0.16**	0.61
	H_E	0.25	0.47	0.70	0.32	0.39	0.59

Summary statistics for six microsatellite loci Manus Basin. n: number of individuals, a: number of alleles, Rs: allelic richness, as: allele size range, H_E: expected heterozygosity, H_O: observed heterozygosity; bold: significant deviation from Hardy-Weinberg expectations after Bonferroni correction for multiple tests.

Table 4. *Chorocaris* sp. 2 and *Munidopsis lauensis* pairwise comparisons of Solwara 8, Solwara 1, South Su, North Fiji, and Lau Basin genetic differentiation.

			Solwara 8	Solwara 1	South Su	Lau Basin
Chorocaris sp. 2	Manus Basin	Solwara 8	--	0.005	0.007	--
		Solwara 1	0.006	--	0.013	--
		South Su	−0.009	0.000	--	--
	North Fiji Basin		**0.334**	**0.372**	**0.339**	--
M. lauensis	Manus Basin	Solwara 8	--	0.07	0.01	**0.12**
		Solwara 1	0.00	--	**0.07**	**0.15**
		South Su	0.00	0.00	--	**0.11**
	Lau Basin		0.00	0.00	0.00	--

Pairwise comparisons of F_{ST} from microsatellites (above the diagonal), φ_{ST} from *COI* sequences (*Chorocaris* sp. 2–616 bp; *Munidopsis lauensis* - 454 bp; below the diagonal). Significant genetic differentiation after sequential Bonferroni correction for multiple tests indicated in bold. No microsatellites were deployed on *Chorocaris* sp. 2 collected from North Fiji Basin.

by introducing divergent genetic diversity into analyses [41], thus identifying and excluding cryptic or misidentified species from population samples is an important first step in any population study where morphological identifications are challenging.

Chorocaris sp. 2: Population structure

The homogeneous distribution of *Chorocaris* sp. 2 haplotypes and microsatellite markers within Manus Basin is consistent with high gene flow among the study sites. Although larvae are likely the primary dispersal vector, the mobility of juvenile and adult shrimp may also allow individuals to travel among neighboring vent sites, reducing the potential for local population structure to emerge.

There is a high frequency of rare, private haplotypes (78.3% are singletons) from Manus and North Fiji Basin samples of *Chorocaris* sp. 2. The excess of rare *COI* haplotypes (significantly negative Fu's F_S) and departure from mutation-drift equilibrium for microsat-

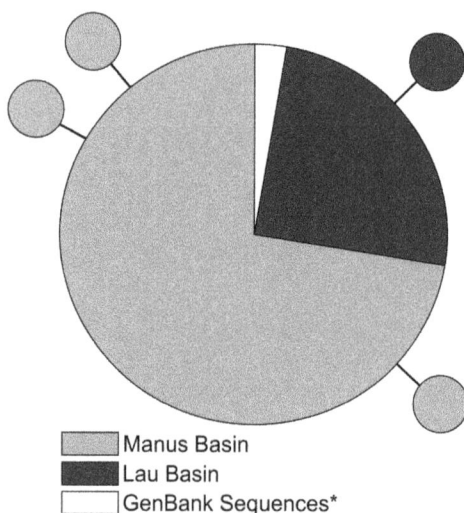

Figure 3. Statistical parsimony network for *Munidopsis lauensis* haplotypes from the western Pacific. Dominant haplotype contains 111 individuals, including four representative sequence recovered from GenBank—Desmos Caldera (Manus Basin; EF157850; [29]), Mariner Vent Field (Lau Basin; EF157851; [29]), Hine Hina (Lau Basin; EF157852; [29]), and Brothers Seamount (New Zealand; EF157853; [29])—indicated with an asterisk. Each node represents 1 bp difference.

ellite markers at Solwara 1 and Solwara 8 suggest that these two populations encountered a recent bottleneck followed by a population expansion. South Su also departed from mutation-drift equilibrium based on *COI* (F_S values are significantly negative but higher than for the other populations) but not for microsatellite markers. This could indicate that the bottleneck encountered by the South Su population was not as strong as bottlenecks affecting other populations, or that the recovery from the bottleneck was faster in South Su. Bottlenecks followed by population expansions have been found in other species [25], including in the *Rimicaris exoculata* shrimp on the Mid-Atlantic Ridge [8,9] and in other hydrothermal vent-associated species from Manus Basin [11,63]. Although null alleles were found at three loci, the presence of null alleles results in inflated estimates of populations differentiation [59]. Given that no genetic differentiation was detected among *Chorocaris* sp. 2 sampled from multiple sites in Manus Basin, the null alleles did not influence the overall outcome.

Despite strong signals of genetic differentiation in *COI* sequences between populations of *Chorocaris* sp. 2 from Manus and North Fiji Basins, the presence of two shared *COI* haplotypes between basins indicates that some migration must occur or have occurred in the recent past or that the populations are experiencing incomplete lineage sorting. The absence of *COI* haplotypes descended from the North Fiji clade in Manus Basin suggests that migration, if it occurs, may be directional, from Manus into North Fiji Basin, though this could be the result of sample bias, given the small number of individuals (n = 9) sampled from North Fiji. A Manus to North Fiji route is consistent with the regional circulation patterns and recent models of larval transport in the southwest Pacific [64].

In contrast to *Chorocaris* sp. 2, *COI* and microsatellite-based population studies of the related *Rimicaris exoculata* from hydrothermal vents along the Mid-Atlantic Ridge found no population structure across more than 5,000 kilometers [8,9]. Haplotype diversity in *R. exoculata* populations (Hd = 0.69 to 1.00) is similar to populations of *Chorocaris* sp. 2 (Hd = 0.70 to 1.00). Teixeira *et al.* [8] suggest that the *R. exoculata* population is the product of a recent founder event followed by demographic expansion along the Mid-Atlantic Ridge, while population structure of *Chorocaris* sp. 2 appears to arise from a barrier to gene flow from basin to basin in the southwestern Pacific.

Munidopsis lauensis: Population structure

COI haplotype diversity in *Munidopsis lauensis* is low (Hd = 0.07–0.09) and is an order of magnitude lower than that observed in

Table 5. *Munidopsis lauensis* microsatellite summary statistics for nine loci Manus and Lau Basin.

Site		Mp8	Mp12	Mp14	Mp15	Mp16	Mp21	Mp24	Mp27	Mp29
Solwara 8	n	19	21	26	23	23	17	29	28	28
	a	2	8	4	4	1	4	8	5	7
	Rs	2.00	7.43	4.00	4.00	1.00	4.00	7.86	5.00	6.61
	as	187–228	103–149	155–242	110–190	207–207	252–314	211–278	197–209	203–246
	H_O	0.68	0.52	0.58	0.26	0.00	0.76	0.45	0.54	0.57
	H_E	0.46	0.45	0.49	0.38	0.00	0.62	0.44	0.49	0.65
Solwara 1	n	28	24	30	25	29	29	29	31	30
	a	2	5	6	4	6	9	6	5	8
	Rs	2.00	4.58	5.85	4.00	5.70	7.37	5.90	4.81	7.66
	as	187–228	103–245	155–244	110–190	207–447	152–314	127–229	197–212	206–246
	H_O	0.32	0.46	0.57	0.72	0.38	0.90	**0.97**	0.32	0.63
	H_E	0.27	0.39	0.74	0.71	0.34	0.72	**0.66**	0.41	0.82
South Su	n	19	19	30	29	28	25	28	29	23
	a	2	2	6	5	6	7	4	4	5
	Rs	2.00	2.00	5.73	4.79	5.29	6.03	4.00	3.97	5.00
	as	187–228	103–143	155–244	110–190	160–446	238–326	129–229	197–206	229–246
	H_O	0.58	0.42	**0.33**	0.41	0.32	0.84	0.25	0.48	0.65
	H_E	0.42	0.34	**0.57**	0.41	0.29	0.74	0.29	0.52	0.60
Lau Basin	n	29	—	35	31	34	—	36	32	37
	a	2	—	4	2	10	—	3	4	3
	Rs	2.00	—	3.74	1.74	8.16	—	2.78	4.00	6.62
	as	187–228	—	155–242	103–110	207–446	—	127–219	197–209	226–244
	H_O	0.17	—	0.40	0.03	0.44	—	0.89	0.34	0.16
	H_E	0.16	—	0.47	0.03	0.43	—	0.51	0.38	0.24

n: number of individuals, a: number of alleles, Rs: allelic richness, as: allele size range, H_E: expected heterozygosity, H_O: observed heterozygosity; bold: significant deviation from Hardy-Weinberg expectations after sequential Bonferroni correction for multiple tests.

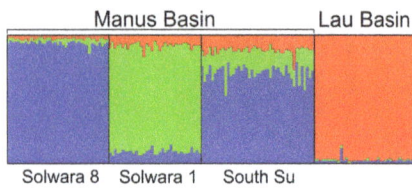

Figure 4. *Munidopsis lauensis* **structure output for seven microsatellite loci shared across Manus and Lau Basin.** Each color represents a different putative population inferred from the distribution of allele frequencies. K = 3 was determined to be the most likely model based on 5 replicates each of model runs from k = 1 to k = 7, with a 1,000,000 step burn-in period followed by 10,000,000 steps. Sampling locations were used as priors for putative population assignments.

other vent-associated species in Manus Basin ([11], Van Dover laboratory, unpublished data). The 96.4% dominance of a single *COI* haplotype in *M. lauensis* contrasts with that of other species from western Pacific deep-sea hydrothermal vents [16] but is consistent with that of *M. polymorpha* from an anchialine pool in the Canary Islands [65]. A low *COI* mutation rate is characteristic of related squat lobsters [60,66]. Alternatively, a recent selective sweep could have reduced the number of haplotypes in the populations [67] though it could also be indicative of a recent population expansion [68]. This interpretation is not supported by the mismatch distribution curves, which indicate a stable population size. For a selective sweep to reduce haplotype diversity in *M. lauensis* populations in both Manus and Lau Basins to a single dominant haplotype, these populations must be or once have been well-connected, otherwise other common haplotypes would have propagated in the isolated populations [69]. The processes that produce low observed haplotype diversity are, as yet, undetermined. However, all interpretation of *M. lauensis* mitochondrial population structure are necessarily constrained by the limited variability of *COI* haplotypes.

Past microsatellite-based studies of galatheid squat lobster population structure have been confounded by the presence of mobile, cryptic, microsatellite-flanking transposable elements in squat lobster genomes [70]. Transposable elements were observed in three squat lobster species (*Munida rugosa*, *Munida sarsi*, and *Galatheae strigosa*), causing inconsistencies and failures in microsatellite amplification and amplification of multiple fragments [70]. The symptoms of these elements were not reported in microsatellites developed for *Munidopsis polymorpha* [71], nor were they observed in the amplification and analysis of *Munidopsis lauensis* for this study.

Microsatellite-based estimates of genetic differentiation in *Munidopsis lauensis* yielded evidence for fine-scale population structure among sites in Manus Basin, but not within sites. *M. lauensis* from Solwara 1 are genetically differentiated from those of South Su (only ~2.5 kilometers apart) and Solwara 8 (~40 km apart), but *M. lauensis* from South Su and Solwara 8 (~40 km apart) are genetically undifferentiated. While family effects—the appearance of genetic differentiation due to a region being colonized by closely related propagules—could explain the observed genetic differentiation between Solwara 1 and the other Manus Basin sites, no indication of significant relatedness among individuals from Solwara 1 was detected. Isolation of *M. lauensis* at Solwara 1 could be the result of physical or hydrographic barriers to gene flow that limit colonization or migration from other sites in Manus Basin. Solwara 1 lies along the northwest flank of a large, submerged, and active volcano (North Su), which physically

separates Solwara 1 from South Su [72]. In addition, the St. George's Undercurrent runs roughly northwest through Manus Basin, passing first over South Su, then over Solwara 1 and Solwara 8 [73]. If *M. lauensis* possesses larva that remain near the sea floor, the path of the St. George's Undercurrent could prevent *M. lauensis* larvae from effectively dispersing up-current, from Solwara 1 southwestward to South Su.

Munidopsis lauensis is genetically differentiated between Manus and Lau Basins based on microsatellite data, but the less variable *COI* sequences do not detect this differentiation. These divergent results from two types of genetic markers highlight one of the challenges in conservation genetics. Because multiple phenomena likely shape the genetic diversity and distribution of *M. lauensis* throughout the western Pacific, no single gene, or suite of similar genetic loci (e.g., mitochondrial or microsatellite), can provide a complete picture of population structure [74]. Our interpretation is that different genes represent different processes in *M. lauensis*. On an evolutionary time-scale, *COI* data suggest that homogenizing processes (gene flow, selective sweeps, or lack of time for mutations to accumulate after a founder event) have reduced the genetic diversity of *M. lauensis*. The lack of a recent population bottleneck detected in the microsatellites supports a selective sweep on *COI*. On an ecologic time-scale, microsatellite data suggest locally differentiated populations and restricted gene flow over relatively short distances (2.5 km); the extent to which locally differentiated populations persist over multiple generations of *M. lauensis* is unclear and requires an assessment of temporal variability among these differentiated populations. A similar phenomenon of divergent estimates of population structure based on differing marker types was observed among species in the tubeworm genus *Escarpia* that inhabit cold seeps in the Gulf of Mexico, Gulf of California, and West Africa; no significant population differentiation was observed based on mitochondrial markers, but microsatellites revealed some significant differentiation among regions [75].

When a suite of microsatellite loci is developed for one particular population, it may not work as well in additional distantly related populations, resulting in null alleles [76]. For *M. lauensis*, this kind of ascertainment bias is apparent in two microsatellite loci, *Mp12* and *Mp21* (as these markers were developed on *M. lauensis* from Manus Basin), which failed to amplify in individuals from Lau Basin. It is possible that these loci are not present in Lau Basin samples. The frequency of alleles for the other seven markers fell within Hardy-Weinberg expectations, with no evidence of null alleles, and suggests that they are not influenced by ascertainment bias. The failure of two microsatellites to amplify, suggests that the observed genetic divergence between *M. lauensis* from Manus and Lau Basin may be even more pronounced.

Isolation of Manus Basin Vent Fauna

There is strong genetic evidence that *Chorocaris* sp. 2 and *Munidopsis lauensis* in Manus Basin are isolated from other back-arc basin vent systems. Three provannid snail species (*Ifremeria nautilei*, *Alviniconcha* sp.1, and *Alviniconcha* sp. 2) also exhibit strong population isolation between Manus Basin and other regional back-arc basins (North Fiji, Lau Basins, Marianna Trough [11,12,20,77]). Barriers to dispersal have been documented for other vent taxa in the eastern Pacific, where the barriers are often associated with geomorphological features (*e.g.* Easter Microplate: [3,7]; Blanco Transform Fault: [22,23]; or hydrodynamic gyres: [3]). Hydrothermal vents in the eastern Pacific are distributed in a roughly linear pattern [17,78], with population differentiation often occurring along a north/south gradient [17,21,23,25,78,79].

The presence of the New Guinea archipelago may create a physical barrier to migration out of Manus Basin [11], while the north-westward path of the St. George's undercurrent may provide a limited pathway for propagules to disperse from Manus eastwards towards North Fiji and Lau Basins [73]. Larval transport models based on an assumption of long-lived, lecithotrophic, deep-sea larvae indicate that, even after 500 days of dispersal, few larvae would be transported into Manus Basin from surrounding regions [64].

Implications for conservation and management

Characterization of patterns of genetic diversity and connectivity within and among populations is a valuable tool for managing and mitigating the effects of anthropogenic disturbance at deep-sea hydrothermal vents [80]. In Manus Basin, Solwara 1 has been identified as a site for mineral extraction, while South Su has been set aside as a refuge [72]. Key vent-associated species shared between South Su and Solwara 1 (*Chorocaris* sp. 2 and *Ifremeria nautilei*) have a high degree of connectivity, providing evidence that South Su may serve as an effective reservoir of genetic diversity for some species. For *Munidopsis lauensis*, however, Solwara 1 and South Su populations are genetically distinct, and, in this case, South Su would not act as an effective reservoir of genetic diversity. Population differentiation between *Munidopsis lauensis* from Solwara 1 and other sites is perplexing. If the squat lobsters are genetically isolated through some as yet-to-be-determined mechanism, it is not clear that this genetic lineage could be sustained in the face of severe population reduction. Genetic differentiation at such a local scale, however, is a surprising outcome for this species and requires further investigation.

There is limited connectivity between the Manus Basin populations *Chorocaris* sp. 2 and *Munidopsis lauensis* and conspecific populations in other western Pacific back-arc basins. This suggests that if maintenance of genetic diversity is an environmental management objective, then management tools must be applied on a regional basis (e.g., within Manus Basin) for these species. Ongoing monitoring of genetic diversity of key taxa before and after mineral extraction would allow managers to assess the impact of the activity on connectivity and population structure and inform best practices.

Supporting Information

Figure S1 Maximum likelihood tree for a subset of *Chorocaris* spp. sampled from Manus and North Fiji Basin. Sequences are 600-base pairs in length. Substitution model is Tamura 3-parameter determined by Find Best Model application in Mega 5. Representatives of *Chorocaris* sp. 1 and sp. 2 were chosen at random. *Chorocaris* sp. 1 and *Chorocaris* sp. 2 indicated by horizontal bars. *Chorocaris vandoverae* (Mariana

Trough; accession # AF125417; [81]) and *Rimicaris exoculata* (Mid-Atlantic Ridge; accession # FN393000) presented for comparison. Bootstrap values greater than 0.50 reported on branches. Scale bar is number of substitutions per base pair.

Figure S2 Observed and expected mismatch curves of pairwise mitochondrial *COI* nucleotide differences for *Chorocaris* sp. 2 sampled from Manus and North Fiji Basin and *Munidopsis lauensis* sampled from Manus and Lau Basin. Each graph represents a comparison between simulated curves for pairwise nucleotide differences and observed pairwise differences for (A) *Chorocaris* sp. 2 under a model of constant population size, (B) *Chorocaris* sp. 2 under a model of population growth and decline, (C) *Munidopsis lauensis* under a model of constant population size, and (D) *M. lauensis* under a model of population growth and decline.

File S1 FASTA format file for all *Chorocaris* spp. *COI* sequences.

File S2 FASTA format file for all *Munidopsis lauensis* *COI* sequences.

File S3 GENPOP format file of all *Chorocaris* spp. microsatellite markers.

File S4 GENPOP format file of all *Munidopsis lauensis* microsatellite markers.

Acknowledgments

We thank Dr. Samantha Smith of Nautilus Minerals, the captain and crew of the M/V *Nor Sky*, the Canyon Offshore ROV team, the *Jason II* ROV team and crew of the R/V *Melville*, and Rebecca Jones and Pen-Yuan Hsing for assistance with field sampling. We thank Bernard Ball for laboratory assistance, Clifford Cunningham for advice and consultation, and Robert Vrijenhoek for providing both samples and guidance. Specimens of *Chorocaris* spp. and *Munidopsis lauensis* from Manus Basin collected for this work are the property of Papua New Guinea, held in trust by Nautilus Minerals, and loaned to Duke University for baseline studies for the Solwara 1 Project.

Author Contributions

Conceived and designed the experiments: ADT SP TFS JC CLVD. Performed the experiments: ADT SP WS FA AJ EAB TFS. Analyzed the data: ADT SP WS FA AJ EAB TFS JC CLVD. Contributed reagents/materials/analysis tools: TFS JC. Wrote the paper: ADT SP CLVD.

References

1. McClain CR, Hardy SM (2010) The dynamics of biogeographic ranges in the deep sea. Proc Biol Sci 277: 3533–3546. doi:10.1098/rspb.2010.1057.
2. Beinart R, Sanders J, Faure B, Sylva S, Lee R, et al. (2012) Evidence for the role of endosymbionts in regional-scale habitat partitioning by hydrothermal vent symbioses. Proc Natl Acad Sci 190: E3241–50.
3. Hurtado LA, Lutz RA, Vrijenhoek RC (2004) Distinct patterns of genetic differentiation among annelids of eastern Pacific hydrothermal vents. Mol Ecol 13: 2603–2615. doi:10.1111/j.1365-294X.2004.02287.x.
4. Coykendall DK, Johnson SB, Karl SA, Lutz RA, Vrijenhoek RC (2011) Genetic diversity and demographic instability in *Riftia pachyptila* tubeworms from eastern Pacific hydrothermal vents. BMC Evol Biol 11: 96. doi:10.1186/1471-2148-11-96.
5. Grassle J (1985) Genetic differentiation in populations of hydrothermal vent mussels (*Bathymodiolus thermophilus*) from the Galapagos Rift and 13°N on the east Pacific rise. Bull Biol Soc Washingt 6: 429–442.

6. Craddock C, Hoeh WR, Lutz RA, Vrijenhoek RC (1995) Extensive gene flow among mytilid (*Bathymodiolus thermophilus*) populations from hydrothermal vents of the eastern Pacific. Mar Biol 124: 137–146. doi:10.1007/BF00349155.
7. Won Y, Young CR, Lutz RA, Vrijenhoek RC (2003) Dispersal barriers and isolation among deep-sea mussel populations (Mytilidae: Bathymodiolus) from eastern Pacific hydrothermal vents. Mol Ecol 12: 169–184. doi:10.1046/j.1365-294X.2003.01726.x.
8. Teixeira S, Cambon-Bonavita M-A, Serrão EA, Desbruyères D, Arnaud-Haond S (2011) Recent population expansion and connectivity in the hydrothermal shrimp *Rimicaris exoculata* along the Mid-Atlantic Ridge. J Biogeogr 38: 564–574. doi:10.1111/j.1365-2699.2010.02408.x.
9. Teixeira S, Serrão EA, Arnaud-Haond S (2012) Panmixia in a Fragmented and Unstable Environment: The Hydrothermal Shrimp *Rimicaris exoculata* Disperses Extensively along the Mid-Atlantic Ridge. PLoS One 7: e38521.

10. Kojima S, Segawa R, Fujiwara Y, Hashimoto J, Ohta S (2000) Genetic Differentiation of Populations of a Hydrothermal Vent-Endemic Gastropod, *Ifremeria nautilei*, between the North Fiji Basin and the Manus Basin revealed by Nucleotide Sequences of Mitochondrial DNA. Zoolog Sci 17: 1167–1174. doi:10.2108/zsj.17.1167.

11. Thaler AD, Zelnio K, Saleu W, Schultz TF, Carlsson J, et al. (2011) The spatial scale of genetic subdivision in populations of *Ifremeria nautilei*, a hydrothermal-vent gastropod from the southwest Pacific. BMC Evol Biol 11: 372. doi:10.1186/1471-2148-11-372.

12. Kojima S, Segawa R, Fijiwara Y, Fujikura K, Ohta S, et al. (2001) Phylogeny of Hydrothermal-Vent – Endemic Gastropods *Alviniconcha* spp. from the Western Pacific Revealed by Mitochondrial DNA Sequences. Biol Bull 200: 298–304.

13. Carney SL, Formica MI, Divatia H, Nelson K, Fisher CR, et al. (2006) Population structure of the mussel "*Bathymodiolus*" *childressi* from Gulf of Mexico hydrocarbon seeps. Deep Sea Res Part I Oceanogr Res Pap 53: 1061–1072.

14. Olu K, Cordes EE, Fisher CR, Brooks JM, Sibuet M, et al. (2010) Biogeography and potential exchanges among the atlantic Equatorial belt cold-seep faunas. PLoS One 5: e11967.

15. Van Dover CL (2000) The Ecology of Deep-Sea Hydrothermal Vents. 1st ed. Princeton: Princeton University Press.

16. Vrijenhoek RC (2010) Genetic diversity and connectivity of deep sea hydrothermal vent metapopulations. Mol Ecol 19: 4391–4411. doi:10.1111/j.1365-294X.2010.04789.x.

17. Marsh AG, Mullineaux LS, Young CM, Manahan DT (2001) Larval dispersal potential of the tubeworm *Riftia pachyptila* at deep-sea hydrothermal vents. Nature 411: 77–80. doi:10.1038/35075063.

18. Moraga D, Jollivet D, Denis F (1994) Genetic differentiation across the Western Pacific populations of the hydrothermal vent bivalve *Bathymodiolus* spp. and the Eastern Pacific (13°N) population of *Bathymodiolus thermophilus*. Deep Sea Res Part I Oceanogr Res Pap 41: 1551–1567. doi:10.1016/0967-0637(94)90060-4.

19. Kyuno A, Shintaku M (2009) Dispersal and differentiation of deep-sea mussels of the genus *Bathymodiolus* (Mytilidae, Bathymodiolinae). J Mar Bio 2009: 1–15. doi:10.1155/2009/625672.

20. Suzuki Y, Kojima S, Sasaki T, Suzuki M, Utsumi T, et al. (2006) Host-symbiont relationships in hydrothermal vent gastropods of the genus *Alviniconcha* from the Southwest Pacific. Appl Environ Microbiol 72: 1388–1393. doi:10.1128/AEM.72.2.1388-1393.2006.

21. Plouviez S, Le Guen D, Lecompte O, Lallier FH, Jollivet D (2010) Determining gene flow and the influence of selection across the equatorial barrier of the East Pacific Rise in the tube-dwelling polychaete *Alvinella pompejana*. BMC Evol Biol 10: 220. doi:10.1186/1471-2148-10-220.

22. Johnson SB, Young CR, Jones WJ, Waren A, Vrijenhoek RC (2006) Migration, Isolation, and Speciation of Hydrothermal Vent Limpets (Gastropoda; Lepetodrilidae) Across the Blanco Transform Fault. Biol Bull 210: 140–157.

23. Young CR, Fujio S, Vrijenhoek RC (2008) Directional dispersal between mid-ocean ridges: deep-ocean circulation and gene flow in *Ridgeia piscesae*. Mol Ecol 17: 1718–1731. doi:10.1111/j.1365-294X.2008.03609.x.

24. Maas P, O'Mullan GD, Lutz RA, Vrijenhoek RC (1999) Genetic and Morphometric Characterization of Mussels (Bivalvia: Mytilidae) From Mid-Atlantic Hydrothermal Vents. Biol Bull 196: 265–272.

25. Plouviez S, Shank TM, Faure B, Daguin-Thiebaut C, Viard F, et al. (2009) Comparative phylogeography among hydrothermal vent species along the East Pacific Rise reveals vicariant processes and population expansion in the South. Mol Ecol 18: 3903–3917. doi:10.1111/j.1365-294X.2009.04325.x.

26. Desbruyères D, Hashimoto J, Fabri M (2006) Composition and biogeography of hydrothermal vent communities in western Pacific back-arc basins. Geophys Monogr 166: 215.

27. Komai T, Segonzac M (2008) Taxonomic Review of the Hydrothermal Vent Shrimp Genera *Rimicaris* Williams & Rona and *Chorocaris* Martin & Hessler. J Shellfish Res 27: 21–41.

28. Cordes EE, Carney SL, Hourdez S, Carney RS, Brooks JM, et al. (2007) Cold seeps of the deep Gulf of Mexico: Community structure and biogeographic comparisons to Atlantic equatorial belt seep communities. Deep Res Part I 54: 637–653. doi:10.1016/j.dsr.2007.01.001.

29. Cubelio SS, Tsuchida S, Watanabe S (2007) Vent Associated *Munidopsis* (Decapoda: Anomura: Galatheidae) from Brothers Seamont, Kermadec Arc, Southwest Pacific, with Description of One New Species. J Crustac Biol 27: 513–519.

30. Baba K (2005) Deep-sea chirostylid and galatheid Crustaceans (Decapoda: Anomura) from the Indo-Pacific, with a list of species. Galathea Rep 20: 1–317.

31. Macpherson E, Segonzac M (2005) Species of genus *Munidopsis* (Decapoda, Anomura, Galatheidae) from the deep Atlantic Ocean, including cold seeps and hydrothermal vent area. Zootaxa 1095: 1–60.

32. Creasey S, Rogers A, Tyler P, Gage J (2000) Genetic and morphometric comparisons of squat lobster, *Munidopsis scobina* (Decapoda: Anomura: Galatheidae) populations, with notes on the phylogeny of the genus *Munidopsis*. Deep Res II 47: 87–118.

33. Galkin SV (1997) Megafauna associated with hydrothermal vents in the Manus Back-Arc Basin (Bismarck Sea). Mar Geol 142: 197–206. doi:10.1016/S0025-3227(97)00051-0.

34. Collins P, Kennedy R, Van Dover C (2012) A biological survey method applied to seafloor massive sulphides (SMS) with contagiously distributed hydrothermal-vent fauna. Mar Ecol Prog Ser 452: 89–107.

35. Erickson KL, Macko S, Van Dover CL (2009) Evidence for a chemoautotrophically based food web at inactive hydrothermal vents (Manus Basin). Deep Sea Res Part II Top Stud Oceanogr 56: 1577–1585. doi:10.1016/j.dsr2.2009.05.002.

36. Walsh PS, Metzger DA, Higuchi R (1991) Chelex 100 as a medium for simple extraction of DNA for PCR-based typing from forensic material. Biotechniques 10: 506–513.

37. Folmer O, Black M, Hoeh W, Lutz R, Vrijenhoek R (1994) DNA primers for amplification of mitochondrial cytochrome c oxidase subunit I from diverse metazoan invertebrates. Mol Mar Biol Biotechnol 3: 294–299.

38. Benson D (1997) GenBank. Nucleic Acids Res 25: 1–6. doi:10.1093/nar/25.1.1.

39. Edgar R (2004) MUSCLE: a multiple sequence alignment method with reduced time and space complexity. BMC Bioinformatics 5: 113.

40. Tamura K, Peterson D, Peterson N, Stecher G, Nei M, et al. (2011) MEGA5: Molecular Evolutionary Genetics Analysis using Maximum Likelihood, Evolutionary Distance, and Maximum Parsimony Methods. Mol Biol Evol 28: 2731–2739. doi:10.1093/molbev/msr121.

41. Bickford D, Lohman DJ, Sodhi NS, Ng PKL, Meier R, et al. (2007) Cryptic species as a window on diversity and conservation. Trends Ecol Evol 22: 148–155. doi:10.1016/j.tree.2006.11.004.

42. Clement M, Posada D, Crandall KA (2000) TCS: a computer program to estimate gene genealogies. Mol Ecol 4: 331–346.

43. Librado P, Rozas J (2009) DnaSP v5: a software for comprehensive analysis of DNA polymorphism data. Bioinformatics 25: 1451–1452. doi:10.1093/bioinformatics/btp187.

44. Excoffier L, Laval G, Schneider S (2005) Arlequin ver. 3.0: an integrated software package for population genetics data analysis. Evol Bioinform Online 1: 47–50.

45. Rice W (1989) Analyzing tables of statistical test. Evolution (N Y) 43: 223–225.

46. Zelnio KA, Thaler AD, Jones RE, Saleu W, Schultz TF, et al. (2010) Characterization of nine polymorphic microsatellite loci in *Chorocaris* sp. (Crustacea, Caridea, Alvinocarididae) from deep-sea hydrothermal vents. Conserv Genet Resour 2: 223–226. doi:10.1007/s12686-010-9243-0.

47. Boyle EA, Thaler AD, Jacobson A, Plouviez S, Dover CL (2013) Characterization of 10 polymorphic microsatellite loci in Munidopsis lauensis, a squat-lobster from the southwestern Pacific. Conserv Genet Resour. 5 (3), 647–649.

48. Rousset F (2008) Genepop'007: a complete reimplementation of the Genepop software for Windows and Linux. Mol Ecol Resour 8: 103–106.

49. Dieringer D, Schlötterer C (2003) Microsatellite analyser (MSA): a platform independent analysis tool for large microsatellite data sets. Mol Ecol Notes 3: 167–169.

50. Goudet J (1995) FSTAT version 1.2: a computer program to calculate F-statistics. J Hered 86: 485–486.

51. Van Oosterhout C, Hutchinson WF, Wills DPM, Shipley P (2004) micro-checker: software for identifying and correcting genotyping errors in microsatellite data. Mol Ecol Notes 4: 535–538. doi:10.1111/j.1471-8286.2004.00684.x.

52. Antao T, Lopes A, Lopes R, Beja-Pereira A, Luikart G (2008) LOSITAN: a workbench to detect molecular adaptation based on a Fst-outlier method. BMC Bioinformatics 9: 323.

53. Beaumont MA, Nichols RA (1996) Evaluating Loci for Use in the Genetic Analysis of Population Structure. Proc R Soc B Biol Sci 263: 1619–1626.

54. Pritchard J, Stephens M, Donnelly P (2000) Inference of Population Structure Using Multilocus Genotype Data. Genetics 155: 945–959.

55. Waples RS, Do C (2008) Ldne: a Program for Estimating Effective Population Size From Data on Linkage Disequilibrium. Mol Ecol Resour 8: 753–756. doi:10.1111/j.1755-0998.2007.02061.x.

56. Tallmon DA, Koyuk A, Luikart G, Beaumont MA (2008) COMPUTER PROGRAMS: onesamp: a program to estimate effective population size using approximate Bayesian computation. Mol Ecol Resour 8: 299–301.

57. Konovalov DA, Manning C, Henshaw MT (2004) kingroup: a program for pedigree relationship reconstruction and kin group assignments using genetic markers. Mol Ecol Notes 4: 779–782. doi:10.1111/j.1471-8286.2004.00796.x.

58. Cornuet JM, Luikart G (1996) Description and power analysis of two tests for detecting recent population bottlenecks from allele frequency data. Genetics 144: 2001–2014.

59. Carlsson J (2008) Effects of microsatellite null alleles on assignment testing. J Hered 99: 616–623. doi:10.1093/jhered/esn048.

60. Jones WJ, Macpherson E (2007) Molecular Phylogeny of the East Pacific Squat Lobsters of the Genus *Munidopsis* (Decapoda: Galatheidae) with the Descriptions of Seven New Species. J Crustac Biol 27: 477–501.

61. Cohen D, Levin SA (1991) Dispersal in patchy environments: the effects of temporal and spatial structure. Theor Popul Biol 39: 63–99.

62. Shulman MJ (1998) What can population genetics tell us about dispersal and biogeographic history of coral-reef fishes? Austral Ecol 23: 216–225. doi:10.1111/j.1442-9993.1998.tb00723.x.

63. Plouviez S, Schultz TF, McGinnis G, Minshall H, Rudder M, et al. (2013) Genetic diversity of hydrothermal-vent barnacles in Manus Basin. Deep Sea Res Part I Oceanogr Res Pap 82: 73–79.

64. Yearsley JM, Sigwart JD (2011) Larval transport modeling of deep-sea invertebrates can aid the search for undiscovered populations. PLoS One 6: e23063. doi:10.1371/journal.pone.0023063.

65. Wilkens H, Parzefall J, Ribowski A (1990) Population Biology and Larvae of the Anchialine Crab *Munidopsis polymorpha* (Galatheidae) from Lanzarote (Canary Islands). J Crustac Biol 10: 667–675.

66. Samadi S, Bottan L, Macpherson E, De Forges BR, Boisselier M-C (2006) Seamount endemism questioned by the geographic distribution and population genetic structure of marine invertebrates. Mar Biol 149: 1463–1475.

67. Ilves KL, Huang W, Wares JP, Hickerson MJ (2010) Colonization and/or mitochondrial selective sweeps across the North Atlantic intertidal assemblage revealed by multi-taxa approximate Bayesian computation. Mol Ecol 19: 4505–4519.

68. Fu YX (1995) Statistical properties of segregating sites. Theor Popul Biol 48: 172–197. doi:10.1006/tpbi.1995.1025.

69. Wares J, Cunningham C (2001) Phylogeography and historical ecology of the North Atlantic intertidal. Evolution (N Y) 55: 2455–2469.

70. Bailie DA, Fletcher H, Prodöhl PA, Prodo PA (2010) High Incidence of Cryptic Repeated Elements in Microsatellite Flanking Regions of Galatheid Genomes and Its Practical Implications for Molecular Marker Development. J Crustac Biol 30: 664–672. doi:10.1651/09-3252.1.

71. Cabezas P, Bloor P, Acevedo I, Toledo C, Calvo M, et al. (2008) Development and characterization of microsatellite markers for the endangered anchialine squat lobster *Munidopsis polymorpha*. Conserv Genet 10: 673–676. doi:10.1007/s10592-008-9611-4.

72. Coffey Natural Systems (2008) Environmental Impact Statement: Nautilus Minerals Niugini Limited, Solwara 1 Project. Queensland, Australia.

73. Zenk W, Siedler G, Ishida A, Holfort J, Kashino Y, et al. (2005) Pathways and variability of the Antarctic Intermediate Water in the western equatorial Pacific Ocean. Prog Oceanogr 67: 245–281. doi:10.1016/j.pocean.2005.05.003.

74. Avise JC (2004) Molecular markers, natural history and evolution. 2nd ed. Sunderland, MA: Sinauer Associates, Inc.

75. Cowart DA, Huang C, Arnaud-Haond S, Carney SL, Fisher CR, et al. (2013) Restriction to large-scale gene flow vs. regional panmixia among cold seep *Escarpia* spp. (Polychaeta, Siboglinidae). Mol Ecol 22: 4147–4162.

76. Schlötterer C (2000) Evolutionary dynamics of microsatellite DNA. Chromosoma 109: 365–371.

77. Denis F, Jollivet D, Moraga D (1993) Genetic separation of two allopatric populations of hydrothermal snails *Alviniconcha* spp.(Gastropoda) from two South Western Pacific back-arc basins. Biochem Syst Ecol 21: 431–440.

78. Thomson RE, Mihály SF, Rabinovich AB, McDuff RE, Veirs SR, et al. (2003) Constrained circulation at Endeavour ridge facilitates colonization by vent larvae. Nature 424: 545–549. doi:10.1038/nature01824.

79. Matabos M, Plouviez S, Hourdez S, Desbruyères D, Legendre P, et al. (2011) Faunal changes and geographic crypticism indicate the occurrence of a biogeographic transition zone along the southern East Pacific Rise. J Biogeogr 38: 575–594. doi:10.1111/j.1365-2699.2010.02418.x.

80. Collins P, Kennedy B, Copley J, Boschen R, Fleming N, et al. (2013) VentBase: Developing a consensus among stakeholders in the deep-sea regarding environmental impact assessment for deep-sea mining–A workshop report. Mar Policy 42: 334–336. doi:10.1016/j.marpol.2013.03.002.

81. Shank TM, Black MB, Halanych KM, Lutz RA, Vrijenhoek RC (1999) Miocene radiation of deep-sea hydrothermal vent shrimp (Caridea: Bresiliidae): evidence from mitochondrial cytochrome oxidase subunit I. Mol Phylogenet Evol 13: 244–254. doi:10.1006/mpev.1999.0642.

7

Humpback Whale Song on the Southern Ocean Feeding Grounds: Implications for Cultural Transmission

Ellen C. Garland[1]*, **Jason Gedamke**[2], **Melinda L. Rekdahl**[3], **Michael J. Noad**[3], **Claire Garrigue**[4], **Nick Gales**[5]

1 National Marine Mammal Laboratory, Alaska Fisheries Science Center, National Marine Fisheries Service, National Oceanic and Atmospheric Administration, Seattle, Washington, United States of America, 2 Ocean Acoustics Program, Office of Science and Technology, National Marine Fisheries Service, National Oceanic and Atmospheric Administration, Silver Spring, Maryland, United States of America, 3 Cetacean Ecology and Acoustics Lab, School of Veterinary Science, University of Queensland, Gatton, Queensland, Australia, 4 Opération Cétacés, Noumea, New Caledonia, 5 Australian Marine Mammal Centre, Australian Antarctic Division, Kingston, Tasmania, Australia

Abstract

Male humpback whales produce a long, complex, and stereotyped song on low-latitude breeding grounds; they also sing while migrating to and from these locations, and occasionally in high-latitude summer feeding areas. All males in a population sing the current version of the constantly evolving display and, within an ocean basin, populations sing similar songs; however, this sharing can be complex. In the western and central South Pacific region there is repeated cultural transmission of song types from eastern Australia to other populations eastward. Song sharing is hypothesized to occur through several possible mechanisms. Here, we present the first example of feeding ground song from the Southern Ocean Antarctic Area V and compare it to song from the two closest breeding populations. The early 2010 song contained at least four distinct themes; these matched four themes from the eastern Australian 2009 song, and the same four themes from the New Caledonian 2010 song recorded later in the year. This provides evidence for at least one of the hypothesized mechanisms of song transmission between these two populations, singing while on shared summer feeding grounds. In addition, the feeding grounds may provide a point of acoustic contact to allow the rapid horizontal cultural transmission of song within the western and central South Pacific region and the wider Southern Ocean.

Editor: Patrick J. O. Miller, University of St Andrews, United Kingdom

Funding: The Antarctic Whale Expedition cruise described in the manuscript was jointly funded by the Australian and New Zealand governments. At the time of the cruise, both Nick Gales and Jason Gedamke were employees of the Australian Government, and along with the other authors, were responsible for study design, data collection and analysis, decision to publish, and preparation of this manuscript. The recording of eastern Australian song was incidental to behavioural response studies being conducted on humpback whales. These studies were funded by the Australian Government (2009) and the E&P Sound and Marine Life Joint Industry Program and the US Bureau of Ocean Energy Management, Regulation and Enforcement (2010). Song collection in New Caledonia was made possible by contributions from Fondation d'Entreprise Total and the Provinces Sud. ECG was supported by a National Research Council Postdoctoral Fellowship. These funding bodies had no role in the design of the current study, nor the decision to publish.

Competing Interests: The authors have declared that no competing interests exist.

* E-mail: Ellen.Garland@noaa.gov

Introduction

Male humpback whales produce a long, stereotyped and constantly evolving vocal breeding display, termed 'song' [1,2]. Within a population, males conform to the current arrangement and content of the song [3–5]. The conformity to a single song type within a population is thought to occur via vocal learning from surrounding males and, when song transmission is examined at the ocean basin scale, is considered one of the best examples of horizontal cultural transmission in a non-human animal [6]. Song similarity has also been documented between populations, although the degree to which this occurs is dependent upon the geographical distance between such populations [6–13]. Thus, song similarity among populations indicates that acoustic contact is likely to have occurred, although there is currently little known about the mechanism(s) through which song transmission is mediated. Therefore, identifying all potential mechanisms of transfer is essential to understanding the dynamics of song transmission within and across regions.

Within an ocean basin, populations in closer proximity to each other typically display a higher degree of song similarity [6–12]. In

the North Pacific, for example, studies of song sharing have shown that the geographically close populations of Hawaii and Mexico shared a higher number of themes, compared to Japan [8–9,11]. Similarly, in the Southern Hemisphere, songs recorded across the South Atlantic were similar across the ocean basin within a year [12]. In more geographically isolated populations, such as in the Indian Ocean (Madagascar and western Australia), little song sharing was found to occur across the ocean basin (shown through a single shared theme [14]).

Typically, humpback whales from different ocean basins sing distinctly different songs (see [3–4,7,15–16]). Despite this, song from Gabon (South Atlantic) was found to be similar in a single year (2003) to song from Madagascar (Indian Ocean) [17]. Similarly, song from the western Australian population (Indian Ocean) has spread into the eastern Australian population (South Pacific) [13], although little is known about how songs are shared between these two populations.

Payne and Guinee [7] hypothesized three possible mechanisms to allow acoustic contact and subsequent song transmission among populations in any ocean basin. The first possibility is movement of individuals from one breeding population to another between seasons (a phenomenon which has been observed repeatedly; see [18–21]). The second is within-season movement of individuals between two breeding populations (rarely observed; see [19–20]). The third is song exchange on shared migration routes and/or on summer feeding grounds in high latitudes.

Although high-latitude song is, compared to that on breeding grounds, relatively uncommon, such singing has been observed in spring, summer or autumn in the North Atlantic [22–24], the North Pacific [25], and off the Western Antarctic Peninsula [26]. It has also been recorded extensively during migration, including off the eastern coast of Australia [16,27], in the North Pacific [28] and the North Atlantic [1,29–30], and off New Zealand [10,31]. Song transmission among populations within the western and central South Pacific region is more likely to occur through the movement of individual males between seasons and/or singing while on shared migratory routes, than by males moving between populations in a single season [6,13], as these movements, although documented, are rare in comparison to both inter-seasonal movements of males or migratory song [20]. Although mixing of populations on the feeding grounds has been reported (between eastern and western Australian whales [32]), singing while on these shared summer feeding grounds has not been reported for the South Pacific populations, or the greater Southern Ocean away from the Western Antarctic Peninsula.

A number of humpback whale populations have been recognised around the world due to strong feeding and/or breeding ground site fidelity [18,33–35]. In the western and central South Pacific region, the International Whaling Commission (IWC) currently recognizes two different breeding stocks (E and F [36]). These are thought to migrate to two corresponding feeding regions in Antarctica (Areas V, 120°E to 180° W, and VI, 180° W to 120° W [36]). Group E is further divided into 'sub-stocks' E1 eastern Australia, E2 New Caledonia and E3 Tonga, while, further to the east, Group F is divided into F1 Cook Islands and F2 French Polynesia [36] (Figure 1). Recovery from commercial whaling, including the severe overexploitation from illegal Soviet whaling [37], has been uneven within the region. The eastern Australian population is recovering strongly with a high rate of population growth [38–39], while other populations within the Oceania region are showing little signs of recovery [40–41].

Linkages between breeding and feeding grounds have been demonstrated in the South Pacific using 'Discovery marks', small metal cylinders shot into whales during whaling and subsequently recovered when whales were later killed [32,42–44]. These provided linkages in the 1950s and early 1960s between Antarctic feeding Area V and Tonga, Norfolk Island, New Zealand [43], and eastern Australia [32]. More recently, genetic analyses (*e.g.*, [45]), photo-identification comparisons (*e.g.*, [46]) and satellite tagging studies (*e.g.*, [47]) have demonstrated connections between feeding and breeding areas. The dedicated Australia-New Zealand Antarctic Whale Expedition to Antarctic Area V in 2010 [48] and project CETA in 2010 [49], highlighted connections between the Balleny Islands region (66–68°S and 162–165°E) and waters off Adelie Land (65–66°S and 140–145°E) and the eastern Australian migratory corridor (six genetic matches, 23 photo-identification recaptures for the Balleny Islands, one photo-identification recapture for Adelie Land), the New Zealand migratory corridor (one genetic match to the Balleny Islands), and the New Caledonian breeding ground (one photo-identification recapture for the Balleny Islands) [45–46]. Opportunistic acoustic recordings were taken as part of the Australia-New Zealand Antarctic Whale Expedition and are analysed here to assess the similarity between humpback whale song recorded in the Southern Ocean and other regions.

Song sharing within the western and central South Pacific is very dynamic (based on a multi-year, multi-population song analysis) [6,50–51]: songs have been documented radiating repeatedly across the region from west to east, from eastern Australia to French Polynesia, usually over a period of two years. The inclusion of song from the feeding grounds into such analyses represents a significant opportunity to acoustically link breeding to feeding grounds, and to explore how mechanisms of song transmission may contribute to the large-scale pattern of song similarity across this vast oceanographic region.

Here we report the first recorded occurrence of humpback whale song in Antarctic Area V, south of the eastern Australian and New Caledonian breeding populations and their shared migratory corridor, New Zealand. Although song was available from only faint singers, we were able to identify the song as being humpback whale in its origin, and then to compare the song from both eastern Australia and New Caledonia in the previous and following breeding seasons, in an attempt to acoustically link breeding to feeding grounds.

Materials and Methods

During the dedicated Australia-New Zealand Antarctic Whale Expedition on *RV Tangaroa* [48], sonobuoy recordings containing presumed humpback whale song were made from sonobuoys deployed in a region 150–200 km southwest of the Balleny Islands, and 90–150 km north of the Antarctic continent. These occurred over a 48 hour period (March 5–7, 2010; Table 1) and spanned roughly a 70 km range, with the southernmost sonobuoy deployed at 69° 18′ S, 166° 16′ E and the northernmost sonobuoy deployed at 68° 43′ S, 166° 53′ E. Recordings were made of underwater sound transmitted from radio-linked sonobuoys (DIFAR 53D, functional audio range of 10 Hz to at least 2.4 kHz; see [52] for further details). Sound was digitized (wav file format, 12 bit, 48 kHz sampling rate) on a National Instruments PCMCIA DAQCard-6062E and recorded onto a laptop running *Ishmael* software [53]. A total of one hour of sound files (~5 minute wav files downsampled to 6 kHz sampling rate) with identifiable song were examined (Table 1). Songs from eastern Australia were recorded at Peregian Beach, Queensland, using moored, radio-linked hydrophone buoys (a brief description is provided below; see [27,54] for additional details on equipment set up). These had High Tech HTI 96 MIN hydrophones with a built-in +40 dB gain pre-amplifier and an additional external custom-built preamplifier (+20 dB). The signals were transmitted using AN/SSQ-47A sonobuoy transmitters and received onshore using a type 8101 sonobuoy receiver. The radio signals were recorded directly to computer (wav file format, 16 bit, 22 kHz sampling rate) running *Ishmael* software [53] using a National Instruments E-series data acquisition card. Songs from the southern lagoon of New Caledonia were recorded using a single High Tech HTI 96 MIN hydrophone with built-in +40 dB gain pre-amplifier and M-Audio Microtrack 24/96 digital recorder (wav file format, 16 bit, 44.1 kHz sampling rate). Ethical and permit approval for this work was obtained from all appropriate organizations (The University of Queensland Animal Ethics Committee, The Australian Federal Government, The Queensland State Government and Direction de l'Environnement Province Sud, New Caledonia).

Songs were viewed in Abode Audition (2.0 and CS5.5) using Blackman-Harris, 75% overlap, 2048 point fast Fourier transform (FFT) for the eastern Australian and New Caledonian recordings and 1024 point FFT for the Antarctic Area V recordings (to ensure a comparable frequency resolution), displaying approximately 30 seconds of song from 0–2.5 kHz. The nature of the Antarctic

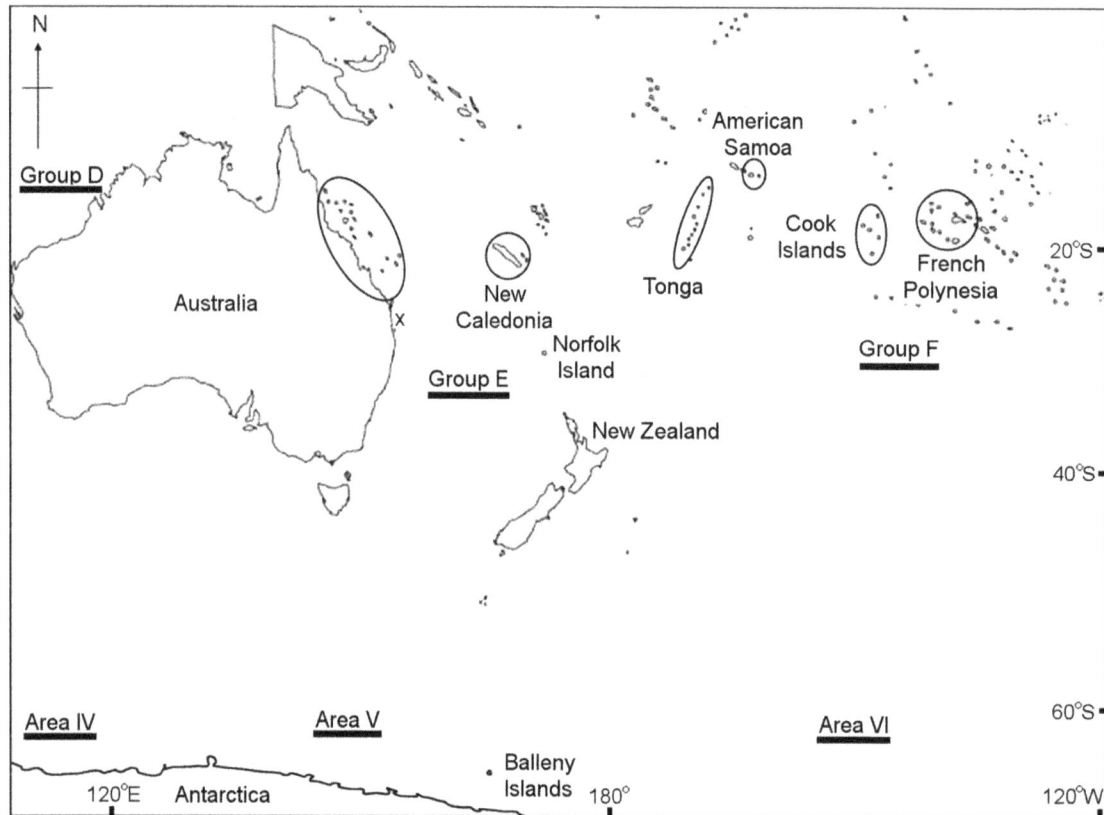

Figure 1. Map of the South Pacific and corresponding Antarctic region. Suggested breeding groups (D, E, and F) and corresponding Antarctic feeding areas (IV, V and VI) are shown. Circles represent the location of the major breeding populations discussed within the region (see text). Recordings in eastern Australia were taken at Peregian Beach, SE Queensland, while animals where on migration (x).

recording permitted only qualitative analysis of the song. First, to identify if the sounds were humpback whale song, all sound units were examined to see if there were any repeating patterns. Each sound type ('unit') was assigned a descriptive name based on the visual and aural qualities of the sound (*e.g.*, 'ascending cry', 'moan', 'purr'; see [6,55] for unit descriptions and qualitative unit identification). Humpback whale song is composed of multiple sounds ('units') that make a stereotyped pattern (a 'phrase'); these are then repeated multiple times to make a 'theme' [1]. A few themes, which are sung in a particular order, comprise a song. Identifying repeated, stereotyped phrases within the vocalization strongly suggests the sound is produced by a male humpback whale and constitutes a 'song'. Multiple potential instances of song were present in some of the Antarctic recording as humpback whale vocalizations/units were recorded; these were not included in further analysis unless a clear phrase pattern was present indicating song (Table 1). Second, once phrases were identified for the Antarctic sample, 2009 and 2010 song from eastern Australia and New Caledonia were examined (Table 2). This was done by comparing the themes for each year and location. For the eastern Australia 2009 and 2010 samples, previous examination of the songs identified all themes that were present within each year of song (M. L. Rekdahl, E. C. Garland and A. Murray unpublished data). For the New Caledonian song for both years, high quality recordings were examined in Adobe Audition.

For each location and year, all themes were described by summarising the units used by each singer for every phrase. In some themes there were alternate forms of phrases where one sound type was replaced by another. To ensure theme classifica-

tion and thus matching was as objective as possible and repeatable, three naïve observers (a bowhead whale song specialist, a killer whale and right whale call specialist, and a fin whale call specialist with humpback whale song experience) were asked to match a number of themes. Ten themes in total (a reference set) were chosen from the song types presented in the current study. These were displayed (along with the test set) in Raven Pro 1.4 (using the same settings as for the analysis) to allow the observers to both visually and aurally assess each theme. Each observer was given the test set containing 20 themes and was asked to assign each sample with a theme number or, in the case of one sample, no matching theme. The observers classified 89%, 90% and 95% of themes in a similar manner, resulting in an average 91% agreement in classification. Mismatches were between themes M1 (eastern Australia 2010) and M2 (eastern Australia 2010), and through the evolution of the eastern Australian 2010 theme H to include 'whoops' in the squeak series. All matches pertaining to the 2009 eastern Australian, New Caledonian 2010 and Antarctic Area V themes resulted in 100% agreement in classification. Humpback whale song themes are known to be highly variable with a large number of sound types that can be combined in many different ways. For example, Oceania song over the period 1998 to 2008 has had at least 93 phrase types [55]. Therefore, the chance of two themes matching by chance is extremely small.

Results

Due to the repetition of a stereotyped sequence of species-typical sounds, the recorded sounds from Antarctic Area V were

Table 1. Summary of recordings that contained suspected humpback whale song in Antarctic Area V. Bold indicates the highest quality recordings. See Figure 2 for corresponding spectrograms.

Date	Time	Length (hr:min:sec)	Humpback vocalizations present	Clear song pattern (Y/N)	Themes identified
5-Mar-10	19:47:33	5:00	Y	N	
	19:52:33	5:00	Y	Y	Theme H
	19:57:33	5:00	Y	N	
	20:02:33	5:00	Y	N	
6-Mar-10	18:45:00	5:00	Y	N	
	19:45:41*	**4:35**	**Y**	**Y**	**Theme I, K & M1**
	19:53:02*	**3:40**	**Y**	**Y**	**Theme H, I & J**
	19:58:50	**2:00**	**Y**	**Y**	**Theme H & K**
	20:20:00	5:00	Y	Y	Theme K
7-Mar-10	06:46:48	5:00	Y	Y	Theme K
	06:51:48	5:00	Y	N	
	08:26:48	5:00	Y	Y	Theme M1
	08:31:48	5:00	Y	N	
	08:36:48	5:00	Y	Y	Theme M1
	17:47:05	5:00	Y	Y	Theme I & M1
	17:52:05	5:00	Y	Y	Theme H & I
	17:57:05	**5:00**	**Y**	**Y**	**Theme H & I**
	18:02:05	5:00	Y	N	
	18:22:05	5:00	Y	Y	Theme K
	18:27:05	5:00	Y	N	
	19:55:30	5:00	Y	Y	Theme M1
Total time examined		1:40:15			
Time with song		1:00:15			

*Spectrograms in Figure 2 were taken from these recordings.

identified as humpback whale song. At least four stereotyped phrases and thus themes (labelled H to M) were identified (Figure 2); these qualitatively matched four themes from the eastern Australian 2009 song (Figure 3) and the same four themes from the 2010 New Caledonian song (Figure 4).

Song Description for Eastern Australia 2009 and 2010, New Caledonia 2010 and Antarctic Area V 2010 Phrases

Four themes from the Antarctic Area V sample were seen repeatedly (H, I, K & M1; Table 1), but no sequence of themes was discernible (Figure 2, Acoustic file S1). Theme J was heard on

a single occasion in the Antarctic Area V sample, and as such is only considered a potential match. Six themes were present in the eastern Australian song in 2009 (Figure 3, Acoustic file S2). Themes were typically sung H, I, J, K, L and M1. In 2010, theme I evolved into theme N, and theme M1 progressed into a second phrase, M2. Thus, the theme sequence was H, N, J, L, M1 and M2 (Figure 5, Acoustic file S4). The New Caledonian 2010 song sequence predominantly consisted of themes H, I, J, K, L and M1 (note description below of an individual singing the 2009 song in New Caledonia in 2010; Figure 4, Acoustic file S3).

New Caledonia 2009 Song Description

Seven themes were present in 2009, some of which contained two phrase types (Figure 6, Acoustic file S5). Themes were typically sung in the order A1, A2, B1, B2, C, D, E, F1 and F2, to form a song. Theme G, the surfacing theme, was inserted after theme D or F2 if the animal surfaced to breathe during the song. Interestingly, the first recording from New Caledonia in 2010 (July 15, 2010) contained this song type; however all background singers in the recording, and all additional singers recorded in 2010, were singing the eastern Australian song type as described above. The sequence for this particular 2010 New Caledonian singer, assessed from two songs, was B1, B2, C, D, E, and F2. This song type qualitatively matches Garland et al.'s Light Green song type [6,51,55].

Table 2. Sample sizes of the number of singers and the total number of songs for eastern Australia and New Caledonia in 2009 and 2010.

Location	Year	Number of singers	Total number of songs
Eastern Australia	2009	6	32
	2010	6	30
New Caledonia	2009	3	14
	2010	6*	14

*One singer sung the New Caledonian 2009 song type.

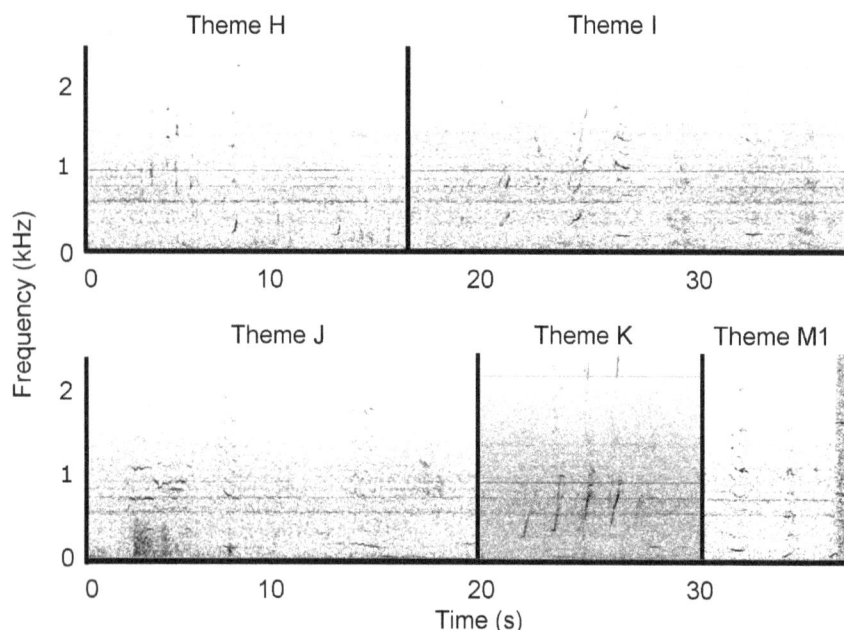

Figure 2. Antarctic Area V 2010 song. A single representative phrase is shown for each theme representing a collection of separately identified themes. While the themes illustrated were taken from an almost continuous ten minute recording session (see Table 1), no clear theme sequence was discernible. Spectrograms were generated in Raven Pro 1.4 (Blackman-Harris window, 1024 FFT size, 75% overlap). Recordings were taken from 5 to 7 March 2010. Audio is provided in acoustic file S1. Note the difference in frequency scale for Figure 2 compared to Figures 3 to 6.

Theme Matching

Four of the available eight themes (H, J, L and M1) matched between the eastern Australia 2009 and 2010 song (Figures 3 and 5). All six themes from the New Caledonian 2010 song (H, I, J, K, L, M1) matched the six themes present in the eastern Australian 2009 song (Figures 3 and 4), and four of the eight possible themes (H, J, L and M1) matched the eastern Australian 2010 song (Figures 4 and 5). The Antarctic sample potentially shared five (out of a possible six) themes with the New Caledonian 2010 and eastern Australian 2009 song (H, I, J, K and M1; Figures 2–4), and

three (out of a possible eight) themes with the eastern Australian 2010 song (H, J and M1; Figures 2 and 5). None of the New Caledonian 2009 themes matched the 2009 and 2010 eastern Australian, 2010 New Caledonian or 2010 Antarctic Area V themes.

Discussion

Here we have presented the first example to our knowledge, of humpback whale song recorded on the Antarctic feeding grounds

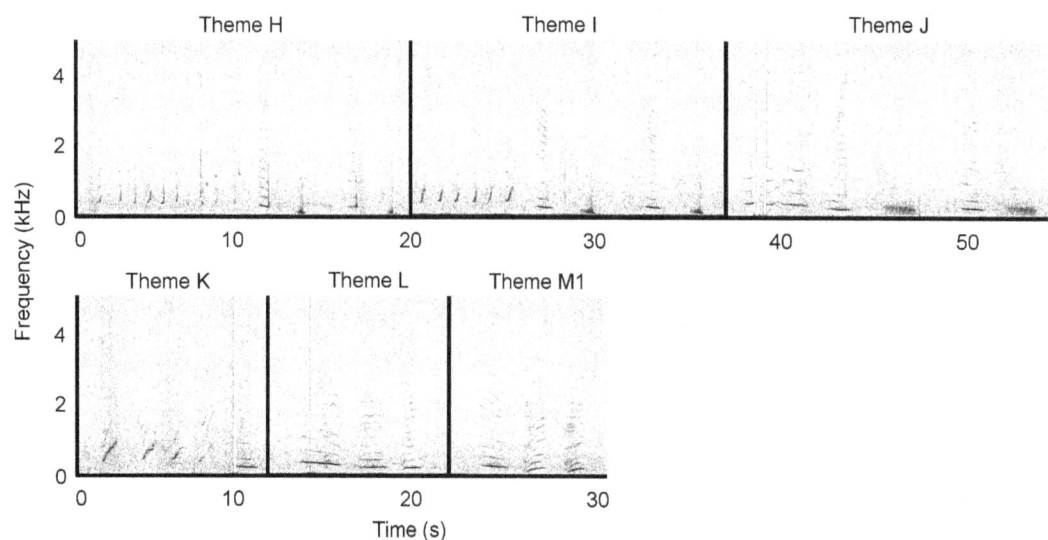

Figure 3. Eastern Australia 2009 song. A single representative phrase is shown for each theme representing a song. Spectrograms were generated in Raven Pro 1.4 (Blackman-Harris window, 2048 FFT size, 75% overlap). Recordings were taken in October 2009. Audio is provided in acoustic file S2. Note difference in frequency scale to Figure 2.

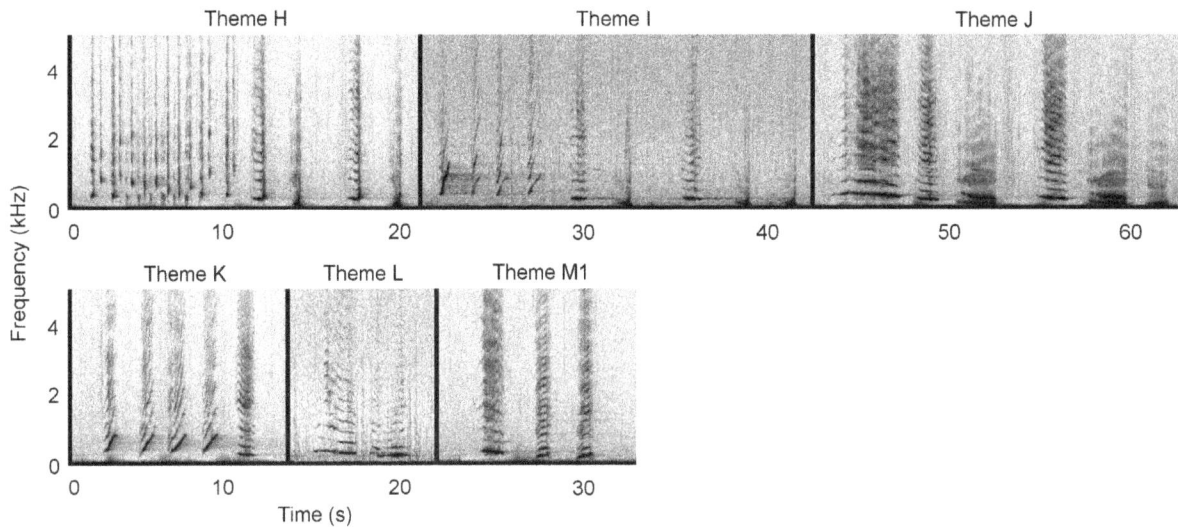

Figure 4. New Caledonia 2010 song. A single representative phrase is shown for each theme representing a song. Spectrograms were generated in Raven Pro 1.4 (Blackman-Harris window, 2048 FFT size, 75% overlap). Recordings were taken from July to September 2010. Audio is provided in acoustic file S3. Note difference in frequency scale to Figure 2.

in the waters of eastern Antarctica. Themes identified from the Antarctic Area V recordings matched song themes from the two closest breeding populations which are likely to feed in this region, eastern Australia and New Caledonia. While feeding grounds along the coasts of North America (Pacific and Atlantic) are more accessible for song recording, the Southern Ocean (particularly away from the Western Antarctic Peninsula) is significantly more challenging as a study area, and this is why there is no previous record of humpback whale singing activity from this region. The opportunistic song recording documented here, and the high density of whales present near the Balleny Islands [48], has highlighted a location where future effort can be concentrated to maximize recordings of feeding ground song.

Payne and Guinee [7] hypothesized three possible mechanisms to allow acoustic contact and subsequent song transmission among populations in any ocean basin: movement of individuals from one breeding population to another between seasons, within-season movement of individuals between two breeding populations, and song exchange on shared migration routes and/or on summer feeding grounds in high latitudes. The similarity of four Antarctic Area V themes to song from the previous year from eastern Australia, and the same year (but following breeding season) in New Caledonia (confirmed by naïve observers and unlikely to be simply by chance), indicates that singing in summer on shared feeding grounds is one possible mechanism through which song can be transmitted between these populations. In addition, a recent photo-identification recapture study has indicated a low level of interchange between eastern Australia and New Caledonia (four out of 1402 individuals) [21], and these populations also share a migration route through New Zealand [56]. Thus, both

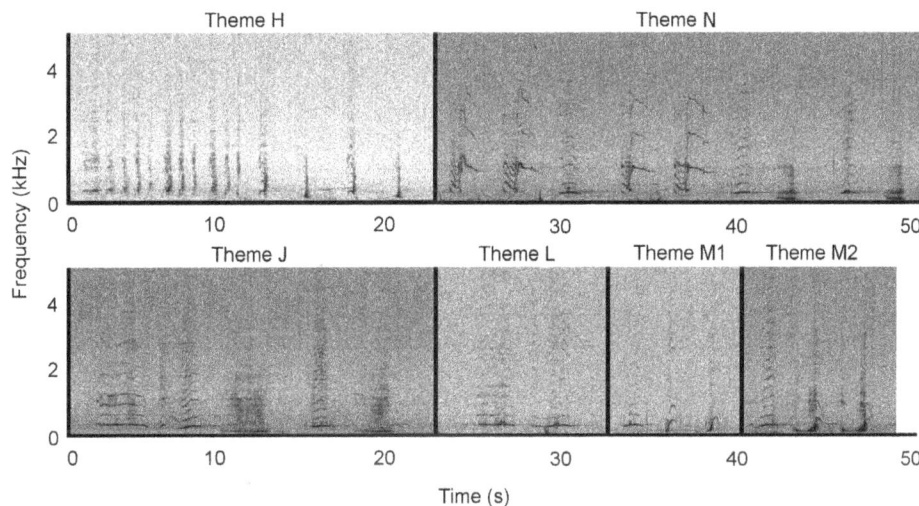

Figure 5. Eastern Australia 2010 song. A single representative phrase is shown for each theme representing a song. Some themes contained 1 and 2 phrase types representing small but stereotyped variations to a theme. Spectrograms were generated in Raven Pro 1.4 (Blackman-Harris window, 2048 FFT size, 75% overlap). Recordings were taken from September to October 2010. Audio is provided in acoustic file S4. Note difference in frequency scale to Figure 2.

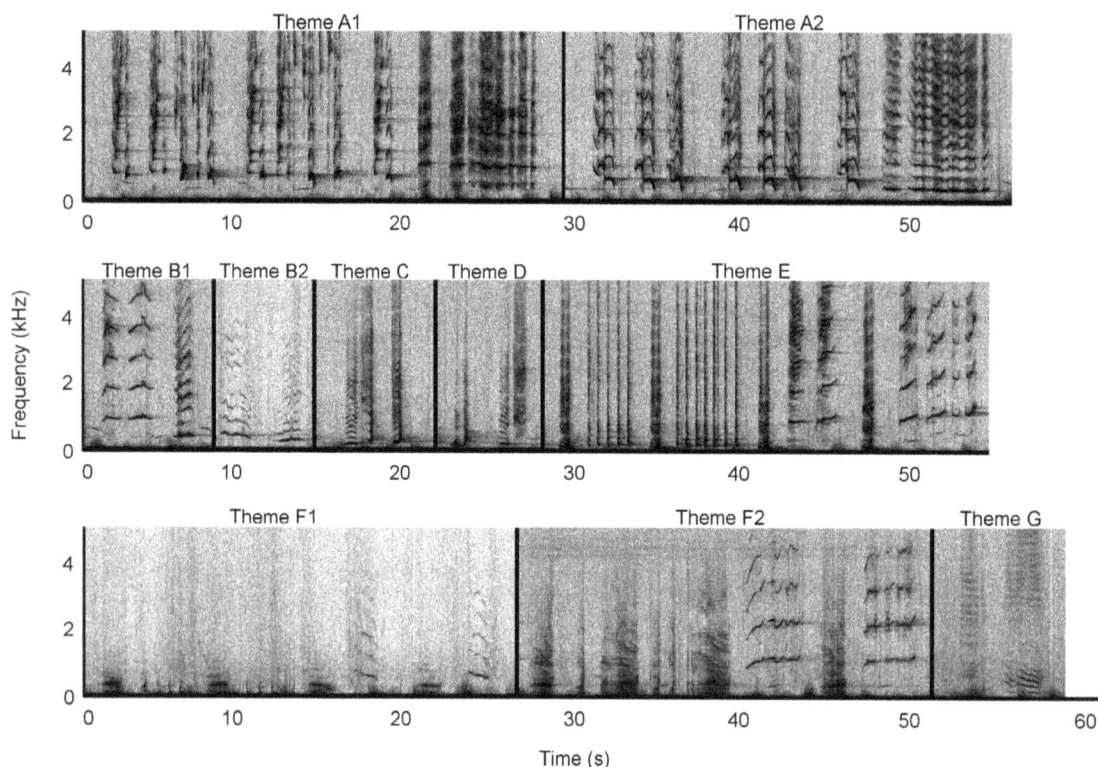

Figure 6. New Caledonia 2009 song. A single representative phrase is shown for each theme representing a song. Some themes contained 1 and 2 phrase types representing small but stereotyped variations to a theme. Spectrograms were generated in Raven Pro 1.4 (Blackman-Harris window, 2048 FFT size, 75% overlap). Recordings were taken from July to August 2009. Audio is provided in acoustic file S5. Note difference in frequency scale to Figure 2.

between-season movement and song exchange on shared migration routes present additional possible mechanisms of song transmission within this region. Interestingly, the eastern Australian population is the largest in the region [39,41], so if similar proportions of each population emigrate to the other in a season, this may explain its undue influence on song within the region [6] as more individuals would be expected to emigrate. While distances between these populations are small compared to populations within the North Pacific (~1,500 km vs. ~4,200 km), song sharing occurs with a one year delay in transmission between eastern Australia and New Caledonia [6,51], leading to within-season differences in song despite being geographically closer than populations within the North Pacific. At this stage, however, we cannot discern among the mechanisms behind the dynamic nature of song transmission in the western South Pacific region (*i.e.*, between-season movement, shared migration routes and summer feeding grounds); we merely present that song relevant to the surrounding populations is being produced on the summer feeding grounds, the pattern of song exchange indicates song was first present in eastern Australia before being recorded in Antarctica and then New Caledonia, and that this is the first time song has been recorded in the waters of eastern Antarctica.

The sample sizes included in this study are relatively small. The sample from New Caledonia in 2009 contained three singers only (Table 2), suggesting the potential for additional variability not captured by this sample. However, the occurrence of the New Caledonian 2009 song in the first recording of the 2010 season in New Caledonia strongly suggests that, based on similar song change events in this population [6,50,51], it is unlikely the eastern

Australian 2009 song was present and simply not recorded in New Caledonia in 2009. The data presented from Antarctic Area V are also a limited sample, but are highly suggestive due to the number of matched themes. The Balleny Islands have historically been connected through Discovery marking with eastern Australia, New Zealand and Norfolk Island (in the 1950s and early 1960s) [32,43]. These locations have additional connections, shown through Discovery marking, photo-identification recaptures, genetic matches and satellite tagging studies with the eastern Australian and New Caledonian populations [32,43,46,56,45,57]. Interestingly, a Discovery mark connection was made between Tonga (1958) and Area V near the Balleny Islands (1957) [42]. Song from Tonga and New Caledonia was typically the same each breeding season over the last decade, indicating a strong acoustic connection between these populations [6,51,55].

The 2009 New Caledonian song type was clearly different to the 2009 and 2010 eastern Australian, 2010 New Caledonian and 2010 Antarctic Area V song type. The New Caledonian 2010 song shared all six themes with the eastern Australian 2009 song. The Antarctic sample included theme I (which evolved into theme N in eastern Australia in 2010), and theme K (which was dropped from the eastern Australian 2010 song), indicating it was more related to the 2009 eastern Australian/2010 New Caledonian song. Due to the time lag in song matching between eastern Australia and New Caledonia [6,50–51] (*i.e.*, the song type does not match within a season), it is likely that song was first learnt by the eastern Australian population and then transmitted at a later point in time to the New Caledonian population (through the movement of animals from eastern Australia to New Caledonia, and/or song sharing on migration, and/or in Antarctica), at which point the two populations' songs diverged. This could occur through the divergence of migratory streams, or the movement of a group of animals away from a shared feeding aggregation (*e.g.*, around the Balleny Islands). Of particular note is the photo-identification recapture between New Caledonia (taken in 2007) and the Balleny Islands, taken on 1st March 2010 as part of the same expedition [46]. This individual was genetically and behaviorally identified as a mature male; it is probable that this male was exposed to the (2009 eastern Australian) song while within the vicinity of the Balleny Islands. Although we cannot comment further as to whether this individual acquired the 2009 eastern Australian song at this specific point in time, the nearly simultaneous presence of this song with a male linked to New Caledonia clearly illustrates the potential for cultural exchange and provides direct evidence of singing while on shared summer feeding grounds in the Southern Ocean Antarctic Area V.

The present study suggests a plausible mechanism of song transmission that has major implications for our understanding of song similarity between seemingly geographically separated populations. The movement of individuals (facilitating song exchange), coupled with song transmission on the feeding grounds, may permit a rapid transfer of song across ocean basins and the circumpolar feeding grounds. Song has been introduced into the eastern Australian population from the western Australian population [13], representing a movement between two discrete breeding populations (IWC stocks D and E). This was suggested to occur through the movement of a few individuals into the population [13] as occasional movement of individuals between breeding Groups D and E has been noted [32]. Within the South Pacific, individuals have been documented feeding in Antarctic Area I where they have not traditionally been thought to feed. One male marked with a Discovery tag in Tonga (1952) was later killed in Area I (1957) [58], and two individuals (one male and one female) recently photo-identified in American Samoa were later recaptured in Area I [59]. This could facilitate the transfer of song from the western and central South Pacific region to another 'stock' if the males concerned sung while on the Area I feeding grounds. Song has recently been recorded on the Western Antarctic Peninsula (Area I) [26] which provides an interesting opportunity to further investigate song transmission between breeding and feeding grounds across this ocean basin.

At the broader scale it is plausible that the movement of individuals, or song transmission on the circumpolar distribution of feeding grounds in the Southern Ocean, may allow the movement of different song types from one area to the next. The movement of different versions (song types) of this acoustic sexual display is a clear example of large-scale population-wide horizontal cultural transmission in a non-human animal [6]. For the future, only through a large, international collaboration of researchers that possess song recordings across the multiple breeding (and now feeding) locations throughout the Southern Hemisphere can we

investigate the intriguing possibility for a song type to undergo a complete circumpolar transmission. Such research would have significant implications for our understanding of population connectivity within the Southern Hemisphere, and would also contribute to the wider understanding of the underlying drivers for population-wide song conformity and cultural traditions in animals.

Supporting Information

Acoustic File S1   Antarctic Area V 2010 song. A single representative phrase is provided for each theme and is the corresponding audio file for Figure 2.

Acoustic File S2   Eastern Australia 2009 song. A single representative phrase is provided for each theme and is the corresponding audio file for Figure 3.

Acoustic File S3   New Caledonia 2010 song. A single representative phrase is provided for each theme and is the corresponding audio file for Figure 4.

Acoustic File S4   Eastern Australia 2010 song. A single representative phrase is provided for each theme and is the corresponding audio file for Figure 5.

Acoustic File S5   New Caledonia 2009 song. A single representative phrase is provided for each theme and is the corresponding audio file for Figure 6.

Acknowledgments

The authors would like to thank Phil Clapham for providing helpful comments on a previous version of this manuscript, Anita Murray for assistance with the 2010 eastern Australian song data, and Jessica Crance, Stephanie Grassia and Jessica Thompson for completing the theme matching test. We thank everyone involved in the Australia-New Zealand Antarctic Whale Expedition, the Humpback Acoustic Research Collaboration (HARC) and the Behavioural Responses of Australian Humpback whales to Seismic Surveys (BRAHSS) projects, and Rémi Dodemont, Veronique Pérard, and all the volunteers that helped in the field in New Caledonia.

Author Contributions

Conceived and designed the experiments: ECG MJN JG NG MLR. Analyzed the data: ECG MLR. Wrote the paper: ECG JG. Provided data: MJN CG NG JG. Drafting the article or revising it critically for important intellectual content: ECG JG MLR MJN CG. Final approval of the version to be published: ECG JG MLR MJN CG NG.

References

1. Payne R, McVay S (1971) Songs of humpback whales. Science 173: 585–597. doi:10.1126/science.173.3997.585.
2. Payne K, Payne R (1985) Large Scale Changes over 19 Years in Song of Humpback Whales in Bermuda. Z Tierpsychol 68: 89–114. doi:10.1111/j.1439-0310.1985.tb00118.x.
3. Winn HE, Winn LK (1978) The song of the humpback whale *Megaptera novaeangliae* in the West Indies. Mar Biol 47: 97–114.
4. Payne R (1978) Behavior and vocalization of humpback whales (Megaptera sp.). In: Norris KS, Reeves RR, editors. Report on a Workshop on Problems Related to Humpback Whales (*Megaptera novaeangliae*) in Hawaii. U.S. Department of Commerce, NTIS PB 280–794. 56–78.
5. Payne K, Tyack P, Payne R (1983) Progressive changes in the songs of humpback whales (*Megaptera novaeangliae*): A detailed analysis of two seasons in

Hawaii. In: Payne R, editor. Communication and Behavior of Whales. Boulder: Westview Press Inc. 9–57.
6. Garland EC, Goldizen AW, Rekdahl ML, Constantine R, Garrigue C, et al. (2011) Dynamic horizontal cultural transmission of humpback whale song at the ocean basin scale. Curr Biol 21: 687–691. doi:10.1016/j.cub.2011.03.019.
7. Payne R, Guinee LN (1983) Humpbacks Whale (*Megaptera novaeangliae*) Songs as an Indicator of "Stocks". In: Payne R, editor. Communication and Behavior of Whales. Boulder: Westview Press Inc. 333–358.
8. Helweg DA, Herman LM, Yamamoto S, Forestell PH (1990) Comparison of Songs of Humpback Whales (*Megaptera novaeangliae*) Recorded in Japan, Hawaii, and Mexico During the Winter of 1989. Scientific Reports of Cetacean Research 1: 1–20.

9. Helweg DA, Frankel AS, Mobley JR Jr, Herman LM (1992) Humpback whale song: our current understanding. In: Thomas JA, Kastelein RA, Supin AY, editors. Marine Mammal Sensory Systems. New York: Plenum Press. 459–483.

10. Helweg DA, Cato DH, Jenkins PF, Garrigue C, McCauley RD (1998) Geographic variation in South Pacific humpback whale songs. Behaviour 135: 1–27.

11. Cerchio S, Jacobsen JK, Norris TF (2001) Temporal and geographical variation in songs of humpback whales, *Megaptera novaeangliae*: synchronous change in Hawaiian and Mexican breeding assemblages. Anim Behav 62: 313–329. doi:10.1006/anbe.2001.1747.

12. Darling JD, Sousa-Lima RS (2005) Songs indicate interaction between Humpback whale (*Megaptera novaeangliae*) populations in the western and eastern South Atlantic Ocean. Mar Mamm Sci 21: 557–566. doi:10.1111/j.1748-7692.2005.tb01249.x.

13. Noad MJ, Cato DH, Bryden MM, Jenner M-N, Jenner KCS (2000) Cultural revolution in whale songs. Nature 408: 537. doi:10.1038/35046199.

14. Murray A, Cerchio S, McCauley R, Jenner CS, Razafindrakoto Y, et al. (2012) Song comparison reveals limited exchange between humpback whales (*Megaptera novaeangliae*) in the southern Indian Ocean. Mar Mamm Sci 28: E41–E57. doi:10.1111/j.1748-7692.2011.00484.x.

15. Winn HE, Thompson TJ, Cummings WC, Hain J, Hudnall J, et al. (1981) Song of the Humpback Whale – Population Comparisons. Behav Ecol Sociobiol 8: 41–46.

16. Cato DH (1991) Songs of Humpback Whales: the Australian perspective. Mem Queensl Mus 30: 277–290.

17. Razafindrakoto Y, Cerchio S, Collins T, Rosenbaum H, Ngouessono S (2009) Similarity of humpback whale song from Madagascar and Gabon indicates significant contact between South Atlantic and southwest Indian Ocean populations. Int Whal Comm: SC61/SH8.

18. Calambokidis J, Steiger GH, Straley JM, Herman LM, Cerchio S, et al. (2001) Movements and population structure of humpback whales in the North Pacific. Mar Mamm Sci 17: 769–794. doi:10.1111/j.1748-7692.2001.tb01298.x.

19. Garrigue C, Aguayo A, Amante-Helweg VLU, Baker CS, Caballero S, et al. (2002) Movements of humpback whales in Oceania, South Pacific. J Cetacean Res Manag 4: 255–260.

20. Garrigue C, Constantine R, Poole M, Hauser N, Clapham P, et al. (2011) Movement of individual humpback whales between wintering grounds of Oceania (South Pacific), 1999 to 2004. J Cetacean Res Manag (Special Issue) 3: 275–281.

21. Garrigue C, Franklin T, Constantine R, Russell K, Burns D, et al. (2011) First assessment of interchange of humpback whales between Oceania and the east coast of Australia. J Cetacean Res Manag (Special Issue) 3: 269–274.

22. Mattila DK, Guinee LN, Mayo CA (1987) Humpback whale songs on a North Atlantic feeding ground. J Mammal 68: 880–883. doi:10.2307/1381574.

23. Clark CW, Clapham PJ (2004) Acoustic monitoring on a humpback whale (*Megaptera novaeangliae*) feeding ground shows continual singing into late spring. Proc R Soc Lond B Biol Sci 271: 1051–1057. doi:10.1098/rspb.2004.2699 1471–2954.

24. Vu ET, Risch D, Clark CW, Gaylord S, Hatch LT, et al. (2012) Humpback whale song occurs extensively on feeding grounds in the western North Atlantic Ocean. Aquat Biol 14: 175–183. doi:10.3354/ab00390.

25. McSweeney DJ, Chu KC, Dolphin WF, Guinee LN (1989) North Pacific humpback whale songs: A comparison on southeast Alaskan feeding ground songs with Hawaiian wintering ground songs. Mar Mamm Sci 5: 139–148.

26. Stimpert AK, Peavey LP, Nowacek DP, Friedlaender AS (2012) Humpback whale song and foraging behavior on an Antarctic feeding ground. PLoS ONE 7: e51214. doi:10.1371/journal.pone.0051214.

27. Noad MJ, Cato DH (2007) Swimming speeds of singing and non-singing humpback whales during migration. Mar Mamm Sci 23: 481–495. doi:10.1111/j.1748-7692.2007.02414.x.

28. Norris TF, McDonald M, Barlow J (1999) Acoustic detections of singing humpbacks whales (*Megaptera novaeangliae*) in the eastern North Pacific during their northbound migration. J Acoust Soc Am 106: 506–514. doi:10.1121/1.427071.

29. Clapham PJ, Mattila DK (1990) Humpback whale songs as indicators of migration routes. Mar Mamm Sci 6: 155–160. doi:10.1111/j.1748-7692.1990.tb00238.x.

30. Charif RA, Clapham PJ, Clark CW (2001) Acoustic detections of singing humpback whales in deep waters off the British Isles. Mar Mamm Sci 17: 751–768. doi:10.1111/j.1748-7692.2001.tb01297.x.

31. Kibblewhite AC, Denham RN, Barnes DJ (1967) Unusual Low-Frequency Signals Observed in New Zealand Waters. J Acoust Soc Am 41: 644–655.

32. Chittleborough RG (1965) Dynamics of two populations of the humpback whale, *Megaptera novaeangliae* (Borowski). Aust J Mar Fresh Res 16: 33–128. doi:10.1071/MF9650033.

33. Baker CS, Herman LM, Perry A, Lawton WS, Straley JM, et al. (1986) Migratory movement and population structure of humpback whales (*Megaptera novaeangliae*) in the central and eastern North Pacific. Mar Ecol Prog Ser 31: 105–119.

34. Baker CS, Palumbi SR, Lambertsen RH, Weinrich MT, Calambokidis J, et al. (1990) Influence of seasonal migration on geographic distribution of mitochon-

35. Clapham PJ (1996) The Social and Reproductive Biology of Humpback Whales: an Ecological Perspective. Mamm Rev 26: 27–49.

36. International Whaling Commission (IWC) (2006) Report of the Workshop on the Comprehensive Assessment of Southern Hemisphere Humpback Whales. Int Whal Comm: SC58/Rep5.

37. Clapham P, Mikhalev Y, Franklin W, Paton D, Baker CS, et al. (2009) Catches of Humpback Whales, *Megaptera novaeangliae*, by the Soviet Union and Other Nations in the Southern Ocean, 1947–1973. Mar Fish Rev 71: 39–43.

38. Paterson RA, Paterson P, Cato DH (2004) Continued increase in east Australian humpback whales in 2001, 2002. Mem Queensl Mus 49: 712.

39. Noad MJ, Dunlop RA, Paton D, Cato DH (2011) Absolute and relative abundance estimates of Australian east coast humpback whales (*Megaptera novaeangliae*). J Cetacean Res Manag (Special Issue) 3: 243–52.

40. Garrigue C, Dodemont R, Steel D, Baker CS (2004) Organismal and 'gametic' capture-recapture using microsatellite genotyping confirm low abundance and reproductive autonomy of humpback whales on the wintering grounds of New Caledonia. Mar Ecol Prog Ser 274: 251–262. doi:10.3354/meps274251.

41. Constantine R, Jackson JA, Steel D, Baker CS, Brooks L, et al. (2012) Abundance of humpback whales in Oceania using photo-identification and microsatellite genotyping. Mar Ecol Prog Ser 453: 249–261. doi:10.3354/meps09613.

42. Dawbin WH (1959) New Zealand and South Pacific Whale Marking and Recoveries to the End of 1958. Norsk Hvalfangst-Tidende 48: 213–238.

43. Dawbin WH (1964) Movements of Humpback Whales Marked in the South West Pacific Ocean 1952 to 1962. Norsk Hvalfangst-Tidende 53: 68–78.

44. Dawbin WH (1966) The Seasonal Migratory Cycle of Humpback Whales. In: Norris KS, editor. Whales, Dolphins, and Porpoises. Berkley: University of California Press. 145–170.

45. Steel D, Schmitt N, Anderson M, Burns D, Childerhouse S, et al. (2011) Initial genotype matching of humpback whales from the 2010 Australia/New Zealand Antarctic Whale Expedition (Area V) to Australia and the South Pacific. Int Whal Comm: SC63/SH10.

46. Constantine R, Allen J, Beeman P, Burns D, Charrassin J-B, et al. (2011) Comprehensive photo-identification matching of Antarctic Area V humpback whales. Int Whal Comm: SC63/SH16.

47. Gales N, Double MC, Robinson S, Jenner C, Jenner M, et al. (2009) Satellite tracking of southbound East Australian humpback whales (*Megaptera novaeangliae*): challenging the feast or famine model for migrating whales. Int Whal Comm: SC61/SH17.

48. Gales N (2010) Antarctic Whale Expedition-Preliminary science field report and summary, *R.V. Tangaroa* Feb/Mar 2010. Int Whal Comm: SC62/O12.

49. Garrigue C, Peltier H, Ridoux V, Franklin T, Charrassin JB (2010) CETA: a new cetacean observation program in East Antarctica. Int Whal Comm: SC62/SH3.

50. Garland EC, Lilley MS, Goldizen AW, Rekdahl ML, Garrigue C, et al. (2012) Improved versions of the Levenshtein distance method for comparing sequence information in animals' vocalisations: tests using humpback whale song. Behaviour 149: 1413–41. doi:10.1163/1568539X-00003032.

51. Garland EC, Noad MJ, Goldizen AW, Lilley MS, Rekdahl ML, et al. (2013) Quantifying humpback whale song sequences to understand the dynamics of song exchange at the ocean basin scale. J Acoust Soc Am 133: 560–569. doi:10.1121/1.4770232.

52. Gedamke J, Robinson SM (2010) Acoustic survey for marine mammal occurrence and distribution off East Antarctica (30–80°E) in January-February 2006. Deep Sea Res Part 2 Top Stud Oceanogr 57: 968–981. doi:10.1016/j.dsr2.2008.10.042.

53. Mellinger DK (2001) Ishmael 1.0 User's Guide. U.S. Department of Commerce, NOAA Technical Memo OAR-PMEL 120: 1–30.

54. Dunlop RA, Cato DH, Noad MJ (2010) Your attention please: increasing ambient noise levels elicits a change in communication behaviour in humpback whales (*Megaptera novaeangliae*). Proc R Soc Lond B Biol Sci 277: 2521–2529. doi:10.1098/rspb.2009.2319.

55. Garland EC (2011) Cultural transmission of humpback whale song and metapopulation structure in the South Pacific Ocean. PhD thesis: University of Queensland. 208 p.

56. Constantine R, Russell K, Gibbs N, Childerhouse S, Baker CS (2007) Photo-identification of humpback whales (*Megaptera novaeangliae*) in New Zealand waters and their migratory connections to breeding grounds of Oceania. Mar Mamm Sci 23: 715–720. doi:10.1111/j.1748-7692.2007.00124.x.

57. Garrigue C, Zerbini AN, Geyer Y, Heide-Jørgensen M-P, Hanaoka W, et al. (2010) Movements of satellite-monitored humpback whales from New Caledonia. J Mammal 91: 109–115. doi:10.1644/09-MAMM-A-033R.1.

58. Brown SG (1957) Whale marks recovered during the Antarctic whaling season 1956/57. Norsk Hvalfangst-Tidende 10: 555–559.

59. Robbins J, Dalla Rosa L, Allen JM, Mattila DK, Secchi ER, et al. (2011) Return movement of a humpback whale between the Antarctic Peninsula and American Samoa: a seasonal migration record. Endanger Species Res 13: 117–121. doi:10.3354/esr00328.

drial DNA haplotypes in humpback whales. Nature 344: 238–240. doi:10.1038/344238a0.

8

Characterization and Function of the First Antibiotic Isolated from a Vent Organism: The Extremophile Metazoan *Alvinella pompejana*

Aurélie Tasiemski[1]*, **Sascha Jung**[2], **Céline Boidin-Wichlacz**[1], **Didier Jollivet**[3], **Virginie Cuvillier-Hot**[1], **Florence Pradillon**[4], **Costantino Vetriani**[5], **Oliver Hecht**[2], **Frank D. Sönnichsen**[6], **Christoph Gelhaus**[7], **Chien-Wen Hung**[8], **Andreas Tholey**[8], **Matthias Leippe**[7], **Joachim Grötzinger**[2], **Françoise Gaill**[9]

1 Université de Lille1-CNRS UMR8198, Laboratoire GEPV, Ecoimmunology of Marine Annelids (EMA), Villeneuve d'Ascq, France, **2** Institute of Biochemistry, Christian-Albrechts-Universität, Kiel, Germany, **3** Université Pierre et Marie Curie-CNRS UMR7144, Laboratoire AD2M, Adaptation et Biologie des Invertébrés en Conditions Extrêmes (ABICE), Station Biologique, Roscoff, France, **4** IFREMER, Centre de Brest, REM/EEP/LEP, Plouzané, France, **5** Department of Biochemistry and Microbiology and Institute of Marine and Coastal Sciences, Rutgers University, New Brunswick, New Jersey, United States of America, **6** Otto Diels Institute for Organic Chemistry, Christian-Albrechts-Universität, Kiel, Germany, **7** Institute of Zoology, Zoophysiology, Christian-Albrechts-Universität, Kiel, Germany, **8** Division of Systematic Proteome Research, Institute for Experimental Medicine, Christian-Albrechts-Universität, Kiel, Germany, **9** Université Pierre et Marie Curie-Muséum National d'Histoires Naturelles CNRS BOREA IRD, Paris, France

Abstract

The emblematic hydrothermal worm *Alvinella pompejana* is one of the most thermo tolerant animal known on Earth. It relies on a symbiotic association offering a unique opportunity to discover biochemical adaptations that allow animals to thrive in such a hostile habitat. Here, by studying the Pompeii worm, we report on the discovery of the first antibiotic peptide from a deep-sea organism, namely alvinellacin. After purification and peptide sequencing, both the gene and the peptide tertiary structures were elucidated. As epibionts are not cultivated so far and because of lethal decompression effects upon Alvinella sampling, we developed shipboard biological assays to demonstrate that in addition to act in the first line of defense against microbial invasion, alvinellacin shapes and controls the worm's epibiotic microflora. Our results provide insights into the nature of an abyssal antimicrobial peptide (AMP) and into the manner in which an extremophile eukaryote uses it to interact with the particular microbial community of the hydrothermal vent ecosystem. Unlike earlier studies done on hydrothermal vents that all focused on the microbial side of the symbiosis, our work gives a view of this interaction from the host side.

Editor: Ramy K. Aziz, Cairo University, Egypt

Funding: This research was supported by the CNRS, the MSER, the Université de Lille1 (BQR 2012), the Région Nord Pas-de-Calais (Emergent 2012) and the Fondation pour la Recherche sur la Biodiversité (VERMER 2013). The funders had no role in study design, data collection and analysis, decision to publish, or preparation of the manuscript.

Competing Interests: The authors have declared that no competing interests exist.

* E-mail: aurelie.tasiemski@univ-lille1.fr

Introduction

Alvinella pompejana is a polychaetous annelid that inhabits active deep-sea hydrothermal vents along the East Pacific Rise, where it colonizes the walls of actively venting high-temperature chimneys [1]. The environment of the worm is characterized by extreme physicochemical gradients, high pressure and bursts of elevated temperatures which can be as high as 105°C [2]. To date, the Pompeii worm is considered as one of the most eurythermal and thermotolerant metazoan known on Earth [3–6].

One of the striking features of this annelid is its association with a unique epibiotic bacterial community that forms cohesive hair-like projections from mucous glands lining the dorsal intersegmental spaces [3]. Numerous studies, including metagenomic analyses, evidenced that the microflora is composed of a multispecies complex of 12 to 15 phylotypes of which >98% are Epsilonproteobacteria, a dominating taxonomic group in hydrothermal vents [7]. These bacteria have been suggested to provide *Alvinella* with a stable source of nutrients and may detoxify the environment of the worm from reactive heavy metals and free hydrogen sulfide [1].

Central theme in beneficial bacterial-host interaction is that hosts must protect themselves against inappropriate colonization and replication of the symbiotic flora [8]. Various mechanisms are employed to control the symbionts without compromising host vitality. Amongst them, beneficial partnership between symbiotic bacteria and the immune reactions of the host has been widely invoked in mammals and insects [8]. The molecular interactions between the two partners of the association seem to modulate host immunity, and in turn the immune system shapes the composition of the microbiota.

Antimicrobial peptides (AMPs) are small sized molecules naturally produced by bacteria, protists, fungi, plants and animals. Their large distribution in nature within both unicellular and multicellular organisms suggests that they are crucial immune effectors which presumably have evolved under positive selection for a long period of time [9]. Recently, Login *et al* demonstrated that coleoptericin A, an AMP produced by the beetle belonging to the *Sitophilus* genus, keeps endosymbionts under control within the bacteriocytes [10]. By comparison with the number of AMPs isolated from terrestrial invertebrates ($\approx$1500), relatively few AMPs ($\approx$40) have been characterized from marine organisms [11]. Yet, marine animals are permanently in close contact with very high densities of microbes (10^5 to 10^7 per ml) suggesting that their immune effectors are effective in microbial growth inhibition and killing [12]. Although AMPs have been found in numerous marine invertebrate taxa such as Cnidarians, Annelids, Mollusks, Arthropods, Tunicates and Echinoderms [11], there is no evidence of active AMPs in organisms living in the deep-sea. To date, the aspect of AMP coevolution under selective pressures associated with the abyssal environment has never been investigated, whereas many life forms, in such an extreme habitat, rely on a symbiotic association.

Here, we describe the nature of the first abyssal AMP found in a symbiotic animal, its common origin with AMPs of coastal annelids as well as the manner in which an extremophile eukaryote uses it to interact with the particular microbial community of the hydrothermal vent ecosystem.

Material and Methods

Biological materials

Animal collection. *Alvinella pompejana* were collected from the bio9 and P Vent sites (EPR 9°50′N, 2.500 m depth) on board of the R/V L'Atalante using the telemanipulated arm of DSV Nautile (MESCAL Cruises 2010, 2012). Animals were brought back to the surface inside an insulated basket and directly dissected upon recovery. Although not subjected to specific property regulations (international water areas), authors have obtained permission to use samples for any analysis from both chief-scientists. This study did not involve endangered or protected species.

Primary cell culture. Freshly harvested coelomic cells were cultured in Leibovitz L-15 medium under sterile conditions on board. For microbial treatment, cells were separately incubated in 500 µL of medium containing 10 µL of killed bacteria, for 12 h. Incubations without bacteria were performed under the same conditions as controls.

Microorganisms. The bacterial strains used in this study are listed in S.1 in Material and Methods S1. Epibionts were scraped with a thin razor from 1 cm^2 of the tegument of *Alvinella* freshly harvested and were suspended in 4 mL of sterile seawater. Primary enrichment cultures were obtained shipboard by adding an aliquot of epibionts or fragments of *Alvinella* tubes to 10 mL of modified SME media prepared as previously described and followed by incubation at 30 and 50°C [13], [14].

Peptide purification and identification

A purification guided assay (see S.2 in Material and Methods S1 for details) was performed from the coelomic liquid of *Alvinella*. After three steps of chromatography (Reverse Phase-HPLC), the purity of the antimicrobial fractions was assessed by mass spectrometry (MS) analyses (DE STR PRO; Applied Biosystems) and homogeneous material was subjected to protein sequencing via Edman degradation (pulse liquid automatic peptide sequenator, Beckman Coulter).

Three dimensional structures

NMR spectroscopy. (see S.3 in Material and Methods S1 for details). The renatured alvinellacin was submitted to NMR measurements on a Bruker Avance III 800 MHz spectrometer. The chemical shift data were deposited in the University of Wisconsin Biological Magnetic Resonance Bank database under the accession number 18085. All spectra were processed with the program NMRPipe [15] and analyzed with the program NMRView [16]. Models of the three dimensional structures of capitellacin were generated using the solution NMR structure of alvinellacin as template. Structure calculations were performed using the program CYANA [17]. The 10 best structures were selected as the final structural ensemble and were deposited (PDB accession code 2LLR).

Assignment of disulfide bridges in alvinellacin by Mass Spectrometry. (see S.4 in Material and Methods S1 for details) Alvinellacin was incubated with the endoproteinase Lys-C. For sequential Lys-C and tryptic digestion, trypsin was added to the Lys-C digestion. During the incubation, aliquots were taken from the digest at different time points in order to monitor the enzyme digestion efficiency. MS experiments were performed on an offline nanoESI-LTQ Orbitrap Velos mass spectrometer with ETD option (Thermo Fisher Scientific, San Jose, CA). Spectrum interpretation and disulfide bridge assignment were performed manually.

Alvinellacin activities

Antimicrobial assays. The minimal inhibitory concentration (MIC) and minimal bactericidal concentration (MBC) of the synthetic peptide (diluted in acidified water 0.05% acetic acid) against bacterial growth were determined by a microdilution susceptibility assay in microtiter plates as previously described [18]. Permeabilization of membranes of viable bacteria and pore-forming activity towards liposomes were measured as previously described in details [19], [20]. Alamethicin, cecropin P1, and magainin II were purchased as synthetic peptides from Sigma.

Shipboard antimicrobial assays against epibionts. Epibionts were scrapped with a thin razor from the tegument of freshly harvested *Alvinella*. They were incubated in presence or absence (control) of the alvinellacin peptide for 4 h and were subsequently fixed in 3% glutaraldedyde on board of the ship. In the laboratory, samples were directly placed on copper grids and counterstained with uranyl acetate and lead citrate. Morphological changes on epibionts were detected and damaged *versus* intact bacteria were counted on a Hitachi H 600 electron microscope.

Gene characterization and gene expression

The nucleotidic sequence coding the preproalvinellacin precursor was obtained by blasting the amino acid sequence of Alvinellacin to the *Alvinella* EST database (TERA 00513) [21].

Gene structure. The complete gene sequence of the preproalvi-nellacin (Genbank accession number KJ489380) was obtained from the cloning of PCR-products coming from the nested amplification of a series of *A. pompejana* gDNA using specific primers targeted on the 5′ and 3′ ends of the cDNA. Used primers are as followed: AP_alvinellacinF starting from the first methio-nine codon: 5′-ATG ACG TAT TCT GTA GTT GTG ACG CTG GTC-3′, AP_alvinellacinR1 (in the 3′UTR region): 5′-TAG GCA GGA CGG AGC CGC CAG ATC A-3′, and AP_alvi-nellacinR2 (starting on codon stop): 5′-CTC AGT GAA ATG AAG CAG GTG AGT TAT G-3′. PCR amplifications were obtained following 40 cycles of 96°C for 45 s, 60°C for 45 s and 72°C for 4 min after a first denaturation of gDNA at 96°C for 4 min and a final elongation of 10 min. Putative splicing sites (ACEs and ISEs) and both mobile and regulatory elements were detected using ACESCAN2 web server (http://genes.mit.edu/

A

B

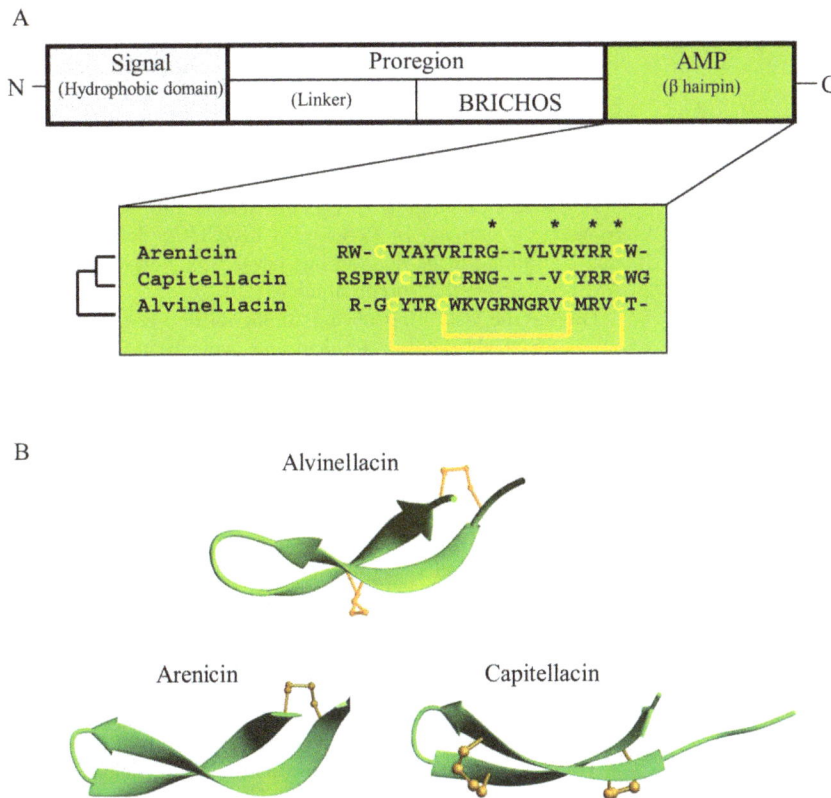

Figure 1. Alvinellacin is evolutionary linked to a family of AMPs from coastal annelids. (A) Molecular organization of the alvinellacin precursor and sequence alignment of alvinellacin with arenicin, an AMP produced by *Arenicola marina*, and capitellacin, a sequence so named by our group, issued from the analysis of the *Capitella teleta* genome. N and C denote N- and C-termini. (B) Comparison of the three-dimensional structures of the three AMPs. Disulfide bonds are depicted as yellow balls and sticks.

acescan2/index.html) and modules of the geneinfinity (http://www.geneinfinity.org/sp/sp_coding.html), webgene (http://www.itb.cnr.it/webgene/) and the TE tools (ergmanlab.smith.man.a-c.uk/?page_id = 295) platforms. The complete gene sequence of preprocapitellacin was obtained by blasting the preproalvinellacin in the *Capitella teleta* genome database (http://genome.jgi-psf.org/Capca1/Capca1.home.html).

Quantitative Reverse Transcription PCR. RNA from cells were extracted (Qiazol, Qiagen) and used for cDNA synthesis with an oligodT according to the protocol of the manufacturer (SuperScript II; Invitrogen). The primers used for quantification were designed with the Primer3 Input software (http://frodo.wi.mit.edu/cgi-bin/primer3/primer3 www.cgi).

-Alvinellacin primers: forward: 5′-TGACATCGTGAAG-GAACTCG-3′; reverse: 5′-CCGTTCCTACCAACTTTCCA-3′

-Ribosomal Protein 26S primers (*Alvinella* database [21]: TERA01523): forward: 5′-CCGGCTAGTTCAAGATGACC-3′; reverse: 5′-AGCTGCTGCCTCCACTATGT-3′.

The RP26S was used as the reference gene. Real Time reactions were conducted on a CFX96 qPCR system (BioRad) using a hot start, then 40 cycles at 94°C, 15 s; 56°C, 30 s; 72°C, 30 s., and a final extension step at 72°C for 3 min. Analysis of relative gene expression data was performed using the ΔΔCt method. For each couple of primers, a plot of the log cDNA dilution versus ΔCt was generated to validate the qPCR experiments (data not shown). Reference and target were amplified in separated wells.

Alvinellacin production sites

Polyclonal antiserum. The alvinellacin antiserum was raised in two New Zealand White rabbits (Saprophyte pathogen-free). The chemically synthesized peptide was coupled to OVA and used for the immunization procedure according to the protocol described previously [22]. The reactivity of the antibody was tested by Dot Immunobinding Assay (DIA) using 1 μL of the RP HPLC fractions [23].

Immunocytochemistry and immunohistochemistry. Cells or tissues were fixed on board in 4% paraformaldehyde. Later, the SHANDON Cytospin 3 was used to spin cell suspension onto poly-lysine slides (8 min, 2,000 rpm). Immunocytochemistry and immunohistochemistry were performed with the rabbit anti-alvinellacin (1:100) and the FITC-conjugated anti-rabbit secondary antibody (1:100; Jackson Immunoresearch Laboratories) according to a protocol already described by our group [21]. Samples were examined using a confocal microscope (Zeiss LSM 510).

Coelomocyte structure. The coelomocytes were collected, and immediately fixed in 3% glutaraldehyde according to the protocol previously described [24]. Coelomocytes were observed on a Hitachi H 600 electron microscope.

Results and Discussion

Nature of *Alvinella* AMP and evolutionary link with AMPs from coastal species

To date, only one AMP isolated from marine species belongs to an AMP family already characterized in terrestrial species [11].

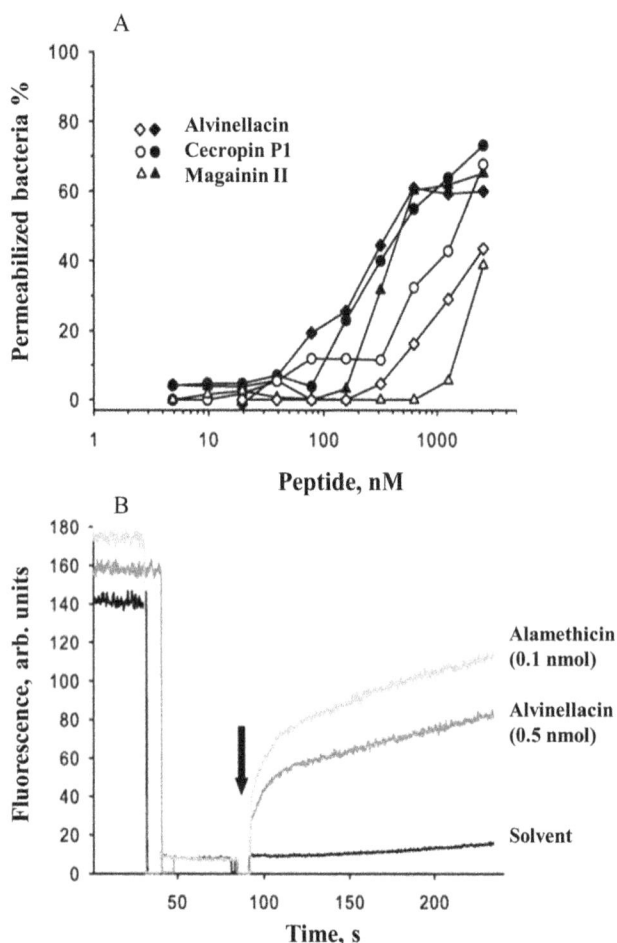

Figure 2. Bactericidal activity of alvinellacin. (A) Membrane permeabilization of viable bacteria induced by alvinellacin compared with that induced by other antimicrobial peptides. The percentage of bacteria with compromised membranes was monitored fluorometrically using the membrane-impermeable SYTOX green dye. Bacteria were incubated with varying concentrations of alvinellacin, cecropin P1 and magainin II at pH 7.4. Induction of membrane permeabilization was monitored against *B. megaterium* after 10 min (closed symbols) and against *E. coli* after 2 h (open symbols). (B) Time course of pore formation induced by alvinellacin. The dissipation of a valinomycin-induced diffusion potential in vesicles of asolectin after addition of alvinellacin (0.5 nmol), control peptide alamethicin (0.1 nmol), and the peptide solvent (0.05% acetic acid) were recorded. Pore-forming activity is reflected by the increase of fluorescence as a function of time. The arrow marks the time point at which the peptide is added.

The majority of marine AMPs presents novel structures and is confined to certain taxa or even species, as observed for AMPs of polychaeta [25]. In order to identify peptides with antibiotic activity from the Pompeii worm, a biochemical approach was combined with the analysis of the *Alvinella* EST database [22]. The anatomy of annelids is characterized by the presence of a coelom, a compartment that includes mobile cells, named coelomocytes that sterilize the coelomic fluid by releasing humoral factors such as antimicrobial peptides (AMPs) [26]. We purified and identified a cationic peptide composed of 22 amino-acid residues which we named alvinellacin, from the coelomocytes of *Alvinella* (Figures S1 and S2). As for most AMPs from all invertebrate phyla, mature alvinellacin is processed from a larger precursor molecule

containing a signal peptide and an anionic proregion [27] (Figure 1A).

Following a BLAST search, the mature peptide did not display any similarity in its primary structure with other known proteins. However, the proregion had ~33% identity to the proregion of AMPs from coastal annelids: arenicin from *Arenicola marina* and capitellacin, a putative peptide inferred from the genome sequence of *Capitella teleta* (Figure S3). Pfam analysis of the proregions revealed the presence of a conserved BRICHOS domain. So far, this 100 amino acids domain has never been reported in other AMP precursors than preproarenicin [28].

Despite the lack of an obvious similarity between the primary structures of alvinellacin, arenicin and capitellacin, we analyzed and compared their three-dimensional structures. Using NMR spectroscopy and mass spectrometry, we determined the tertiary structure of alvinellacin (Figures S4 and S5, Tables S1 and S2) and compared it to the solved and predicted structures of arenicin and capitellacin, respectively (Figure 1B). Alvinellacin and capitellacin are stabilized by two disulfide bonds, whereas arenicin possesses only one cystine. Like capitellacin, alvinellacin folds into a double-stranded antiparallel beta-sheet resembling the structure of arenicin [29]. Consequently, the three AMP precursors *i.e.* preproalvinellacin, preprocapitellacin and preproarenicin harbor the conserved pattern of almost all the BRICHOS containing proteins: a hydrophobic domain (here, the signal peptide), a linker region, the BRICHOS domain itself and a C terminal region with β-sheet propensities (here, the AMP) (Figure 1A) [28].

To date, proregions of AMP precursor are essentially known to be implicated in cell chemotaxy and/or protection against the cytotoxic activities of certain AMPs [30]. The BRICHOS domain has been found as a constituent of proteins associated with a wide variety of human diseases such as dementia, respiratory distress and cancer [31].

Recent data evidence that BRICHOS participates in the complex post-translational processing of proteins, and functions as an intramolecular chaperone domain that can bind β hairpin motifs and prevents them from β sheet aggregation and amyloid fibril formation [28]. Because of their strand-loop-strand structure, it seems reasonable that alvinellacin like the two other AMPs interacts with BRICHOS. Coastal and, even more, hydrothermal annelids are naturally submitted to strong hypoxic and thermal stresses. We hypothesize that the presence of the BRICHOS domain might be an evolution-driven adaptation of the worms to warrant the correct folding of their AMP under extreme conditions such as hypoxia and/or eurythermality. All these suggestions should be experimentally tested: BRICHOS might also have a novel function in *A. pompejana* that remains undiscovered.

As a conserved gene structure constitutes a convincing evidence for evolutionary relatedness between protein families, we also characterized the complete gene sequence of alvinellacin and compared it to the capitellacin gene [32] (Figure S6). Both genes display a 5 introns/6 exons structure with nearly all conserved intron-splicing positions. Given the taxonomic position of *Capitella* and *Alvinella* [33], their gene structure along with the proregion sequence identity and the three-dimensional peptide structure, strongly indicate that alvinellacin and capitellacin presumably together with arenicin, share an ancient origin and are evolutionary correlated since hundred millions of years. Further detailed comparisons of the proregions showed a high level of amino acid changes in the first part of the propieces; that may be also attributable to an adaptive 'hot spot' of mutations (functional change) in the face of the very long period of time since divergence between the two polychaeta species. The low amino acid

Table 1. Antimicrobial activity of alvinellacin.

	MIC, μM	MBC, μM
Gram-negative bacteria		
Escherichia coli D31	0.012–0.024	0.048
Escherichia coli D31 (300 mM NaCl)	0.012–0.024	0.048
Escherichia coli D31 (500 mM NaCl)	0.012–0.024	0.048
Pseudomonas sp.*	0.001–0.003	0.012
Vibrio diabolicus*	0.048–0.096	>0.19
Vibrio MPV19	0.012–0.024	0.024
Gram-positive bacteria		
Bacillus megaterium	0.012–0.024	0.024
Bacillus megaterium (300 mM NaCl)	0.024–0.048	0.048
Bacillus megaterium (500 mM NaCl)	0.048–0.096	0.096
Staphylococcus aureus	0.048–0.096	>0.19

Assays were performed against bacteria routinely used for antimicrobial assays or having a medical incidence, and against the scarce hydrothermal strains (asterisk*) cultivable under the conditions of a microbial assay. The minimal inhibitory concentration (MIC) and the minimal bactericidal concentration (MBC) are expressed as final concentration in μM. > denotes no activity detected at the given concentration. The MBC and MIC values are the same, indicating that the bacterial growth inhibition is due to the killing of bacteria.

conservation in the AMP sequences compared to the proregions suggests that they might have evolved independently. To the best of our knowledge, AMP proregions are not known to interfere with components of the external environment as AMPs do by interacting with microbes. Thus, the mature AMP presumably evolved to respond to the specific microbial communities (hydrothermal or coastal habitats) as well as to the specific lifestyle (symbiotic or not) of the worms while the proregion did not. This is consistent with the observation that the C-terminal propeptides of the interstitial collagen of *Alvinella* and *Arenicola* are similar, while the helical domain of the mature protein, which is located in the extracellular matrix and is presumably more exposed to environmental conditions, is not. [34].

Alvinellacin in the first line of defense towards environmental microbes

In general, microbial invasion into the host causes bacterial infection which prompts an immune response such as the release of AMPs to eliminate invaders. Since alvinellacin was isolated from the coelomocytes, these cells are likely to produce and secrete the AMP into the coelomic fluid where it exerts its antibacterial activities. The presence of a signal peptide in the alvinellacin precursor (Figure 1A) together with the results obtained by immunocytochemistry (see below) corroborates this assumption. The antimicrobial activity of alvinellacin was then evaluated (Figure 2 and Table 1). As the worm's coelomic fluid composition is not very different from seawater, assays were performed at salt concentrations mimicking this environment [35]. Under these conditions, alvinellacin's activity was constant primarily against Gram-negative bacteria. This may represent an adaptation of the worm to its associated microorganisms, which have been shown to be predominantly Gram-negative ε-proteobacteria [3,36]. We then wondered whether an exposure to various microorganisms might have differential impacts on the synthesis of alvinellacin. Usually, to investigate the immune response of an organism, animals are submitted to experimental infections and variations of immune markers are quantified. Since *Alvinella* precludes *in vivo* investigation because of lethal decompression effects upon

sampling [6,37], we developed an *ex vivo* model by establishing primary culture of cells obtained from freshly harvested animals.

Despite being subjected to 250 bars decompression, cells were alive and morphologically preserved (Figure 3A). Primary cell cultures were then initiated and maintained on board. To mimic a systemic infection, coelomocytes were incubated in presence of either *Alvinella* epibionts or of different vent bacterial strains (Figure 3B). Quantitative RT-PCR experiments showed a selective induction of the gene encoding alvinellacin upon exposure of the worm to the vent bacteria. Both epibionts, which are highly represented by ε-proteobacteria [7], and bacterial enrichments obtained shipboard from *Alvinella* tubes using culture conditions that support growth of vent ε-proteobacteria, appear to be better inducers than pure cultures of γ-proteobacteria. These results show that *Alvinella* coelomocytes can sense different microorganisms and that alvinellacin synthesis might be the outcome of the adaptation of *Alvinella*'s immune defense system against the specific microorganisms present in its environment. To date, the role of archaea in activating the host's immune system and the ability of its immune receptors to detect their presence has never been investigated. Interestingly, *Alvinella* tubes are inhabited by an extremely dense population of archaea related to the Thermococcales, including members of the two major genera, *Thermococcus* and *Pyrococcus*, the former being more prevalent in *Alvinella* tubes than the latter [38,39]. We investigated the ability of these hyperthermophilic microbes to induce the expression of the alvinellacin gene in our cell cultures. Remarkably, only archaea belonging to the *Thermococcus* genus induced the expression of the alvinellacin gene. These data indicate that *Alvinella* can selectively recognize specific archaea and in turn induces an immune response, suggesting for the first time the existence of pattern recognition receptors in an eukaryote organism able to recognize and discriminate archaeal microbe-associated molecular patterns [40].

Alvinellacin controls and shapes the epibiotic flora

While a key role of AMPs in fighting infections is well described, very recent studies also evidenced that these effector molecules can be employed to regulate/control the symbiotic microflora [10,30].

A

B

Alvinellacin mean normalized expression

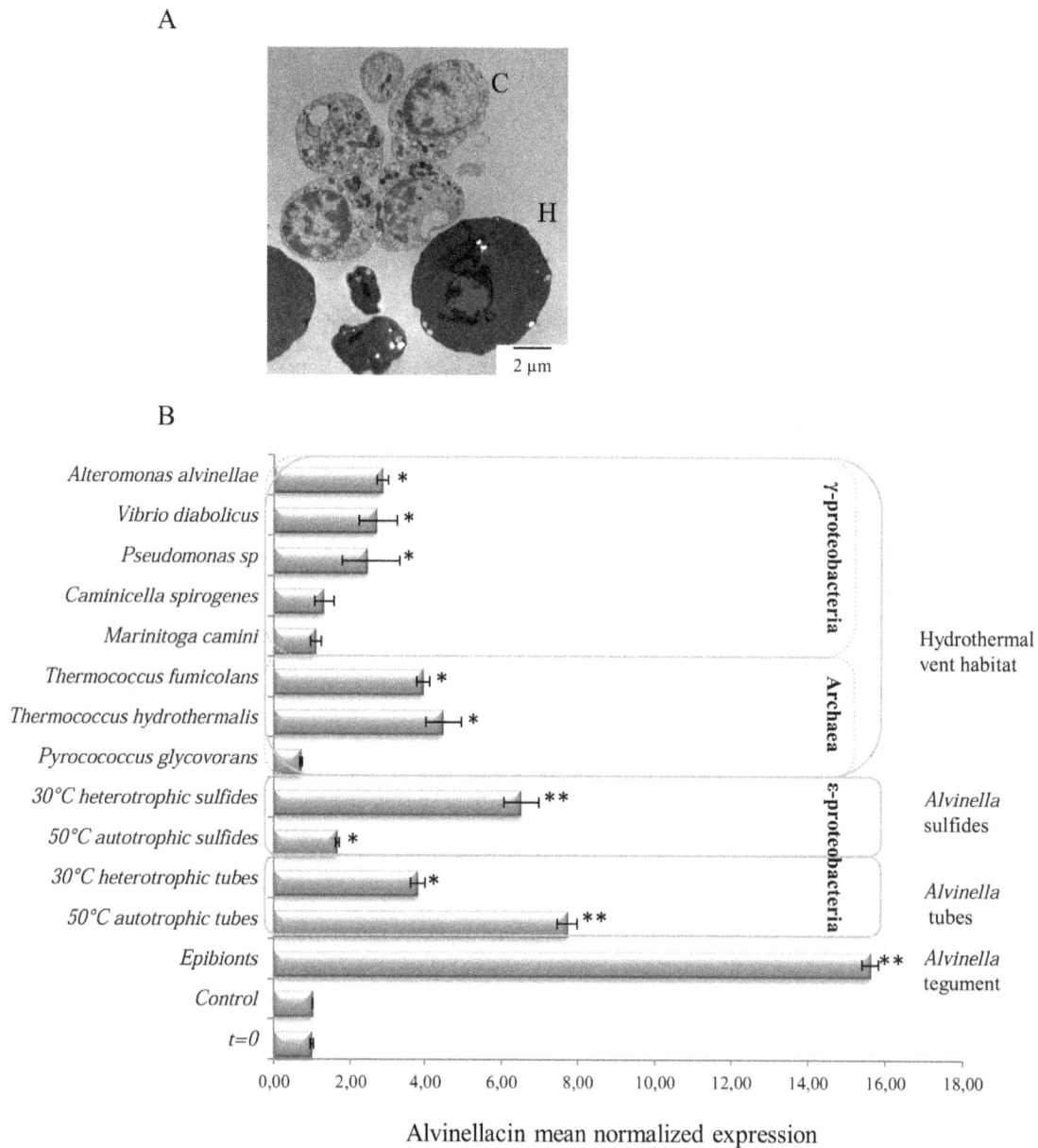

Figure 3. Selective induction of the gene encoding alvinellacin in cœlomocytes exposed to vent bacteria. (A) Electron-microscopic image showing the intact structure of *Alvinella* cœlomocytes despite 250 bars decompression. C: cœlomocytes. H: blood cells. (B) RT-qPCR on primary culture of coelomocytes incubated in the presence (t = 12 h) or not (control) of *Alvinella* epibionts, enrichment cultures obtained from *Alvinella* samples, and different pure cultures of bacteria and archaea isolated from hydrothermal vents. Graphics represent the results of two independent experiments; p-values from Student's tests were calculated *versus* the control treatment, based on the experimental measures performed in triplicates (*p<0.05).

The strong and vital relationship between *Alvinella* and its epibionts prompted us to investigate such an alternative function of alvinellacin. Immunohistochemistry experiments showed that alvinellacin is expressed constitutively by epithelial cells of the tegument associated with the epibiotic microflora, i.e. the dorsal but not the ventral epidermis (Figures 4A *vs* 4B). This observation supports the idea that alvinellacin may prevent bacterial entrance and/or keep epibionts under control. Accordingly, we determined the antimicrobial potency of alvinellacin against epibionts (Figure 5). As epibionts have not been cultivated so far, we carried out a shipboard antimicrobial assay aimed at detecting epibiont-cell damage in response to exposure to alvinellacin. Interestingly,

alvinellacin significantly targeted epibionts that correspond to filamentous bacteria (epibiont types 5 to 9). In particular, alvinellacin killed 100% of the two most abundant morphotypes within this group (types 6 and 7). In contrast, the presence/absence of alvinellacin on epibionts types 1 and 4 did not have distinguishable effects and other epibionts (types 2 and 3) are not affected by alvinellacin. Altogether, the data suggest that alvinellacin controls epibiosis by selectively killing the most dominant part of the filamentous bacteria found on the dorsal part of the worm.

That is reminiscent of the role of the defensin HD5 in shaping the composition of the symbiotic microflora of the digestive tract in

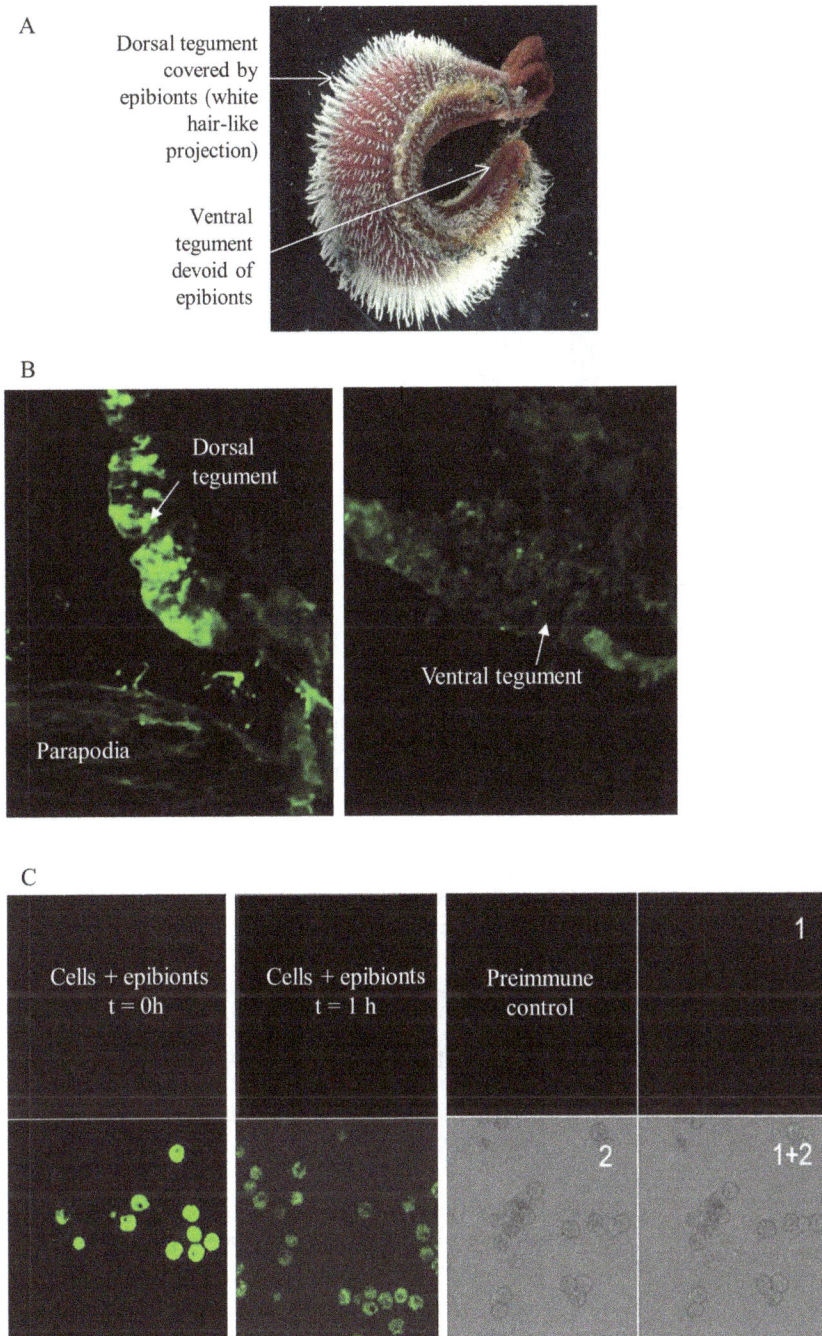

Figure 4. Alvinellacin is produced by tissues or cells in contact with epibionts. (A) Picture of *Alvinella* showing the distribution of epibionts. (B) Immunohistochemistry data evidence that alvinellacin peptide accumulates in tissue hosting epibionts *i.e.* the dorsal but not the ventral tegument. (C) Accidental entrance of epibionts stimulates the secretion of alvinellacin by circulating cells. Images of immunodetection of alvinellacin in coelomocytes incubated with epibionts. After one hour of exposure, the signal was reduced evidencing an extracellular secretion of the peptide. Control is performed with preimmune serum 1: FITC fluorescence, 2: transmission, 1+2: overlay.

mammals [41]. We hypothesize that high production of alvinellacin by epidermal cells selects and shapes the epibiotic microflora and prevents microbiota from over proliferating and subsequently penetrating the underlying tissue. To test this hypothesis, an accidental invasion was simulated by incubating coelomocytes with epibionts (Figure 4C). The distribution of alvinellacin-immune reactivity was compared by immunofluorescence in unchallenged *versus* challenged cells. Under basal conditions (t = 0), the AMP was strongly detectable inside the cells, suggesting that this active compound is stored after synthesis. One hour after the bacterial infestation, the immune staining inside the cells faded, evidencing that alvinellacin is secreted rapidly when the cells are challenged by microorganisms. The induction of transcription observed by RT-qPCR in the cells incubated for 12 h with epibionts probably contributes to the renewal of the alvinellacin peptide stock (Figure 3B). Overall, these

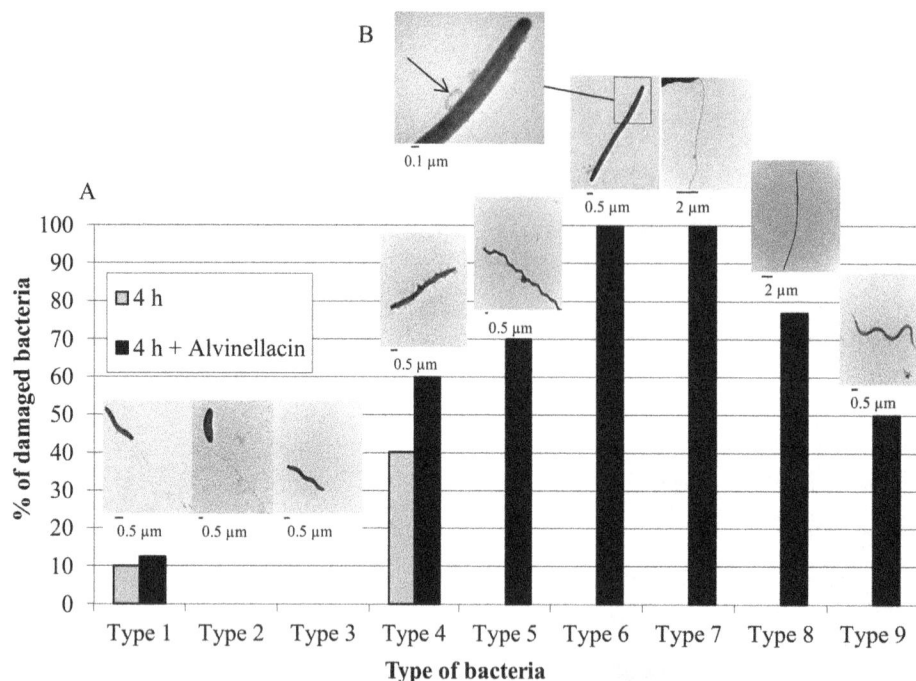

Figure 5. Alvinellacin activities against epibionts. (A) Freshly collected epibionts were incubated alone or in the presence of alvinellacin for 4 h, and after fixation observed under the electron microscope. The number of damaged bacteria was estimated among nine different morphotypes (numbered from 1 to 9) clearly distinguishable in all our preparations. (B) Bacterial lesions are visible at high magnification as the formation of membrane blebs and the release of cytoplasmic material (arrow).

results reinforce the role of alvinellacin in keeping symbionts under control.

Conclusion

Altogether, the data indicate the production of an original AMP from a deep-sea animal that endorses a durable relationship with Epsilonproteobacteria and possibly archaea in the face of the hostile vent habitat. Alvinellacin appears to act as a first line of defence against microbial invasion. The specificity of the gene induction along with the selective anti epibiotic activity and the expression in tissues exposed to the environment suggest that alvinellacin is actively participating in the surveillance of the epibiotic community. The conservation of the proregion and the gene structure of alvinellacin with AMPs of coastal annelids, suggest a common origin of the molecules. To draw a decisive conclusion regarding the gene evolution of alvinellacin, we plan to search for related genes in more than 30 annelid species living in various habitats. Such phylogenetic analysis will aim at determining whether the amino acid sequences of the antimicrobial part of the precursor diverged between species in order to face (i) contrasted temperatures, (ii) different microbial environments and/or (iii) to allow the establishment of epibioses.

Supporting Information

Figure S1 Alvinellacin purification and molecular identification. Material eluting at 60% acetonitrile (ACN) upon solid phase extraction was loaded onto a C18 column (250×4 mm, Vydac). Elution was performed with a linear gradient of acetonitrile in acidified water (dotted line), and absorbance was monitored at 225 nm. Each individually collected fraction was tested for its antimicrobial activity (white bar) and its immunore-

activity to the alvinellacin Ab by DIA (grey bar). Fractions containing antimicrobially active alvinellacin were further purified by two additional RP-HPLC purification steps. Asterisk shows the active final fraction containing alvinellacin.

Figure S2 MS spectrum of native alvinellacin. Analysis of purified alvinellacin by MALDI TOF-MS shows a m/z value of 2,600.35 MH+ which perfectly matches the theoretical mass of the peptide including two disulfide bonds.

Figure S3 Sequence alignments of the precursors of alvinellacin, capitellacin, and two arenicin isoforms.

Figure S4 Intact protein MS spectrum of alvinellacin measured by nanoESI-Orbitrap MS. (A) Full range MS survey spectrum. (B) Zoom-in of the [M+5H]5+ charge state species in a. A small species (indicated as asterisk) found next to the major component was identified as the methionine oxidation product of alvinellacin. The experimentally determined monoisotopic MW of alvinellacin was 2,599.2221 Da. (C) Display of theoretical MW (2,599.2067 Da) of alvinellacin and its isotope distribution at charge state 5. The results indicated that all four cysteines are involved in the formation of disulfide bonds.

Figure S5 Time-course analysis of the proteolytic cleavage of alvinellacin. The products of alvinellacin digestion were analyzed by nanoESI-Orbitrap MS. (A) Peptide MS survey spectra of alvinellacin digested with Lys-C at 35 °C (overnight). (B) Subsequent digestion of the Lys-C-digest with trypsin after

30 min; (C) after 2 h; (D) after 18 h at 37°C. The identities of the peptides are summarized in Table S2.

Figure S6 Alvinellacin and capitellacin gene structures.
(A) As opposed to CDS (648 bp), the alvinellacin gene is rather long (1949 bp from the initial methionine to the stop codon) with a 5 introns/6 exons structure and a first large intron of 442 bp. Introns are all inserted in phase 0 with the exception of the last one in phase 1. (B) Alignment of the translated regions of the alvinellacin and capitellacin genes. The intron splicing positions (triangles) are nearly conserved. The BRICHOS domains are shaded and the AMP sequences are in bold type.

Material and Methods S1

Table S1 Structural statistics for the 10 best structures of alvinellacin showing the lowest target functions. None of the distance constraints was violated by more than 0.5 Å in any structure.

Table S2 Disulfide-connected peptide fragments of alvinellacin observed after proteolytic cleavage. Peptides

with oxidized cysteines were successively digested using the proteases Lys-C and trypsin. The resulting peptides were analyzed by offline nanoESI-Orbitrap MS/MS as shown in Figure S2. The results unambiguously indicated two disulfide linkages between C1–C4 and C2–C3.

Acknowledgments

We thank the captain of the crew of the R/V Atalante, the DSV Nautile group (IFREMER), along with N. Lebris and F. Lallier chief scientists of the MESCAL 2010 and 2012 cruises. We also thank C. Slomianny and N. Barois for access to the confocal laser microscope and electron microscope, respectively, M.A. Cambon for deep sea microbial strains and H. Ließegang for technical help during membrane-activity testing.

Author Contributions

Conceived and designed the experiments: A. Tasiemski ML JG SJ FG. Performed the experiments: A. Tasiemski CBW DJ VCH SJ CV CWH A. Tholey CG ML JG FDS OH. Analyzed the data: A. Tasiemski SJ CV DJ JG ML. Contributed reagents/materials/analysis tools: FP. Wrote the paper: A. Tasiemski SJ CV DJ JG ML.

References

1. Le Bris N, Gaill F (2007) How does the annelid *Alvinella pompejana* deal with an extreme hydrothermal environment? Rev Environ Sci Biotechnol 6: 197–221.
2. Girguis PR and Lee RW (2006) Thermal preference and tolerance of alvinellids. Science 312: 231.
3. Cary SC, Cottrell MT, Stein JL, Camacho F, Desbruyeres D (1997) Molecular Identification and Localization of Filamentous Symbiotic Bacteria Associated with the Hydrothermal Vent Annelid *Alvinella pompejana*. Appl Environ Microbiol 63: 1124–1130.
4. Cary SC, Shank T and Stein J (1998) Worms basks in extreme temperatures. Nature 391: 545–546.
5. Chevaldonné P, Desbruyères D, Childress JJ (1992) Some like it hot. and some even hotter. Nature 359: 593–594.
6. Ravaux J, Hamel G, Zbinden M, Tasiemski A, Boutet I, et al. (2013) Thermal limit for metazoan life in question: *in vivo* heat tolerance of the Pompeii worm. PLoSONE 8.
7. Grzymski JJ, Murray AE, Campbell BJ, Kaplarevic M, Gao GR, et al. (2008) Metagenome analysis of an extreme microbial symbiosis reveals eurythermal adaptation and metabolic flexibility. Proc Natl Acad Sci U S A 105: 17516–17521.
8. Dale C, Moran NA (2006) Molecular interactions between bacterial symbionts and their hosts. Cell 126: 453–465.
9. Tennessen JA (2005) Molecular evolution of animal antimicrobial peptides: widespread moderate positive selection. J Evol Biol 18: 1387–1394.
10. Login FH, Balmand S, Vallier A, Vincent-Monegat C, Vigneron A, et al. (2011) Antimicrobial peptides keep insect endosymbionts under control. Science 334: 362–365.
11. Sperstad SV, Haug T, Blencke HM, Styrvold OB, Li C, et al. (2011) Antimicrobial peptides from marine invertebrates: challenges and perspectives in marine antimicrobial peptide discovery. Biotechnol Adv 29: 519–530.
12. Austin B (1988) Marine Microbiology. Cambridge University Press.
13. Stetter KO, Konig H, Stackebrandt E (1983) *Pyrodictium* gen.nov., a New Genus of Submarine Disc-Shaped Sulphur Reducing Archaebacteria Growing Optimally at 105 degrees C. Syst Appl Microbiol 4: 535–551.
14. Vetriani C, Speck MD, Ellor SV, Lutz RA, Starovoytov V (2004) *Thermovibrio ammonificans* sp. nov., a thermophilic, chemolithotrophic, nitrate-ammonifying bacterium from deep-sea hydrothermal vents. Int J Syst Evol Microbiol 54: 175–181.
15. Delaglio F, Grzesiek S, Vuister GW, Zhu G, Pfeifer J, et al. (1995) NMRPipe: a multidimensional spectral processing system based on UNIX pipes. J Biomol NMR 6: 277–293.
16. Johnson BA (2004) Using NMRView to visualize and analyze the NMR spectra of macromolecules. Methods Mol Biol 278: 313–352.
17. Güntert P (2004) Automated NMR structure calculation with CYANA. Methods Mol Biol 278: 353–378.
18. Fedders H, Leippe M (2008) A reverse search for antimicrobial peptides in *Ciona intestinalis*: identification of a gene family expressed in hemocytes and evaluation of activity. Dev Comp Immunol 32: 286–298.
19. Herbst R, Ott C, Jacobs T, Marti T, Marciano-Cabral F, et al. (2002) Pore-forming polypeptides of the pathogenic protozoon *Naegleria fowleri*. J Biol Chem 277: 22353–22360.
20. Leippe M, Ebel S, Schoenberger OL, Horstmann RD, Muller-Eberhard HJ (1991) Pore-forming peptide of pathogenic *Entamoeba histolytica*. Proc Natl Acad Sci U S A 88: 7659–7663.
21. Gagniere N, Jollivet D, Boutet I, Brelivet Y, Busso D, et al. (2010) Insights into metazoan evolution from *Alvinella pompejana* cDNAs. BMC Genomics 11: 634.
22. Baert JL, Britel M, Slomianny MC, Delbart C, Fournet B, et al. (1991) Yolk protein in leech. Identification, purification and characterization of vitellin and vitellogenin. Eur J Biochem 201: 191–198.
23. Salzet M, Bulet P, Wattez C, Malecha J (1994) FMRFamide-related peptides in the sex segmental ganglia of the Pharyngobdellid leech *Erpobdella octoculata*. Identification and involvement in the control of hydric balance. Eur J Biochem 221: 269–275.
24. Boidin-Wichlacz C, Vergote D, Slomianny C, Jouy N, Salzet M, et al. (2012) Morphological and functional characterization of leech circulating blood cells: role in immunity and neural repair. Cell Mol Life Sci 69: 1717–1731.
25. Tasiemski A (2008) Antimicrobial peptides in annelids. Invertebrate Survival Journal 166.
26. Salzet M, Tasiemski A, Cooper E (2006) Innate immunity in lophotrochozoans: the annelids. Curr Pharm Des 12: 3043–3050.
27. Zasloff M (2002) Antimicrobial peptides of multicellular organisms. Nature 415: 389–395.
28. Willander H, Hermansson E, Johansson J, Presto J (2011) BRICHOS domain associated with lung fibrosis, dementia and cancer—a chaperone that prevents amyloid fibril formation? FEBS J 278: 3893–3904.
29. Andra J, Jakovkin I, Grotzinger J, Hecht O, Krasnosdembskaya AD, et al. (2008) Structure and mode of action of the antimicrobial peptide arenicin. Biochem J 410: 113–122.
30. Maroti G, Kereszt A, Kondorosi E, Mergaert P (2011) Natural roles of antimicrobial peptides in microbes, plants and animals. Res Microbiol 162: 363–374.
31. Sanchez-Pulido L, Devos D, Valencia A (2002) BRICHOS: a conserved domain in proteins associated with dementia, respiratory distress and cancer. Trends Biochem Sci 27: 329–332.
32. Zhu S, Gao B (2012) Evolutionary origin of beta-defensins. Dev Comp Immunol.
33. Struck TH, Paul C, Hill N, Hartmann S, Hosel C, et al. (2011) Phylogenomic analyses unravel annelid evolution. Nature 471: 95–98.
34. Sicot FX, Mesnage M, Masselot M, Exposito JY, Garrone R, et al. (2000) Molecular adaptation to an extreme environment: origin of the thermal stability of the pompeii worm collagen. J Mol Biol 302: 811–820.
35. Hourdez S, Lallier FH, De Cian MC, Green BN, Weber RE, et al. (2000) Gas transfer system in *Alvinella pompejana* (Annelida polychaeta, Terebellida): functional properties of intracellular and extracellular hemoglobins. Physiol Biochem Zool 73: 365–373.

36. Nakagawa S, Takaki Y, Shimamura S, Reysenbach AL, Takai K, et al. (2007) Deep-sea vent epsilon-proteobacterial genomes provide insights into emergence of pathogens. Proc Natl Acad Sci U S A 104: 12146–12150.

37. Cottin D, Ravaux J, Leger N, Halary S, Toullec JY, et al. (2008) Thermal biology of the deep-sea vent annelid *Paralvinella grasslei: in vivo* studies. J Exp Biol 211: 2196–2204.

38. Bonch-Osmolovskaya EA, Perevalova AA, Kolganova TV, Rusanov, II, Jeanthon C, et al. (2011) Activity and distribution of thermophilic prokaryotes in hydrothermal fluid, sulfidic structures, and sheaths of alvinellids (East Pacific Rise, 13 degrees N). Appl Environ Microbiol 77: 2803–2806.

39. Moussard H, Moreira D, Cambon-Bonavita MA, Lopez-Garcia P, Jeanthon C (2006) Uncultured Archaea in a hydrothermal microbial assemblage: phylogenetic diversity and characterization of a genome fragment from a euryarchaeote. FEMS Microbiol Ecol 57: 452–469.

40. Herwald H, Egesten A (2011) The discovery of pathogen recognition receptors (PRR). J Innate Immun 3: 435–436.

41. Salzman NH, Hung K, Haribhai D, Chu H, Karlsson-Sjoberg J, et al. (2009) Enteric defensins are essential regulators of intestinal microbial ecology. Nat Immunol 11: 76–83.

9

Evolutionary Dynamics in the Southwest Indian Ocean Marine Biodiversity Hotspot: A Perspective from the Rocky Shore Gastropod Genus *Nerita*

Bautisse Postaire[1,2]*, J. Henrich Bruggemann[1,2], Hélène Magalon[1,2], Baptiste Faure[1,3]

1 Laboratoire d'ECOlogie MARine, Université de la Réunion, FRE3560 INEE-CNRS, Saint Denis, La Réunion, France, **2** Labex CORAIL, Perpignan, France, **3** Biotope, Service Recherche et Développement, Mèze, France

Abstract

The Southwest Indian Ocean (SWIO) is a striking marine biodiversity hotspot. Coral reefs in this region host a high proportion of endemics compared to total species richness and they are particularly threatened by human activities. The island archipelagos with their diverse marine habitats constitute a natural laboratory for studying diversification processes. Rocky shores in the SWIO region have remained understudied. This habitat presents a high diversity of molluscs, in particular gastropods. To explore the role of climatic and geological factors in lineage diversification within the genus *Nerita*, we constructed a new phylogeny with an associated chronogram from two mitochondrial genes [cytochrome oxidase subunit 1 and 16S rRNA], combining previously published and new data from eight species sampled throughout the region. All species from the SWIO originated less than 20 Ma ago, their closest extant relatives living in the Indo-Australian Archipelago (IAA). Furthermore, the SWIO clades within species with Indo-Pacific distribution ranges are quite recent, less than 5 Ma. These results suggest that the regional diversification of *Nerita* is closely linked to tectonic events in the SWIO region. The Reunion mantle plume head reached Earth's surface 67 Ma and has been stable and active since then, generating island archipelagos, some of which are partly below sea level today. Since the Miocene, sea-level fluctuations have intermittently created new rocky shore habitats. These represent ephemeral stepping-stones, which have likely facilitated repeated colonization by intertidal gastropods, like *Nerita* populations from the IAA, leading to allopatric speciation. This highlights the importance of taking into account past climatic and geological factors when studying diversification of highly dispersive tropical marine species. It also underlines the unique history of the marine biodiversity of the SWIO region.

Editor: Sharyn Jane Goldstien, University of Canterbury, New Zealand

Funding: Funding was provided by the French Agence Nationale de Recherche (programme BIOTAS, no. ANR-06-BDIV-002) and the EU-funded program RUN-EMERGE. We acknowledge the les Territoires Australes et Antarctiques Françaises for logistic support during our fieldwork. We further thank the Affaires Maritimes and the Réserve Naturelle Marine de La Réunion for sampling permits. The funders had no role in study design, data collection and analysis, decision to publish, or preparation of the manuscript.

Competing Interests: The authors have declared that no competing interests exist.

* E-mail: bautisse.postaire@univ-reunion.fr

Introduction

The heterogeneity of species distribution is primarily due to latitudinal, longitudinal and altitudinal gradients in environmental conditions [1]. However, such gradients do not explain the presence of highly species-rich regions, or "biodiversity hotspots". Marine and terrestrial biodiversity hotspots have been defined as restricted regions with high species richness, high endemism and facing habitat loss due to human activities [2,3]. Over the past two decades, 34 terrestrial et 10 marine biodiversity hotspots have been identified [2,4]. The Indo-Australian archipelago (IAA), as defined by [5], represents the most important marine biodiversity hotspot, with species richness declining gradually westward and eastward [6,7]. Two main theories have been proposed to explain this distribution of marine biodiversity: the centre-of-origin and the centre-of-overlap hypothesis [8]. The first proposes that this region generates novel species arising in sympatry or allopatry between islands that subsequently migrate outwards to peripheral regions. The second theory proposes that the IAA, located at the limit of several biogeographical regions, accumulates species

formed by allopatric speciation outside the IAA. It should be noted that these hypotheses are not mutually exclusive and patterns may vary among taxa (reviewed in [9]).

Shallow marine habitats are strongly affected by tectonics, eustatic sea-level changes, physical disturbances and run-off from land, impacts that continually modify habitat availability over various time scales [10,11]. Furthermore, climatic variation over geologic time changes environmental conditions and ocean currents at global and regional scales [10]. These abiotic changes alter gene flow and species distribution and may eventually lead to the formation of new species due to vicariance and/or new selective pressures [12]. Several studies argued that part of the origin of the IAA marine hotspot could be explained by abiotic factors, such as sea level fluctuation and oceanic circulation, tectonic activity and temperature changes [13–18]. Recent studies further pointed out the role of habitat, with coral reefs enhancing species diversification [17,19].

The Southwest Indian Ocean (SWIO) region, corresponding to the Western Indian marine ecoregion [20] comprises a main landmass, Madagascar, and several island archipelagos such as

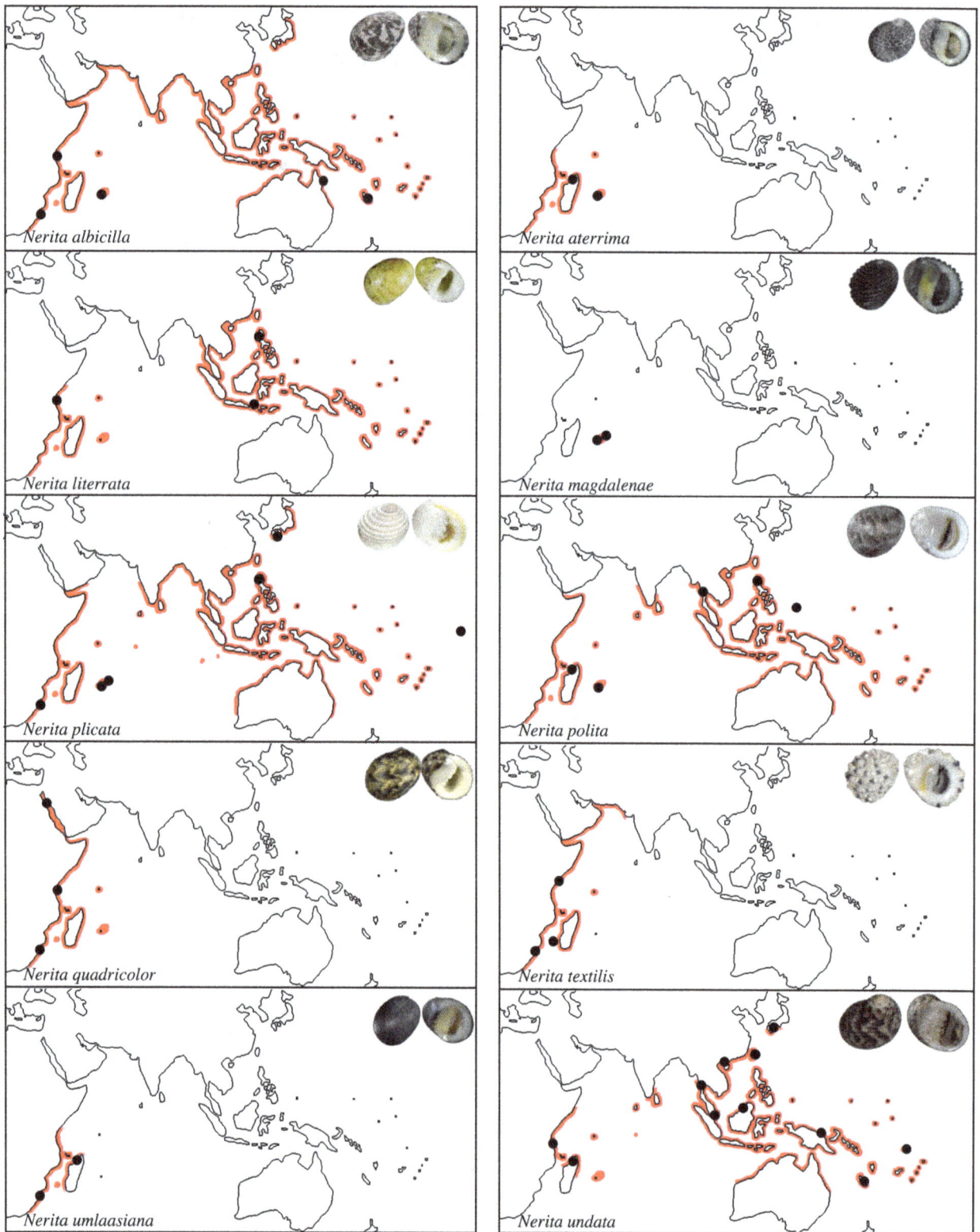

Figure 1. Geographic distribution of *Nerita* species living in the South-Western Indian Ocean, adapted from [16]**.** Grey dots indicate sampling sites from [16]; grey squares indicate sampling sites of new haplotypes.

Comoros, Mascarenes and Seychelles, each with different origins and ages. While Madagascar and the Seychelles are fragments of eastern Gondwana, the Mascarene archipelago was formed during

the past 10 Ma by the volcanic activity of a mantle plume, which has been active since the late Cretaceous [21]. This region hosts a high proportion of endemics and is highly threatened by human activities, hence its classification as a marine biodiversity hotspot [2,22]. It further is one of the regions where terrestrial and marine biodiversity hotspots coincide. The study of the SWIO terrestrial biota revealed a high relatedness between SWIO and Asian species, leading to the hypothesis of the presence of discontinuous land bridges connecting islands and continents to explain present species distributions [23–25]. The extent of such ephemeral bridges varied with sea level and is thought to have facilitated the colonisation of SWIO by terrestrial species following a stepping-stone model [24,26,27]. The study of SWIO marine taxa, focusing on reef-associated species, also highlighted the relatedness of SWIO and IAA faunas, but also revealed the presence of numerous cryptic species [8,17,28,29]. In contrast, a recent study argued that the northern Mozambique Channel hosts scleractinian corals that are relicts of the West Tethys biodiversity [30]. While these studies revealed the complex and contrasted evolutionary histories of SWIO marine taxa, they concerned mainly reef-associated biota, leaving aside other marine ecosystems.

Rocky shores represent only 3000 km^2 area in the Indian Ocean (excluding the Western Australian coast) [31]. In spite of the limited extent, this habitat is extremely interesting for studying species evolution. Rocky shores are characterised by strong gradients in environmental conditions at small scales, varying within meters and hours, and so exert strong selection pressures enhancing species diversification over time [32]. Several biogeographical studies revealed the high variability of genetic structures and speciation patterns of rocky shore species, with some taxa showing only weak imprint of past variations in environmental conditions while others displaying distribution patterns that are highly correlated to the geological history [33–35]. Differential dispersal capacities provides an intuitive explanation for these discrepancies among species: population structure of species with long-lived planktonic larva show little imprint of environmental changes due to high connectivity, as speciation occurs at ocean basin scales and exclusively by allopatric processes [36]. However, attributing the observed biogeographic patterns and population genetic structures to the specificities of larval life alone is too simplistic, as the ecological requirements of the adult stage also determine the distribution of marine species [37,38]. Knowing the geographic range, the ecological requirements and life history traits, in addition to geological history, is challenging but often necessary to accurately resolve the evolutionary history of a species.

The Neritidae (Rafinesque) (Neritimorpha [39]) family is composed of tropical gastropods inhabiting a wide variety of habitats: open seawater, rocky shores, brackish and fresh waters,

mangroves, mud and sand. However, the genus *Nerita* (Linnaeus), with more than 60 known extant species is almost restricted to tropical rocky shores and has already been subject to phylogeographic studies [16,40,41]. Niche changes have played a relatively minor role in the diversification of the genus [16]. Species diversity in this genus peaks in the IAA region, but many species present large biogeographical ranges, from the SWIO to Hawaii [16]. Members of the genus have planktotrophic larvae that live for weeks up to months before settling [42,43]. Those long-lived larvae are expected to limit the number of regional endemics due to high connectivity between populations, which is generally observed in *Nerita*. Contrasting with this general pattern, the SWIO rocky shores present a high proportion of endemics. Among the ten *Nerita* species present, one is endemic to the Mascarene archipelago, *N. magdalenae* (Gmelin), two are endemic to the SWIO, *N. aterrima* (Gmelin) and *N. umlaasiana* (Krauss), while two are restrained to the western coasts of the Indian Ocean, *N. textilis* (Gmelin) and *N. quadricolor* (Gmelin) ([16], Figure 1). The phylogeny of the genus is almost resolved and several clades restricted to the SWIO have been identified within species presenting an Indo-Pacific distribution. Further study of the diversification of the genus have set its origin at the end of the Paleocene and showed a constant diversification rate since then [38]. According to this study, species living in the SWIO originated during two distinct periods, at the beginning (±22 Ma; *N. magdalenae*, *N. textilis* and *N. aterrima*) and the end of the Miocene (±9 Ma; *N. quadricolor* and *N. umlaasiana*).

Time-calibrated phylogenies contain information on the temporal diversification of clades, permitting the exploration of macroevolutionary processes such as variation of speciation and extinction rates over time [44]. When biogeographical information of extant lineages is also considered, a chronogram allows understanding how past geological and climatic events influenced present-day species richness and distribution patterns [45]. According to [40], the IAA is a centre-of-overlap, conclusion supported by the observation that no recent lineage of *Nerita* species has originated within this region. However, this study did not fully explore the diversification patterns of SWIO *Nerita* species and clades.

The present study focuses on the origins and diversification of SWIO *Nerita* species and complements previous phylogenies. It participates to the global effort in understanding how the unique biodiversity of the SWIO region appeared and was maintained through geological times. Many questions remain unanswered: Are local species formed via regional or more global speciation processes? Do they present trans-ocean origins? Are they ancient

Table 1. Number of newly sampled specimens per location.

	Reunion Island	Mauritius	Europa Island	Madagascar
Nerita albicilla (Linnaeus 1758)	4			
Nerita aterrima (Gmelin 1791)	6			
Nerita magdalenae (Gmelin 1791)	3	2		
Nerita plicata (Linnaeus 1758)	6	2		
Nerita polita (Linnaeus 1758)	3			
Nerita quadricolor (Gmelin 1791)			3	
Nerita textilis (Gmelin 1791)			2	
Nerita umlaasiana (Krauss 1848)				1

Each specimen was sequenced for both markers (CO1 and 16S).

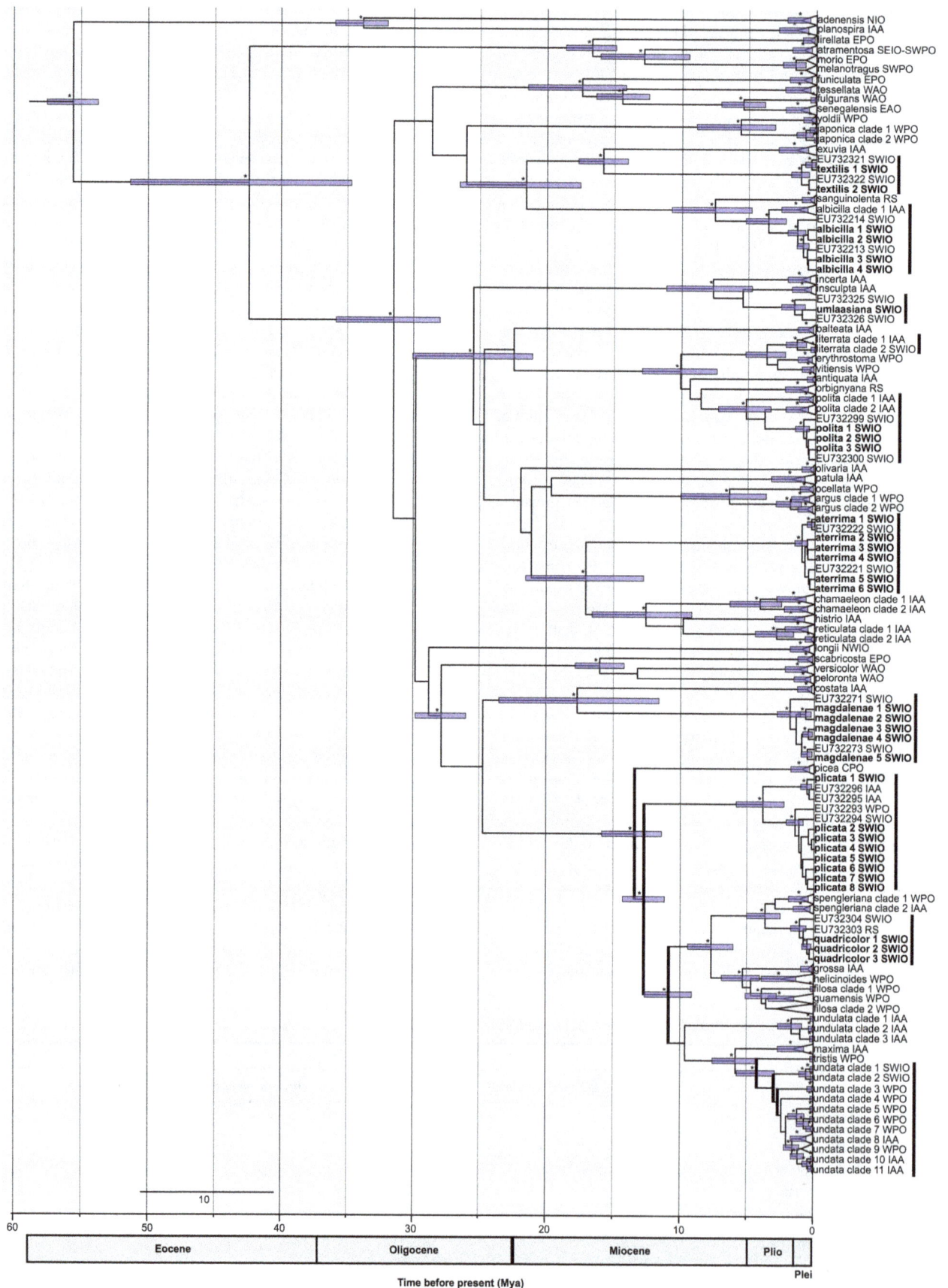

Figure 2. Maximum clade credibility tree. Branch lengths are proportional to time in million years (Ma) and are estimated with Bayesian relaxed lognormal molecular clocks for each marker on the concatenated dataset. Clades with significant increase of diversification rates are indicated by thick black lines (p<0.05). Asterisks on branches represent clades supported by the Bayesian analysis (posterior probability >0.95). Bars represent the 95% HPD interval around the dated nodes. NWIO: Northwest Indian Ocean; SEIO: Southeast Indian Ocean; SWIO: Southwest Indian Ocean; IAA: Indo-Australian archipelago; SWPO: Southwest Pacific Ocean; WPO: West Pacific Ocean; CPO: Central Pacific Ocean; EPO: East Pacific Ocean; RS: Red Sea; WAO: West Atlantic Ocean; EAO: East Atlantic Ocean. Each triangle tips represent two sequences from Genbank. Bold names represent new sequences. Thick black lines identify species present in the SWIO.

or relatively recent species? To answer such questions, we used up-to-date methods of chronogram reconstruction and macroevolution studies to (1) reveal variations in the diversification history of *Nerita* species, and (2) study the potential influence of extrinsic geological and climatic factors on the diversification of the genus.

Materials and Methods

Sampling

We thank the Affaires Maritimes, the Réserve Naturelle Marine de La Réunion and the Territoires Australes et Antarctiques Françaises for sampling permits and the latter also for logistic support during fieldwork at the Eparse Islands. We collected shells and tissues of eight *Nerita* morpho-species from different locations within the SWIO (32 individuals in total, Table 1, Figure 1). Species were identified *a posteriori* using online databases and taxonomic works [46,47]. To detect possible cryptic species, we collected and sequenced at least three individuals per morpho-species. Samples from Madagascar and Europa Island were collected between 2008 and 2010 and samples from Reunion and Mauritius were collected in 2011. Each individual (shell) was photographed alive. Samples previous to 2010 were stored in 90% ethanol at room temperature and samples of later campaigns were preserved after dissection in 70% ethanol and stored at −20°C.

Sequencing

DNA of each individual was extracted from ~10 mg of foot tissue with DNeasy Blood & Tissue Kit (Qiagen) following the manufacturer's protocol. Extraction quality was assessed visually on a 0.8% agarose gel stained with GelRed Nucleic Acid Stain, 10000X in DMSO (Gentaur). We amplified and sequenced fragments of two mitochondrial genes, cytochrome oxidase subunit 1 (CO1) and 16 s rRNA (16S). Each PCR reaction was conducted in 25 µl: 8,5 µl of ultra-pure water, 12.5 µl (0.625U) of AmpliTaq mix (Applied Biosystems), 1 µl of each primer (10 µM) and 2 µl of template DNA (final concentration: 1.6 ng/µl). Each gene was amplified with the following sets of primers: CO1 with HCO2198 (5′-TAA ACT TCA GGG TGA CCA AAA AAT CA-3′) and LCO1490 (5′-GGT CAA CAA ATC ATA AAG ATA TTG G-3′) [48] and 16S with 16Sar (5′-CGC CTG TTT ATC AAA AAC AT-3′) and 16Sbr (5′-CCG GTC TGA ACT CAG ATC ACG T-3′) [49]. The PCR profile for both markers consisted of an initial denaturation step (5 min at 95°C), 5 cycles of denaturation (30 s at 94°C), annealing (30 s at 46°C) and elongation (1 min at 72°C), 30 cycles of denaturation (30S at 94°C), annealing (30 s at 51°C) and elongation (1 min at 72°C), and a final extension step (10 min at 72°C). PCR products were visualized on a 1% agarose gel stained with GelRed Nucleic Acid Stain, 10000X in DMSO (Gentaur). PCR products were sequenced in both directions at Genoscreen (www.genoscreen.fr) by the Sanger method (3730XL, Applied Biosystems).

Tree Construction

The 64 new sequences were checked and edited using Geneious 6.0 (created by Biomatters, available on from http://www.

geneious.com/) and deposited in Genbank (Table S1). To complete our data set, sequences for CO1 and 16S genes were downloaded from Genbank (Table S1). In fine, 370 *Nerita* sequences were used (185 for each marker). The complete data set was aligned using the ClustalW algorithm (default options) and confirmed by eye. Using Modeltest v3.7 [50,51] and AIC, GTR+I+Γ was the best evolution model. Lack of significant divergence in the phylogenies of the two markers permitted conducting the analysis on the combined data set [40]. Three methods of phylogenetic tree construction were used: (1) Neighbor-joining (NJ, 1000 bootstrap permutations) was calculated using Geneious 6.0 (2) Maximum Likelihood (ML) was implemented using the PHYML Geneious plug-in [52] and (3) Bayesian analyses were performed using MrBayes v3.2 (MB, 30.10^6 generations, 8 chains, 3 runs, temperature to 0.2) [53]. Bayesian analyses were conducted on Titan, Reunion University's cluster. Each tree was rooted by the outgroup *Bathynerita naticoidea* [54].

Estimation of Nodes Ages

We used BEAST v1.7.5 [55] to reanalyse the combined dataset of CO1 and 16S sequences incorporating an uncorrelated relaxed, lognormal molecular clock for both markers. This method allows the construction of phylogeny and estimation of divergences times at the same time, calculating the 95% Highest Posterior Density (HPD) of each node. The chronogram was calibrated as in [40]. We used a GTR+Γ+I substitution model for each marker. We used the birth-death model for the tree reconstruction [56], which assumes constant speciation and extinction rates over time (exponential increase of species number over time) with a faster accumulation of lineages when considering recent past geological time (accumulation of lineages that are not extinct yet).

Analyses were undertaken with three independent chains of 1.10^8 generations, sampling every 10^4 generations. The convergence of all parameters was assessed using Tracer v1.5 [57]. Final tree was produced by constructing a consensus tree from the data set of accepted trees (burnin: 25%, total accepted: 7500) with the maximum clade credibility option and median node height using TreeAnnotator v1.7.5 [58]. Final tree editing was performed using FigTree v1.4 [59].

Diversification Analysis

Tree imbalance assessment. We assessed the imbalance of the tree using β. The β parameter compares the observed nodal imbalance among clades to the equal-rates Markov model [60]. Under the theoretical model, each node has the same probability of splitting, and β should not be distinguishable from 0. Strong negative or positive values indicate variations in the splitting probability among lineages indicating whether lineages within a tree have diversified with different rates [60]. We performed this analysis for the whole genus using *apTreeshape* [61] package under R [62].

To explore the evolution of the genus through time, lineage-through-time (LTT) plots were constructed. Based on the chronogram, it represents the evolution of the number of lineages against the node ages and thus the change in diversification rate

Table 2. Determination of the number of rate shifts for *Nerita*.

Rs	T0	Tu	Div	T1	Tu	Div	T2	Tu	Div	p value	Log-likelihood
0	0	0.49	0.068								−204.656
1	0	0.99	0.00013	24	0.57	0.058				0.0403	−202.554
2	0	0.98	0.0042	6	0.18	0.006	24	6.10^{-7}	0.13	0.0709	−200.923

Log-likelihood values were tested against the model with one more shift. Rs : rate shifts, Tn : n rate shift, Tu : turnover, Div: diversification.

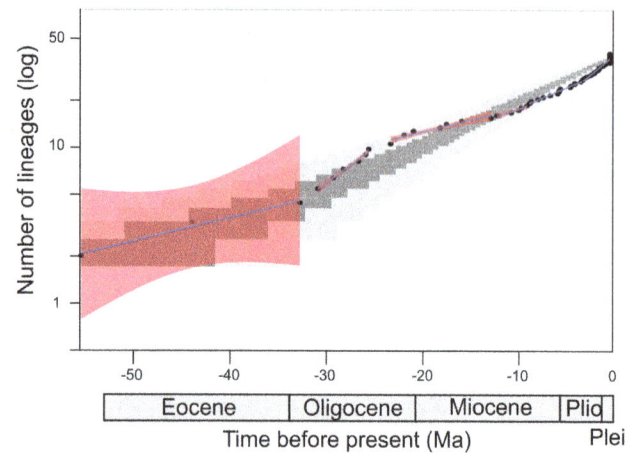

Figure 3. LTT plot of the maximum clade credibility tree. Grey zone represent the 50% (dark grey) and 95% (light grey) null distribution of LTT plots generated under the Yule model. Black dots represent observed values. Blue lines represent successive linear models fitted to the data and confidence intervals (red).

through time. One thousand random trees were calculated with the same parameters (root height, number of taxa, diversification and extinction rates) as our data to assess a null distribution of LTT curves. Under a constant birth-death model, a straight line is expected with slope equal to b–d (b: speciation, d: extinction) [56,63].

Diversification rate shifts. Departures from a constant rate of diversification (null hypothesis) were analysed using the constant rate test. This test uses the gamma (γ) statistic to compare the positions of nodes of the studied phylogeny to those from a theoretical phylogeny generated under a constant rate of diversification [63] using the package *ape* [64]. Negative values of γ imply a reduction of the speed of accumulation of lineages and thus of diversification rate; positive values imply an acceleration of lineage accumulation. The gamma statistic assumes that all extant lineages have been sampled and that speciation and extinction occur with equal probability among lineages. To account for incomplete sampling, we used the MCCR test [63]. We simulated a thousand phylogenies with 100 taxa under the pure birth model. We compared the γ value of the original phylogeny to the distribution of values of the random phylogenies. To test the second assumption of the constant rate test, we used the relative cladogenesis (RC) test with a Bonferroni correction implemented the *geiger* package [65]. This test calculates the probability that a particular lineage existing at a time t will have n descendants under a constant rate birth-death model at the present time. The null hypothesis of this test is that all ancient branches have the same proportion of extant species [66]. It allows identifying branches producing more descendant than the others at the same time interval.

When using phylogenies with a non-null extinction rate [67], the power of the γ statistic for detecting changes in diversification rates is lower than likelihood methods (e.g., ΔAIC_{RC}) (Table S2). The ΔAIC_{RC} test statistic included in the *laser* package [68] compares data fits to rate-constant diversification models (Yule and birth-death, AIC_{RC}) and several rate-variable models (Yule-2-rate, Yule-3-rate, logistic density-dependent DDL and exponential density-dependent DDX, AIC_{RV}) with likelihood ratio tests and AIC scores. Under DDL model, the speciation rate slows in relation to a logistic growth depending on: the initial speciation rate, the number of lineages at a specific point and a parameter

Table 3. Parameters of the linear models fitted to the LTT for each geological period.

Geological period (in Ma)	Slope (lineage per Ma)	Intercept
Eocene to early Oligocene (−50, −30)	**1.11**	0.020
Early Oligocene to early Miocene (−30, −20)	2.44	0.059
Early Miocene to mid-Miocene (−20, −15)	1.79	0.036
Mid-Miocene to late Miocene (−15, −10)	1.60	0.023
Late Miocene (−10, −5)	1.63	0.026
Pliocene to nowadays (−5, 0)	1.77	0.051

similar to the carrying capacity. Under the DDX model, the speciation rate at a certain point of time is function of initial speciation rate, the number of lineage at this specific point and the magnitude of the rate change as the number of lineage increases (a rate change equals to 0 implies constant speciation through time) [69]. In the other two rate-variable models, the speciation rate under a pure birth model within a clade is assumed to change before one or several breakpoints [68]. For example, the Yule-2-rate assumes a first speciation rate, a final speciation rate, and inserts a breakpoint in time when rate shifts (optimized during model fitting). According to [67], rate-constant models can be rejected with confidence when the difference between AIC_{RC} and AIC_{RV} is close to 4 for small phylogenies ($n_{taxa} = 30$) and 5.5 for large phylogenies ($n_{taxa} = 100$). We assessed the critical values of the ΔAIC_{RC} test by simulating 1000 phylogenies of 74 taxa under the pure birth model (Table S3).

We used the *TreePar* package 2.5 [70] to highlight changes in diversification rates (speciation-extinction), i.e. the slope of the LTT, through time by adjusting the starting time of the analysis and taking into account incomplete sampling. The control of these two parameters allows controlling the possibility that recent evolution and incomplete sampling may hide diversification phases during the early stages of the phylogeny. The sampling parameter represents the proportion of extant lineages present in the phylogeny (this parameter set to 1 indicates that all lineages have been sampled). This method needs the specification of the boundaries of the period studied (here from 34 to 3 Ma, covering Oligocene, Miocene and Pliocene), the grid (here 1 Ma) and the number of shifts to include. On each point of the grid, a shift is inserted and rates are estimated. Once the best shift point is determined, it is fixed and the analysis restarts to insert another shift point. Likelihood ratio tests were used to compare the current model to a model with one more shift point.

Sliding window analysis. In order to explore potential bursts of diversification through time, we performed a sliding-window analysis [71]. Assuming that all non-sampled lineages were formed during the Pleistocene (e.g. 2.6 Ma to Present), we set a window of 5 Ma wide and a step of 1 Ma, and calculated a diversification rate for each step using the chronogram from BEAST from 56 to 2.6 Ma. In [72], the diversification rate is given as $-\ln (Nb/Nt)/\Delta t$, where Nb is the number of lineages at the beginning of the period, Nt the number of lineages at the end of the period and Δt the length of the period in million years.

Results

Molecular Dating

Our phylogenetic analysis was fully consistent with the previous published phylogeny of *Nerita* [40]. No new clades within *Nerita* species were discovered. The tree generated with an uncorrelated relaxed lognormal molecular clock for both markers indicates divergence times (Figure 2). The relaxed molecular clock estimated that the genus *Nerita* originated 55.7 Ma ago (95% HPD: 53.7 to 57.6 Ma). Even though some nodes are poorly supported, the majority of nodes ages were estimated with high confidence. Our chronogram, using two independent molecular clocks, dated all diversification events more recently than previous findings, though with the same topology and confidence of nodes. Estimated ages of *Nerita* species present in the SWIO region are all younger than 20 Ma: *N. aterrima* appeared 17.26 Ma (95% HPD: 12.76 to 22.71 Ma); *N. magdalenae* 17.9 Ma (95% HPD: 11.27 to 23.8 Ma); *N. quadricolor* 3.59 Ma (95% HPD: 2.4 to 4.99 Ma); *N. textilis* 15.86 Ma (95% HPD: 13.99 to 17.65 Ma); *N. umlaasiana* 5.39 Ma (no 95% HPD given by BEAST). For species with Indo-Pacific range, individuals collected in the SWIO show divergent and strongly supported monophyletic clades (except *N. polita*): *N. albicilla* SWIO diverged from the IAA clade 3.64 Ma (95% HPD: 2.18 to 5.63 Ma); *N. polita* 3.73 Ma (no 95% HPD given by BEAST); *N. litterata* 1.13 Ma (no 95% HPD given by BEAST); two *N. undata* clades 2.67 Ma (no 95% HPD given by BEAST).

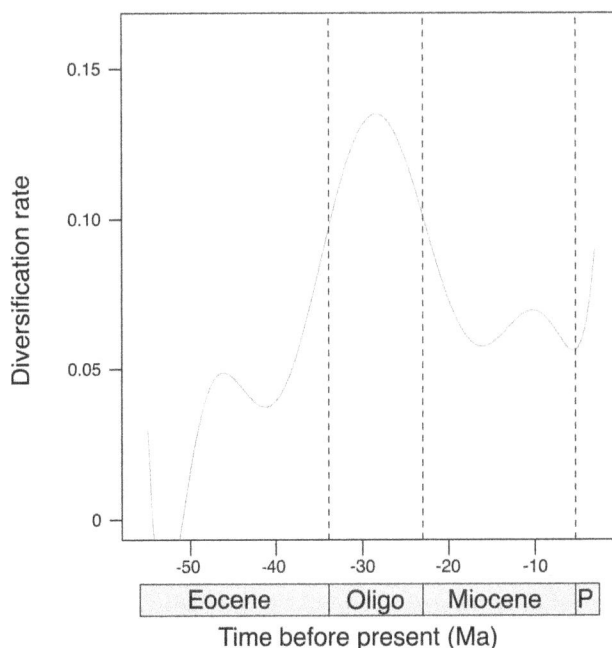

Figure 4. Sliding-window analysis of the net diversification rate of the genus *Nerita* through geological periods. Oligo: Oligocene; P: Pliocene.

Evolution of Lineages and Tree Shape

The chronogram obtained with BEAST suggested imbalance in the diversification of *Nerita* as it is more imbalanced than expected under the Yule model ($\beta = -1.07$, 95% confidence interval: -1.43 to -0.56). The RC test highlighted an increase in diversification rate starting at 13.15 Ma (95% HPD: 11.39 to 15.83 Ma, $p = 0.0076$, $\alpha = 0.05$) and running all along the phylogeny to the *N. undata* species complex.

Studies of diversification rate shifts through time selected the Yule-3-rate model as the best fit model for the whole *Nerita*, but did not reject the Yule model ($\gamma = 0.63$; critical value $\gamma = 1.96$, $\alpha = 0.05$).

The *TreePar* analyses detected a reduction of diversification rates situated at 24 Ma, decreasing from 0.058 to 0.00013 (Table 2).

The LTT plot is consistent with the *TreePar* analyses, showing a faster accumulation of lineages during the Oligocene and a later slowdown (Figure 3, Table 3). However, the LTT plot remains within the 95% confidence interval of null distributions of LTT plots under the Yule model.

The sliding-window analysis is also consistent with previous results, showing that the diversification rate was higher during the Oligocene. It decreased during the Miocene and remained relatively constant for 10 Ma (Figure 4) to finally increase again during more recent geological times.

Discussion

Biogeography: Global Patterns and Timing of Diversification

Our results provide another example of extensive cryptic diversity in species with Indo-Pacific distribution ranges. *Nerita* species from IAA present numerous robust divergent clades within this region. This pattern has been interpreted as a consequence of intense diversification within this region, producing species that further disperse and colonize peripheral islands in the Indian and Pacific oceans [73]. Long distance colonization events may lead to allopatric and peripatric speciation with the modification of gene flow through time due to changes in biotic and abiotic conditions. Allopatry seems to be the most frequent speciation mechanism of marine species [28,33,37,74], but other processes like disruptive selection, habitat or resource choice, may occur at smaller geographic scales and lead to sympatric sister species, particularly in gastropods [75].

The first fossil record of *Nerita* is from the late Paleocene (56 Ma) [76] and the average genus diversification rate varied over time according to the *TreePar* analysis. Our results suggest that diversification was higher during the Oligocene and early Miocene compared to later geologic periods when net diversification rate decreased from 0.058 to 0.00013. During Oligo-Miocene, the oceanic circulation of the southern hemisphere was greatly modified by the northward movement of the Australian and South-American plates and the southward migration of the Antarctic plate [77]. These led to the formation of the circumpolar current, the formation of the Antarctic ice sheet [78–81] and global cooling, lowering sea levels and expanding emerged land masses and coastlines [82]. Furthermore, the collision of the Australian and Asian plates led to the emergence of new landmasses and tropical habitats suitable for colonization by shallow water species: the IAA [83]. The IAA formation modified equatorial currents, constraining seaways between the Pacific and Indian oceans. These changes in marine habitats induced massive species extinctions in many taxa and important changes in faunal compositions [84]. Based on our results, *Nerita* gastropods do not appear to have suffered from these global environmental changes

since their diversity increased rapidly during this period (Figure 4). Increased availability of suitable habitats and the high spatial and temporal heterogeneity of the environment during Oligocene and early Miocene have likely modified the distribution and connectivity of populations and boosted diversification by increasing the opportunities for allopatric and peripatric speciation. The pattern we found is congruent with that of three other intertidal or shallow marine gastropod genera: *Conus*, *Echinolittorina* and *Turbo* [15]. The observed slowdown of diversification in these genera during late Miocene and Pliocene was interpreted as a consequence of limited speciation opportunities, due to the progressive filling of newly created niches [15]. Diversification rates in *Nerita* have likewise decreased slightly since the end of the Oligo-Miocene period. Contrastingly, the *N. undata* complex (originating from the IAA) presents a higher probability of diversification as shown by the RC test ($p = 0.0076$, $\alpha = 0.05$). Originating during mid-Pliocene, this clade has diversified since this period in 11 robust genetic clades: 2 in the SWIO and 9 in the IAA. The detection of different geographically restricted lineages within species with Indo-Pacific distributions, like *N. albicilla* or *N. undata*, suggests that dispersal occurs at relatively small geographic scales, despite a high dispersal potential due to long-lived planktotrophic larvae. The spatial and temporal heterogeneity of the IAA region may enhance species diversification at small geographic scales by constantly modifying connectivity between populations for species with benthic adult and planktonic larval stages which are dependent on ocean currents and available habitats for settlement [85,86]. Our results, like in other gastropod genera, support the centre-of-origin hypothesis for *Nerita*, the IAA presenting significantly more diversification events during the Oligocene.

Isolation by Distance as a Driving Process of Diversification in the Indian Ocean

The genus *Nerita* being almost restricted to tropical rocky shores, its distribution is partly correlated to the existence of these particular habitats. Although the cryptic diversity of IAA *Nerita* can be explained by global climatic variations and environmental modifications over geological time, these factors do not explain the high proportion of endemic and cryptic lineages found within the SWIO. In this region, the volcanic activity (geological hotspot) started more than 65 Ma ago and created a North-South oriented chain of islands across the Indian Ocean: Laccadive islands (emergence: 65-60 Ma), Maldivian islands (60-50 Ma), Chagos archipelago (50-49 Ma), Mascarene plateau (48-31 Ma), Mauritius island (8 Ma) and finally Reunion Island (2 Ma) [87,88]. During sea level low stands, these islands represented large landmasses, particularly during the Miocene. Various terrestrial clades used those multiple islands as stepping-stones to colonize the SWIO, while subsequent sea level rises facilitated secondary isolation and speciation [24]. Thus, colonization events of SWIO by Asian and IAA species favoured terrestrial speciation due to the action of geologic/climatic events throughout this period.

This model seems to apply to Indian Ocean *Nerita* species as well. Our hypothesis is supported by the old asynchronous divergence, ranging from 21 to 5 Ma (the Mascarene plateau emerged during this period) of three endemic species from their closest parent species living in the IAA region: *N. aterrima*, *N. magdalenae*, *N. umlaasiana*. Nowadays, there is little connectivity between the western and eastern Indian Ocean populations for a wide range of marine taxa [34,89]. However, throughout the Miocene and assuming no major changes in ocean currents compared to nowadays, the intermittent emergence of volcanic landmasses and new rocky shores in SWIO may have permitted larval colonization from IAA *Nerita* populations. Without constant larval input due to sea level variations changing distances between populations of IAA and SWIO, newly settled populations diverged and formed new species by allopatry or peripatry. Ecological transition has not played a role in the formation of SWIO *Nerita* species as all sister species-pairs occupy the same ecological niches: lower littoral for *N. albicilla*-*N. sanguinolenta*; mid-littoral for *N. textilis*-*N. exuvia*; upper littoral for the pairs *N. magdalenae*-*N. costata* and *N. quadricolor*-*N. spengleriana*; supra-littoral for *N. insculpta*-*N. umlaasiana* [16]. Ecological conservatism have been identified in other intertidal gastropods, e.g., sister species of *Echinolittorina* remain allopatric for millions of years without changing their habitat preferences [33]. Ecological conservatism during diversification has also been documented in other taxa, such as coral reef fishes [90]. For a temperate terrestrial gastropod (*Arion subfuscus*), the habitat fidelity over time (as evidenced by the persistence of allopatry) has even contributed to the increase of lineage accumulation during the past glacial maximum [91].

Therefore SWIO *Nerita* endemics followed a "terrestrial" diversification pattern in the region and formed due to the synergy of several abiotic factors: the presence of an active geological hotspot and sea level variations, favouring colonization of *Nerita* populations from the IAA and subsequent genetic isolation. Changes in ocean circulation may have also played a role, but modelling currents at small geographical scale through geological time seems presently not feasible.

Conclusions

This study joins the increasing number of publications linking geological history and diversification processes to explore biogeographic patterns. It brings attention to the role of geologic events and climatic variations modifying colonization opportunities. It further highlights the strong influence of allopatric and peripatric speciation processes in establishing intertidal gastropods diversity patterns, depending on the presence of islands to maintain their presence across oceanic basins and thus exhibiting a "terrestrial diversification pattern". The link between the IAA and SWIO

intertidal biodiversity is here evidence and it would be interesting to compare this pattern with other SWIO rocky shore taxa, in order to assess the prevalence of our results among a wider variety of organisms.

Supporting Information

Table S1 List of species used in this study, sampling zone, biogeographic region and Genbank accession numbers for CO1 and 16S genes. CPO: Central Pacific Ocean; EAO: East Atlantic Ocean; EPO: East Pacific Ocean; IAA: Indo-Australian archipelago; NIO: North Indian Ocean; RS: Red Sea; SEIO: Southeast Indian Ocean; SWIO: Southwest Indian Ocean; SWPO: Southwest Pacific Ocean; WPO: West Pacific Ocean; WAO: West Atlantic Ocean;.

Table S2 Tests used on the final chronogram with assumptions and alternative hypotheses.

Table S3 Diversification models tested with corresponding lineage-through-time plots.

Acknowledgments

Computations have been performed on the supercomputer facilities of Université de la Réunion. We would like to thank Matthieu Bober, Guy Hoarau and late Maurice Jay for their taxonomic work on the collections of the Université de la Réunion.

Author Contributions

Conceived and designed the experiments: BP JHB HM BF. Performed the experiments: BP JHB HM BF. Analyzed the data: BP JHB HM BF. Contributed reagents/materials/analysis tools: BP JHB HM BF. Wrote the paper: BP JHB HM BF.

References

1. Barnes RSK (2010) Regional and latitudinal variation in the diversity, dominance and abundance of microphagous microgastropods and other benthos in intertidal beds of dwarf eelgrass, *Nanozostera* spp. Mar Biodiv 40: 95–106. doi:10.1007/s12526-010-0036-1.

2. Roberts CM, McClean CJ, Veron JEN, Hawkins JP, Allen GR, et al. (2002) Marine biodiversity hotspots and conservation priorities for tropical reefs. Science 295: 1280–1284. doi:10.1126/science.1067728.

3. Mittermeier RA, Hawkins F, Rajaobelina S, Langrand O (2005) Wilderness Conservation in a Biodiversity Hotspot. International Journal of Wilderness 11: 42–45.

4. Myers N, Mittermeier RA, Mittermeier CG, da Fonseca GA, Kent J (2000) Biodiversity hotspots for conservation priorities. Nature 403: 853–858. doi:10.1038/35002501.

5. Lohman DJ, de Bruyn M, Page T, Rintelen von K, Hall R, et al. (2011) Biogeography of the Indo-Australian Archipelago. Annu Rev Ecol Evol Syst 42: 205–226. doi:10.1146/annurev-ecolsys-102710–145001.

6. Hughes TP, Bellwood DR, Connolly SR (2002) Biodiversity hotspots, centres of endemicity, and the conservation of coral reefs. Ecology Letters 5: 775–784.

7. Hoeksema BW (2007) Delineation of the Indo-Malayan centre of maximum marine biodiversity: the Coral Triangle. *Biogeography, Time, and Place: Distributions, Barriers, and Islands*. Biogeography. 117–178.

8. Hubert N, Meyer CP, Bruggemann JH, Guérin F, Komeno RJL, et al. (2012) Cryptic diversity in Indo-Pacific coral-reef fishes revealed by DNA-barcoding provides new support to the centre-of-overlap hypothesis. PLoS ONE 7: e28987. doi:10.1371/journal.pone.0028987.

9. Briggs JC, Bowen BW (2013) Marine shelf habitat: biogeography and evolution. Journal of Biogeography 40: 1023–1035. doi:10.1111/jbi.12082.

10. Paulay G, Meyer C (2002) Diversification in the tropical Pacific: comparisons between marine and terrestrial systems and the importance of founder speciation. Integrative and Comparative Biology 42: 922–934.

11. DiBattista JD, Berumen ML, Gaither MR, Rocha LA, Eble JA, et al. (2013) After continents divide: comparative phylogeography of reef fishes from the Red Sea and Indian Ocean. Journal of Biogeography 40: 1170–1181. doi:10.1111/jbi.12068.

12. Santini F, Winterbottom R (2002) Historical biogeography of Indo-western Pacific coral reef biota: is the Indonesian region a centre of origin? Journal of Biogeography 29: 189–205.

13. Benzie JAH, Williams ST (1997) Genetic Structure of Giant Clam (*Tridacna maxima*) Populations in the West Pacific is Not Consistent with Dispersal by Present-Day Ocean Currents. Evolution 51: 768–783.

14. Alfaro ME, Santini F, Brock CD (2007) Do Reefs Drive Diversification in Marine Teleosts? Evidence from the Pufferfish and their Allies (Order Tetraodontiformes). Evolution 61: 2104–2126. doi:10.1111/j.1558-5646.2007.00182.x.

15. Williams ST, Duda TF Jr (2008) Did Tectonic Activity Stimulate Oligomiocene Speciation in the Indo-West Pacific? Evolution 62: 1618–1634. doi:10.1111/j.1558-5646.2008.00399.x.

16. Frey MA (2010) The relative importance of geography and ecology in species diversification: evidence from a tropical marine intertidal snail (*Nerita*). Journal of Biogeography 37: 1515–1528. doi:10.1111/j.1365-2699.2010.02283.x.

17. Cowman PF, Bellwood DR (2011) Coral reefs as drivers of cladogenesis: expanding coral reefs, cryptic extinction events, and the development of biodiversity hotspots. Journal of Evolutionary Biology 24: 2543–2562. doi:10.1111/j.1420-9101.2011.02391.x.

18. Cabezas P, Sanmartín I, Paulay G, Macpherson E, Machordom A (2012) Deep Under the Sea: Unraveling the Evolutionary History of the Deep-Sea Squat Lobster Paramunida (Decapoda, Munididae). Evolution 66: 1878–1896. doi:10.1111/j.1558-5646.2011.01560.x.

19. Cowman PF, Bellwood DR (2013) The historical biogeography of coral reef fishes: global patterns of origination and dispersal. Journal of Biogeography 40: 209–224. doi:10.1111/jbi.12003.

20. Spalding MD, Fox HE, Allen GR, Davidson N, Ferdaña ZA, et al. (2007) Marine Ecoregions of the World: A Bioregionalization of Coastal and Shelf Areas. BioScience 57: 573–583. doi:10.1641/B570707.

21. Peng ZX, Mahoney JJ (1995) Drillhole lavas from the northwestern Deccan Traps, and the evolution of Reunion hotspot mantle. Earth and Planetary Science Letters 134: 169–185.

22. Bellard C, Leclerc C, Courchamp F (2013) Impact of sea level rise on the 10 insular biodiversity hotspots. Global Ecology and Biogeography 23: 203–212. doi:10.1111/geb.12093.

23. Masters JC, de Wit MJ, Asher RJ (2006) Reconciling the Origins of Africa, India and Madagascar with Vertebrate Dispersal Scenarios. Folia Primatol 77: 399–418. doi:10.1159/000095388.

24. Warren BH, Strasberg D, Bruggemann JH, Prys-Jones RP, Thébaud C (2010) Why does the biota of the Madagascar region have such a strong Asiatic flavour? Cladistics 26: 526–538. doi:10.1111/j.1096-0031.2009.00300.x.

25. Agnarsson I, Kuntner M (2012) The generation of a biodiversity hotspot: biogeography and phylogeography of the western Indian Ocean islands. Current Topics in Phylogenetics and Phylogeography of Terrestrial and Aquatic Systems. 33–82.

26. Yoder AD, Nowak MD (2006) Has Vicariance or Dispersal Been the Predominant Biogeographic Force in Madagascar? Only Time Will Tell. Annu Rev Ecol Evol Syst 37: 405–431. doi:10.1146/annurev.ecolsys.37.091305.110239.

27. Vences M, Wollenberg KC, Vieites DR, Lees DC (2009) Madagascar as a model region of species diversification. Trends in Ecology & Evolution 24: 456–465. doi:10.1016/j.tree.2009.03.011.

28. Malay MC, Paulay G (2010) Peripatric Speciation Drives Diversification and Distributional Pattern of Reef Hermit Crabs (Decapoda: Diogenidae: Calcinus). Evolution 64: 634–662. doi:10.1111/j.1558-5646.2009.00848.x.

29. Hoareau TB, Boissin E, Paulay G (2013) The Southwestern Indian Ocean as a potential marine evolutionary hotspot: perspectives from comparative phylogeography of reef brittle-stars. Journal of Biogeography 40: 2167–2179.

30. Obura DO (2012) Evolutionary mechanisms and diversity in a western Indian Ocean center of diversity. Proceedings of the 12th International Coral Reef Symposium.

31. Wafar M, Venkataraman K, Ingole B, Ajmal Khan S, LokaBharathi P (2011) State of Knowledge of Coastal and Marine Biodiversity of Indian Ocean Countries. PLoS ONE 6: e14613. doi:10.1371/journal.pone.0014613.

32. Lubchenco J (1980) Algal Zonation in the New England Rocky Intertidal Community: An Experimental Analysis. Ecology 61: 333–344.

33. Williams ST, Reid DG (2004) Speciation and diversity on tropical rocky shores: a global phylogeny of snails of the genus Echinolittorina. Evolution 58: 2227–2251.

34. Kirkendale LA, Meyer CP (2004) Phylogeography of the Patelloida profunda group (Gastropoda: Lottidae): diversification in a dispersal-driven marine system. Mol Ecol 13: 2749–2762. doi:10.1111/j.1365-294X.2004.02284.x.

35. Waters JM, King TM, O'Loughlin PM, Spencer HG (2005) Phylogeographical disjunction in abundant high-dispersal littoral gastropods. Mol Ecol 14: 2789–2802. doi:10.1111/j.1365-294X.2005.02635.x.

36. Bohonak AJ (1999) Dispersal, gene flow, and population structure. Quarterly Review of Biology 74: 21–45.

37. Paulay G (2006) Dispersal and divergence across the greatest ocean region: Do larvae matter? Integrative and Comparative Biology 46: 269–281. doi:10.1093/icb/icj027.

38. Weersing K, Toonen RJ (2009) Population genetics, larval dispersal, and connectivity in marine systems. Mar Ecol Prog Ser 393: 1–12. doi:10.3354/meps08287.

39. Bouchet P, Rocroi JP (2005) Classification and nomenclator of gastropod families. Malacologia 47: 397.

40. Frey MA, Vermeij GJ (2008) Molecular phylogenies and historical biogeography of a circumtropical group of gastropods (Genus: Nerita): Implications for regional diversity patterns in the marine tropics. Molecular Phylogenetics and Evolution 48: 1067–1086. doi:10.1016/j.ympev.2008.05.009.

41. Castro LR, Colgan DJ (2010) The phylogenetic position of Neritimorpha based on the mitochondrial genome of Nerita melanotragus (Mollusca: Gastropoda). Molecular Phylogenetics and Evolution 57: 918–923. doi:10.1016/j.ympev.2010.08.030.

42. Underwood AJ (1975) Comparative studies on the biology of Nerita atramentosa Reeve, Bembicium nanum (Lamarck) and Cellana tramoserica (Sowerby) (gastropoda: Prosobranchia) in S.E. Australia. Journal of Experimental Marine Biology and Ecology 18: 153–172.

43. Lewis JB (1960) The fauna of rocky shores of Barbados, West Indies. Can J Zool 38: 391–435. doi:10.1139/z60-043.

44. Antonelli A, Sanmartín I (2011) Why are there so many plant species in the Neotropics? Taxon 60: 403–414.

45. Condamine F, Silva-Brandão K (2012) Biogeographic and diversification patterns of Neotropical Troidini butterflies (Papilionidae) support a museum model of diversity dynamics for Amazonia. BMC Evol Biol 12. doi:doi:10.1186/1471-2148-12-82.

46. Drivas J, Jay M (1988) Coquillages de La Réunion et de l'île Maurice. Times Editions/Les Editions de Pacifique. 159 pp.

47. Jay M, Drivas J, Hoareau G, Martin J-C (2014) Mollusques de l'île de la Réunion. http://vieoceane.free.fr/mollusques.

48. Folmer O, Black M, Hoeh W, Lutz R, Vrijenhoek R (1994) DNA primers for amplification of mitochondrial cytochrome c oxidase subunit I from diverse metazoan invertebrates. Mol Marine Biol Biotechnol 3: 294–299.

49. Palumbi SR (1996) Nucleic acids II: The polymerase chain reaction. Molecular Systematics. 205–247.

50. Posada D, Buckley TR (2004) Model Selection and Model Averaging in Phylogenetics: Advantages of Akaike Information Criterion and Bayesian Approaches Over Likelihood Ratio Tests. Systematic Biology 53: 793–808. doi:10.1080/10635150490522304.

51. Posada D, Crandall KA (1998) Modeltest: testing the model of DNA substitution. Bioinformatics 14: 817–818.

52. Guindon S, Dufayard JF, Lefort V, Anisimova M, Hordijk W, et al. (2010) New Algorithms and Methods to Estimate Maximum-Likelihood Phylogenies: Assessing the Performance of PhyML 3.0. Systematic Biology 59: 307–321. doi:10.1093/sysbio/syq010.

53. Ronquist F, Teslenko M, van der Mark P, Ayres DL, Darling A, et al. (2012) MrBayes 3.2: Efficient Bayesian Phylogenetic Inference and Model Choice Across a Large Model Space. Systematic Biology 61: 539–542. doi:10.1093/sysbio/sys029.

54. Aktipis SW, Giribet G (2010) A phylogeny of Vetigastropoda and other "archaeogastropods": re-organizing old gastropod clades. Invertebrate Biology 129: 220–240. doi:10.1111/j.1744-7410.2010.00198.x.

55. Drummond AJ, Suchard MA, Xie D, Rambaut A (2012) Bayesian Phylogenetics with BEAUti and the BEAST 1.7. Molecular Biology and Evolution 29: 1969–1973. doi:10.1093/molbev/mss075.

56. Nee S (2006) Birth-Death Models in Macroevolution. Annu Rev Ecol Evol Syst 37: 1–17. doi:10.1146/annurev.ecolsys.37.091305.110035.

57. Rambaut A, Drummond AJ (2007) Tracer 1.5. University of Edinburgh, Edinburgh, UK Available at: http://tree bio ed ac uk/software/tracer.

58. Rambaut A, Drummond AJ (2007) TreeAnnotator v1.7.5. University of Edinburgh, Edinburgh, UK Available at: http://tree bio ed ac uk/software/treeannotator.

59. Rambaut A (2012) FigTree v1.4. University of Edinburgh, Edinburgh, UK Available at: http://tree bio ed ac uk/software/figtree.

60. Aldous DJ (2001) Stochastic models and descriptive statistics for phylogenetic trees, from Yule to today. Statistical Science: 23–34.

61. Bortolussi N, Durand E, Blum M, Francois O (2006) apTreeshape: statistical analysis of phylogenetic tree shape. Bioinformatics 22: 363–364. doi:10.1093/bioinformatics/bti798.

62. Team RC (2013) R: A language and environment for statistical computing. Available: http://www.R-project.org.

63. Pybus OG, Harvey PH (2000) Testing macro-evolutionary models using incomplete molecular phylogenies. Proceedings of the Royal Society B: Biological Sciences 267: 2267–2272. doi:10.1098/rspb.2000.1278.

64. Paradis E, Claude J, Strimmer K (2004) APE: Analyses of Phylogenetics and Evolution in R language. Bioinformatics 20: 289–290. doi:10.1093/bioinformatics/btg412.

65. Harmon LJ, Weir JT, Brock CD, Glor RE, Challenger W (2007) GEIGER: investigating evolutionary radiations. Bioinformatics 24: 129–131. doi:10.1093/bioinformatics/btm538.

66. Nee S, Mooers AO, Harvey PH (1992) Tempo and mode of evolution revealed from molecular phylogenies. PNAS USA 89: 8322–8326.

67. Rabosky DL (2006) Likelihood methods for detecting temporal shifts in diversification rates. Evolution 60: 1152–1164.

68. Rabosky DL (2006) LASER: a maximum likelihood toolkit for detecting temporal shifts in diversification rates from molecular phylogenies. Evolutionary bioinformatics online: 247–250.

69. Rabosky DL, Lovette IJ (2009) Problems detecting density-dependent diversification on phylogenies: reply to Bokma. Proceedings of the Royal Society B: Biological Sciences 276: 995–997.

70. Stadler T (2011) Mammalian phylogeny reveals recent diversification rate shifts. PNAS 108: 6187–6192. doi:10.1073/pnas.1016876108.

71. Meredith RW, Janecka JE, Gatesy J, Ryder OA, Fisher CA, et al. (2011) Impacts of the Cretaceous Terrestrial Revolution and KPg extinction on mammal diversification. Science 334: 521–524. doi:10.1126/science.1211028.

72. Nagalingum NS, Marshall CR, Quental TB, Rai HS, Little DP, et al. (2011) Recent synchronous radiation of a living fossil. Science 334: 796–799. doi:10.1126/science.1209926.

73. Briggs JC (1999) Coincident biogeographic patterns: Indo-west Pacific ocean. Evolution: 326–335.

74. Palumbi SR (1994) Genetic Divergence, Reproductive Isolation, and Marine Speciation. Annu Rev Ecol Syst 25: 547–572.

75. Krug PJ (2011) Patterns of Speciation in Marine Gastropods: A Review of the Phylogenetic Evidence for Localized Radiations in the Sea. American Malacological Bulletin 29: 169–186. doi:10.4003/006.029.0210.

76. Woods AJC, Saul LR (1986) New Neritidae from southwestern north America. Journal of Paleontology: 636–655.

77. Potter PE, Szatmari P (2009) Global Miocene tectonics and the modern world. Earth-Science Reviews 96: 279–295.

78. Kennett JP, Houtz RE, Andrews PB, Edwards AR, Gostin VA, et al. (1975) 44. Cenozoic Paleoceanography in the Southwest Pacific Ocean, Antarctic Glaciation, and the Development of the Circum-Antarctic Current. Deep Sea Drilling Project Reports and Publications 29: 1155–1169.

79. Lawver LA, Gahagan LM (2003) Evolution of Cenozoic seaways in the circum-Antarctic region. Palaeogeography 198 : 11–37. doi:10.1016/S0031-0182(03)00392-4.

80. Huber M, Brinkhuis H, Stickley CE, Döös K, Sluijs A, et al. (2004) Eocene circulation of the Southern Ocean: Was Antarctica kept warm by subtropical waters? Paleoceanography 19: 12. doi:10.1029/2004PA001014.

81. Barker PF, Thomas E (2004) Origin, signature and palaeoclimatic influence of the Antarctic Circumpolar Current. Earth-Science Reviews 66: 143–162. doi:10.1016/j.earscirev.2003.10.003.

82. Haq BU, Hardenbol J, Vail PR (1987) Chronology of fluctuating sea levels since the Triassic. Science 235: 1156–1167.

83. Wilford GE, Brown PJ (1994) Maps of late Mesozoic-Cenozoic Gondwana break-up: some palaeogeographical implications. History of the Australian vegetation Cretaceous to recent. Cambridge University Press. 5–13.

84. Ivany LC, Patterson WP, Lohmann KC (2000) Cooler winters as a possible cause of mass extinctions at the Eocene/Oligocene boundary. Nature 407: 887–890.

85. Meyer CP, Geller JB, Paulay G (2005) Fine scale endemism on coral reefs: archipelagic differentiation in turbinid gastropods. Evolution 59: 113–125.

86. Claremont M, Williams ST, Barraclough TG, Reid DG (2011) The geographic scale of speciation in a marine snail with high dispersal potential. Journal of Biogeography 38: 1016–1032. doi:10.1111/j.1365-2699.2011.02482.x.

87. Duncan RA, Hargraves RB (1990) 40Ar/39Ar geochronology of basement rocks from the Mascarene Plateau, the Chagos Bank, and the Maldives Ridge. Proceedings of the Ocean Drilling Program 115: 43–52.

88. Duncan RA, Storey M (1992) The life cycle of Indian Ocean hotspots. Synthesis of Results from Scientific Drilling in the Indian Ocean 70: 91–103.

89. Ridgway T, Riginos C, Davis J, Hoegh-Guldberg O (2008) Genetic connectivity patterns of Pocillopora verrucosa in southern African Marine Protected Areas. Mar Ecol Prog Ser 354: 161–168. doi:10.3354/meps07245.

90. Hubert N, Paradis E, Bruggemann H, Planes S (2011) Community assembly and diversification in Indo-Pacific coral reef fishes. Ecology and Evolution 1: 229–277. doi:10.1002/ece3.19.

91. Pinceel J, Jordaens K, van Houtte N, De Winter AJ, Backeljau T (2004) Molecular and morphological data reveal cryptic taxonomic diversity in the terrestrial slug complex Arion subfuscus/fuscus (Mollusca, Pulmonata, Arionidae) in continental north-west Europe. Biological Journal of the Linnean Society 83: 23–38.

10

A New Barrier to Dispersal Trapped Old Genetic Clines That Escaped the Easter Microplate Tension Zone of the Pacific Vent Mussels

Sophie Plouviez[1,2,3]*, **Baptiste Faure**[1,2,4,5], **Dominique Le Guen**[1,2], **François H. Lallier**[1,2], **Nicolas Bierne**[4,5], **Didier Jollivet**[1,2]

1 Université Pierre et Marie Curie-Paris 6, Laboratoire Adaptation et Diversité en Milieu Marin, Station Biologique de Roscoff, Roscoff, France, 2 CNRS UMR 7144, Station Biologique de Roscoff, Roscoff, France, 3 Division of Marine Science and Conservation, Nicholas School of the Environment, Duke University, Beaufort, North Carolina, United States of America, 4 Université Montpellier 2, Montpellier, France, 5 CNRS UMR 5554, Institut des Sciences de l'Evolution, Station Méditerranéenne de l'Environnement Littoral, Sète, France

Abstract

Comparative phylogeography of deep-sea hydrothermal vent species has uncovered several genetic breaks between populations inhabiting northern and southern latitudes of the East Pacific Rise. However, the geographic width and position of genetic clines are variable among species. In this report, we further characterize the position and strength of barriers to gene flow between populations of the deep-sea vent mussel *Bathymodiolus thermophilus*. Eight allozyme loci and DNA sequences of four nuclear genes were added to previously published sequences of the cytochrome *c* oxidase subunit I gene. Our data confirm the presence of two barriers to gene flow, one located at the Easter Microplate (between 21°33'S and 31°S) recently described as a hybrid zone, and the second positioned between 7°25'S and 14°S with each affecting different loci. Coalescence analysis indicates a single vicariant event at the origin of divergence between clades for all nuclear loci, although the clines are now spatially discordant. We thus hypothesize that the Easter Microplate barrier has recently been relaxed after a long period of isolation and that some genetic clines have escaped the barrier and moved northward where they have subsequently been trapped by a reinforcing barrier to gene flow between 7°25'S and 14°S.

Editor: Donald James Colgan, Australian Museum, Australia

Funding: This work was supported by the GDR Ecchis and the ANR-06-BDV-005 (Deep Oases: coord. D. Desbruyès). The 'Bivalvomix'GIS programme (coord. N. Bierne) partly supported the sequencing costs of the B. azoricus cDNA library construction together with the NoE "Marine Genomics Europe". S. Plouviez was supported by a PhD grant from the Université Pierre et Marie Curie and the National Science Foundation under Grant Number OCE-1031050 (coord. C. Van Dover). B. Faure was supported by the NoE "Marine Genomics Europe" (Fish & Shellfish node). The funders had no role in study design, data collection and analysis, decision to publish, or preparation of the manuscript.

Competing Interests: The authors have declared that no competing interests exist.

* E-mail: sophie.plouviez@duke.edu

Introduction

Genetic structure is easier to detect and understand in a one-dimensional system than in a two-dimension space [1]. However, even in a one-dimension space, detecting a genetic cline with a correlation between genetic differentiation and geographical distance does not always mean that populations are following an isolation-by-distance (IBD) model. Such a correlation can also be due to the presence of barriers to dispersal (e.g., [2]) or secondary contacts between previously isolated populations under expansion (e.g., [3]).

When detecting a genetic cline, one should consider the possibility that the location of this cline may be due to the presence of a natural barrier to dispersal because clines are expected to be trapped by such a barrier [4], [5]. Genetic clines depend on the relative impact of dispersal, selection (e.g., [6], [7]) and the recent demographic history of populations (e.g., [8]), and a natural barrier to dispersal impacts the balance among these parameters. As a result, clines often typify adaptive gradients [9], [10] or hybrid zones (i.e., regions containing recombinant individuals between genetically differentiated populations).

The geographic region in which a balance between dispersal and selection is maintained is defined as a tension zone. Tension zones tend to stabilize over natural barriers to dispersal [5], [7], [11], maintaining a cline around that barrier. They can also couple with a local adaptation cline and be stabilized at an environmental boundary [12]. If the origin of stabilized clines is difficult to establish from a single gene (e.g., isolation-by-distance without natural barriers to dispersal, hybrid zone), comparing allele frequencies between genes and gene divergences can help explain its emergence and, if a barrier is present, to identify its position more precisely.

Deep-sea hydrothermal vents are patchily distributed along mid-ocean ridges and back-arc basins. Along the East Pacific Rise (EPR), they follow a one-dimension pattern, ideal for testing an isolation-by-distance model of populations [13]. However, vent displacements along the ridge and eruptive phases leading to local faunal extinctions together with transform faults which likely impede gene flow are able to seriously alter expectations of such population models [14].

Comparative phylogeographic analyses of deep-sea hydrothermal vent species have previously shown the presence of a genetic break between the northern and southern regions of the EPR [15], [16], [17]. These studies established a shared vicariant event among species and suggested the emergence of a barrier to dispersal near the equator about 1.5 to 2 Mya [18]. However, the width and position of the barrier was not matching among species: some such as the tube-dwelling polychaete *Alvinella pompejana* displayed an abrupt separation of the populations across the equator [3] whereas others showed genetic patterns closer to an isolation-by-distance model [15]. Interestingly, the deep-sea mussel *Bathymodiolus thermophilus* exhibited a smooth clinal distribution of mitochondrial lineages along the EPR (13°N to 21°33′S) [15], [17] and the presence of a cryptic species, *B.* aff. *thermophilus* further south (31°S–32°S) [15].

Bathymodiolus species have been supposed to be long-distance dispersers because of their planktotrophic mode of larval development [19] and the possibility of larvae reaching the upper (photic) layers of the water column [20]. The wide-dispersal capabilities of *Bathymodiolus* have been confirmed in several population genetic studies showing the absence of genetic differentiation across the Atlantic [21] and across the Gulf of Mexico (Mississippi Canyon and Alaminos Canyon, 550 km apart) [22]. Such dispersal characteristics could strongly favour population connectivity among geographically isolated sites. However, hybrid zones resulting in an abrupt change of allele frequencies over relatively short distances have also been observed in both the Atlantic [23] and the Pacific [24].

The likely cause of the clines observed at the EPR has been a source of debate. Although Plouviez *et al.* [17] suggested the observed genetic shifts might be the consequence of a natural barrier to dispersal that separated two interacting *Bathymodiolus* units, Audzijonyte & Vrijenhoek [25] proposed that the differentiation observed could simply be obtained under IBD. However, Johnson *et al.* [24] recently refuted IBD by describing a tension zone localised at the Easter Microplate between *B. thermophilus* and *B.* aff. *thermophilus* (renamed *B. antarcticus* by the authors). One of the loci, S-Adenosyl Homocysteine Hydrolase (SAHH), displayed a discordant cline position with fixed substitutions around the geographic zone previously identified by Plouviez *et al.* [17] to be a barrier to gene flow. The small likelihood of discovering fixed alleles over such a small spatial scale under IBD, and the fact that mitochondrial DNA exhibited an abrupt shift in allele frequency at the exact same position [17] prompted a reinvestigation of the hypothesis of a second barrier to gene flow.

In the present study, we obtained allozyme and DNA sequence datasets for four nuclear genes including the SAHH marker and analysed this new dataset together with mitochondrial haplotypes previously obtained from the EPR mussels to further investigate how genetic diversity is structured across the EPR and more specifically to test for the homogeneity of gene divergences across identified barriers. We obtained evidence for the existence of two barriers, including the one positioned at 7°25′S–14°S. We thus propose that a pair of (semi-)permeable barriers along the EPR is likely responsible for the geographically discordant clinal distribution of alleles in this region.

Materials and Methods

Ethics statement

No specific permits were required to perform field studies described in this article. No specific permissions were required to access geographic localities and sample specimens (sampling sites belong to international waters). The locations are not privately owned or protected in any way. The field studies did not involve endangered or protected species.

Collection

Bathymodiolus thermophilus specimens were sampled from seven deep-sea hydrothermal vent fields along the East Pacific Rise (EPR) from 9°50′N to 21°33′S (Table 1) using the tele-manipulated arm of the manned submersible Nautile operated from the oceanographic vessels Le Nadir and L'Atalante during three oceanographic cruises: at 9°50′N during HOT 1996 and Mescal 2010 and from latitudes 7°25′S to 21°33′S during BIOSPEEDO 2004. During the three cruises, all fresh specimens were measured and dissected on board and tissues (mantle and muscle) were preserved in 80% alcohol. In addition, the anterior muscle of each individual was also frozen in liquid nitrogen for allozyme analyses during BIOSPEEDO 2004 and Mescal 2010.

Allozyme genotyping

Eight enzyme loci were genotyped for each individual of *B. thermophilus* collected from 7°25′S to 21°33′S (Table 1) following the protocols of Boutet *et al* [26]: Phosphoglucomutase (*Pgm*, E.C. 5.4.2.2), Mannose phosphate isomerase (*Mpi*, E.C. 5.3.1.8), Octopine deshydrogenase (*Odh*, 1.5.1.11), Leucine amino peptidase *(Lap*, 3.4.11.1), Glucose phosphate isomerase (*Gpi*, 5.3.1.9), Malate deshydrogenase-1 and -2 (*Mdh 1* and *2*, 1.1.1.37), and Hexokinase-1 (*Hk*, 2.7.1.1). Alleles were numbered according to their relative mobility from the most frequent allele (labelled as 100) previously determined for the Atlantic species *B. azoricus*, this species was used as a reference (see [26]).

The program Genetix 4.05.2 [27] was used to perform population genetic analyses on allozyme data. For each locus, allele frequencies, heterozygosities and Weir & Cockerham [28] f statistic (departure from Hardy-Weinberg equilibrium tested by a 1000-permutations test) were estimated for each population along the EPR. The overall genetic differentiation across populations was estimated at all loci using Weir & Cockerham's θ estimator [28] and 1000 permutations used to determine significance. The isolation-by-distance model was tested with a Mantel Spearman test with 5000 permutations using Genepop 4.0.10 [29].

DNA sequencing

Genomic DNA was extracted using a CTAB-PVP extraction procedure following Jolly *et al.* [30]. Mitochondrial lineages of *B. thermophilus* were identified from cytochrome oxidase I gene (mtCOI) sequences previously obtained by Plouviez *et al.* ([17], Table 1). Sequences from three nuclear genes (GenBank accession numbers KC858658- KC858846, see Table S1) were obtained from the same individuals using the mark-recapture (MR) cloning technique developed by Bierne *et al.* [31] with two times the capture effort. MR-cloning was chosen over direct sequencing of PCR products because of the presence of insertions/deletions in intronic regions and to have access to linkage disequilibrium among polymorphic sites without the use of computer algorithms to determine allelic phase. Primers developed by Faure *et al.* [23] were used for amplification of introns in the S-Adenosyl Homocysteine Hydrolase (SAHH) and Lysozyme (Lyso) genes, previously obtained from EST sequences from a *Bathymodiolus azoricus* cDNA library. The SAHH gene contains a poly-A tract varying in length among individuals, the length polymorphism in this region was not included in the analyses due to the high potential error rate resulting from PCR and cloning. Specific primers for the Sulfotransferase (Sulfo) gene were also designed from the cDNA sequence data (BtSulfo-F: 5′-TCTTTAAAGT-CAGGATCACATTGG-3′, BtSulfo-R: 5′-TAAGGCAAAGTG-GAACAACGAGACCGC-3′). Sequences from Sulfotransferase were sorted in two paralogous genes called Sulfo1 and Sulfo2 from individual allele recaptures, respectively. Because of the low

Table 1. Location and sample size of populations sampled for allozymes (n_{allo}), nuclear genes (n_{SAHH}, n_{Lyso}, n_{Sulfo1}, $n_{EF1\alpha}$) and mitochondrial (n_{mtCOI}) gene.

Locality	Site name	Latitude Longitude	Depth	n_{allo}	n_{Sulfo1}	n_{Lyso}	n_{SAHH}	$n_{EF1\alpha}$	n_{mtCOI}*
GR									
1°N	Mussel Bed	0°48′ N 86°09′ W	2486	-	-	-	-	-	12
1°N	Rose Garden	0°48′ N 86°14′ W	2460	-	-	-	-	-	12
EPR									
13°N	-	12°48′ N 103°56′ W	2630	-	-	-	-	-	12
11°N	-	11°25′ N 103°47′ W	2515	-	-	-	-	-	12
9°50′N [‡]	East Wall	9°50′ N 104°17′ W	2530	66	9	11	6	4	45
7°25′S [‡]	Last Hope	7°25′ S 107°47′ W	2735	9	16	19	25	25	48
11°S	-	11°18′ S 110°32′ W	2669	-	-	-	-	-	12
14°S [‡]	Lucky Eric	13°59′ S 112°29′ W	2623	38	12	14	12	-	30
17°25′S [‡]	Oasis	17°25′ S 113°12′ W	2575	69	12	10	11	-	21
17°35′S [‡]	Ms Wormwood	17°35′ S 113°15′ W	2595	23	12	15	10	-	60
18°33′S [‡]	Animal Farm	18°33′ S 113°24′ W	2636	33	12	12	11	-	30
21°33′S [‡]	Gromit	21°33′ S 114°18′ W	2800	108	8	14	19	14	27
PAR									
31°S	-	31°09′ S 111°55′ W	2332	-	-	-	-	-	12
32°S	-	31°51′ S 112°02′ W	2331	-	-	-	-	-	12
38°S	Foundation hotspot	37°70′ S 110°87′ W	2200	-	2	2	2	2	2

GR, Galapagos Rift; EPR, East Pacific Rise; PAR, Pacific Antarctic Rise. Depth is given meters. Sulfo 1, Sulfotransferase paralogue 1; Lyso, Lysozyme; SAHH, S-Adenosyl Homocysteine Hydrolase; EF1α, Elongation Factor 1α; mtCOI, cytochrome oxidase I. n_{SAHH}, n_{Lyso}, n_{Sulfo1}, $n_{EF1\alpha}$, total number of recaptured individuals for each nuclear gene. *, sequences from [15], [17]. [‡], populations used for the Monmonier analysis.

number of sequences recaptured from Sulfo2, only Sulfo1 sequences were used in this study.

Faure et al. [23] used a similar MR-cloning approach on two individuals from 37°70′S for the three nuclear genes studied in the present paper. Faure et al. [32] also used a MR-cloning approach for a fourth nuclear gene (Elongation Factor 1α, EF1α), for which four of our populations were sampled and already sequenced prior to this population analysis. This gene was thus included in our analyses. In Johnson et al. [24], a study done in parallel to ours, the SAHH and EF1α genes were also used but alleles were obtained by direct sequencing. Because of the presence of insertion/deletion in the intronic region, Johnson et al. [24] were able to analyze only

a quarter of the sequence length we obtained by MR-cloning. Consequently, this did not allow us to include these new datasets into our SAHH and EF1α analyses.

DNA sequences obtained from the MR-cloning method were visualized and edited using CodonCode Aligner 2.0.6 (http://www.codoncode.com/aligner/). Sequence alignments were initially performed with ClustalW [33] and improved manually. The number of individuals for which at least one of the two alleles was recaptured (number of recaptured individuals) varied from one population to another is indicated in Table 1. Because of the random nature of the recapture, it was not possible to distinguish 'true' homozygotes from heterozygotes with the recapture effort.

Consequently, and because our main results is based on the coalescence theory, only the most recaptured allele was retained from each individual to avoid sample bias when performing demographic analyses and genetic diversity estimations (see: Table 1). Multiple recaptures allowed us to discard intra-individual *in vitro* recombinants and putative artefactual/somatic mutations. Recombinants between different individuals (1–2% of the dataset for each population) from the same PCR set were detected (and removed) based on abnormal combinations of the 5′-tails.

Polymorphism and divergence from DNA sequences

For each locus and locality, nucleotide diversity (π_n) and Watterson's theta (θ_w) were estimated using DnaSP 4.10.3 [34]. Phylogenetic relationships among alleles were estimated using the median joining algorithm of the Network software (version 4.5.0.0; www.fluxus-engineering.com) [35] to detect potentially divergent clades. The geographic distribution of divergent clades was then examined to locate potential barriers to gene flow by plotting synthetic clade-specific allele frequency distributions for each locality. To test for gene divergence homogeneity across barriers, divergence time between clades was estimated using the formula $T = D/2r$ under the assumption of a local molecular clock (tested using the BEAUti/BEAST 1.4.8 package [36] with parameters previously described in [17]), where D is the average net divergence between geographic clades and r the mutation rate per site per million years [37]. The initial separation between *B. thermophilus* and *B. azoricus* (across the Isthmus of Panama, set to about 8–12 Mya for deep-sea fauna; [38]) was used as calibration point using a published dataset [23]. This calibration of deep-sea fauna separation across the Isthmus of Panama has been used successfully in other species [39].

Neutrality of loci was also tested using a multi-loci Hudson-Kreitman-Aguadé test (HKA, [40]) performed separately for each divergent clade, with one sequence of *B. azoricus* as an outgroup, via the software HKA (J. Hey's web page: http://lifesci.rutgers.edu/~heylab/HeylabSoftware.htm#HKA). This test compares polymorphism and divergence at several loci to detect if at least one of these loci displays a departure from neutral evolution. This test was preferred to the McDonald-Kreitman [41] test because of the absence of fixed non-synonymous mutations between the two deep-sea mussel species.

Statistical evidence for a barrier to gene flow at 7°25′S-14°S

A Monmonier algorithm was implemented using Barrier 2.2 [42] that compares matrices of multigene genetic distances (F-statistics, ϕ_{st}, estimated with DnaSP 4.10.3 [34]) and geographic distances under the assumption of gene flow breaks. This Bayesian program was used to determine the geographic position of a potential barrier along the East Pacific Rise (insertions/deletions were coded as presence/absence). As the software Barrier 2.2 does not hold missing ϕ_{st} values in the matrix, the EF1α gene and some of the populations for which, at least one gene was missing were discarded from the analysis (see Table 1). The 38°S population was not included because of its small sample size. Localities from each side of the 7°25′S–14°S barrier were grouped together to test for genetic differentiation across this barrier using ϕ_{st} [43] computed using DnaSP 4.10.3 (1000-permutations [34]).

Migration rates and demographic history of populations

Migration rates across the 7°25′S–14°S barrier and the effective size of the southern and northern populations were estimated by fitting an isolation with migration model (IMa2 program, [44])

using previously described parameters [3], with the following exceptions: upper bounds of uniform priors set at $\theta = 50$ (population size), $m = 5$ (migration rate) and $t = 30$ (divergence time). Demographic and migration parameters were calibrated using divergence across Panama to inform mutation rates of loci (geometric mean among loci following a strict molecular clock) and a generation time of 2 years as previously estimated by Faure *et al.* [23].

Hybrid/introgressed individuals detected using SAHH RFLP analysis

A *Hinf I* restriction site polymorphism fixed between the two main sets of SAHH alleles was identified and used to check for the occurrence of putative introgressed/hybrid individuals between mitochondrial lineages by looking for individuals that possess one SAHH allele from each of the two distinct clades (called N/S individuals). An RFLP analysis was then performed using the *Hinf I* site to detect N/S individuals by screening all individuals from the *Bathymodiolus* collection. Incubation of PCR-products was done at 37°C for 1.5 hour in a 20 µl total volume containing 17 µl of PCR product, 1X buffer (supplied by the manufacturer) and 10 U of *Hinf I* (Ozyme™).

Results

Geographic distribution of allozymes

A series of differentiation tests were performed on *B. thermophilus* populations located along the EPR. Among the eight allozyme loci, four (*Mpi*, *Odh*, *Mdh 2* and *Hk*) were nearly monomorphic with the most frequent allele occurring at a frequency greater than 95% in all populations. The remaining four loci (*Pgm*, *Lap*, *Gpi* and *Mdh 1*) exhibited enough polymorphism to investigate their allelic distribution over the range of the EPR. Genetic differentiation was very low and not statistically significant overall ($\theta = 0.016$, P value > 0.05), showing the absence of any strong genetic structure at allozyme loci along this portion of the EPR. The Mantel Spearman test showed a slight but significant correlation (P value $= 0.05$) between the genetic distance ($\theta/(1-\theta)$) and the geographic distance between vent fields (Fig. 1).

Polymorphism and divergence from DNA sequences

Population structure across the Easter Microplate was found for two genes (Fig. 2, 3), mtCOI (4.4% divergence) and EF1α (1.0% divergence), with two divergent clades corresponding to EPR and

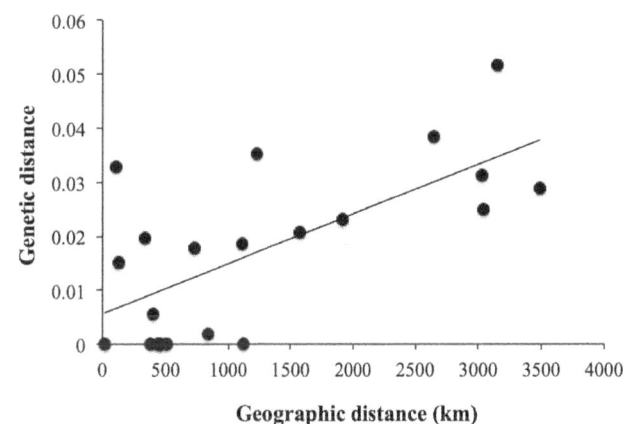

Figure 1. Relationship between pairwise genetic distances (θ/(1-θ)) from allozymes and geographic distances (in kilometres).

Pacific-Antarctic Ridge (PAR, i.e., 31°S-38°S) populations, respectively. None of the three other nuclear genes (i.e., SAHH, Lyso, Sulfo1) were divergent across the Easter Microplate (Fig. 2, 3). Conversely, networks and geographic distribution of alleles along the ridge (Fig. 2, 3) revealed a pronounced geographic differentiation along the EPR for these three nuclear genes between the 9°50'N-7°25'S and 14°S-21°33'S regions in concordance with mtCOI, but not EF1α.

Nuclear gene networks exhibited two clades separated by a pronounced net divergence (Fig. 2). The SAHH gene revealed the presence of two clades (2% divergence) well established from each part of the previously suggested 7°25'S-14°S break (Fig. 2, Fig. 3, [17]). SAHH clade 1 contains the sequences from most individuals sampled at 9°50'N or 7°25'S whereas clade 2 corresponds to individuals sampled only from the southern EPR sites. Two 0.3%-divergent clades were also found using the Lyso gene but are more difficult to attribute to a particular geographic area (Fig. 2, 3). However, a 1-bp deletion was only found in the intron of sequences sampled in the southern EPR sites below the latitude of 7°25'S (Fig. 3), indicating that the 7°25'S-14°S barrier is playing a role in impeding the spread of new mutations. The Sulfo1 gene also displayed a clear geographic structure between north and south of the barrier with a 0.5% divergence between the two clades (Fig. 2).

Statistical evidence for a 7°25'S–14°S barrier

The Monmonier analysis identified a barrier between 7°25'S and 14°S for all tested loci (Fig. 3) and ϕ_{st} values were significantly different from zero for all sampled genes across this barrier (Table 2). An Isolation with Migration (IMa2) analysis was performed between the southern and northern populations across this barrier. The marginal posterior probability distribution of migration rate across the barrier overlapped with zero, indicating that the absence of migration cannot be ruled out. If present, migration across this barrier could have occurred in both directions, possibly slightly orientated from north to south (Table 3). The estimated effective population sizes of the present populations were both greater than the ancestral (N_A) population size (Table 3), indicating that present-day populations of the vent mussel may be expanding, possibly at a higher rate in the south (N_N being slightly higher than N_S). However, expansion cannot be confirmed because of the overlap range of Highest Posterior Density among north, south and ancestral populations.

Gene divergence homogeneity across the barrier

Under the verified assumption of a molecular clock (BEAST analysis) and using a 8–12 Mya calibration time obtained from the splitting of B. thermophilus and B. azoricus across the Isthmus of Panama, divergence times between clades 1 and 2 were estimated at 0.6–0.9 Mya for mtCOI, conforming to geological estimates of the ages of the transform faults in the 7°25'S-14°S area (Fig. 4). In contrast, divergence times between mtCOI clades 2 and 3 (3.0–4.5 Mya) as well as between Sulfo1 (4.4–6.6 Mya) and between SAHH (4.0–6.1 Mya) clades conformed to geological estimates for Easter Microplate formation (Fig. 4). Divergence times between clades for the Lyso (0.4–0.6 Mya) and EF1α (3.1–4.7 Mya) genes have to be interpreted with caution because of mutation rate heterogeneity among clades (departure from a strict molecular clock).

The multi-locus HKA test showed a significant departure from neutral evolution (P < 0.02) among genes within clade 1 (no departure among genes within clade 2), indicating that at least one gene could be under selection. Sulfo1 displayed the highest divergence to polymorphism deviation when compared to other genes. The removal of Sulfo1 (possible outlier) from the dataset led the HKA test to refit the model of neutral evolution (P > 0.31). This gene displayed a polymorphism/divergence ratio of 8.66: a value that is more than eight times greater than expected (1.05) with a very low divergence between B. thermophilus and B. azoricus, indicative of balancing or strong purifying selection.

Detection of introgressed individuals using the SAHH gene

The RFLP analysis of the SAHH marker revealed that only 13 individuals south of the 7°25'S–14°S barrier displayed at least one nuclear allele corresponding to the northern clade. Except for one individual showing two northern-type alleles in the southern populations, all of these individuals possessed one allele of each clade. These N/S individuals were mainly located at 14°S and decreased abruptly in frequency further south (Fig. 5). No N/S individual was detected at 7°25'S or further north. N/S individuals of SAHH had Sulfo1 alleles from clade 2 only but they were not associated with a specific mtCOI clade: the SAHH northern-type allele being found in association with haplotypes from both the mtCOI clades 1 and 2, indicating that these individuals are most probably introgressed rather than first generation hybrids which should result in cyto-nuclear disequilibrium [45], [46].

Discussion

The unpredictable nature of fluid circulation and the high level of fragmentation found at deep-sea hydrothermal vents should result in species with high dispersal capabilities in order to promote long-distance (re)colonization of new vent sites [47]. In accordance with this expectation, most species from the genus Bathymodiolus have high dispersal capabilities [20], [21], [22]. Data from B. thermophilus supports this hypothesis in showing a complete lack of differentiation at allozyme markers between the northern EPR (13°N, 11°N, 9°N) and the Galapagos sites [48] (but see [49]). Having high dispersal capabilities, however, does not necessarily imply a lack of geographic structure because gene flow is likely impacted by physical barriers to dispersal, great distances without suitable habitat and/or hybridization fronts [23].

Previous and present works strongly support the hypothesis that the population structure of Bathymodiolus thermophilus is impacted by two barriers to gene flow along the EPR: the so-called Easter Microplate barrier ([15], [24], present manuscript) and the 7°25'S–14°S barrier. This 7°25'S–14°S barrier was first suggested by Plouviez et al. [17], but then challenged by Audzijonyte & Vrijenhoek [25] who proposed that frequency changes in the two most divergent mtCOI clades found in the mussel populations along the EPR were most likely due to isolation-by-distance (Mantel Spearman test performed on the Won et al.'s dataset found significant [15]) and not an actual barrier. Johnson et al.'s study [24], together with our present work, refutes the hypothesis that isolation-by-distance alone might be responsible for the genetic patterns observed along the EPR, instead these studies suggest that a tension zone across the Easter Microplate might explain the clinal distribution of alleles. Johnson et al. [24] proposed the occurrence of a hybrid zone at 23°S and confirmed that alleles from the Pacific-Antarctic mussel B. antarcticus were likely to introgress in the South EPR populations of B. thermophilus but did not discuss the possible presence of a barrier to gene flow further north (7°25'–14°S).

In the present manuscript, we confirm the existence of such a barrier for the deep-sea mussel and argue that the observed positive correlation of ϕ_{st} with geographic distances at the mtCOI locus (significant Mantel Spearman tests) likely reflects the

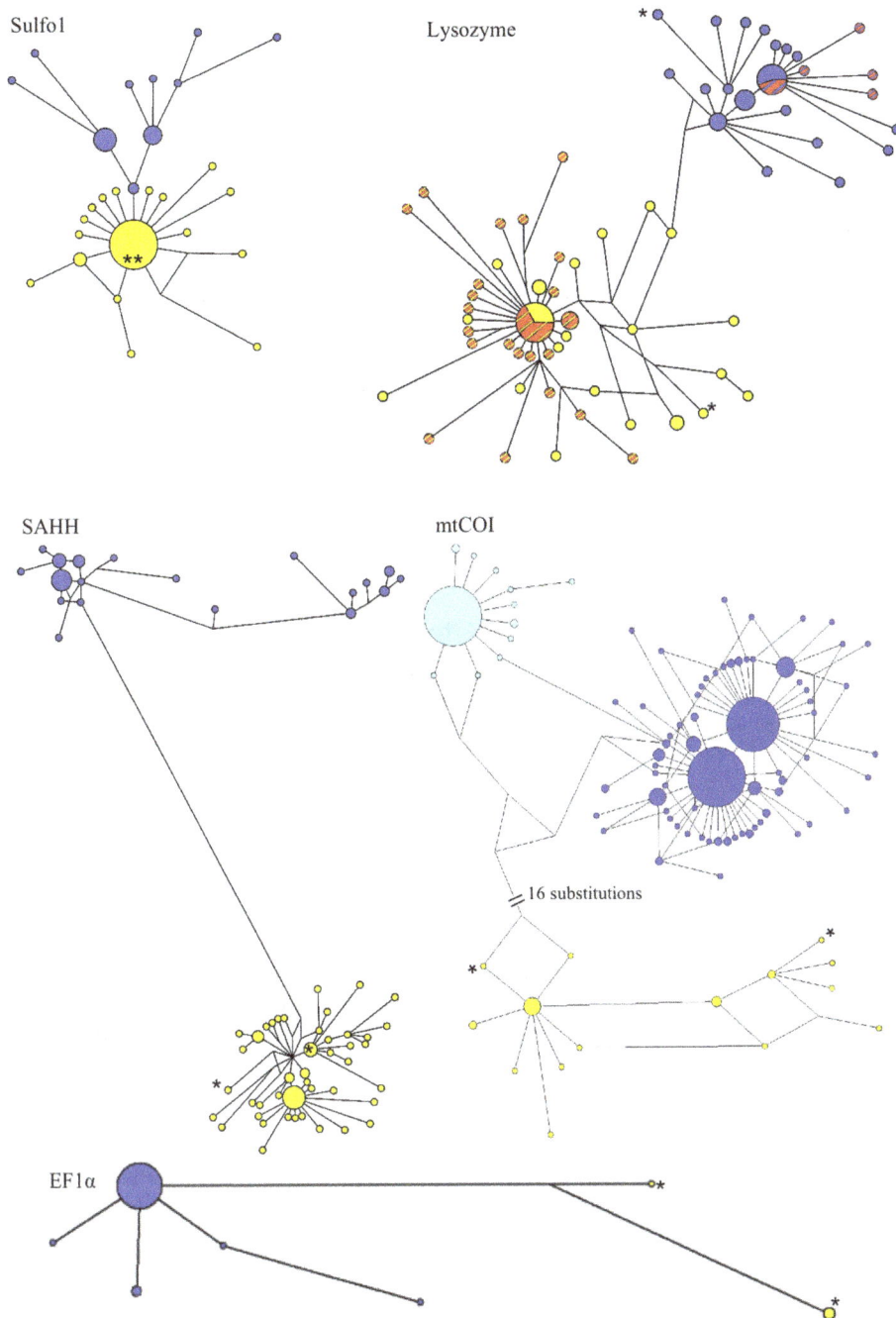

Figure 2. Median Joining Networks on the three nuclear genes and the mitochondrial cytochrome oxidase I gene. For each gene, the sizes of haplotype/allele circles and lengths of connecting lines are proportional to the number of individuals and the number of mutations that separate two linked haplotypes/alleles, respectively (length is not reflected in the 16-substitution link indicated on the mtCOI network). Colours represent divergent clades used for mapping the geographic distribution of alleles in Figure 3. For the nuclear genes, dark blue circles correspond to clades 1 and yellow circles to clades 2 in the manuscript. For mtCOI, light blue circles = clade 1, dark blue circles = clade 2, yellow circles = clade 3. For the Lysozyme gene, position of the 1-bp deletion in the network is represented by red stripes within both the yellow and the blue circles. *, position of individuals from 38°S in the network. Sulfo 1, Sulfotransferase paralogue 1; Lyso, Lysozyme; SAHH, S-Adenosyl Homocysteine Hydrolase; EF1α, Elongation Factor 1α; mtCOI, cytochrome oxidase I.

occurrence of a tension zone that originated at the Easter Microplate (as proposed by Johnson *et al.* [24]), but which subsequently moved northward and became captured by a second area of restricted larval exchanges (responsible for the cline), the 7°25′S–14°S barrier to gene flow. As for many genetic barriers it is semipermeable and affects loci differentially.

Genetic differentiation of B. thermophilus is triggered by the recent formation of a series of transform faults at 7°25′S–14°S

The Monmonier analysis and coalescence analyses statistically detected the presence of a barrier to gene flow at 7°25′S–14°S using nuclear and COI genes. Sulfo1 and SAHH loci both showed

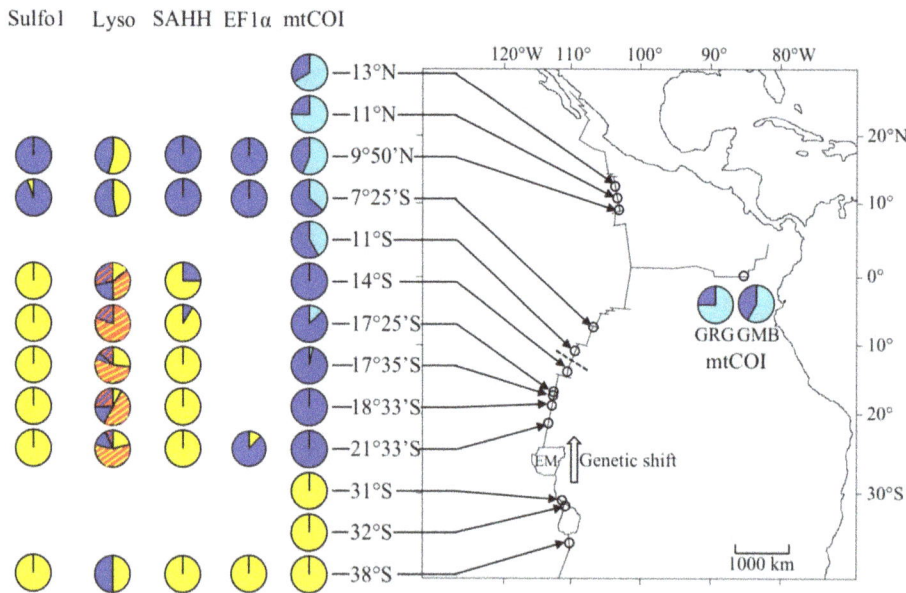

Figure 3. Geographic distribution of divergent alleles for three nuclear genes and the mitochondrial cytochrome oxidase I gene.
Colours match divergent clades identified in Figure 2. Stripes indicate the presence of the "deletion" in the Lysozyme gene on either the yellow clade (red stripes on yellow background) or the blue clade (red stripes on blue background). The dashed line depicts the recent barrier to gene flow identified by the Monmonier analysis. The white-block arrow represents the hypothesized northward genetic shift of the tension zone for some genes. Sulfo 1, Sulfotransferase paralogue 1; Lyso, Lysozyme; SAHH, S-Adenosyl Homocysteine Hydrolase; EF1α, Elongation Factor 1α; mtCOI, cytochrome oxidase I; EM, Easter Microplate; GRG, Galapagos Rose Garden; GMB, Galapagos Mussel Bed.

two sets of divergent alleles distributed in the 9°50′N–7°25′S and 14°S–21°33′S regions, with an almost reciprocal monophyly. The gene encoding Lyso exhibited two sets of divergent alleles that were geographically interspersed over the whole EPR and PAR, indicating that they were able to cross both the Easter Microplate and the 7°25′S–14°S barriers. However, the 1-bp deletion allele found at a high frequency in populations from 14°S-21°33′S was not observed from 9°50′S to 7°25′S nor within individuals from 38°S. The sample size (2 individuals) is far too low to exclude the absence of this deletion at 38°S. The absence of the deletion in the 9°50′N and 7°25′S individuals, however, strongly suggests that the spread of this new deletion has been blocked since its first occurrence in the southern populations (at least across the 7°25′S–14°S barrier). This deletion is however old enough to provide recombinant alleles at a high frequency between the two Lyso divergent clades, as the deletion is present at the tip of both divergent clades (see Fig. 2). The geographic distribution of this deletion on the Lyso gene suggests that this newly-formed barrier is now impermeable or only sporadically permeable to gene flow. The multigenic estimation of migration rates using a Bayesian approach and the mitochondrial and nuclear sequences (IMa2) also fit the relative isolation of mussels located at 7°25′S or 9°50′N when compared to the more southern populations. The IMa2 analysis indeed showed the absence or the very limited number of mussel migrants across the 7°25′S–14°S barrier. Although sequences of *B. thermophilus* from 11°S were not available for the other genes, we hypothesize this barrier is located, more precisely, between 11°S and 14°S based on haplotype frequencies of the mtCOI gene (ϕ_{st}).

The barrier was however not sufficient to create an allele frequency shift at allozyme markers, highlighting an apparent discrepancy among marker types (allozymes vs. DNA sequences). Discrepancies in the genetic differentiation observed with allozyme and DNA markers have regularly been suggested to result from selection at allozyme loci, sometimes claimed to be under balancing selection [50], [51] and sometimes under disruptive selection [52], [53]. However, subsequent analyses often refute the hypothesis of selection on allozymes by demonstrating insufficient sampling [54], [55]. Here, EF1α is sufficient to show that a low level of differentiation is observed at a DNA marker in the area sampled for allozyme analysis. Furthermore, Johnson *et al.* [24] did not observe any genetic differentiation north of the Easter Microplate at the three DNA markers that are not in common between the two surveys. We conclude that there is no real discrepancy between the two categories of markers and that the few allozyme loci analyzed simply fell in genomic regions unlinked to isolation genes. These loci are thus able to cross the barrier freely and to organize themselves according to geographic distance, producing an IBD pattern. Moreover, the barrier may not be detectable on some of the allozyme loci (*Mpi, Odh, Mdh 2* and *Hk*) because of their low level of polymorphism.

Geologic and hydrodynamic conditions from the equator to 15°S are consistent with limited gene flow observed for multiple species between north and south EPR [3], [15], [16], [17]. In terms of geology, dispersal in this region can be impeded by the equatorial triple junction between the EPR and the Galapagos Rift, as well as the Gofar/Discovery multiple transform fault complex near 4°S [56], [57], [58]. When looking at local hydrodynamism in the region, surveys of He-3 plumes produced by the venting activity along the EPR also indicated the occurrence of a westward flow centered at 15°S [59], [60]. This flow, possibly linked to the anticyclonic circulation of water masses in the eastern Pacific [61], creates strong cross-axis currents able to produce a hydrodynamic barrier at these latitudes. The co-occurrence of these geologic and hydrodynamic features represents ideal conditions for the establishment of new barriers to gene flow.

Table 2. Summary statistics of nucleotide polymorphism according to locality for nuclear genes and mitochondrial gene.

Locus	Locality	h	S	$\theta_W \times 100$	$\pi_n \times 100$	ϕ_{st}
Sulfo1						0.474 ***
	9°50′N	4	3	0.511	0.617	
	7°25′S	9	12	1.674	1.196	
	14°S	3	4	0.613	0.309	
	17°25′S	5	6	0.920	0.463	
	17°35′S	5	5	0.767	0.386	
	18°33′S	5	4	0.613	0.379	
	21°33′S	4	4	0.714	0.546	
	38°S	1	-	-	-	
Lyso						0.108 ***
	9°50′N	11	24	0.665	0.692	
	7°25′S	18	23	0.534	0.592	
	14°S	10	30	0.766	0.566	
	17°25′S	9	29	0.832	0.568	
	17°35′S	11	25	0.625	0.415	
	18°33′S	10	26	0.699	0.579	
	21°33′S	13	34	0.868	0.542	
	38°S	2	7	-	-	
SAHH						0.717***
	9°50′N	6	18	0.918	0.893	
	7°25′S	6	16	0.494	0.687	
	14°S	10	34	1.479	1.140	
	17°25′S	9	41	1.649	0.955	
	17°35′S	8	11	0.439	0.341	
	18°33′S	8	8	0.344	0.270	
	21°33′S	17	36	1.270	0.594	
	38°S	2	6	-	-	
EF1α						0.033[NS]
	9°50′N	1	0	-	-	
	7°25′S	3	2	0.101	0.030	
	21°33′S	6	12	0.699	0.395	
	38°S	1	-	-	-	

Locus	Locality	h	S	$\theta_W \times 100$	$\pi_n \times 100$	ϕ_{st}
mtCOI						0.280***
	GMB	7	11	0.700	0.749	
	GRG	6	11	0.700	0.612	
	13°N	4	8	0.509	0.624	
	11°N	6	11	0.700	0.612	
	9°50′N	16	18	0.751	0.687	
	7°25′S	21	22	0.907	0.715	
	11°S	6	8	0.509	0.682	
	14°S	14	11	0.534	0.294	
	17°25′S	8	12	0.553	0.366	
	17°35′S	24	27	1.113	0.345	
	18°33′S	14	13	0.631	0.286	
	21°33′S	12	10	0.499	0.266	
	31°S	8	9	0.573	0.533	

Table 2. Cont.

Locus	Locality	h	S	$\theta_W \times 100$	$\pi_n \times 100$	ϕ_{st}
	32°S	7	9	0.573	0.533	
	38°S	2	4	-	-	

h, number of different alleles across individuals; S, number of segregating sites; θ_W, Watterson's theta; π_n, nucleotide diversity. ϕ_{st} values correspond to levels of differentiation between populations from 9°N-7°25'S and populations from 14°S-21°33'S. ***, P value < 0.001; NS, P value > 0.05. Sulfo 1, Sulfotransferase paralogue 1; Lyso, Lysozyme; SAHH, S-Adenosyl Homocysteine Hydrolase; EF1α, Elongation Factor 1α; mtCOI, cytochrome oxidase I.

A dynamic effect of a couple of physical barriers to gene flow

Based on tectonic plate history, the Easter Microplate barrier is older than the 7°25'S–14°S barrier. The Easter Microplate originated from the progressive offsetting of the Pacific-Antarctic Ridge (PAR) along with the 'old' EPR (Nazca Pacific spreading centre) about 3.88 Mya (anomaly 3). This offsetting progressively expanded with the formation of an overlapping spreading center (OSC: see [62]). Ridge offsetting is known to have a profound impact on gene flow [17], [63], [64] and to isolate populations from each other. In the specific case of the Easter Microplate, species divergence time was thus expected to coincide with (or to be slightly older than) anomaly 3 (i.e., 3.88 Mya) at the Easter Microplate barrier. In comparison, transform faults responsible for offsetting the ridge axis between 9°N and 17°S initiated 1-2 Mya [56], [57], [58]. If the two geographical barriers were both impermeable to dispersal since their occurrence, one would expect to find reciprocal monophylies for nearly all genes, with a greater divergence, at the older barrier (i.e., the Easter Microplate barrier) and incomplete lineage sorting at the most recent barrier (i.e., the 7°25'S–14°S barrier).

Contrasting results among genes across the Easter Microplate are in accordance with the presence of a tension zone at the latitude of Easter Island as proposed by Johnson et al. [24] who studied the PAC/EPR population connectivity between 21°S to 38°S in more detail. The present results on mtCOI integrating both Won et al. [15] and Plouviez et al. [17] datasets are consistent with geological evidence, indicating that gene flow is either blocked or greatly impeded by both the Easter Microplate and the 7°25'S–14°S barriers. MtCOI haplotypes indeed show a reciprocal monophyly with a 2.1–4.3 Mya divergence time across the Easter Microplate [15], while populations display only haplotype frequency differences across the 7°25'S and 14°S barrier (significant φ_{st}, [17] and Table 2). Absence of detectable differentiation on the EF1α gene across the more recent 7°25'S–14°S barrier (Table 2) while populations were divergent across the older Easter Microplate barrier was expected considering nuclear

genes generally evolve slower than mitochondrial genes. In contrast, the other nuclear gene structure did not fit this expectation; showing no differentiation across the Easter Microplate, whereas two of them (SAHH, Sulfo1) displayed clear but unexpected patterns of reciprocal monophyly across the more recent barrier (7°25'S–14°S). Such an inconsistency among genes might be explained by the presence of a tension zone extending further north toward 7°25'S associated with the relaxation of the Easter Microplate barrier.

The hypothesis of a relaxation of the barrier at the Easter Microplate is also supported by the geological formation of the microplate itself. If initiated by the offsetting of the PAR and 'old' EPR, the plate formation was achieved when the ends of the two expanding ridges joined together following a clockwise rotation with the setting up of insular volcanic arcs (Orongo and Pito rifts) about 2.5–1.75 Mya (end of anomaly 2a: see [65]). This junction between the PAR and EPR is likely to have provided stepping-stones for episodic bursts of larvae between the PAR and the southern EPR (using vents located on active seamounts: e.g., the Pito Seamount [66]).

We propose that the progressive ridge-offsetting at 7°25'–14°S played an important role in enhancing barriers to dispersal by efficiently trapping endogenously-induced genetic clines associated with the reconnection of B. thermophilus and B. antarcticus, which could have subsequently escaped from the relaxing barrier at the Easter Microplate. Divergence times between clades for the SAHH and Sulfo1 nuclear genes across the 7°25'S–14°S barrier (4–6.6 Mya) were similar to those estimated at the Easter Microplate boundary between the PAC and SEPR clades of mtCOI and EF1α. This convergence in divergence times found at two distinct spatial locations, as well as the fact that nuclear mutation rates should be lower than that of the mtCOI (as found in the Atlantic Bathymodiolus sp. [23]) fits this "dual relaxing/enhancing barriers" hypothesis. Indeed, the fixation of alleles in the two observed-divergent clades associated with the SAHH and Sulfo1 loci (and possibly alleles from the Lyso gene as well) might have been initiated by the separation of populations during the

Table 3. IMa2 estimates and the 95% Highest Posterior Density (HPD) intervals of migration and demographic parameters across the 7°25'S-14°S barrier to gene flow.

	θ_N	θ_S	θ_A	m_{N-S}	m_{S-N}	N_N	N_S	N_A	M_{N-S}	M_{S-N}
Estimate	5.125	5.975	3.575	0.373	0.003	642647	749232	448285	1.555	0.057
L-HPD	2.825	3.575	0.775	0.013	0.000	354239	448285	97181	0.000	0.000
H-HPD	11.720	44.380	41.230	0.989	1.103	1470250	5564379	5169386	11.010	3.343

L-HPD, lower 95% Highest Posterior Density; H-HPD, higher 95% Highest Posterior Density. θ, demographic parameter estimated by IMa2. N, calibration of θ in number of individuals. m, migration parameters estimated by IMa2. M, calibration of m in number of individuals. $_N$, north of the barrier; $_S$, south of the barrier; $_A$, ancestral population; forward in time, $_{N-S}$, migration from north to south; $_{S-N}$, migration from south to north.

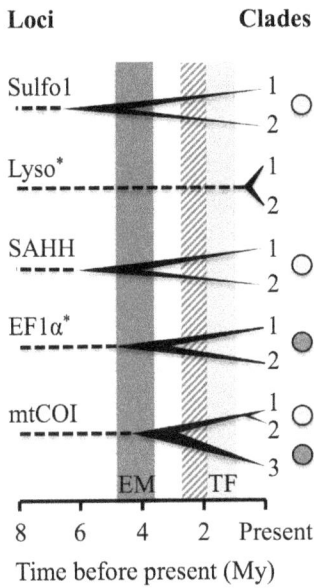

Figure 4. Bifurcated trees showing the correspondence between divergence times between sister clades and times of geological formations. Clades correspond in those identified in Fig. 2. Grey boxes represent the estimated time of the two barriers: EM, time since the first offsetting of overlapping faults leading to the Easter Microplate (dark grey); TF, time since the formation of Gofar/Discovery transform faults at 7°25′S–14°S latitude (light grey). Striped box represents the estimated time elapsed since the junction between the Pacific Antarctic Ridge and the East Pacific Rise. Grey circles correspond to the geographic position at which the divergence is observed, EM: dark grey and TF: light grey. Dashed lines represent the roots of the trees with the outgroup and calibration point (i.e. *Bathymodiolus azoricus*, Mid-Atlantic Ridge, about 8–12 Mya). Intervals of estimated divergence, due to the range of the calibration point date, are represented by horizontal thickness of tree nodes. Sulfo1, Sulfotranferase paralogous gene 1; Lyso, Lysozyme; SAHH, S-Adenosyl Homocysteine Hydrolase; mtCOI, mitochondrial cytochrome oxidase 1; EF1α, Elongation Factor 1α. *Lysozyme and EF1α estimates of divergence are indicated but have to be interpreted with caution because these loci did not follow a strict molecular clock.

Easter Microplate formation before migrating northward and being trapped by the emerging 7°25′S–14°S barrier.

To conclude, we have shown that the clinal distribution of mtCOI haplotypes along the EPR was in fact due to highly dynamic historical/geological processes. A system of two genetic barriers to gene flow, in which one would be progressively relaxed (i.e., Easter Microplate) and the other enforced (i.e., 7°25′S–14°S transform faults), can explain the discordant distribution of genetic clines. A volcanic arc joining PAR and EPR probably played an important role in setting up of a secondary contact zone (and subsequent northward spreading of alleles/genetic clines) between individuals across the Easter Microplate. The proposed relaxation/enforcement scenario of barriers is consistent with the occurrence of a biogeographic transition zone along the southern EPR (resulting from the overlap of the North-EPR and the South-Easter Microplate biogeographic provinces) to explain the high diversity of vent fauna species observed between 17°25′S and 21°33′S [67]. Studies on motion of hybrid/tension zones are in their infancy (but see e.g., [12], [68], [69], [70]). Tension zone movement is expected theoretically although they are expected to move toward areas of lower population density [5], [71]. Evidence for the movement of hybrid zones has been obtained directly following known tension zones over time (e.g., [68]) or indirectly using molecular and/or phenotypic traits (e.g., [70], review in [69]). Theoretical analyses have demonstrated the ability of low-migration areas (i.e., lack of suitable habitats/larval transport disruption) to trap genetic incompatibilities previously generated by other historical/ecological/biogeographical processes (see: [12]) and might apply to the case of *B. thermophilus*, which, in turn represents one of the first empirical observation of such theoretical predictions. The theory of genetic barriers and hybrid zones explains semi-permeable barriers to gene flow and the structure observed, possibly indicating selection against hybrids. Of course one could also speculate that different loci respond to different ecological pressures at the two positions. However, not only are the possible ecological differences unidentified but getting by chance a series of outlier loci (or hichhicked introns) under differential ecological selection out of four genes is highly unlikely. This therefore suggests that allele frequency shifts are linked to tension zones and are thus able to move. Together with what we know of the geological history of the EPR, the hypothesis of an

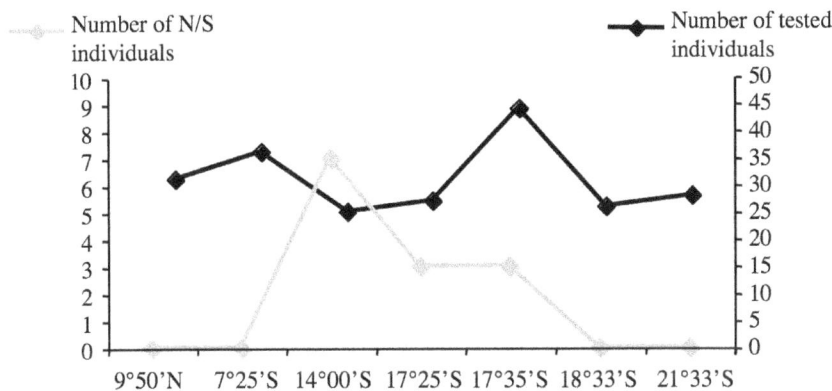

Figure 5. Localisation of individuals with both north and south type alleles in SAHH using RFLP analyses. N/S individuals, individuals that present one allele from the northern clade (blue clade in Figure 2) and one from the southern clade (yellow clade in Figure 2). Left Y-axis is the number of N/S individuals. Right Y-axis is the total number of individuals.

interchange of clines from the southern barrier to the northern barrier explains the geographic discordance of cline positions.

Supporting Information

Table S1 GenBank accession numbers of each unique sequence and their geographic distribution. Numbers correspond to the number of individuals having the accession number in a given population. For example, accession numbers KC858658 through KC858662 all have a single individuals recovered in the 9°50′N population.

Acknowledgments

We thank the chief scientists and 'Nautile' and 'Alvin' crews for their technical support and efforts during the oceanographic expeditions: HOT96, BIOSPEEDO2004, SO-157 Foundation III and MESCAL2010.

We acknowledge the chief scientists from those oceanographic cruises: F. Gaill, D. Jollivet, P. Stoffers, F. Lallier and N. Le Bris. We are very grateful to C. Daguin-Thiébaut and F. Viard for collecting/preserving mussels during the BIOSPEEDO cruise and N. Dubilier and C. Borowski who provided the 38°S samples. We are also truly indebted to the sequencing genomic plateforms: GENOMER (Station Biologique de Roscoff, France) and GENOSCOPE (Evry, France) for the sequencing of the mtCOI fragment and our clone collections, respectively. We also want to thank Arnaud Tanguy for his contribution to the *B. azoricus* cDNA library construction, which provided exonic sequences for the nuclear genes. We would like to thank Thomas Schultz, Cindy Van Dover and the anonymous reviewers for their valuable comments and editorial suggestions on the manuscript.

Author Contributions

Conceived and designed the experiments: SP BF NB DJ. Performed the experiments: SP BF DLG. Analyzed the data: SP. Contributed reagents/materials/analysis tools: BF FHL DJ. Wrote the paper: SP NB DJ.

References

1. Kimura M, Weiss GH (1964) The stepping stone model of population structure and the decrease of genetic correlation with distance. Genetics 49: 561–576.
2. Fontaine MC, Baird SJE, Piry S, Ray N, Tolley KA, et al. (2007) Rise of oceanographic barriers in continuous populations of a cetacean: the genetic structure of harbour porpoises in Old World waters. BMC Biol 5: 30.
3. Plouviez S, Le Guen D, Lecompte O, Lallier FH, Jollivet D (2010) Determining gene flow and influence of selection across the equatorial barrier of the East Pacific Rise in the tube-dwelling polychaete *Alvinella pompejana*. BMC Evol Biol 10: 220.
4. Hewitt GM (1975) A sex chromosome hybrid zone in the grasshopper *Podisma pedestris* (Orthoptera: Acrididae). Heredity 35: 375–387.
5. Barton NH (1979) The dynamics of hybrid zones. Heredity 43: 341–359.
6. Slatkin M (1973) Gene flow and selection in a cline. Genetics 75: 733–756.
7. Barton NH, Hewitt GM (1985) Analysis of hybrid zones. Annu Rev Ecol Evol Syst 16: 113–148.
8. Castric V, Bernatchez L (2003) The rise and fall of isolation by distance in the anadromous brook charr (*Salvelinus fontinalis* Mitchill). Genetics 163: 983–996.
9. Lewontin RC, Krakauer J (1973) Distribution of gene frequency as a test of the theory of the selective neutrality of polymorphism. Genetics 74: 175–195.
10. Olson MS, Levsen N (2012) Classic clover cline clues. Mol Ecol 21: 2315–2317.
11. Hewitt GM (1988) Hybrid zones, natural laboratories for evolutionary studies. Trends Ecol Evol 3: 158–167.
12. Bierne N, Welch J, Loire E, Bonhomme F, David P (2011) The coupling hypothesis: why genome scans may fail to map adaptation genes. Mol Ecol 20: 2044–2072.
13. Vrijenhoek RC (1997) Gene flow and genetic diversity in naturally fragmented metapopulations of deep-sea hydrothermal vent animals. J Hered 88: 285–293.
14. Jollivet D, Chevaldonné P, Planque B (1999) Hydrothermal-vent alvinellid polychaete dispersal in the eastern Pacific. 2. A metapopulation model based on habitat shifts. Evolution 53: 1128–1142.
15. Won Y, Young CR, Lutz RA, Vrijenhoek RC (2003) Dispersal barriers and isolation among deep-sea mussel populations (Mytilidae: *Bathymodiolus*) from eastern Pacific hydrothermal vents. Mol Ecol 12: 169–184.
16. Hurtado L, Lutz R, Vrijenhoek RC (2004) Distinct patterns of genetic differentiation among annelids of eastern Pacific hydrothermal vents. Mol Ecol 13: 2603–2615.
17. Plouviez S, Shank TM, Faure B, Daguin-Thiébaut C, Viard F, et al. (2009) Comparative phylogeography among hydrothermal vent species along the East Pacific Rise reveals vicariant processes and population expansion in the South. Mol Ecol 18: 3903–3917.
18. Remington CL (1968) Suture-zones of hybrid interaction between recently joined biotas. Evol Biol 2: 321–428.
19. Lutz RA, Jablonski D, Turner RD (1984) Larval development and dispersal at deep-sea hydrothermal vents. Science 226: 1451–1454.
20. Arellano SM, Young CM (2009) Spawning, development and the duration of larval life in a deep-sea cold-seep mussel. Biol Bull 216: 149–162.
21. Olu-Le Roy L, von Cosel R, Hourdez S, Carney SL, Jollivet D (2007) Amphi-Atlantic cold-seep *Bathymodiolus* species complexes across the equatorial belt. Deep Sea Res Part 1 54: 1890–1911.
22. Carney SL, Formica MI, Divatia H, Nelson K, Fisher CR, et al. (2006) Population structure of the mussel "*Bathymodiolus*" *childressi* from Gulf of Mexico hydrocarbon seeps. Deep Sea Res Part 1 53: 1061–1072.
23. Faure B, Jollivet D, Tanguy A, Bonhomme F, Bierne N (2009) Speciation in the deep sea: multi-locus analysis of divergence and gene flow between two hybridizing species of hydrothermal vent mussels. PLoS One 4: e6485. doi:10.1371/journal.pone.0006485.

24. Johnson SB, Won Y-J, Harvey JBJ, Vrijenhoek RC (2013) A hybrid zone between *Bathymodiolus* mussel lineages from eastern Pacific hydrothermal vents. BMC Evol Biol 13: 21.
25. Audzijonyte A, Vrijenhoek RC (2010) When gaps are really gaps: statistical phylogeography of hydrothermal vent invertebrates. Evolution 64: 2369–2384.
26. Boutet I, Tanguy A, Le Guen D, Piccino P, Hourdez, et al. (2009) Global depression in gene expression as a response to rapid thermal changes in vent mussels. Proc R Soc Lond B 276: 3071–3079.
27. Belkhir K, Borsa P, Chikhi L, Raufaste N, Bonhomme F (2004) Genetix 4.05, logiciel sous Windows TM pour la génétique des populations. Laboratoire Génome, Populations, Interactions, Adaptations, UMR 5000, Université de Montpellier 2. Available at: www.genetix.univ-montp2.fr/genetix/genetix.htm
28. Weir B, Cockerham C (1984) Estimating F-statistics for the analysis of population structure. Evolution 38: 1358–1370.
29. Raymond M, Rousset F (1995) GENEPOP, version 1.2: population genetics software for exact tests and ecumenicism. J Hered 86: 248–249.
30. Jolly MT, Viard F, Gentil F, Thiébaut E, Jollivet D (2006) Comparative phylogeography of two coastal polychaete tubeworms in the Northeast Atlantic supports shared history and vicariant events. Mol Ecol 15: 1841–1855.
31. Bierne N, Tanguy A, Faure M, Faure B, David E, et al. (2007) Mark-recapture cloning: a straightforward and cost-effective cloning method for population genetics of single copy nuclear DNA sequences in diploids. Mol Ecol Notes 7: 562–566.
32. Faure B, Bierne N, Tanguy A, Bonhomme F, Jollivet D (2007) Evidence for a slightly deleterious effect of intron polymorphisms at the EF1α gene in the deep-sea hydrothermal vent bivalve *Bathymodiolus*. Gene 406: 99–107.
33. Thompson JD, Higgins DG, Gibson TJ (1994) CLUSTAL W: improving the sensitivity of progressive multiple sequence alignment through sequence weighting, position specific gap penalties and weight matrix choice. Nucleic Acids Res 22: 4673–4680.
34. Rozas J, Sanchez-DelBarrio JC, Messeguer X, Rozas R (2003) DnaSP, DNA polymorphism analyses by the coalescent and other methods. Bioinformatics 19: 2496–2497.
35. Bandelt HJ, Forster P, Rohl A (1999) Median-joining networks for inferring intraspecific phylogenies. Mol Biol Evol 16: 37–48.
36. Drummond AJ, Rambaut A (2007) BEAST: Bayesian evolutionary analysis by sampling trees. BMC Evol Biol 7: 214.
37. Kumar S, Balczarek KA, Lai Z-C (1996) Evolution of the hedgehog gene family. Genetics 142: 965–972.
38. Burton KW, Ling H-F, O'Nions RK (1997) Closure of the Central American Isthmus and its effect on deep-water formation in the North Atlantic. Nature 386: 382–385.
39. Stiller J, Rousset V, Pleijel F, Chevaldonné P, Vrijenhoek RC, et al. (2013) Phylogeny, biogeography and systematics of hydrothermal vent and methane seep *Amphisamytha* (Ampharetidae, Annelida), with descriptions of three new species. Syst Biodivers 11: 35–65.
40. Hudson RR, Kreitman M, Aguade M (1987) A test of neutral molecular evolution based on nucleotide data. Genetics 116: 153–159.
41. McDonald JH, Kreitman M (1991) Adaptive protein evolution at the Adh locus in *Drosophila*. Nature 351: 652–654.
42. Manni F, Guérard E, Heyer E (2004) Geographic patterns of (genetic, morphologic, linguistic) variation: how barriers can be detected by "Monmonier's algorithm". Hum Biol 76: 173–190.
43. Hudson RR, Slatkin M, Maddison WP (1992) Estimation of levels of gene flow from DNA-sequence data. Genetics 132: 583–589.
44. Hey J, Nielsen R (2007) Integration within the Felsenstein equation for improved Markov chain Monte Carlo methods in population genetics. Proc Nat Acad Sci USA 104: 2785–2790.

45. Won Y, Hallam SJ, O'mullan GD, Vrijenhoek RC (2003) Cytonuclear disequilibrium in a hybrid zone involving deep-sea hydrothermal vent mussels of the genus *Bathymodiolus*. Mol Ecol 12: 3185–3190.

46. Bierne N, Borsa P, Daguin C, Jollivet D, Viard F, et al. (2003) Introgression patterns in the mosaic hybrid zone between *Mytilus edulis* and *M. galloprovincialis*. Mol Ecol 12: 447–461.

47. Mullineaux LS, Adams DK, Mills SW, Beaulieu SE (2010) Larvae from afar colonize deep-sea hydrothermal vents after a catastrophic eruption. Proc Natl Acad Sci USA 107: 7829–7834.

48. Craddock C, Hoeh WR, Lutz RA, Vrijenhoek RC (1995) Extensive gene flow among mytilid (*Bathymodiolus thermophilus*) populations from hydrothermal vents of the eastern Pacific. Mar Biol 124: 137–146.

49. Grassle JF (1985) Genetic differentiation in populations of hydrothermal vent mussels (*Bathymodiolus thermophilus*) from the Galapagos Rift and 13°N on the East Pacific Rise. Bull Biol Soc Wash 6: 429–442.

50. Karl SA, Avise JC (1992) Balancing selection at allozyme loci in oysters: implications from nuclear RFLPs. Science 256: 100–102.

51. Pogson GH, Mesa KA, Boutilier RG (1995) Genetic population structure and gene flow in the Atlantic cod *Gadus morhua*: a comparison of allozyme and nuclear RFLP loci. Genetics 139: 375–385.

52. Lemaire C, Allegrucci G, Naciri M, Bahri-Sfar L, Kara H, et al. (2000) Do discrepancies between microsatellite and allozyme variation reveal differential selection between sea and lagoon in the sea bass (*Dicentrarchus labrax*)? Mol Ecol 9: 457–467.

53. Riginos C, Sukhdeo K, Cunningham CW (2002) Evidence for selection at multiple allozyme loci across a mussel hybrid zone. Mol Biol Evol 19: 347–351.

54. McDonald JH, Verrelli BC, Geyer LB (1996) Lack of geographic variation in anonymous nuclear polymorphisms in the American oyster, *Crassostrea virginica*. Mol Biol Evol 13: 1114–1118.

55. Bierne N, Daguin C, Bonhomme F, David P, Borsa P (2003) Direct selection on allozymes is not required to explain heterogeneity among marker loci across a *Mytilus* hybrid zone. Mol Ecol 12: 2505–2510.

56. Kureth CL, Rea DK (1981) Large-scale oblique features in an active transform fault, the Wilkes fracture zone near 9°S on the East Pacific Rise. Mar Geophys Res 5: 119–137.

57. Naar DF, Hey RN (1989) Speed limit for oceanic transform faults. Geology 17: 420–422.

58. Francheteau J, Armijo R, Cheminee JL, Hekinian R, Lonsdale P, et al. (1990) 1 Ma East Pacific Rise oceanic-crust and uppermost mantle exposed by rifting in Hess Deep (equatorial Pacific-ocean). Earth Planet Sci Lett 101: 281–295.

59. Lupton JE, Craig H (1981) A major He-3 source at 15°S on the East Pacific Rise. Science 214: 13–18.

60. Lupton JE (1998) Hydrothermal helium plumes in the Pacific ocean. J Geophys Res Oceans 103: 15853–15868.

61. Fujio SZ, Imasato N (1991) Diagnostic calculation for circulation and water mass movement in the deep Pacific. J Geophys Res Oceans 96: 759–774.

62. Tebbens SF, Cande SC (1997) Southeast Pacific tectonic evolution from Oligocene to present. J Geophys Res Solid Earth 102: 12601–12084.

63. Johnson SB, Young CR, Jones WJ, Warén A, Vrijenhoek RC (2006) Migration, isolation, and speciation of hydrothermal vent limpets (Gastropoda; Lepetodrilidae) across the Blanco transform fault. Biol Bull 210: 140–157.

64. Young CR, Fujio S, Vrijenhoek RC (2008) Directional dispersal between midocean ridges: deep-ocean circulation and gene flow in *Ridgeia piscesae*. Mol Ecol 17: 1718–1731.

65. Larson RL, Searle RC, Kleinrock MC, Schouten H, Bird RT, et al. (1992) Roller-bearing tectonic evolution at the Juan Fernandez Microplate. Nature 356: 571–576.

66. Naar DF, Hekinian R, Segonzac M, Francheteau J, Armijo R, et al. (2004) Vigorous venting and biology at Pito seamount, Easter Microplate. Geophys Monograph Ser 148: 305–318.

67. Matabos M, Plouviez S, Hourdez S, Desbruyères D, Legendre P, et al. (2011) Faunal changes and geographic crypticism indicate the occurrence of biogeographic transition zone along the southern East Pacific Rise. J Biogeogr 38: 575–594.

68. Dasmahapatra KK, Blum MJ, Aiello A, Hackwell S, Davies N, et al. (2002) Inferences from a rapidly moving hybrid zone. Evolution 56: 741–753.

69. Buggs RJA (2007) Empirical study of hybrid zone movement. Heredity 99: 301–312.

70. Gay L, Crochet P-A, Bell DA, Lenormand T (2008) Comparing clines on molecular and phenotypic traits in hybrid zones: a window on tension zone models. Evolution 62: 2789–2806.

71. Barton NH, Turelli M (2011) Spatial waves of advance with bistable dynamics: cytoplasmic and genetic analogues of Allee effects. Am Nat 178: 48–75. doi: 10.1086/661246.

11

Marine Litter Distribution and Density in European Seas, from the Shelves to Deep Basins

Christopher K. Pham[1,2]*, Eva Ramirez-Llodra[3,4], Claudia H. S. Alt[5], Teresa Amaro[6], Melanie Bergmann[7], Miquel Canals[8], Joan B. Company[3], Jaime Davies[9], Gerard Duineveld[10], François Galgani[11], Kerry L. Howell[9], Veerle A. I. Huvenne[12], Eduardo Isidro[1,2], Daniel O. B. Jones[12], Galderic Lastras[8], Telmo Morato[1,2], José Nuno Gomes-Pereira[1,2], Autun Purser[13], Heather Stewart[14], Inês Tojeira[15], Xavier Tubau[8], David Van Rooij[16], Paul A. Tyler[5]

1 Center of the Institute of Marine Research (IMAR) and Department of Oceanography and Fisheries, University of the Azores, Horta, Portugal, 2 Laboratory of Robotics and Systems in Engineering and Science (LARSyS), Lisbon, Portugal, 3 Institut de Ciències del Mar (ICM-CSIC), Barcelona, Spain, 4 Norwegian Institute for Water Research (NIVA), Marine Biology section, Oslo, Norway, 5 Ocean and Earth Science, University of Southampton, National Oceanography Centre, Southampton, United Kingdom, 6 Norwegian Institute for Water Research, Bergen, Norway, 7 Alfred-Wegener-Institut, Helmholtz-Zentrum für Polar- und Meeresforschung, Bremerhaven, Germany, 8 GRC Geociències Marines, Departament d'Estratigrafia, Paleontologia i Geociències Marines, Facultat de Geologia, Universitat de Barcelona, Campus de Pedralbes, Barcelona, Spain, 9 Marine Biology & Ecology Research Centre, Marine Institute, Plymouth University, Plymouth, United Kingdom, 10 Netherlands Institute for Sea Research (NIOZ), Texel, The Netherlands, 11 Institut Français de Recherche pour l'Exploitation de la Mer (IFREMER), Bastia, France, 12 National Oceanography Centre, University of Southampton Waterfront Campus, Southampton, United Kingdom, 13 OceanLab, Jacobs University Bremen, Bremen, Germany, 14 British Geological Survey, Murchison House, Edinburgh, United Kingdom, 15 Portuguese Task Group for the Extension of the Continental Shelf (EMEPC), Paço de Arcos, Portugal, 16 Renard Centre of Marine Geology (RCMG), Department of Geology and Soil Science, Ghent University, Gent, Belgium

Abstract

Anthropogenic litter is present in all marine habitats, from beaches to the most remote points in the oceans. On the seafloor, marine litter, particularly plastic, can accumulate in high densities with deleterious consequences for its inhabitants. Yet, because of the high cost involved with sampling the seafloor, no large-scale assessment of distribution patterns was available to date. Here, we present data on litter distribution and density collected during 588 video and trawl surveys across 32 sites in European waters. We found litter to be present in the deepest areas and at locations as remote from land as the Charlie-Gibbs Fracture Zone across the Mid-Atlantic Ridge. The highest litter density occurs in submarine canyons, whilst the lowest density can be found on continental shelves and on ocean ridges. Plastic was the most prevalent litter item found on the seafloor. Litter from fishing activities (derelict fishing lines and nets) was particularly common on seamounts, banks, mounds and ocean ridges. Our results highlight the extent of the problem and the need for action to prevent increasing accumulation of litter in marine environments.

Editor: Andrew Davies, Bangor University, United Kingdom

Funding: This research was supported by the European Community's Seventh Framework Programme (FP7/2007^2013) under the HERMIONE project, Grant agreement (GA) no. 226354. The authors would like to acknowledge further funds from the Condor project (supported by a grant from Iceland, Liechtenstein, Norway through the EEA Financial Mechanism (PT0040/2008)), Corazon (FCT/PTDC/MAR/72169/2006; COMPETE/QREN), CoralFISH (FP7 ENV/2007/1/21314 4), EC funded PERSEUS project (GA no. 287600), the ESF project BIOFUN (CTM2007-28739-E), the Spanish projects PROMETEO (CTM2007-66316-C02/MAR) and DOS MARES (CTM2010-21810-C03-01), la Caixa grant "Oasis del Mar", the Generalitat de Catalunya grant to excellence research group number 2009 SGR 1305, UK's Natural Environment Research Council (NERC) as part of the Ecosystems of the Mid-Atlantic Ridge at the Sub-Polar Front and Charlie-Gibbs Fracture Zone (ECOMAR) project, the Marine Environmental Mapping Programme (MAREMAP), the ERC (Starting Grant project CODEMAP, no 258482), the Joint Nature Conservation Committee (JNCC), the Lenfest Ocean Program (PEW Foundation), the Department for Business, Enterprise and Regulatory Reform through Strategic Environmental Assessment 7 (formerly the Department for Trade and Industry) and the Department for Environment, Food and Rural Affairs through their advisors, the Joint Nature Conservation Committee, the offshore Special Areas for Conservation programme, BELSPO and RBINS-OD Nature (Belgian Federal Government) for R/V Belgica shiptime. The footage from the HAUSGARTEN observatory was taken during expeditions ARK XVIII/1, ARK XX/1, ARK XXII/1, ARK XXIII/2 and ARK XXVI/2 of the German research icebreaker "Polarstern". The authors also acknowledge funds provided by FCT-IP/MEC to LARSyS Associated Laboratory and IMAR-University of the Azores (R&DU #531), Thematic Area E, through the Strategic Project (PEst-OE/EEI/LA0009/2011^2014, COMPETE, QREN) and by the Government of Azores FRCT multiannual funding. CKP was supported by the doctoral grant from the Portuguese Science Foundation (SFRH/BD/66404/2009; COMPETE/QREN). AP was supported by Statoil as part of the CORAMM project. MB would like to thank Antje Boetius for financial support through the DFG Leibniz programme. JNGP was supported by the doctoral grant (M3.1.2/F/062/2011) from the Regional Directorate for Science, Technology and Communications (DRCTC) of the Regional Government of the Azores. ERLL was supported by a CSIC-JAE-postdocotral grant with co-funding from the European Social Fund. Publication fees for this open access publication were supported by IFREMER. The funders had no role in study design, data collection and analysis, decision to publish, or preparation of the manuscript.

Competing Interests: The authors have declared that no competing interests exist.

* E-mail: phamchristopher@uac.pt

Introduction

Litter disposal and accumulation in the marine environment is one of the fastest growing threats for the world's oceans health.

Marine litter is defined as "any persistent, manufactured or processed solid material discarded, disposed of or abandoned in the marine and coastal environment"[1]. The issue has been

Table 1. Sampling locations, date and methods for the collection of data on litter along with litter densities (mean number of items ha^{-1} and kg ha^{-1} ± standard errors).

Location	Year	Method	N° of samples	Mean depth (m)	Density (n ha^{-1})	Density (kg ha^{-1})	Area covered (ha)
ATLANTIC							
Continental slopes							
North Faroe-Shetland Channel	2006	TC	19	657	0.3±0.2	-	2.3
North-East Faroe-Shetland Channel	2006	TC	11	501	1.9±1.0	-	1.2
Continental shelf							
Norwegian Margin	2007	SUB	9	304	9.7±3.8	-	0.6
Submarine canyons							
Dangeard & Explorer Canyons	2007	TC	44	578	7.2±2.7	-	3
Nazaré Canyon	2007	ROV	13	3144	4.2±1.6	-	9
Lisbon Canyon	2007	ROV	1	1602	66.2	-	1
Setúbal Canyon	2007	ROV	1	2194	24.6	-	0.9
Cascais Canyon	2007	ROV	1	4574	10.6	-	1
Guilvinec Canyon	2008-2010	ROV	8	661	31.9±28.1	-	4.1
Whittard Canyon	2010	ROV-TC	11	2668	1.4±0.4	-	12.4
Seamounts, banks and mounds							
Anton Dohrn Seamount	2005-2009	TC	24	992	1.9±1.0	-	2.2
Condor Seamount	2010-2011	ROV	48	258	14.6±3.0	-	5.6
Josephine Seamount	2012	ROV	4	1455	5.7±3.3	-	0.9
Hatton Bank	2005-2011	ROV-TC	52	706	1.9±0.8	-	4
Rockall Bank	2005-2011	ROV-TC	29	702	0.7±0.5	-	2.4
Rosemary Bank	2006	TC	14	577	3.3±2.3	-	1.1
Pen Duick Alpha/Beta Mound	2009	ROV	7	534	2.5±1.7	-	1.1
Darwin Mounds	2011	ROV	7	1007	9.7±2.9	-	1.8
Ocean ridges							
North Charlie Gibbs Fracture Zone	-	ROV	24	2300	0.4±0.3	-	2.4
South Charlie Gibbs Fracture Zone	-	ROV	24	2600	2.9±1.4	-	2.4
Wyville-Thomson Ridge	2006	TC	15	670	10.9±4.3	-	1.2
MEDITERANEAN							
Continental slopes							
Calabrian Slope (Central Med.)	2009	Trawl	4	1400	-	0.6±0.4	18.9
Western Mediterranean Slope	2009	Trawl	8	1500	-	4±1.8	56
Crete-Rhodes Ridge (E. Med.)	2009	Trawl	8	1500	-	1.1±0.3	37.9
Blanes slope (NW Med.)	2009	Trawl	94	1387	-	1.2±0.4	407
Continental shelf							
Gulf of Lion (NW Med.)	2009	Trawl	52	85	0.4±0.1	-	276.4

Table 1. Cont.

Location	Year	Method	N° of samples	Mean depth (m)	Density (n ha⁻¹)	Density (kg ha⁻¹)	Area covered (ha)
Submarine canyons							
Blanes Canyon (NW Med.)	2009–2011*	ROV-Trawl	4 (13)	1496(1431)	32.1±11.9	0.7±0.2	2(33.9)
Gulf of Lion Canyons (NW Med.)	2009	Trawl	11	510	0.4±0.1	-	126.5
Deep basins							
Algero-Balearic Basin (W. Med.)	2009	Trawl	3	2883	-	1.8±1.5	16
Crete-Rhodes Ridge (E. Med.)	2009	Trawl	2	3000	-	1.2±0.3	2.8
Calabrian Basin (Central Med.)	2009	Trawl	3	2967	-	1.7±0.6	12.5
ARCTIC							
Continental slope							
HAUSGARTEN, station IV	1999–2011	TC-ROV	10	2450	13.6±7.9	-	72.2

*Numbers in parentheses refer to trawl surveys. ROV = remotely operated vehicle; TC = towed camera system; TRAWL = Otter Trawl or Maireta System; SUB = manned submersible.

highlighted by the United Nations Environment Program [1] and was included in the 11 Descriptors set by Europe's Marine Strategy Framework directive (2008/56/EC) (MSFD) [2]. The MSFD requires each Descriptor in all European marine waters not to deviate from the undisturbed state and reach Good Environmental Status (GES) by 2020.

With an estimated 6.4 million tonnes of litter entering the oceans each year [1], the adverse impacts of litter on the marine environment are not negligible. Besides the unquestionable aesthetic issue, litter can be mistaken for food items and be ingested by a wide variety of marine organisms [3–8]. Entanglement in derelict fishing gear is also a serious threat, particularly for mammals [9–11], turtles [12] and birds [13] but also for benthic biota such as corals [14,15]. High mortality of fish through "ghost fishing" is another consequence of derelict fishing gear in the marine environment [16]. Moreover, floating litter facilitates the transfer of non-native marine species (e.g. bryozoans, barnacles) to new habitats [17,18]. Barnes et al. [19] estimated that the dispersal of alien species through marine litter more than doubles the rate of natural dispersal processes, especially during an era of global change.

Although the type of litter found in the world's oceans is highly diverse, plastics are by far the most abundant material recorded [20–22]. Because of their persistence and hydrophobic nature, their impact on marine ecosystems is of great concern. Plastics are a source of toxic chemicals such as polychlorinated biphenyls (PCBs) and dioxins that can be lethal to marine fauna [23]. Furthermore, the degradation of plastics generates microplastics which, when ingested by organisms, can deliver contaminants across trophic levels [24–27].

Litter type, composition and density vary greatly among locations and litter has been found in all marine habitats, from surface water convergence in the pelagic realm (fronts) down to the deep sea where litter degradation is a much slower process [21]. The spatial distribution and accumulation of litter in the ocean is influenced by hydrography, geomorphological factors [21,28], prevailing winds and anthropogenic activities [29]. Hotspots of litter accumulation include shores close to populated areas, particularly beaches [30], but also submarine canyons, where litter originating from land accumulates in large quantities [28,31].

In Europe, much has been written on the abundance and distribution of litter on the coastline and in surface waters [32–41]. As more areas of Europe's seafloor are being explored, benthic litter is progressively being revealed to be more widespread than previously assumed [15,28,29,31,42–52]. The sources of litter accumulating on the seafloor are variable, depending upon interactions between distances from shore [31,45], oceanographic and hydrographic processes [47] and human activities such as commercial shipping [29] and leisure craft [43].

Early studies used trawling to quantify litter abundance on the seafloor [53], whilst more recent studies have demonstrated the potential of remotely operated vehicles (ROV), manned submersibles or towed cameras to study litter in the deep sea [15,31,43,47,54,55]. However, understanding spatial patterns in litter abundance and distribution in the deep sea is challenging, owing to the lack of standardization in the sampling and analytical methodologies used. Furthermore, the high cost of sampling in the deep sea has limited our ability to perform standardized surveys across large areas to understand fully the extent of this pollution issue.

The problem of marine litter on the deep seafloor was addressed by the EU-FP7 project HERMIONE, recognising the need to use the surveys conducted by all partners (although designed for other purposes) to gather data on litter in the deep sea. This paper

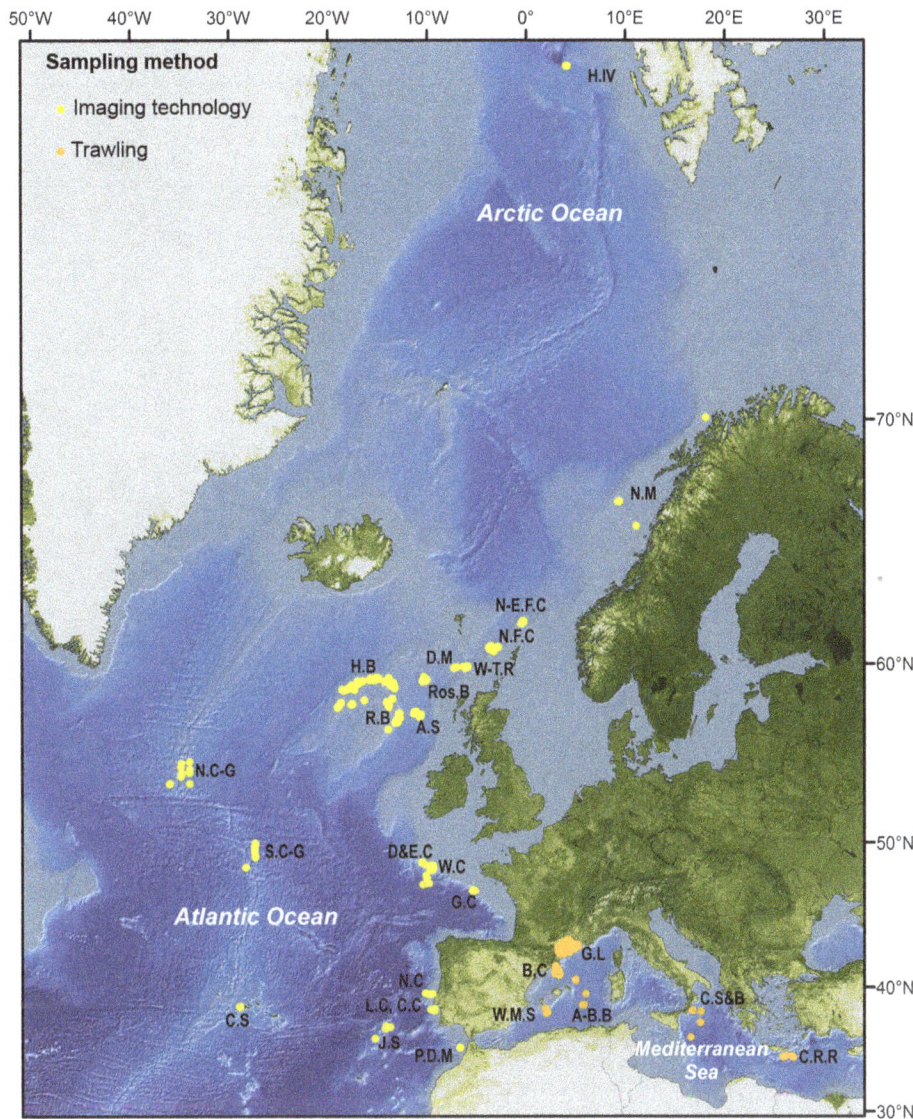

Figure 1. Locations of the study sites sampled with imaging technology (ROVs, manned submersible, towed camera systems) and trawling. A-B.B = Algero-Balearic Basin (W. Med.), A.S = Anton Dohrn Seamount, B.C = Blanes Canyon (NW Med.), C.C = Cascais Canyon, C.S = Condor Seamount, Calabrian Slope & Basin = C.S&B, Crete-Rhodes Ridge = C.R.R, D&E.C = Dangeard & Explorer Canyons, D.M = Darwin Mounds, G.L.C = Gulf of Lion canyons (NW Med.), G.L = Gulf of Lion, G.C = Guilvinec Canyon, H.B = Hatton Bank, H.IV = HAUSGARTEN, station IV, J.S = Josephine Seamount, L.C = Lisbon Canyon, N.C = Nazaré Canyon, N.C-G = North Charlie Gibbs Fracture Zone, N-E.F.C = North-East Faroe-Shetland Channel, N.F.C = North Faroe-Shetland Channel, N.W = Norwegian margin, P.D.M = Pen Duick Alpha/Beta Mound, R.B = Rockall Bank, Ros.B = Rosemary Bank, S.C = Setúbal Canyon, S.C-G = South Charlie Gibbs Fracture Zone, W.C = Whittard Canyon, W.M.S = Western Mediterranean slope, W-T.R = Wyville-Thomson Ridge.

presents the results on the distribution and densities of marine litter obtained during these surveys, with additional data provided by the UK's Mapping the Deep project as well as other previous projects. It provides a unique large-scale analysis of litter on the seafloor across different physiographic settings and depths.

Materials and Methods

Study areas

Data were gathered from surveys conducted during research cruises led by various European institutions between 1999 and 2011. A total of 32 sites in the northeastern Atlantic Ocean, Arctic Ocean and Mediterranean Sea were surveyed (Table 1; Figure 1). Surveyed sites were located on continental shelves and slopes,

submarine canyons, seamounts, banks, mounds, ocean ridges and deep basins, at depths ranging from 35 to 4500 meters (Table 1).

Sampling methods

Sampling methods included both imaging technology (still photograph and video) and fishing trawls (Figure 1; Table 2). The Atlantic sites were surveyed uniquely using imaging technology, whilst sites located in the Mediterranean Sea were primarily investigated by trawling (except for some ROV transects in the Blanes submarine canyon). Video footage was collected by different ROVs (*Genesis, Isis, Liropus, Luso, Lynx, SP* and *Victor 6000*), manned submersible (*JAGO*, GEOMAR) and towed camera systems (Seatronics and the HD-video hopper video system). Still photographs were taken with the Ocean Floor

Table 2. Information on each platform used to collect video and photographs for the collection of data on litter densities and distribution on the seafloor of European waters.

Sampling platform	Name	Format	N° of samples	Total area surveyed (m²)	Field of view (m)	References
Manned submersible	*Jago*	video	13	5561	1.5	[95]
ROVs	*Luso*	video	8	35587	3.6–4.4	[15]
	Sp	video	44	29749	2.3	[15]
	Isis	video	64	167308	2.0	[31]
	Genesis	video	20	86700	2.6	[96]
	Liropus	video	4	19867	3.0	[97]
	Lynx	video	19	3750	1.0	[98]
	Victor 6000	video	6	421840	10.0	[46]
Towed camera systems	Seatronics	video	194	158528	1.5	[99]
	HD video hopper system	video	6	21490	3.0	[100]
	Ocean Floor Observation System	photographs	2882	8570	0.8–11.6	[43]

Further technical information about each platform can be found in the indicated references.

Observation System (OFOS) at the HAUSGARTEN observatory, station IV. Technical details about each platform can be found elsewhere (see Table 2). Trawl samples were collected using two different gears: a net (GOC 73) with a 20 mm-diamond stretched mesh size at the cod-end [56] and an otter trawl Maireta System (OTMS), with a cod-end mesh size of 40 mm and an outer cover of 12 mm [29,57].

Analysis of image data

Protocols for video analysis varied slightly according to the platform used, but followed the same general outline. The entire footage was visualised and the number of litter items and depth recorded. Each litter item was classified into six different categories: plastic (all plastic with exception of fishing line and net), derelict fishing gear (fishing line or net), metal, glass, clinker (residue of burnt coal). Because of the low densities found at all sites, paper and cardboard, fabric, wood and unidentified items were grouped in the same category (other items). Although fishing lines and nets are mostly made of plastic, fishing gear was considered as a separate litter category because of our knowledge on its source and social implications and the particular impacts of this type of litter, such as ghost fishing and entanglement.

For each dive (sample), the area covered was calculated by multiplying the linear distance on the seafloor (off bottom footage were excluded from the analysis) by the average width of view of each of the platforms (Table 2).

For data derived from still photographs (OFOS), all images along each transect (taken at 30 s to 50 s-intervals) were analysed for the presence of litter items. Parallel laser points on the images allowed calculations of the area for each image; ranging between 0.8 and 11.6 m². For OFOS, each image was considered to be a separate sample, while for video data, each dive was considered a single sample.

Trawl data

Hauls in the Gulf of Lion (shelf and submarine canyons) were performed with a bottom trawl equipped with a GOC 73 net [56]. After trawling, litter items were counted and classified into the different categories (see above).

Trawling at the other Mediterranean sites was performed using an otter trawl Mareita System (OTMS). All litter items were separated and classified into different categories (see above) and weighed, after excess water and mud had been removed. The use of weight rather than number to quantify litter was based on the high abundance of broken plastics (from whole plastic bags to very small (<0.5 cm) pieces of plastics) and broken glass, which impeded the quantification of single items without overestimating abundances of certain categories over others [29].

Data analysis

For each sample (video and still photographs), litter density was estimated as items of litter hectare^{-1} (ha; 10,000 m²) of seafloor surveyed. For trawl data where litter was measured in weight, litter density was estimated as kg of litter ha^{-1}. Sites were grouped into 6 different groups according to physiographic characteristics (Table 1); (1) continental shelves; (2) continental slopes (excluding submarine canyons); (3) submarine canyons; (4) seamounts, banks and mounds; (5) ocean ridges and (6) deep basins. Tests for investigating differences among litter densities across physiographic settings were done separately according to the unit in which litter density was estimated (number ha^{-1} or weight ha^{-1}). For both cases, the data were not normally distributed but variances were equal, therefore, the non-parametric Kruskal-Wallis rank sum test followed by a multiple comparison test (Dunn's pairwise comparison) were performed using the statistical package R. Variation in litter composition between physiographic settings were tested for significance using ANOSIM (Analysis of similarity) in PRIMER v6 software [58]. Bray-Curtis similarity [59] was calculated on log(x+1) transformation of the percentage contribution of litter type for each of the physiographic settings, across the entire data set. A similarity percentage analysis (SIMPER) was applied to identify the discriminating feature of the dissimilarities and similarities between physiographic settings.

Results

Litter density

Litter was found at all sites and all depths (from 35 m down to 4500 m) sampled. Most common litter items included plastic bags,

Figure 2. Litter items on the seafloor of European waters. A = Plastic bag entrapped by a small drop stone harbouring sponges (*Cladorhiza gelida, Caulophacus arcticus*), shrimps (*Bythocaris* sp.) and a crinoid (*Bathycrinus carpenterii*) recorded by an OFOS at the HAUSGARTEN observatory (Arctic) at 2500 m; B = Litter recovered within the net of a trawl in Blanes open slope at 1500 m during the PROMETO V cruise on board the R/V "García del Cid"; C = "Heineken" beer can in the upper Whittard canyon at 950 m water depth with the ROV Genesis; D = Plastic bag in Blanes Canyon at 896 m with the ROV "Liropus"; E = "Uncle Benn's Express Rice" packet at 967 m in Darwin Mound with the ROV "Lynx" (National Oceanography Centre, UK); F = Cargo net entangled in a cold-water coral colony at 950 m in Darwin Mound with the ROV "Lynx" (National Oceanography Centre, UK).

glass bottles and derelict fishing lines and nets (Figure 2). Locations with highest litter densities (>20 items ha^{-1}) included the Lisbon Canyon, the Blanes Canyon, the Guilvinec Canyon, and the Setúbal Canyon (Table 1; Figure 3). Sites with intermediate litter density (between 10 and 20 items ha^{-1}) were found on the Condor Seamount, the Wyville-Thomson Ridge, the continental slope of the HAUSGARTEN observatory and the Cascais Canyon (Figure 3). Low densities (between 2 and 10 items ha^{-1}) were recorded on the Darwin Mounds, off the Norwegian margin, in Dangeard and Explorer Canyons, on the Josephine Seamount, in the Nazaré Canyon, on the Rosemary Bank, south of the Charlie-Gibbs Fracture Zone and on the Pen Duick Alpha and Beta Mounds (Figure 3). The lowest litter density (<2 items ha^{-1}) was found on the Hatton Bank, the continental slope on the northern side of the Faroe-Shetland Channel, on the Anton Dohrn Seamount, in the Whittard Canyon, on the Rockall Bank, north of the Charlie-Gibbs Fracture Zone, and in the Gulf of Lion (in both the continental shelf and submarine canyons). Sites with higher litter density were found principally closer to shore (Figure 4), but there were exceptions, such as the samples from the Gulf of Lion where litter densities were low (Table 1).

The sites sampled by trawling in the Mediterranean revealed a relatively even distribution of litter but with a higher density on the continental slope, south of Palma de Mallorca (western Mediterranean) with a mean ($\pm$SE) of 4.0$\pm$1.8 kg of litter ha^{-1} as opposed to densities ranging between 0.7 and 1.8 kg of litter ha^{-1} at the other sites (Figure 5).

When grouping all sites into physiographic settings, there were significant differences in litter density (items ha^{-1}) between the various groups (Kruskal-Wallis $\chi^2 = 26.68$; p<0.01; DF = 4). Multiple comparisons tests indicated that litter density in submarine canyons was significantly higher than those from all other physiographic settings, reaching an average ($\pm$ SE) of 9.3$\pm$2.9 items ha^{-1} (Figure 6a). Litter density on seamounts, mounds and banks was similar to the densities found on the continental slopes with mean ($\pm$ SE) densities of 5.6$\pm$1.0 and 4.1$\pm$2.1 items ha^{-1}, respectively (Figure 6a). Mean ($\pm$ SE) litter density for continental shelves and ocean ridges was 2.2$\pm$0.8 and 3.9$\pm$1.3 items ha^{-1}, respectively (Figure 6a). For Mediterranean sites, where litter density was quantified by weight rather than number of items, no significant differences were found in litter density between the three different physiographic settings (Kruskal-Wallis $\chi^2 = 3.88$; p = 0.144; DF = 2). However, litter density in deep basins was slightly higher (1.55$\pm$0.57 kg ha^{-1}) compared to continental slopes (1.36$\pm$0.34 kg ha^{-1}) and submarine canyons (0.71$\pm$0.25 kg ha^{-1}) (Figure 6b).

Litter composition

There was a high variability in the composition of litter across the different sites (Table 3). A total of 546 litter items were encountered throughout all sites surveyed with imaging technology. Plastic and derelict fishing gear were the most abundant litter items. Plastic represented 41% of the litter items, whilst derelict fishing gear accounted for 34% of the total. Clinker, glass and metal were least common (1, 4 and 7%, respectively). Items classified as "other items" accounted for 13% of the litter items encountered in sites surveyed by imaging technology and included wood, paper/cardboard, clothing, pottery, and unidentified material. Analysis of litter density from trawl surveys found plastic to be the most common litter type to be recovered (found in 98% of the trawls), followed by clinker (73%), fabric (48%), derelict fishing gear (33%), metal (31%) and glass (28%).

Results from ANOSIM showed that there were significant differences in litter composition between physiographic settings (1-way ANOSIM; Global R = 0.32; p<0.001), the analysis also showed some settings to be similar (Table S1). There were no significant differences between litter composition in submarine canyons and continental shelves (R = 0.01; p = 0.58). According to SIMPER analysis (Table S2), the similarity in composition between submarine canyons and continental shelves was mostly driven by plastic. Plastic was the dominant litter category for both settings (Figure 7). Litter composition on ocean ridges and on seamounts, banks and mounds did not show significant differences in litter composition (R = 0.17; p = 0.06), due to a predominance of derelict fishing gear (Figure 7). Finally, litter composition found on continental slopes was similar to deep basins (R = −0.11; p = 0.87). Clinker and plastic were the categories contributing most to the similarities between these two physiographic settings.

Discussion

The occurrence of litter on the seafloor has been far less investigated than in surface waters or on beaches, principally because of the high cost and the technical difficulties involved in sampling the seafloor at bathyal and abyssal depths [21,60].

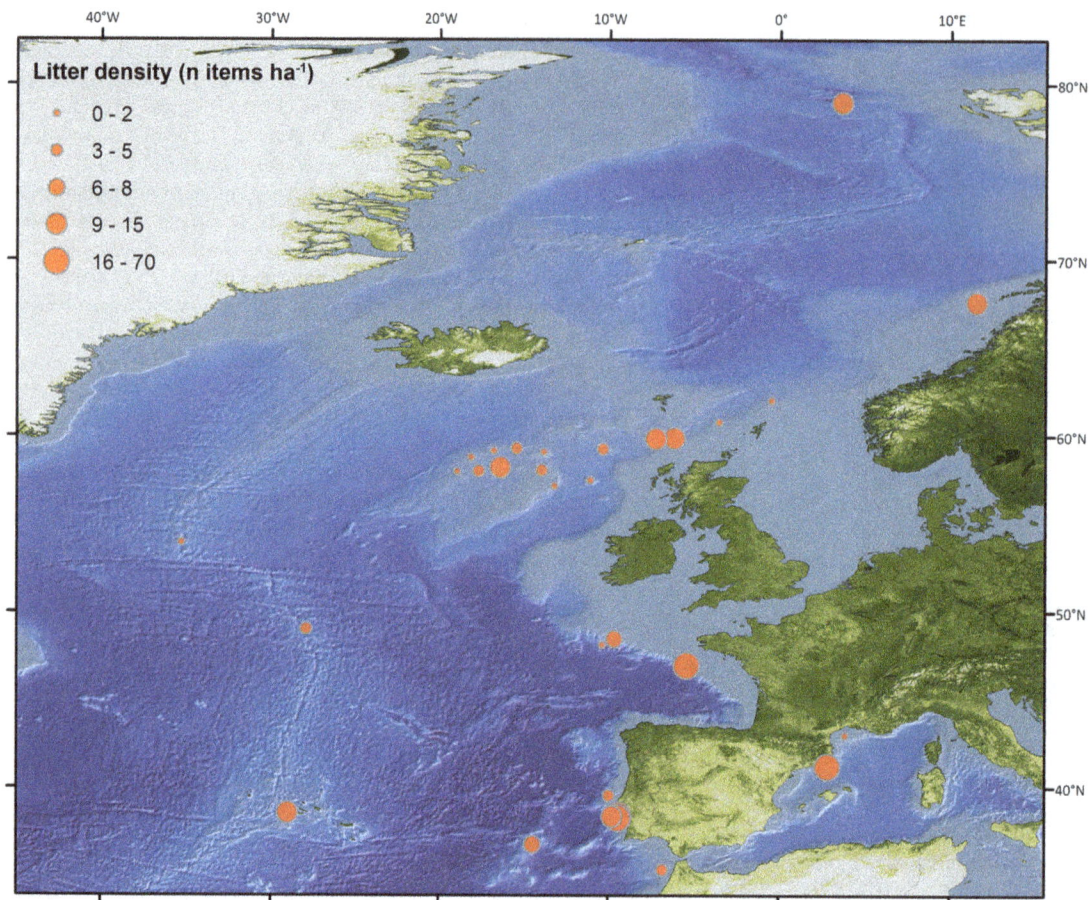

Figure 3. Litter densities (number of items ha^{-1}) in different locations across European waters obtained with ROVs, towed camera systems, manned submersible and trawls.

Considering such limitations and poor knowledge on litter accumulation in deep waters, every survey is of great value for obtaining information on litter density and distribution. In the present study, we integrated data collected during numerous cruises over a large regional scale into a single analysis, providing insight on the density and composition of litter across a wide variety of seafloor settings and over a large geographical area in European waters. Although standardisation of the data permitted comparisons between sites, dissimilarities in the sampling equipment implies that the results should be treated with caution. Furthermore, differences in the areas of the seafloor surveyed between locations may lead to overestimations or underestimations of the litter density. Also, studying litter from trawls introduces the issue of quantification units (number vs. weight), with no correct solution. When using number of items, certain litter categories may be overestimated such as plastic or glass that can break into many small pieces. As a counterpart, if weight is used, the abundance of litter type with different weights (e.g. heavy clinker vs. light plastic) cannot be compared. Ideally, both units for litter quantification will help to understand better trends, but the EU Marine Strategy Framework Directive stresses that for monitoring litter in the marine environment, number is mandatory whilst weight is only recommended [2].

Litter was found at all the locations surveyed, from sites close to population centres such as the Gulf of Lion or the Lisbon Canyon to as far as the South Charlie-Gibbs Fracture Zone on the Mid-Atlantic Ridge, located at about 2000 km from land. Litter was found from shallow waters (35 meters in Gulf of Lion) down to 4500 meters (Cascais Canyon). Such records were not surprising, as litter is known to be present in all seas and oceans of the planet, as remote as the Southern Ocean [21] and at depths as deep as 7216 m in the Ryuku trench, south of Japan [61]. The range of litter densities found on our study sites was within the same order of magnitude to the ones found on the seabed in other parts of the globe (North America [55,62,63], China [54], Japan [64,65]) and for other locations in Europe [28,44,45,47,48]. On the other hand, macro litter densities on the seabed were higher than reported for surface waters [32,66–69]. At the surface, floating litter tends to accumulate in frontal areas but eventually reaches the seabed when heavily covered by fouling organisms [70] or loaded with sediments. Contrary to a common notion that most plastic items float at the sea surface it has been estimated that 70% of the plastic sinks to the seafloor [23]. This results in macro litter accumulation on the seabed rather than in the open sea [21]. For example, on the seafloor of the Mediterranean Sea, our data showed much higher litter densities (0.4 to 48 litter items ha^{-1}) than that estimated to float at the surface (0.021 items ha^{-1}; [1]). Alternatively, floating litter may be transported for considerable distances and get washed ashore [71,72]. Litter density on the coastline is typically higher than on the seafloor given that there is an additional input of waste coming from inland sources (e.g. man-made drainage systems, recreational usage, rivers, winds, etc.) [71,73]. On European coasts, litter densities can exceed 30,000 litter items per linear km [1,41,74], while much higher densities

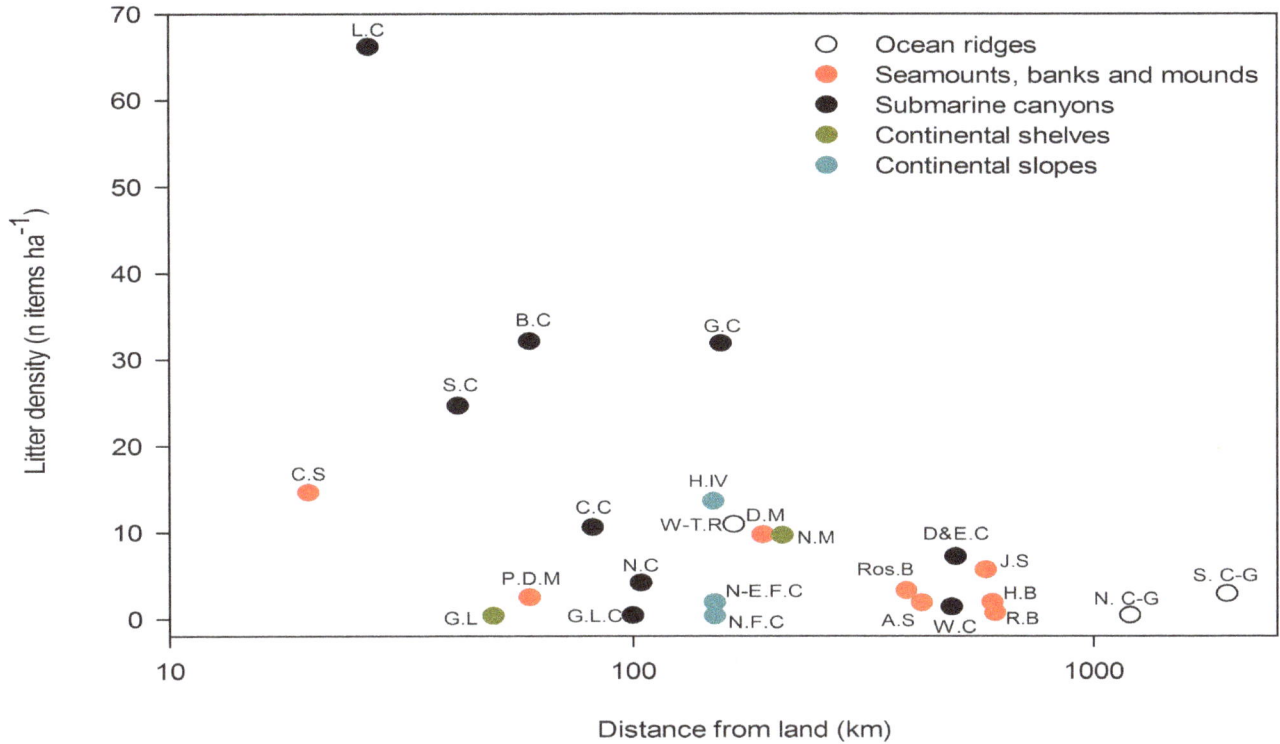

Figure 4. Litter densities (number of items ha^{-1}) in different locations across European waters according to their closest distances from land. x axis is in a Log$_{10}$ scale. A.S = Anton Dohrn Seamount, B.C = Blanes Canyon (NW Med.), C.C = Cascais Canyon, C.S = Condor Seamount, D&E.C = Dangeard & Explorer Canyons, D.M = Darwin Mounds, G.L.C = Gulf of Lion canyons (NW Med.), G.L = Gulf of Lion, G.C = Guilvinec Canyon, H.B = Hatton Bank, H.IV = HAUSGARTEN, station IV, J.S = Josephine Seamount, L.C = Lisbon Canyon, N.C = Nazaré Canyon, N.C-G = North Charlie Gibbs Fracture Zone, N-E.F.C = North-East Faroe-Shetland Channel, N.F.C = North Faroe-Shetland Channel, N.W = Norwegian margin, P.D.M = Pen Duick Alpha/Beta Mound, R.B = Rockall Bank, Ros.B = Rosemary Bank, S.C = Setúbal Canyon, S.C-G = South Charlie Gibbs Fracture Zone, W.C = Whittard Canyon, W-T.R = Wyville-Thomson Ridge.

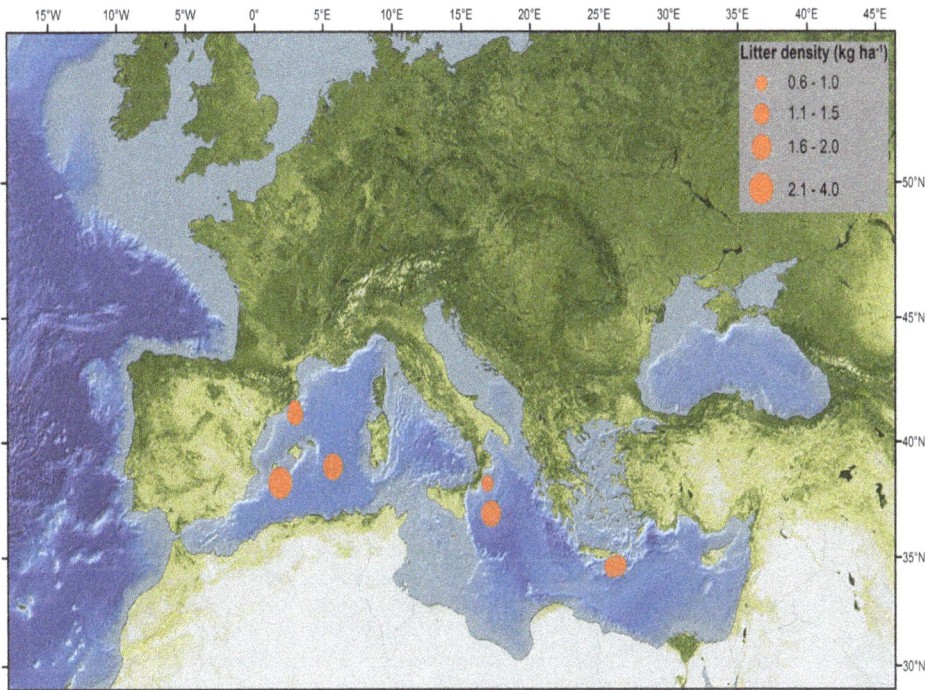

Figure 5. Litter densities (kg ha^{-1}) in different locations across the Mediterranean Sea obtained from trawl surveys.

A

B

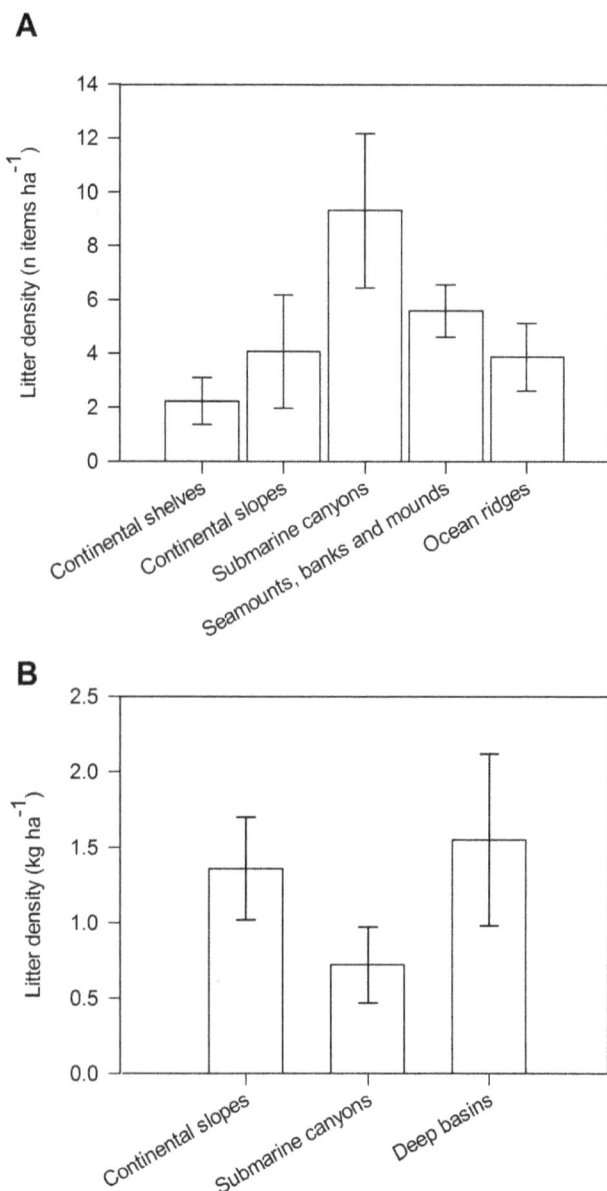

Figure 6. Mean litter density (± standard error) in A = number of items ha^{-1} and B = in kg of items ha^{-1}, across different physiographic settings in European waters.

continental shelf down into Monterey Canyon [78]. Such phenomena may explain why continental shelves were the settings with overall lowest litter density, whilst submarine canyons had the highest litter concentration. Litter levels on seamounts, banks, mounds and ocean ridges were characterised by intermediate levels when compared to other physiographic settings. They are typically located far away from coastal areas where the main anthropogenic activities include fishing [79] and seabed mining [80,81]. The presence of litter on these settings is of concern because they harbor Vulnerable Marine Ecosystems (VMEs) (such as cold-water corals and hydrothermal vents) that have reduced capacity to recover from disturbance events and for which conservation is a global priority [82].

The types of accumulated litter can provide an indication on the human activities impacting a particular location. However, one must be cautious and consider the differences in the buoyancy and longevity of the different types of litter. For example, while some plastics sink to the seafloor, others float on the surface and are able to travel great distances before eventually sinking far from their initial dumping locations, following biofouling and degradation [23]. On the other hand, glass, metal and clinker will sink rapidly and are expected to be recovered from the seafloor close to sites where they were initially released. Cardboard and fabrics (of organic origin) will break down quickly, implying that such items will not reach the deep ocean with the frequency of more resistant materials such as plastic and negatively buoyant items such as glass, metal and clinker. Although it is difficult to determine the exact source of the litter observed on the seafloor, the dominant litter category can be used as an indicator to separate ocean and terrestrial sources [15,29,31,78]. Plastic (other than derelict fishing gear) was the most abundant litter category in submarine canyons, continental shelves and continental slopes. The predominance of plastics in submarine canyons reaffirms that litter accumulation in these habitats comes from coastal and land sources and that submarine canyons act as conduits for litter transport from continental shelves into deeper waters [21,28,29,31,47,78]. Therefore, submarine canyons can be considered to be accumulation zones of land-based marine litter in the deep sea. In fact, submarine canyons are areas where macrophyte detritus that originates from coastal areas accumulates in high quantities. This results in a localised increase of organic matter and high abundances of associated fauna, dominated by deposit and suspension-feeding invertebrates [83–85]. Since some deposit-feeders (e.g. holothurians) have been shown to select plastic fragments over sediment grains under laboratory conditions [7], the accumulation of plastics in submarine canyons could have detrimental effects for these ecologically important deep-sea organisms. Furthermore, plastic fragments contain a wide variety of persistent organic pollutants (POPs) that may accumulate in the consumer's tissues and can be transferred upwards in the trophic webs to predators, including humans [86].

Derelict fishing gear was the main litter item found on seamounts, banks, mounds and ocean ridges implying that, unlike submarine canyons, fishing activities are the major source of litter at those settings. Seamounts and banks are targeted by commercial fishing activities as they are often highly productive areas supporting dense aggregations of commercially valuable fish and shellfish [87]. At other locations where recreational [55,88] and commercial [28,54,62,89] fishing activities are intense, derelict fishing gear dominated the litter on the seabed. It was beyond the scope of this study to evaluate the impacts caused by derelict fishing gear, but numerous studies have shown diverse impacts including ghost fishing [16,90] and entanglement by sessile invertebrates such as corals [15], as well as causing damage to

have been reported for beaches in Indonesia [75] or on the beaches along Armação dos Buzios, Rio de Janeiro, Brazil [76]. However, comparisons between studies are challenging considering differences in the size of the litter items sampled and the sampling methodology used [77].

Our data showed a general increase in litter density in locations closer to the shore, a pattern previously reported for the French Mediterranean coast [47] and off California [55]. Nevertheless, low litter densities in some near-shore sites (e.g. Gulf of Lion or Faroe-Shetland channel) suggest that many other factors (such as geomorphology, hydrography and human activity) affect litter distribution and accumulation rates [29]. In the Gulf of Lion, Galgani et al. [47] suggested that low litter density on the shelf was caused by strong water flow from the Rhone River, transporting litter down south to deeper waters. A similar situation occurs in Monterey Bay where sediment and litter are being swept off the

Table 3. Composition of litter (%) in different locations on the seafloor of European waters.

Location	Derelict fishing gear	Glass	Metal	Plastic	Other items	Clinker
ATLANTIC						
Continental slopes						
North Faroe-Shetland Channel	100.0	0.0	0.0	0.0	0.0	0.0
North-East Faroe-Shetland Channel	100.0	0.0	0.0	0.0	0.0	0.0
Continental shelf						
Norwegian Margin	80.0	0.0	0.0	20.0	0.0	0.0
Submarine canyons						
Dangeard & Explorer Canyons	72.2	0.0	0.0	16.7	11.1	0.0
Nazaré Canyon	37.1	0.0	17.1	25.7	20.0	0.0
Lisbon Canyon	9.2	0.0	1.5	86.2	3.1	0.0
Setúbal Canyon	8.7	4.3	4.3	30.4	52.2	0.0
Cascais Canyon	9.1	0.0	0.0	54.5	36.4	0.0
Guilvinec Canyon	43.8	0.0	0.0	43.8	6.3	6.3
Whittard Canyon	28.6	7.1	14.3	42.9	0.0	7.1
Seamounts, banks and mounds						
Anton Dohrn Seamount	0.0	0.0	100.0	0.0	0.0	0.0
Condor Seamount	85.5	14.5	0.0	0.0	0.0	0.0
Josephine Seamount	42.9	28.6	14.3	0.0	14.3	0.0
Hatton Bank	87.5	0.0	12.5	0.0	0.0	0.0
Rockall Bank	33.3	0.0	66.7	0.0	0.0	0.0
Rosemary Bank	66.7	0.0	33.3	0.0	0.0	0.0
Pen Duick Alpha/Beta Mound	75.0	0.0	25.0	0.0	0.0	0.0
Darwin Mounds	10.0	0.0	15.0	60.0	15.0	0.0
Ocean ridges						
North Charlie Gibbs Fracture Zone	0.0	0.0	100.0	0.0	0.0	0.0
South Charlie Gibbs Fracture Zone	0.0	28.6	28.6	28.6	14.3	0.0
Wyville-Thomson Ridge	85.7	0.0	14.3	0.0	0.0	0.0
MEDITERANEAN						
Continental slopes						
Calabrian Slope (Central Med.)	13.2	0.0	8.4	36.2	26.6	15.5
Western Mediterranean Slope	21.6	0.6	0.2	12.1	0.6	64.9
Crete-Rhodes Ridge (E. Med.)	1.6	9.3	6.0	17.0	20.5	45.5
Blanes slope (NW Med.)	2.3	7.9	8.4	12.6	11.6	57.1
Continental shelf						
Gulf of Lion (NW Med.)	0.0	0.0	0.0	88.9	11.1	0.0
Submarine canyons						
Blanes Canyon (NW Med.)	3 (0.2)	3 (4.9)	6 (2.2)	78 (76.3)	9 (1.7)	0 (14.7)
Gulf of Lion Canyons (NW Med.)	0.0	0.0	0.0	67.3	32.7	0.0
Deep basins						
Algero-Balearic Basin (W. Med.)	16.5	0.8	29.6	14.0	2.1	37.0
Crete-Rhodes Ridge (E. Med.)	0.0	9.7	25.0	19.5	7.2	38.5
Calabrian Basin (Central Med.)	0.5	6.7	0.7	5.9	36.1	50.1
ARCTIC						
Continental slope						
HAUSGARTEN, station IV	2.5	2.5	2.5	60	32.5	0

*Numbers in parentheses refer to trawl surveys.

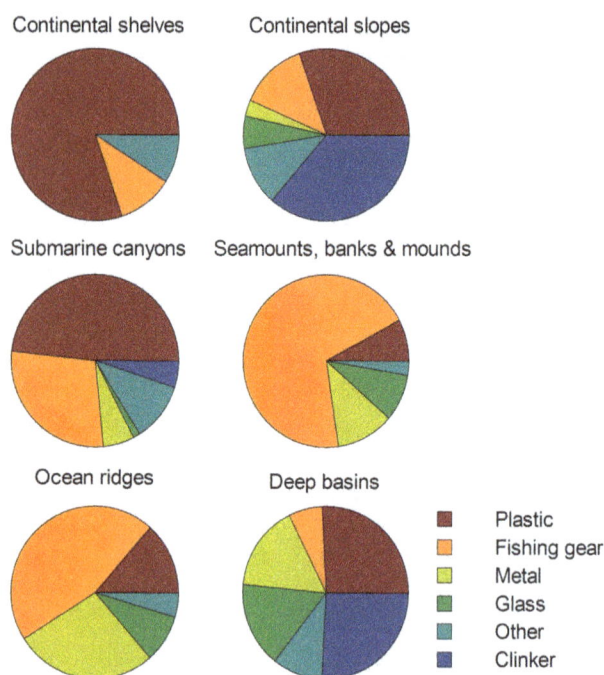

Figure 7. Litter composition in different physiographic settings across European waters.

fishing equipment [91]. Discarded trawl gear can also have a compounding effect by trapping more mobile litter resulting in a litter 'depot' that has a greater impact than single pieces of litter [31]. Since most fishing equipment (lines and nets) is made mostly of highly resistant plastics, such negative effects will likely persist for a long time. Sites located in deep basins and continental slopes were dominated by clinker. Clinker, the residue of burnt coal, was commonly dumped from steam ships from the late 18th century and well into the 20th century. In the Mediterranean Sea, its occurrence on the deep seafloor has been shown to coincide with such shipping routes [29]. However, it is important to acknowledge that in this study, deep basins and continental slopes were principally sampled by trawling and it is difficult to determine if the differences in litter composition with other physiographic settings are the results of differences in the sampling methodology, particularly since clinker is difficult to identify from underwater footage. Indeed, clinker was present in non-quantitative trawls undertaken at HAUSGARTEN (Bergmann, unpublished data), but could not be detected on images from the seafloor. Similarly, a high abundance of clinker was recovered from trawl surveys in Blanes Canyon that could not be identified in analysis of ROV footage from the same area (Table 3). Given that most of the clinker present on the seafloor was dumped over 100 years ago, sedimentation will have buried it, which would explain the differences in clinker quantification between images and trawl data. The deep seafloor is a passive accumulation area for litter, integrating information over long-time periods. If trawls are able to recover heavy clinker deposited on the seafloor over a century ago, these gears must be retrieving at the same time all of the lighter and most recent litter items, such as plastic for example, that have been accumulating only in the last 50 years. Overall, the composition of litter found on the seafloor showed some dissimilarity with the composition found on the coasts or in surface waters. Although plastics are dominant in all settings [70],

some areas of the seafloor investigated here and elsewhere [28,44,45,54,78] harbour significant quantities of non-buoyant litter such as glass, metal and clinker, directly dumped from ships but that are seldom found in surface waters [41,68] or on the coasts [41,72]. The coasts and surface waters are a source of litter items for the open seas and all this litter, sooner or later, will sink to the seafloor where it accumulates.

The most common method used to provide data on benthic marine litter has been trawling, typically as a parallel objective to surveys directed to fish or benthic organism sampling [53]. With the recent development of optical methods fitted to platforms such as submersibles, ROV and drop-down systems, the use of underwater imaging technology has greatly increased our ability to quantify deep-sea litter. Both methods (imaging technology and trawling) have distinct assets for studying benthic litter that should be used in conjunction to best understand the dynamics of pollution on the seafloor. Video surveys can provide data for areas where topography is complex (e.g seamounts or canyon walls), habitats made by structure-building organisms (e.g. cold-water corals), or dynamic systems (e.g. hydrothermal vents and cold seeps), that cannot be accessed with a trawl [53]. Furthermore, imaging is a non-intrusive method that does not remove benthic organisms or damage the environment. On the other hand, a trawl has the advantages of recovering litter items of very small size (e.g. small plastic fragments) or that are buried in the sediments (e.g. clinker), which otherwise would not be detected through imaging technology. In addition, litter items collected with a trawl can be analysed in the laboratory to obtain further important information, such as state of degradation or colonisation by fouling organisms [92]. Such data will help understand sinking processes of plastic, facilitate the identification of their location of arrival into the ocean and provide information on the impacts of litter on marine organisms.

The large quantities of litter reaching the deep ocean floor is a major issue worldwide, yet little is known about its sources, patterns of distribution, abundance and, particularly, impacts on the habitats and associated fauna [1]. At present, density of litter in the deep sea is lower than found on some heavily polluted beaches [33,93], but unlike the coastal zone, only a tiny fraction of the (deep) seafloor has been surveyed to date. Furthermore, microplastic accumulation may become an important component of pollution in deep-sea ecosystems [94] that urgently needs to be evaluated. Our results for European waters show that litter sources are distinct across different physiographic settings and that their abundance is variable, most probably guided by a complex set of interactions between physiography, anthropogenic activities and hydrography. It is important that in the future, large-scale assessments are done in a standardised manner to understand fully the scale of the problem and set the necessary actions to prevent the accumulation of litter in the marine environment.

Supporting Information

Table S1 Results of analyses of similarity (ANOSIM) evaluating variation in the composition of litter among physiographic settings. RIDGE: ocean ridges; CANY: submarine canyons; SHELF: continental shelves; SLOPE: continental slopes; SBM: seamounts, banks and mounds; BASIN: deep basins.

Table S2 Similarity percentage analysis (SIMPER) of litter composition for each pooled physiographic settings (based on similarities revealed by ANOSIM) and the contribution of litter category to group similarity.

Acknowledgments

The authors would like to thank the captains, crews and scientific parties of all cruises for their help and support during the data collection. PT would like to thank Gideon Mordecai for analytical work and Doug Masson. Finally, the authors would like to thank Martin Thiel and two other anonymous reviewers, whose suggestions and comments greatly improved the manuscript. This is publication number 33575 of the Alfred-Wegener-Institut Helmholtz-Zentrum für Polar- und Meeresforschung.

Author Contributions

Conceived and designed the experiments: CKP ERL CHSA TA MB MC JBC JD GD FG KLH VAIH EI DOBJ GL TM JNGP AP HS IT XT DVR PT. Performed the experiments: CKP ERL CHSA TA MB MC JBC JD GD FG KLH VAIH EI DOBJ GL TM JNGP AP HS IT XT DVR PT. Analyzed the data: CKP. Wrote the paper: CKP.

References

1. UNEP (2009) Marine Litter: A Global Challenge. Nairobi. 232 p.
2. Galgani F, Hanke G, Werner S, De Vrees L (2013) Marine litter within the European Marine Strategy Framework Directive. ICES J. Mar. Sci. 70: 1055–1064.
3. Moser ML, Lee DS (1992) A 14-year survey of plastic ingestion by western north-atlantic seabirds. Colon. Waterbirds 15: 83–94.
4. Ryan PG (1988) Effects of ingested plastic on seabird feeding: Evidence from chickens. Mar. Pollut. Bull. 19: 125–128.
5. Bjorndal KA, Bolten AB, Lagueux CJ (1994) Ingestion of marine debris by juvenile sea turtles in coastal Florida habitats. Mar. Pollut. Bull. 28: 154–158.
6. Tomás J, Guitart R, Mateo R, Raga JA (2002) Marine debris ingestion in loggerhead sea turtles, *Caretta caretta*, from the Western Mediterranean. Mar. Pollut. Bull. 44: 211–216.
7. Graham ER, Thompson JT (2009) Deposit- and suspension-feeding sea cucumbers (Echinodermata) ingest plastic fragments. J. Exp. Mar. Biol. Ecol. 368: 22–29.
8. Carson HS (2013) The incidence of plastic ingestion by fishes: From the prey's perspective. Mar. Pollut. Bull. 74: 170–174.
9. Neilson JL, Straley JM, Gabriele CM, Hills S (2009) Non-lethal entanglement of humpback whales (*Megaptera novaeangliae*) in fishing gear in northern Southeast Alaska. J. Biogeogr. 36: 452–464.
10. Williams R, Ashe E, O'Hara PD (2011) Marine mammals and debris in coastal waters of British Columbia, Canada. Mar. Pollut. Bull. 62: 1303–1316.
11. Allen R, Jarvis D, Sayer S, Mills C (2012) Entanglement of grey seals *Halichoerus grypus* at a haul out site in Cornwall, UK. Mar. Pollut. Bull. 64: 2815–2819.
12. Carr A (1987) Impact of nondegradable marine debris on the ecology and survival outlook of sea turtles. Mar. Pollut. Bull. 18: 352–356.
13. Schrey E, Vauk GJM (1987) Records of entangled gannets (*Sula bassana*) at Helgoland, German Bight. Mar. Pollut. Bull. 18: 350–352.
14. Chiappone M, Dienes H, Swanson DW, Miller SL (2005) Impacts of lost fishing gear on coral reef sessile invertebrates in the Florida Keys National Marine Sanctuary. Biol. Conserv. 121: 221–230.
15. Pham CK, Gomes-Pereira JN, Isidro EJ, Santos RS, Morato T (2013) Abundance of litter on Condor seamount (Azores, Portugal, Northeast Atlantic). Deep-Sea Res. Part II-Top. Stud. Oceanogr. 98: 204–208.
16. Brown J, Macfadyen G (2007) Ghost fishing in European waters: Impacts and management responses. Mar. Pol. 31: 488–504.
17. Winston JE (1982) Drift plastic—An expanding niche for a marine invertebrate? Mar. Pollut. Bull. 13: 348–351.
18. Barnes DKA, Milner P (2005) Drifting plastic and its consequences for sessile organism dispersal in the Atlantic Ocean. Mar.Biol. 146: 815–825.
19. Barnes DKA (2002) Biodiversity: Invasions by marine life on plastic debris. Nature 416: 808–809.
20. Derraik JGB (2002) The pollution of the marine environment by plastic debris: a review. Mar. Pollut. Bull. 44: 842–852.
21. Barnes DKA, Galgani F, Thompson RC, Barlaz M (2009) Accumulation and fragmentation of plastic debris in global environments. Philos. Trans. R. Soc. Lond. B Biol. Sci. 364: 1985–1998.
22. Sheavly SB, Register KM (2007) Marine debris & plastics: environmental concerns, sources, impacts and solutions. J. Polym. Environ. 15: 301–305.
23. Engler RE (2012) The complex interaction between marine debris and toxic chemicals in the ocean. Environ. Sci. Technol. 46: 12302–12315.
24. Andrady AL (2011) Microplastics in the marine environment. Mar. Pollut. Bull. 62: 1596–1605.
25. Cole M, Lindeque P, Fileman E, Halsband C, Goodhead R, et al. (2013) Microplastic ingestion by zooplankton. Environ. Sci. Technol. 47: 6646–6655.
26. Farrell P, Nelson K (2013) Trophic level transfer of microplastic: *Mytilus edulis* (L.) to *Carcinus maenas* (L.). Environ. Pollut. 177: 1–3.
27. Murray F, Cowie PR (2011) Plastic contamination in the decapod crustacean *Nephrops norvegicus* (Linnaeus, 1758). Mar. Pollut. Bull. 62: 1207–1217.
28. Galgani F, Leaute JP, Moguedet P, Souplet A, Verin Y, et al. (2000) Litter on the sea floor along European coasts. Mar. Pollut. Bull. 40: 516–527.
29. Ramirez-Llodra E, Company JB, Sardà F, De Mol B, Coll M, et al. (2013) Effects of natural and anthropogenic processes in the distribution of marine litter in the deep Mediterranean sea. Prog. Oceanogr. 118: 273–287.
30. Corcoran PL, Biesinger MC, Grifi M (2009) Plastics and beaches: A degrading relationship. Mar. Pollut. Bull. 58: 80–84.
31. Mordecai G, Tyler PA, Masson DG, Huvenne VAI (2011) Litter in submarine canyons off the west coast of Portugal. Deep Sea Res. Part II Top. Stud. Oceanogr. 58: 2489–2496.
32. Aliani S, Griffa A, Molcard A (2003) Floating debris in the Ligurian Sea, north-western Mediterranean. Mar. Pollut. Bull. 46: 1142–1149.
33. Ariza E, Jiménez JA, Sardá R (2008) Seasonal evolution of beach waste and litter during the bathing season on the Catalan coast. Waste Manag. 28: 2604–2613.
34. Gabrielides GP, Golik A, Loizides L, Marino MG, Bingel F, et al. (1991) Man-made garbage pollution on the Mediterranean coastline. Mar. Pollut. Bull. 23: 437–441.
35. Kornilios S, Drakopoulos PG, Dounas C (1998) Pelagic tar, dissolved/dispersed petroleum hydrocarbons and plastic distribution in the Cretan Sea, Greece. Mar. Pollut. Bull. 36: 989–993.
36. McCoy FW (1988) Floating megalitter in the eastern Mediterranean. Mar. Pollut. Bull. 19: 25–28.
37. Morris RJ (1980) Floating plastic debris in the Mediterranean. Mar. Pollut. Bull. 11: 125.
38. Shiber JG (1982) Plastic pellets on Spain's 'Costa del Sol' beaches. Mar. Pollut. Bull. 13: 409–412.
39. Thompson RC, Olsen Y, Mitchell RP, Davis A, Rowland SJ, et al. (2004) Lost at sea: Where is all the plastic? Science 304: 838.
40. Vauk GJM, Schrey E (1987) Litter pollution from ships in the German Bight. Mar. Pollut. Bull. 18: 316–319.
41. Van Cauwenberghe L, Claessens M, Vandegehuchte MB, Mees J, Janssen CR (2013) Assessment of marine debris on the Belgian Continental Shelf. Mar. Pollut. Bull. 73: 161–169.
42. Anastasopoulou A, Mytilineou C, Smith CJ, Papadopoulou KN (2013) Plastic debris ingested by deep-water fish of the Ionian Sea (Eastern Mediterranean). Deep-Sea Res. Part I Oceanogr. Res. Pap. 74: 11–13.
43. Bergmann M, Klages M (2012) Increase of litter at the Arctic deep-sea observatory HAUSGARTEN. Mar. Pollut. Bull. 64: 2734–2741.
44. Galgani F, Burgeot T, Bocquene G, Vincent F, Leaute JP, et al. (1995b) Distribution and abundance of debris on the continental shelf of the Bay of Biscay and in Seine Bay. Mar. Pollut. Bull. 30: 58–62.
45. Galgani F, Jaunet S, Campillo A, Guenegen X, His E (1995a) Distribution and abundance of debris on the continental shelf of the north-western Mediterranean Sea. Mar. Pollut. Bull. 30: 713–717.
46. Galgani F, Lecornu F (2004) Debris on the seafloor at "Hausgarten". Reports on Polar and Marine Research 488: 260–262.
47. Galgani F, Souplet A, Cadiou Y (1996) Accumulation of debris on the deep sea floor off the French Mediterranean coast. Mar. Ecol. Prog. Ser. 142: 225–234.
48. Galil BS, Golik A, Tuerkay M (1995) Litter at the bottom of the sea: A sea bed survey in the eastern Mediterranean. Mar. Pollut. Bull. 30: 22–24.
49. Katsanevakis S, Katsarou A (2004) Influences on the distribution of marine debris on the seafloor of shallow coastal areas in Greece (Eastern Mediterranean). Water, Air, & Soil Pollution 159: 325–337.
50. Kidd RB, Huggett QJ (1981) Rock debris on abyssal plains in the northeast Atlantic - a comparison of epibenthic sledge hauls and photographic surveys. Oceanol. Acta 4: 99–104.
51. Revill AS, Dunlin G (2003) The fishing capacity of gillnets lost on wrecks and on open ground in UK coastal waters. Fish Res. 64: 107–113.
52. Stefatos A, Charalampakis M, Papatheodorou G, Ferentinos G (1999) Marine debris on the seafloor of the Mediterranean Sea: examples from two enclosed gulfs in Western Greece. Mar. Pollut. Bull. 38: 389–393.
53. Spengler A, Costa MF (2008) Methods applied in studies of benthic marine debris. Mar. Pollut. Bull. 56: 226–230.
54. Lee D-I, Cho H-S, Jeong S-B (2006) Distribution characteristics of marine litter on the sea bed of the East China Sea and the South Sea of Korea. Est. Coast. Shelf Sci. 70: 187–194.
55. Watters DL, Yoklavich MM, Love MS, Schroeder DM (2010) Assessing marine debris in deep seafloor habitats off California. Mar. Pollut. Bull. 60: 131–138.
56. Fiorentini L, Dremière P-Y, Leonori I, Sala A, Palumbo V (1999) Efficiency of the bottom trawl used for the Mediterranean international trawl survey (MEDITS). Aquat. Living Resour. 12: 187–205.
57. Sardà F, Cartes JE, Company JB, Albiol A (1998) A Modified Commercial Trawl Used to Sample Deep-Sea Megabenthos. Fish. Sci. 64: 492–493.
58. Clarke KR, Clarke RK, Gorley RN (2006) Primer V6: User Manual - Tutorial: Plymouth Marine Laboratory.
59. Bray JR, Curtis JT (1957) An Ordination of the Upland Forest Communities of Southern Wisconsin. Ecol. Monogr. 27: 325–349.
60. Ramirez-Llodra E, Tyler PA, Baker MC, Bergstad OA, Clark MR, et al. (2011) Man and the last great wilderness: Human impact on the deep sea. PLoS ONE 6: e22588.

61. Miyake H, Shibata H, Furushima Y (2011) Deep-sea litter study using deep-sea observation tools. In: Omori K, Guo X, Yoshie N, Fujii N, Handoh IC et al., editors. Interdisciplinary Studies on Environmental Chemistry-Marine Environmental Modeling and Analysis: Terrapub. pp. 261–269.

62. Hess NA, Ribic CA, Vining I (1999) Benthic marine debris, with an emphasis on fishery-related items, surrounding Kodiak Island, Alaska, 1994–1996. Mar. Pollut. Bull. 38: 885–890.

63. Keller AA, Fruh EL, Johnson MM, Simon V, McGourty C (2010) Distribution and abundance of anthropogenic marine debris along the shelf and slope of the US West Coast. Mar. Pollut. Bull. 60: 692–700.

64. Kanehiro H, Tokai T, Matuda K (1996) The distribution of litter in fishing ground of Tokyo Gulf. Fish. Eng. 32: 211–217.

65. Kuriyama Y, Tokai T, Tabata K, Kanehiro H (2003) Distribution and composition of litter of Tokyo Gulf and its age analysis. Nippon Suisan Gakkaishi 69: 770–781.

66. Hinojosa IA, Thiel M (2009) Floating marine debris in fjords, gulfs and channels of southern Chile. Mar. Pollut. Bull. 58: 341–350.

67. Thiel M, Hinojosa I, Vasquez N, Macaya E (2003) Floating marine debris in coastal waters of the SE-Pacific (Chile). Mar. Pollut. Bull. 46: 224–231.

68. Thiel M, Hinojosa IA, Joschko T, Gutow L (2011) Spatio-temporal distribution of floating objects in the German Bight (North Sea). J. Sea Res. 65: 368–379.

69. Zhou P, Huang CG, Fang HD, Cai WX, Li DM, et al. (2011) The abundance, composition and sources of marine debris in coastal seawaters or beaches around the northern South China Sea (China). Mar. Pollut. Bull. 62: 1998–2007.

70. Derraik JGB (2002) The pollution of the marine environment by plastic debris: a review. Mar. Pollut. Bull. 44: 842–852.

71. Moore SL, Gregorio D, Carreon M, Weisberg SB, Leecaster MK (2001) Composition and distribution of beach debris in Orange County, California. Mar. Pollut. Bull. 42: 241–245.

72. Topcu EN, Tonay AM, Dede A, Ozturk AA, Ozturk B (2013) Origin and abundance of marine litter along sandy beaches of the Turkish Western Black Sea Coast. Mar. Environ. Res. 85: 21–28.

73. Silva-Iniguez L, Fischer DW (2003) Quantification and classification of marine litter on the municipal beach of Ensenada, Baja California, Mexico. Mar. Pollut. Bull. 46: 132–138.

74. Martinez-Ribes L, Basterretxea G, Palmer M, Tintore J (2007) Origin and abundance of beach debris in the Balearic Islands. Sci. Mar. 71: 305–314.

75. Willoughby NG, Sangkoyo H, Lakaseru BO (1997) Beach litter: an increasing and changing problem for Indonesia. Mar. Pollut. Bull. 34: 469–478.

76. Oigman-Pszczol SS, Creed JC (2007) Quantification and classification of marine litter on beaches along Armacao dos Buzios, Rio de Janeiro, Brazil. J. Coast. Res. 23: 421–428.

77. Ryan PG, Moore CJ, van Franeker JA, Moloney CL (2009) Monitoring the abundance of plastic debris in the marine environment. Philosophical Transactions of the Royal Society B-Biological Sciences 364: 1999–2012.

78. Schlining K, von Thun S, Kuhnz L, Schlining B, Lundsten L, et al. (2013) Debris in the deep: Using a 22-year video annotation database to survey marine litter in Monterey Canyon, central California, USA. Deep-Sea Res. Part I Oceanogr. Res. Pap. 79: 96–105.

79. Halpern BS, Selkoe KA, Micheli F, Kappel CV (2007) Evaluating and ranking the vulnerability of global marine ecosystems to anthropogenic threats. Conserv. Biol. 21: 1301–1315.

80. Halfar J, Fujita RM (2007) Danger of deep-sea mining. Science 316: 987.

81. He G, Ma W, Song C, Yang S, Zhu B, et al. (2011) Distribution characteristics of seamount cobalt-rich ferromanganese crusts and the determination of the size of areas for exploration and exploitation. Acta Oceanol. Sin. 30: 63–75.

82. Davies AJ, Roberts JM, Hall-Spencer J (2007) Preserving deep-sea natural heritage: Emerging issues in offshore conservation and management. Biol. Conserv. 138: 299–312.

83. Amaro T, Bianchelli S, Billett DSM, Cunha MR, Pusceddu A, et al. (2010) The trophic biology of the holothurian *Molpadia musculus*: implications for organic matter cycling and ecosystem functioning in a deep submarine canyon. Biogeosciences 7: 2419–2432.

84. Pagès F, Martín J, Palanques A, Puig P, Gili JM (2007) High occurrence of the elasipodid holothurian *Penilpidia ludwigi* (von Marenzeller, 1893) in bathyal sediment traps moored in a western Mediterranean submarine canyon. Deep-Sea Res. Part I Oceanogr. Res. Pap. 54: 2170–2180.

85. Vetter EW, Dayton PK (1998) Macrofaunal communities within and adjacent to a detritus-rich submarine canyon system. Deep Sea Res. Part II Top. Stud. Oceanogr. 45: 25–54.

86. Wright SL, Thompson RC, Galloway TS (2013) The physical impacts of microplastics on marine organisms: A review. Environ. Pollut. 178: 783–492.

87. Clark MR, Koslow JA (2008) Impacts of fisheries on seamounts. In: Pitcher TJ, Morato T, Hart PJB, Clark MR, Haggan N, et al., editors. Seamounts: Ecology, Fisheries & Conservation: Blackwell Publishing Ltd. pp. 413–441.

88. Moore SL, Allen MJ (2000) Distribution of anthropogenic and natural debris on the mainland shelf of the southern California Bight. Mar. Pollut. Bull. 40: 83–88.

89. Cho D-O (2011) Removing derelict fishing gear from the deep seabed of the East Sea. Mar. Pol. 35: 610–614.

90. Carr HA, Harris J (1997) Ghost-fishing gear: have fishing practices during the past few years reduced the impact? In: Coe J, Rogers D, editors. Marine Debris: Springer New York. pp. 141–151.

91. Nash AD (1992) Impacts of marine debris on subsistence fishermen: An exploratory study. Mar. Pollut. Bull. 24: 150–156.

92. Sanchez P, Maso M, Saez R, De Juan S, Muntadas A, et al. (2013) Baseline study of the distribution of marine debris on soft-bottom habitats associated with trawling grounds in the northern Mediterranean. Sci. Mar. 77: 247–255.

93. Martins J, Sobral P (2011) Plastic marine debris on the Portuguese coastline: A matter of size? Mar. Pollut. Bull. 62: 2649–2653.

94. Van Cauwenberghe L, Vanreusel A, Mees J, Janssen CR (2013) Microplastic pollution in deep-sea sediments. Environ. Pollut. 182: 495–499.

95. Purser A, Orejas C, Gori A, Tong RJ, Unnithan V, et al. (2013) Local variation in the distribution of benthic megafauna species associated with cold-water coral reefs on the Norwegian margin. Cont. Shelf Res. 54: 37–51.

96. Van Rooij D, Blamart D, De Mol L, Mienis F, Pirlet H, et al. (2011) Cold-water coral mounds on the Pen Duick Escarpment, Gulf of Cadiz: The MiCRO-SYSTEMS project approach. Mar. Geol. 282: 102–117.

97. Tubau X, Canals M, Lastras G, Company JB, Rayo X (2012) The PROMARES-OASIS DEL MAR shipboard party; Marine litter in the deep sections of the North Catalan submarine canyons from ROV video-inspection. 3rd Annual Hermione Meeting, Faro (Portugal), Abstr. Vol., p. 21.

98. Huvenne VAI (2011) Benthic habitats and the impact of human activities in Rockall Trough, on Rockall Bank and in Hatton Basin. National Oceanography Centre, Cruise Report No. 04, RRS James Cook Cruise 60, 133 pp.

99. Bullimore RD, Foster NL, Howell KL (2013) Coral-characterized benthic assemblages of the deep Northeast Atlantic: defining "Coral Gardens" to support future habitat mapping efforts. ICES J. Mar. Sci. 70: 511–522.

100. Lavaleye MSS (2011) CoralFISH-HERMIONE cruise report of Cruise 64PE345 with RV Pelagia Texel-Vigo, 28 Sept – 14 Oct 2011 to Belgica Mound Province (CoralFISH & HERMIONE) and Whittard Canyon (HERMIONE). NIOZ-cruise report. pp. 47.

12

Spatial Differences in East Scotia Ridge Hydrothermal Vent Food Webs: Influences of Chemistry, Microbiology and Predation on Trophodynamics

William D. K. Reid[1]*, Christopher J. Sweeting[1], Ben D. Wigham[2], Katrin Zwirglmaier[3¤], Jeffrey A. Hawkes[4], Rona A. R. McGill[5], Katrin Linse[3], Nicholas V. C. Polunin[1]

1 School of Marine Science and Technology, Newcastle University, Newcastle upon Tyne, United Kingdom, **2** Dove Marine Laboratory, School of Marine Science and Technology, Newcastle University, Cullercoats, United Kingdom, **3** British Antarctic Survey, Natural Environment Research Council, High Cross, Madingley Road, Cambridge, United Kingdom, **4** Ocean and Earth Science, University of Southampton, National Oceanography Centre Southampton, Southampton, United Kingdom, **5** Natural Environment Research Council Life Sciences Mass Spectrometry Facility, Scottish Universities Environmental Research Centre, East Kilbride, United Kingdom

Abstract

The hydrothermal vents on the East Scotia Ridge are the first to be explored in the Antarctic and are dominated by large peltospiroid gastropods, stalked barnacles (*Vulcanolepas* sp.) and anomuran crabs (*Kiwa* sp.) but their food webs are unknown. Vent fluid and macroconsumer samples were collected at three vent sites (E2, E9N and E9S) at distances of tens of metres to hundreds of kilometres apart with contrasting vent fluid chemistries to describe trophic interactions and identify potential carbon fixation pathways using stable isotopes. $\delta^{13}C$ of dissolved inorganic carbon from vent fluids ranged from $-4.6‰$ to $0.8‰$ at E2 and from $-4.4‰$ to $1.5‰$ at E9. The lowest macroconsumer $\delta^{13}C$ was observed in peltospiroid gastropods ($-30.0‰$ to $-31.1‰$) and indicated carbon fixation via the Calvin-Benson-Bassham (CBB) cycle by endosymbiotic gamma-Proteobacteria. Highest $\delta^{13}C$ occurred in *Kiwa* sp. ($-19.0‰$ to $-10.5‰$), similar to that of the epibionts sampled from their ventral setae. *Kiwa* sp. $\delta^{13}C$ differed among sites, which were attributed to spatial differences in the epibiont community and the relative contribution of carbon fixed via the reductive tricarboxylic acid (rTCA) and CBB cycles assimilated by *Kiwa* sp. Site differences in carbon fixation pathways were traced into higher trophic levels e.g. a stichasterid asteroid that predates on *Kiwa* sp. Sponges and anemones at the periphery of E2 assimilated a proportion of epipelagic photosynthetic primary production but this was not observed at E9N. Differences in the $\delta^{13}C$ and $\delta^{34}S$ values of vent macroconsumers between E2 and E9 sites suggest the relative contributions of photosynthetic and chemoautotrophic carbon fixation (rTCA v CBB) entering the hydrothermal vent food webs vary between the sites.

Editor: Simon Thrush, National Institute of Water & Atmospheric Research, New Zealand

Funding: The research was funded by the Natural Environment Research Council (NERC) through ChEsSO consortium grant NE/DO1249X/1 and studentships NE/F010664/1 (WDKR) and NE/H524922 (JAH). Sample analysis was funded via NERC Life Sciences Mass Spectrometry Facilities grant LSMSFBRIS043_04/10_R_09/10. The funders had no role in the study design, data collection and anlaysis, decision to publish, or preparation of the manuscript.

Competing Interests: The authors have declared that no competing interests exist.

* E-mail: wdkreid@gmail.com

¤ Current address: Technische Universitaet Muenchen, Wissenschaftszentrum Weihenstephan, Limnologische Station Iffeldorf, Iffeldorf, Germany

Introduction

Deep-sea hydrothermal vents are chemically reducing habitats occurring on mid-ocean and back-arc spreading centres, seamounts, volcanic hotspots and off-axis ridge settings [1,2,3]. They are distinct from the surrounding deep sea with respect to environmental conditions, the energy sources sustaining life and their biological communities [4,5]. High densities of organisms are found to thrive at the interface where hot, mineral-rich fluids discharge from the seafloor and mix with colder, oxygenated seawater. The hot fluids emitted from the seafloor may differ in pH and are enriched in reduced gases (e.g. H_2S, CH_4, H_2) and metals (e.g. Fe^{2+}, Cu, Mn) relative to seawater [6]. Microorganisms oxidise the reduced species in vent fluids and utilise the energy released to fix CO_2 or other single carbon compounds (e.g. CO, CH_4) into cellular material [7]. This results in microbial chemosynthesis replacing photosynthetic primary production at the base of the food chain [7].

Sulfide oxidation appears to be the principal energy acquisition pathway, which microorganisms use to drive carbon fixation [3,7,8]. The most important carbon fixation pathways at the base of the metazoan hydrothermal vent food webs are the Calvin-Benson-Bassham (CBB) and reductive tricarboxylic acid (rTCA) cycles [9,10,11]. Methane oxidation (methanotrophy) is a further carbon fixation process at hydrothermal vents with CH_4 of thermogenic, biogenic or magmatic origin available depending on the host substrate [3,12]. Epipelagic photosynthetic primary production may also provide some nutrition to vent macroconsumers, although the relative contribution to vent fauna is thought to be negligible [12,13]. Macroconsumers utilise the vent organic carbon through endo- and episymbiotic relationships, consumption of free-living microorganisms either from various surfaces or the water column and indirectly through predation and scavenging [14,15,16].

The relative contributions of different carbon sources and complexity of hydrothermal vent food webs vary globally depending on the species present, the geological host substrate and the vent fluid chemistry [17,18,19]. The first Antarctic hydrothermal vent communities were discovered recently on the East Scotia Ridge (ESR), a back-arc spreading centre in the Atlantic sector of the Southern Ocean [20,21]. The two basalt-hosted vent fields occur on the ridge segments E2 and E9, which lack the characteristic alvinocarid shrimps, bathymodiolid mussels and siboglinid worms found at Atlantic, Indian and Pacific hydrothermal vents, respectively [21]. Instead, biomass at the ESR vents is dominated by anomuran crabs (*Kiwa* sp.), stalked barnacles (*Vulcanolepas* sp.) and large peltospiroid gastropods [22], indicating a new biogeographic province [21]. Furthermore, there are differences in the end-member vent fluid chemistry between the E2 and E9 vent fields as well as within field between northern (E9N) and southern (E9S) areas of E9 [21].

Stable isotopes of carbon ($^{13}C/^{12}C$ expressed as $\delta^{13}C$), nitrogen ($^{15}N/^{14}N$ expressed as $\delta^{15}N$) and sulfur ($^{34}S/^{32}S$ expressed as $\delta^{34}S$) have been used to examine hydrothermal vent community trophodynamics [23,24]. $\delta^{13}C$ can be used to characterise the various carbon sources utilised by vent macroconsumers [25]. This is done by comparing the expected carbon fractionation between dissolved inorganic carbon (DIC) and the macroconsumer's tissue. Enzymatic reactions catalysed by the ribulose-1,5-biphosphate carboxylase/oxygenase form I (RuBisCO form I) of the CBB cycle (22‰ to 30‰: [26,27,28]) exhibit greater fractionation than those of the rTCA cycle (2‰ to 14‰: [29,30,31]). Once organic material is incorporated into the macroconsumer food web, carbon trophic discrimination ($\Delta^{13}C$) is small, ranging from 0 to 1.5‰ between the food source and consumer [32]. $\delta^{34}S$ also identifies energy sources (sulfur trophic discrimination, −1‰ to 2‰: [32]). The large difference in $\delta^{34}S$ between seawater sulphate and sulfides at hydrothermal vents [33] results in organic matter of photosynthetic (~16‰ to 19‰) and chemosynthetic (−9‰ to 10‰) origin having distinctive $\delta^{34}S$ values [34,35]. The greater trophic discrimination (2‰ to 5‰) in $\delta^{15}N$ between consumer and food source provides information on the trophic position of an organism relative to a primary consumer [32]. Therefore, the isotopic value of a vent macroconsumer is the product of the following factors: (1) the inorganic substrate and its isotopic value used by the chemoautotroph; (2) the isotopic discrimination processes occurring during metabolic reactions involving inorganic substrates to create organic compounds (e.g. CBB or rTCA cycles) by the chemoautotroph; (3) food source-macroconsumer trophic interactions (e.g. endosymbiont-host, predator-prey) that occur as a function of (1) and (2); and (4) the physiology associated with the macroconsumer's isotopic trophic discrimination.

The goal of the present research was to investigate intra- and inter-site patterns in the trophic assemblages of macroconsumers occurring at hydrothermal vents on the ESR using $\delta^{13}C$, $\delta^{34}S$ and $\delta^{15}N$. Specifically, the aims were to: (1) compare $\delta^{13}C_{DIC}$ among vent sites and thus establish difference in the isotopic inorganic substrates used by chemoautotrophs; (2) compare $\delta^{13}C$, $\delta^{15}N$ and $\delta^{34}S$ between vent and benthic non-vent fauna to assess any photosynthetic inputs into the hydrothermal vent food web; (3) investigate differences in trophic structures among the three sites; and (4) assess which species are driving any differences in trophic structure. The investigation provides a unique opportunity to examine differences in trophic structure at the scale of tens of metres to 100s of kilometres in a newly discovered hydrothermal vent biogeographical province.

Materials and Methods

Ethics Statement

Permits for the fieldwork were granted by the United Kingdom Foreign and Commonwealth Office. This study met the ethical requirements of the affiliated research institutions for research utilising animal tissues. No animal husbandry or laboratory controlled experiments were part of the research that required permits from the UK Home Office. The fish were collected at a water depth of 2500 m, which meant that they were dead when they arrived on deck as a result of changes in pressure. This was the case with the majority of the animals dissected within this study. The research also adhered to the Inter Ridge code of conduct for sampling hydrothermal vents (http://www.interridge.org/IRStatement).

Study Sites

The E2 and E9 vent fields are situated approximately 440 km apart at 56° 05.35′S, 30° 19.20′W and 60° 02.50′ S, 29° 58.93′ W, respectively (Fig. 1). E2 is at a depth of ~2600 m and seafloor topography is complex with a series of terraced features and lobed pillow basalts filling a major north-south steep-sided fissure [21]. The main high-temperature and diffusive venting occurred at an intersection between this fissure and an east-west running fault or scarp [21]. E9 was located at ~2400 m depth and its topography was relatively flat with sheet lava, a series of lava drain back features and collapsed pillow basalts. A series of north-south fissures were found with venting mainly occurring on the most western [21,22]. The end-member fluid chemistry exiting chimneys differed between the northern and southern sections of E9 [21], therefore E9N and E9S are here considered to be separate sites. Ambient seabed water temperatures were 0.0°C at E2 and between −0.1°C and −1.3°C at E9 [21].

Sample Collection and Ship-board Processing

Samples were collected onboard the R.R.S *James Cook* during the 2010 austral summer (7 January to 21 February) using the remotely operated vehicle (ROV) *Isis*. High temperature and diffuse flow fluids were collected for DIC using titanium samplers, equipped with an inductively coupled link high temperature sensor. The nozzle of the titanium sampler was inserted into the chimney orifice for high temperature fluid samples and once the temperature reading became stable the fluid was collected. For diffuse flow samples, a circular titanium housing was placed over the area of diffuse venting to minimise the entrainment of seawater. Once the diffuse flow was visible exiting the top of the housing, the titanium sampler was inserted into the opening and the diffuse flow sample was collected once the temperature reading was stable. On board, an aliquot for stable isotope analysis of DIC was sampled to exclude air and poisoned with mercury chloride.

Vent macroconsumers were collected by suction sampler or scoop with species separated into a series of acrylic chambers or perspex boxes to avoid predation or contamination. Six species were collected at all three sites. No female *Kiwa* sp. were collected from E9N or E9S. Fish and pycnogonids were collected using large collapsible and small metal baited traps deployed from the ROV. Non-vent macroconsumers were collected from metres to tens of metres away from active venting where there were no obvious signs of hydrothermal influence, i.e. no bacterial mat, and where temperature was consistent with local Antarctic bottom water. Non-vent samples were collected on separate dives from those for vent fauna to avoid contamination. Only one non-vent species was collected from E2 and sampling was limited to the areas adjacent to E9N because of ROV operational time

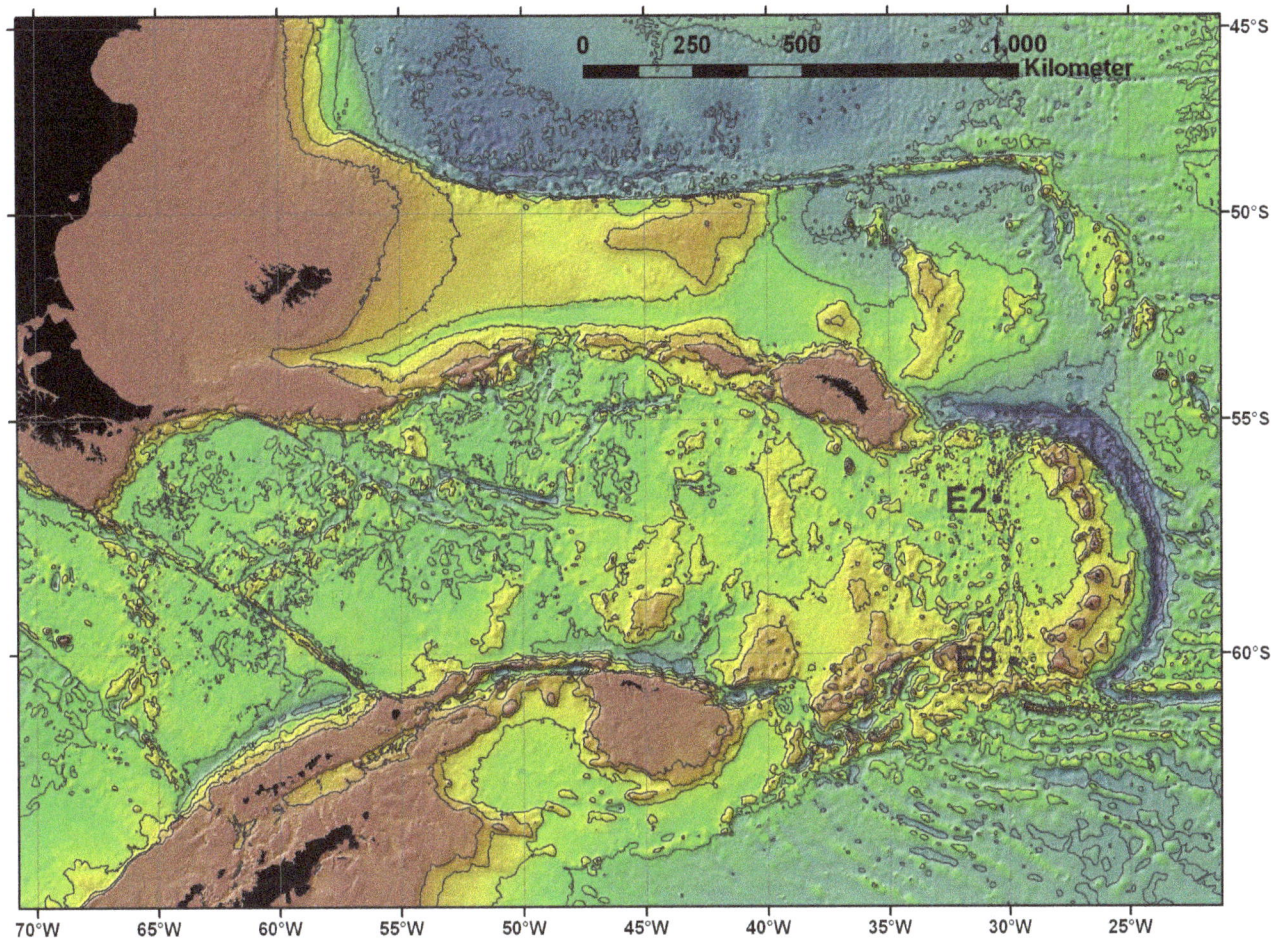

Figure 1. Bathymetric map illustrating positions of the E2 and E9 vent sites (black circles). The vent sites are located at the northern and southern ends of the East Scotia Ridge (ESR), located in the Atlantic sector of the Southern Ocean. The map shows the position of the ESR in relation to South America and the Antarctic Peninsula.

constraints. Potential food sources were collected by scraping material from rocks collected by ROV manipulators and epibionts from the ventral setae of the decapod *Kiwa* sp. Particulate suspended material was collected from the acrylic chambers, which was sampled incidentally during faunal collection. Samples were sorted on board to the lowest possible taxonomic resolution. The majority of the vent species are undescribed to date.

Faunal samples were frozen at −80°C whole or after dissection, depending on their size, for stable isotope analysis. Muscle was removed from the chelipeds of *Kiwa* sp., foot dissected from Peltospiroidea sp., tube feet removed from the asteroids Stichasteridae sp. and *Freyella* cf *fragilissima* and tentacles removed from the anemones. Legs were removed from the pycnogonids *Colossendeis* cf. *concedis* and *C*. cf. *elephantis*, while *Sericosura* spp. was sampled whole. The gastropods Provannidae sp. 1 and 2, *Lepetodrilus* sp., and juvenile Peltospiroidea sp. (<7 mm shell length), and the stalked barnacle *Vulcanolepas* sp. were removed from their shells and sampled whole. White muscle tissue was dissected from the anterior dorso-lateral region of the zoarcid fish.

Sample Processing Onshore

Each end-member and diffuse flow DIC sample was prepared for isotopic analysis by removing a 1 mL water sample and transferring it into a separate vial. The headspace was flushed with

helium, phosphoric acid was injected into the vial and then the contents were vortex mixed. The samples were then left to react for 24 hours to ensure complete conversion of all DIC to CO_2 for isotopic analysis. The CO_2 was then analysed by continuous-flow isotope ratio mass spectrometry (IRMS) using a Europa Scientific 20–20 IRMS by Iso-Analytical (Crewe, United Kingdom). Samples were run in duplicate and the mean is reported. An internal reference gas (IA-R060, $\delta^{13}C = −36.08‰ ± SD\ 0.13$) was used to determine the $\delta^{13}C_{DIC}$ values and is traceable to the International Atomic Energy Agency standard, NBS-19. Concentrations of CH_4 in the water samples were insufficient for isotope analysis.

Faunal tissue samples were freeze dried and ground to a homogenous powder using a pestle and mortar. Aliquots of fauna, particulate suspended material and material scraped from rocks were tested for carbonates prior to analysis with 0.1 N HCl. If the sample effervesced, this indicated carbonates were present and it was subsequently acidified by further addition of HCl until the effervescence ceased. Samples were re-dried at 50°C for 48 hours. If the sample did not effervesce, no acidification was carried out. Aliquots for $\delta^{13}C$ analysis were not lipid extracted. Any confounding lipid effects due to metabolic processes would not affect the interpretation of the ultimate carbon sources of the vent fauna described by $\delta^{13}C$ because of the large differences in the $\delta^{13}C$ values of trophic end-members.

Approximately 0.7 mg of powder was weighed into a tin capsule for carbon and nitrogen IRMS. For sulfur, 2 mg of sample and 4 mg of the catalyst vanadium pentoxide were weighed into each tin capsule. Dual stable carbon and nitrogen isotope ratios were measured by continuous-flow IRMS using a Costech Elemental Analyser interfaced with Thermo Finnigan Delta Plus XP (Natural Environment Research Council, Life Sciences Mass Spectrometry Facility, SUERC, East Kilbride, United Kingdom). Two laboratory standards were analysed for every ten samples in each analytical sequence. These alternated between paired alanine standards, differing in $\delta^{13}C$ and $\delta^{15}N$, and an internal laboratory gelatin standard. Sulfur was analysed by Iso-Analytical using a SERCON Elemental Analyser coupled to a Europa Scientific 20–20 IRMS. Laboratory standards of barium sulphate (two sets of differing $\delta^{34}S$) and silver sulfide were used for calibration and drift correction. An internal standard of whale baleen was used for quality control (n = 28, 16.34‰ ± SD 0.21). Stable isotope ratios were expressed in delta (δ) notation as parts per thousand/permil (‰). All internal standards are traceable to the following international standards: v-PDB (Pee Dee Belemnite), AIR (atmospheric nitrogen) and NBS-127 (barium sulphate), IAEA-S-1 (silver sulfide) and IAEA-SO-5 (barium sulphate). An external reference material of freeze dried and ground deep-sea fish white muscle (*Antimora rostrata*) was also analysed ($\delta^{13}C$, n = 24, −18.94‰ ± SD 0.09; $\delta^{15}N$, n = 24, 13.11‰ ± SD 0.38; $\delta^{34}S$, n = 30, 18.20‰, ± SD 0.59).

Data Analysis

Data were assessed for normality using a Shapiro-Wilk test before statistical tests examining spatial patterns in trophic structure and species stable isotope values. Homogeneity, or otherwise, of variances is ecologically informative, for example in identifying distinct energy sources at the base of the food web [36]. Inter-site differences in trophic structure were examined using a Fligner-Killeen test for homogeneity of variance to assess differences in the spread of the mean stable isotope values of each species. Inter-site differences in species were analysed using a one-way ANOVA followed by Tukey's honest significant difference (HSD) when variance was homogeneous among sites. Welch's ANOVA followed by t-tests were used when there was heterogeneity of variance among sites because it uses adjusted degrees of freedom to protect against Type I errors when variances are unequal [37]. A Bonferroni correction (p = 0.05/n) was used for multiple comparisons. When data were not normally distributed, a two sample Wilcoxon test was used. All statistics were preformed in R version 12.13.1 [38].

Results

Dissolved Inorganic Carbon Stable Isotope Values

Mean (± SD) $\delta^{13}C_{DIC}$ of high temperature and diffuse flow fluids are summarised in Table 1. $\delta^{13}C_{DIC}$ of high temperature samples collected at two E2 locations were −4.7‰ (±0.0) (max temperature 351.0°C) and −2.5‰ (±0.1) (max temperature 323.0°C). At E9N, $\delta^{13}C_{DIC}$ from separate orifices of the same chimney structure were −4.6‰ (±0.0) (max temperature 380.2°C) and −4.5‰ (±0.0) (max temperature 357.0°C). No high temperature fluids were collected from E9S for $\delta^{13}C_{DIC}$ analysis because the pressure was too high within the titanium samples to safely and accurately collect a representative sample. Diffuse flow samples from amongst *Kiwa* sp. and anemones, at E2, had $\delta^{13}C_{DIC}$ values of 0.8‰ (±0.1) (max temperature 19.9°C) and 0.2‰ (±0.2) (max temperature 3.5°C), respectively. A single diffuse flow sample collected from amongst an aggregation of *Kiwa*

sp. at E9N had a $\delta^{13}C_{DIC}$ value of 1.5‰ (±0.1) (max temperature 12.6°C). At E9S, diffuse flow samples from amongst *Kiwa* sp. had a $\delta^{13}C_{DIC}$ value of 0.9‰ (±0.1) (max temperature 19.9°C), while a sample taken from a mixed aggregation of *Kiwa* sp. and peltospiroid gastropods had $\delta^{13}C_{DIC}$ value of 0.1‰ (±0.1) (max temperature 5.0°C).

Comparison between Vent and Benthic Non-vent Macroconsumers at E9N

At E9N, mean $\delta^{13}C$ and $\delta^{15}N$ values of vent fauna overlapped with non-vent benthic fauna (Welch's t-test, $\delta^{13}C$ DF = 10.59, t = 0.66, p = 0.52; Welch's t-test, $\delta^{15}N$ DF = 10.42, t = −0.30, p = 0.76; Fig. 2, Tables 2 & 3) while mean $\delta^{34}S$ values differed between non-vent benthic fauna and vent fauna (Welch's t-test, DF = 12.56, t = −9.08, p<0.01) (Fig. 3, Tables 2 & 3).

Intra- and Inter-site Differences in Community Trophodynamics

Eleven, ten and seven species were collected at E2, E9N and E9S respectively for stable isotope analysis (Table 3). The ranges of mean $\delta^{13}C$ values of the vent fauna differed amongst the three sites (Fligner-Killeen test, DF = 2, χ^2 = 6.46, p<0.05). E2 had the narrowest $\delta^{13}C$ range (−29.9‰ to −19.0‰), whereas at E9N and E9S $\delta^{13}C$ ranged from −31.4‰ to −9.9‰ and −30.0‰ to −10.5‰, respectively (Fig. 2). Across the three sites Peltospiroidea sp. had the lowest values while *Kiwa* sp. had the highest $\delta^{13}C$ values (Fig. 2, Table 3), and *Lepetodrilus* sp., *Vulcanolepas* sp., *Pacmanactis* sp. and *Colossendeis* spp. all had intermediate $\delta^{13}C$ values (Fig. 2, Table 3). However, there was no overall difference in mean $\delta^{13}C$ values among sites for the combined data across species (Welch's ANOVA, DF = 2.00, F = 0.59, p = 0.56). The range and mean $\delta^{34}S$ values (Fig. 3, Table 3) did not differ among sites (Fligner-Killeen test, DF = 2, χ^2 = 0.84, p = 0.65; ANOVA, DF = 2, 26, F = 1.94, p = 0.16), however *Kiwa* sp. had the lowest $\delta^{34}S$ at E2 and E9S while *Lepetodrilus* sp. had the lowest $\delta^{34}S$ values at E9N (Fig. 3, Table 3). The highest vent fauna $\delta^{34}S$ values were in *Pacmanactis* sp. (E2), *Vulcanolepas* sp. (E9N) and *Sericosura* spp. (E9S) (Fig. 3, Table 3). Neither the range nor the mean $\delta^{15}N$ values differed among sites (Fligner-Killeen test, DF = 2, χ^2 = 0.40, p = 0.83; ANOVA, DF = 2, 26, F = 1.19, p = 0.31). The provannid gastropods at E2 and E9S had the lowest $\delta^{15}N$ values while Peltospiroidea sp. had the lowest values at E9N (Fig. 2, Table 3).

Table 1. $\delta^{13}C$ values of dissolved inorganic carbon (DIC) sampled from high temperature and diffuse flow venting from the E2 and E9 ridge segments of the East Scotia Ridge, Southern Ocean.

Site	Temperature (°C)	$\delta^{13}C$ DIC
E2	351.0	−4.7 (0.0)
	323.0	−2.5 (0.1)
	19.9	0.8 (0.1)
	3.5	0.2 (0.2)
E9N	380.2	−4.7 (0.0)
	357.0	−4.7 (0.0)
	12.6	1.5 (0.1)
E9S	19.9	0.9 (0.1)
	5.0	0.1 (0.1)

Standard deviations are in parentheses.

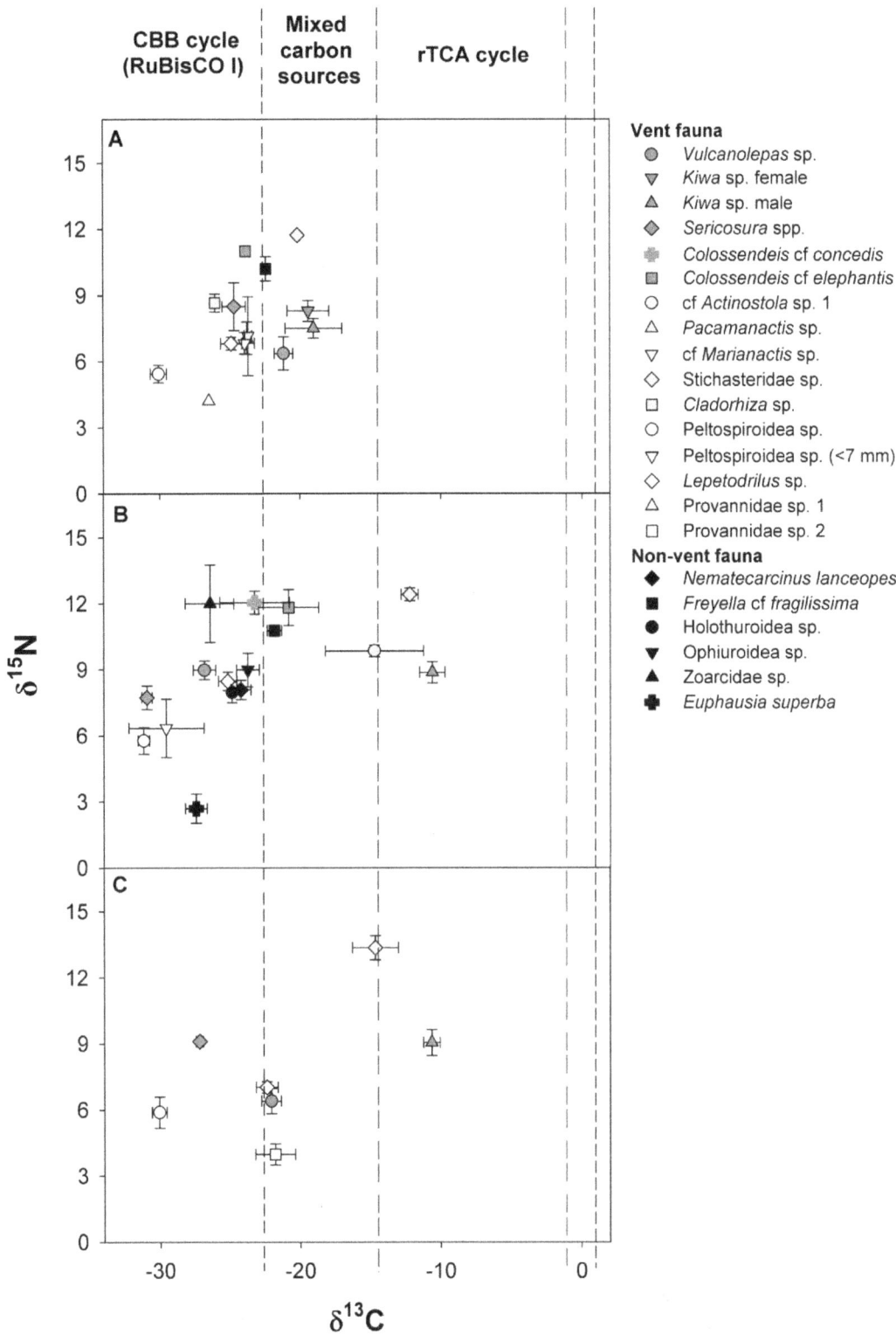

Figure 2. δ^{13}C and δ^{15}N values of macroconsumers collected from the East Scotia Ridge, Southern Ocean. The values represent means (± standard deviations) for hydrothermal vent and non-vent macroconsumers from the three sample sites: (a) E2, (b) E9N and (c) E9S. Dashed vertical lines represent potential ranges of δ^{13}C values indicative of carbon sources sustaining macroconsumers at the ESR: triple dashed line represents the Calvin-Benson-Bassham (CBB) cycle utilising form I RuBisCO, double dashed line represents the reductive tricarboxylic acid (rTCA) cycle, mixed carbon sources occur between the triple and double dashed line and the continuous dashed line represents the approximate δ^{13}C values of the dissolved inorganic carbon from the diffuse flow areas.

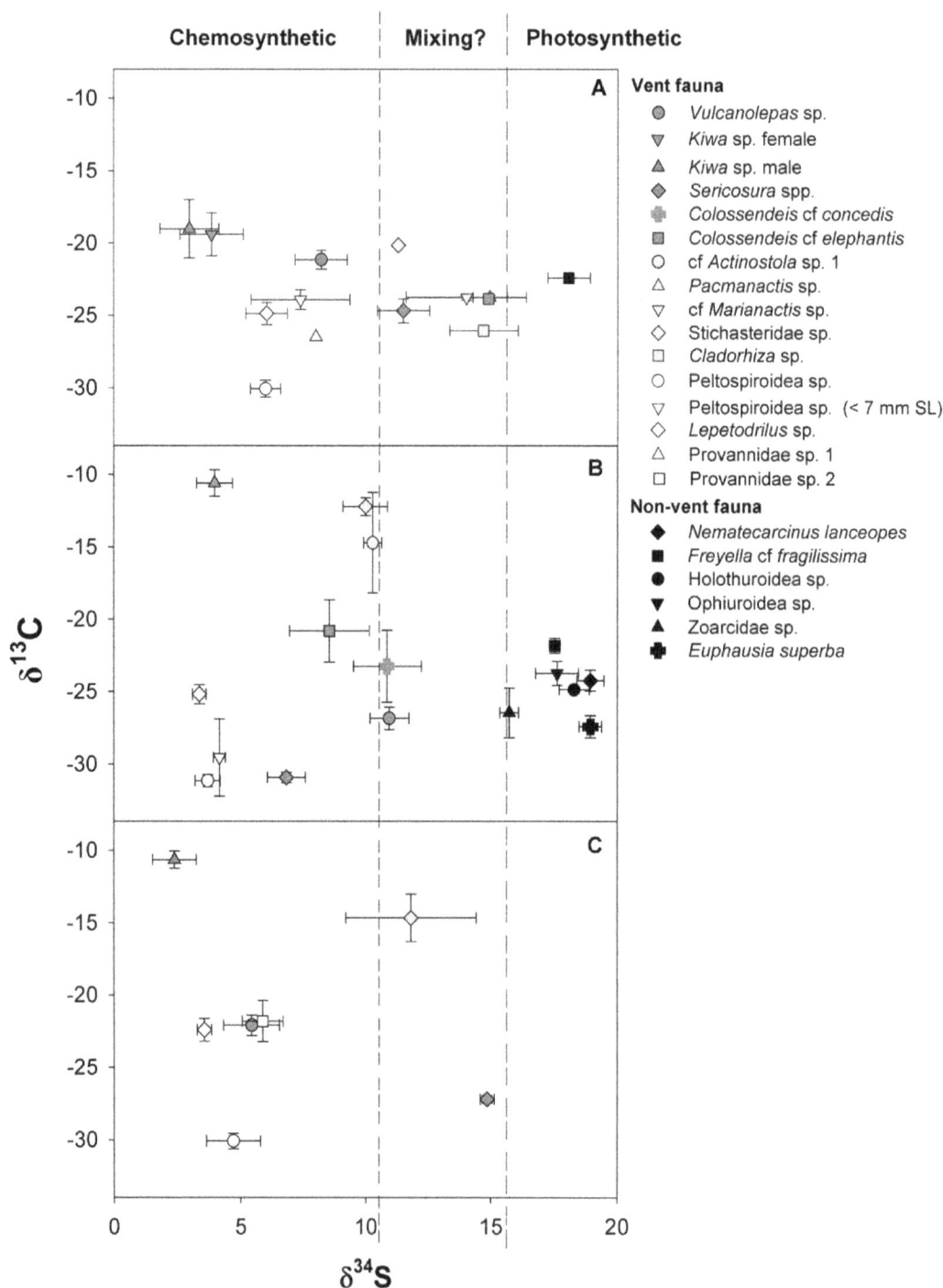

Figure 3. δ¹³C and δ³⁴S values of macroconsumers collected from the East Scotia Ridge, Southern Ocean. The values represent means (± standard deviations) for hydrothermal vent and non-vent macroconsumers from the three sample sites: (a) E2, (b) E9N and (c) E9S. δ³⁴S values between the triple and double dashed lines represent potential areas of isotopic mixing between chemosynthetic and photosynthetic food sources.

The stichasterid sp. consistently had the highest $\delta^{15}N$ values relative to the other vent fauna at each site (Fig. 2, Table 3).

Spatial Differences in Macroconsumer Trophodynamics

Vulcanolepas sp. exhibited spatial differences in $\delta^{13}C$, $\delta^{15}N$ and $\delta^{34}S$ but there was no consistent pattern in isotopic differences among sites (Table 4). Male and female *Kiwa* sp. at E2 did not

differ in $\delta^{13}C$ but males were lower in $\delta^{15}N$ and $\delta^{34}S$ than females (Table 5). Male *Kiwa* sp. showed spatial differences in each stable isotope (Table 4). $\delta^{13}C$ of the males showed a greater range (Fligner-Killeen test, $DF = 2$, $\chi^2 = 10.91$, $p < 0.01$) and lower values at E2 than E9N and E9S (Table 3 & 4). The epibionts attached to the ventral surface of male *Kiwa* sp. also exhibited a greater spread of $\delta^{13}C$ values at E2 than E9S (F-test, $DF = 4$, 3, $F = 244.46$,

Table 2. Mean $\delta^{13}C$, $\delta^{15}N$ and $\delta^{34}S$ values (‰) of non-vent deep-sea fauna collected from the E2 and E9 ridge segments of the East Scotia Ridge, Southern Ocean. Standard deviations are in parentheses.

Taxonomic group	Species	Site	N	$\delta^{13}C$	$\delta^{34}S$	$\delta^{15}N$
Crustacea						
Decapoda	*Nematocarcinus lanceopes*	E9	5	−24.2 (0.7)	18.9 (0.5)	8.1 (0.4)
	Euphausia superba	E9	3	−27.4 (0.8)	19.0 (0.1)	2.7 (0.7)
Echinodermata						
Asteroidea	*Freyella* cf *fragilissima*	E2	3	−22.4 (0.3)	18.0 (0.8)	10.2 (0.6)
	Freyella cf *fragilissima*	E9	2	−21.8 (0.5)	17.5 (0.2)	10.8 (0.2)
Holothuroidea	Holothuroidea sp.	E9	3	−24.9 (0.0)	18.3 (0.6)	8.0 (0.4)
Ophiuroidea	Ophiuroidea sp.	E9	3	−23.7 (0.8)	17.6 (0.9)	9.0 (0.8)
Vertebrata						
Osteichthys	Zoarcidae sp.	E9	4	−26.5 (1.7)	15.7 (0.4)	12.0 (1.8)

Table 3. Mean $\delta^{13}C$, $\delta^{15}N$ and $\delta^{34}S$ values (‰)of hydrothermal vent fauna collected from the E2 and E9 ridge segments of the East Scotia Ridge, Southern Ocean.

Taxonomic group	E2				E9N				E9S			
	N	$\delta^{13}C$	$\delta^{34}S$	$\delta^{15}N$	N	$\delta^{13}C$	$\delta^{34}S$	$\delta^{15}N$	N	$\delta^{13}C$	$\delta^{34}S$	$\delta^{15}N$
Cirripedia												
Vulcanolepas sp.	22	−21.1 (0.6)	8.2 (1.0)	6.3 (0.7)	23	−26.9 (0.8)	11.0 (0.8)	9.0 (0.4)	23	−22.1 (0.8)	5.4 (1.1)	6.4 (0.6)
Decapoda												
Kiwa sp. female	20	−19.4 (1.5)	3.9 (1.3)	8.2 (0.5)	0	–	–	–	0	–	–	–
Kiwa sp. Male	18	−19.0 (2.0)	3.0 (1.2)	7.5 (0.5)	22	−10.6 (0.9)	4.0 (0.7)	8.9 (0.5)	30	−10.7 (0.6)	2.4 (0.9)	9.1 (0.6)
Pycnogonida												
Sericosura spp.	6	−24.7 (0.9)	11.9 (0.4)	8.5 (1.3)	9	−30.9 (0.5)	6.8 (0.8)	7.7 (0.5)	2	−27.2 (0.3)	14.9 (0.3)	9.1 (0.1)
Colossendeis cf *concedis*	0	–	–	–	6	−23.3 (2.5)	10.9 (1.4)	12.1 (0.5)	0	–	–	–
Colossendeis cf *elephantis*	1	−23.8	14.9	11.0	3	−20.8 (2.2)	8.5 (1.6)	11.8 (0.8)	0	–	–	–
Anthozoa												
cf *Actinostola* sp. 1	0	–	–	–	4	−14.7 (3.5)	10.3 (0.4)	9.9 (0.3)	0	–	–	–
Pacmanactis sp.	5	−23.8 (0.2)	14.9 (0.7)	7.1 (0.7)	0	–	–	–	0	–	–	–
cf *Marianactis* sp.	5	−23.7 (0.3)	14.0 (2.4)	7.2 (1.8)	0	–	–	–	0	–	–	–
Asteroidea												
Stichasteridae sp.	1	−20.2	11.3	12.3	5	−12.2 (0.6)	10.0 (0.9)	12.4 (0.4)	5	−14.7 (1.6)	11.8 (2.6)	13.4 (0.6)
Gastropoda												
Peltospiroidea sp.	19	−30.1 (0.6)	6.0 (0.6)	5.4 (0.4)	22	−31.2 (0.4)	3.7 (0.5)	5.8 (0.6)	15	−30.1 (0.5)	4.7 (1.1)	5.9 (0.7)
Peltospiroidea sp (<7 mm)	4	−23.9 (0.7)	7.4 (2.0)	6.8 (0.5)	5	−29.6 (2.7)	4.2 (0.2)	6.4 (1.3)	0	–	–	–
Provannidae sp. 1	1	−26.5	8.0	4.2	0	–	–	–	0	–	–	–
P-rovannidae sp. 2	0	–	–	–	0	–	–	–	4	−21.8 (1.4)	5.9 (0.8)	4.0 (0.5)
Lepetodrilus sp.	5	−24.9 (0.8)	6.4 (0.5)	6.8 (0.3)	4	−25.2 (0.7)	3.4 (0.3)	8.5 (0.4)	4	−22.4 (0.8)	3.6 (0.3)	7.0 (0.3)
Cladorhizidae												
Cladorhiza sp.	5	−26.1 (0.4)	14.7 (1.4)	8.7 (0.4)	0	–	–	–	0	–	–	–
Potential food sources												
Particulate suspended material	3	−23.2 (5.4)	10.0 (1.1)	−0.1 (4.9)	0	–	–	–	0	–	–	–
Rock scrapings	0	–	–	–	1	−23.2	0.8	2.4	1	−31.1	–	1.9
K-iwa n. sp episymbiont	5	−18.9 (5.3)	7.5 (0.3)	3.3 (1.5)	0	–	–	–	5	−9.9 (0.3)	6.6 (0.2)	5.2 (0.8)

Standard deviations are in parentheses and - indicates no data.

p<0.01) as well as lower δ^{13}C but higher δ^{15}N values at E2 than E9S (Table 3 & 5). *Sericosura* spp. δ^{13}C and δ^{34}S values varied amongst sites but δ^{15}N values scarcely did (Table 4). δ^{13}C and δ^{34}S values were lowest at E9N but highest at E2 for δ^{13}C and E9S for δ^{34}S (Table 3). Peltospiroidea sp. showed spatial differences in δ^{13}C and δ^{34}S but not in δ^{15}N (Table 4). δ^{34}S values differed among all sites but E9N δ^{13}C values were lower than those at E9S and E2 (Table 4). Stichasteridae sp. revealed differences between all sites for δ^{13}C and δ^{15}N but for δ^{34}S only between E2 and E9N (Table 5).

Discussion

This study described the trophic structure of a new vent biogeographical province recently discovered on the ESR in the Southern Ocean [21]. In addressing this aim, the study shared the challenges of preceding work in characterising energy sources, separating isotopic overlap and mixing of energy sources, and following energy sources into subsequent predator-prey relationships. However, the tri-isotope approach and integration of both vent chemistry and microbiology, here, provided a more holistic understanding of vent trophic ecology at within- and among-vent field scales.

Intra-site Trophic Interactions and Energy Sources

Scarcity of Δ^{13}C estimates between inorganic carbon and cellular biomass for primary producers at hydrothermal vents [27,30] makes interpretation of the origin of organic carbon fixed within the hydrothermal vent system and assimilated by macro-consumers tentative for species not within a symbiotic or known predator-prey relationship. Diffuse flow δ^{13}C$_{DIC}$ of approximately 1‰ at the ESR vent fields suggests ESR vent macroconsumers with δ^{13}C values <−22‰ are assimilating carbon fixed via the CBB cycle because the net fractionation associated with fixing inorganic into organic carbon for RuBisCO form I ranges from −22‰ to −30‰ [26,27]. Peltospiroidea sp. housed an endosymbiotic gamma-Proteobacteria (K. Zwirglmaier unpublished data) and is within the δ^{13}C range expected for carbon fixed via RuBisCO form I at all three locations. Molluscs containing a single strain of endosymbiotic gamma-Proteobacteria living in other biogeographical vent provinces include some species of bathymodiolid mussels, vesicomyid clams and *Ifremeria* gastropods, all of which have δ^{13}C values between −37‰ and −27‰ [39,40]. Other species of ESR vent macroconsumers, which had δ^{13}C values <−22‰ included *Vulcanolepas* sp. (E2 and E9S), *Sericusora* spp., E2 anemones and *Lepetodrilus* sp. These species consume free-living bacteria [14,23] so organic carbon fixed via other carbon

fixation pathways cannot be ruled out as part of their assimilated diet.

Vent macroconsumers inhabiting the hottest areas of the hydrothermal vent tolerable to metazoan life, including rimicarid shrimps, polychaetes *Alvinella* spp. and *Riftia pachyptila* and some alvinoconchid gastropods, tend to assimilate rTCA-fixed carbon from their diet [11,41,42] and have δ^{13}C values >−16‰ [14,25,43,44]. As δ^{13}C$_{DIC}$ is approximately 1‰ at the ESR sites, vent macroconsumers utilising carbon fixed via the rTCA cycle would have had δ^{13}C values >−13‰; assuming a −2‰ to −14‰ net fractionation between the inorganic substrate and organic product catalysed by the enzymes involved in the rTCA cycle [29,30]. *Kiwa* sp. living at E9N and E9S, along with its epibionts, had δ^{13}C values that were >−12‰ and are found in areas close to discharging vent fluids [22]. This potentially indicates the epibionts living on *Kiwa* sp. ventral setae were fixing carbon via the rTCA cycle. *Kiwa* sp. was also ^{15}N-enriched by between 3.8‰ and 4.2‰ relative to its epibionts, suggesting the epibionts were an important food source. A similar episymbiotic relationship between the ESR kiwid is therefore hypothesised to that of *Kiwa puravida*, for which lipid, stable isotope and behavioural analyses indicate the harvesting of epibiont bacteria [45]. Stichasteridae sp. (~−13‰) and cf *Actinostola* sp. (~−14‰) also appeared to be assimilating carbon indicative of the rTCA cycle at E9N and E9S.

Several vent macroconsumers fell within the range of δ^{13}C values indicative of mixed carbon sources. Those within the δ^{13}C −22‰ to −15‰ range may consume free-living bacteria or are predators or scavengers that utilise a number of trophic pathways. At the ESR hydrothermal vents, *Lepetodrilus* sp., Provannidae sp. 2, *Vulcanolepas* sp., *Kiwa* sp., Stichasteridae sp. and *Colossendeis* cf. *elephantis* fell into this range at one or more sites. Related species of *Lepetodrilus* sp., Provannidae sp. 2 and *Vulcanolepas* sp. are all thought to consume free-living bacteria at other vents sites [14,23]. Such feeding can result in consuming heterogeneous bacterial communities, which have multiple pathways for carbon fixation and elemental cycling [9,46,47]. The biological cycling of carbon is very complex at hydrothermal vents because of the multiple single carbon substrates for carbon fixation (e.g. CO_2, CH_4, CO), spatial variability in the δ^{13}C value of the substrate and various microbial primary producers associated with different carbon fixation pathways [7,11,48]. Furthermore, the incorporation of photosynthetic derived carbon as particulate or dissolved organic matter is possible and may provide some nutrition to vent macroconsumers [12,13]. Therefore, complex isotopic mixes of food sources are available to these species.

Table 4. Results of ANOVA and *post-hoc* Tukey honest significant differences tests for the differences in stable isotope values of vent fauna among the three sites on the East Scotia Ridge.

Species	δ^{13}C				δ^{34}S				δ^{15}N			
	DF	F	p	Post-hoc	DF	F	p	Post-hoc	F	DF	p	Post-hoc
Vulcanolepas sp.	2, 63	403.18	<0.01	E9N<E9S<E2	2, 63	176.16	<0.01	E9S<E2 = E9S	2, 63	138.26	<0.01	E2 = E9S<E9N
Kiwa sp. male	2, 31.36	147.29	<0.01	E2< E9S = E9N*	2, 66	19.52	<0.01	E2<E9S<E9N	2, 66	52.64	<0.01	E2< E9N = E9S
Sericosura spp.	2, 15	215.00	<0.01	E9N<E9S<E2	2, 15	100.61	<0.01	E9N<E2< E9S	2, 15	3.39	0.06	NA
Peltospiroidea sp.	2, 52	29.50	<0.01	E9N<E9S = E2	2, 52	49.26	<0.01	E9N<E9S<E2	2, 52	2.90	0.06	NA
Lepetodrilus sp.	2, 10	17.41	<0.01	E2 = E9N<E9S	2, 10	31.99	<0.01	E9N = E9S<E2	2, 10	32.10	<0.01	E2 = E9S<E9N

*Welch's ANOVA with *post hoc* analysis by t-test with Bonferroni correction (p = 0.05/3 = 0.017).

Table 5. Results of t-tests for between-sites differences in stable isotope values of vent fauna at the East Scotia Ridge.

Species	Comparison	$\delta^{13}C$			$\delta^{34}S$			$\delta^{15}N$		
		DF	t	p	DF	t	p	DF	t	p
Kiwa sp.	E2 female v male	36	−0.50	0.62	36	2.23	<0.05	36	5.13	<0.01
Kiwa sp. Epibionts	E2 v E9S	4.05	−3.81	<0.05*	6.93	4.57	<0.01*	2	na	<0.05ᵠ
Stichasteridae sp.	E2 v E9N	4	28.92	<0.01	4	−3.29	<0.05	4	5.14	<0.01
	E2 v E9S	4	7.28	<0.01	4	0.385	0.59	4	5.94	<0.01
	E9N v E9S	8	3.64	<0.01	8	−1.76	0.11	8	−3.49	<0.05
Colossendeis cf *elphantis*	E2 v E9N	2	2.45	0.13	2	−6.93	<0.05	2	1.71	0.23

*Welch's t-test,
ᵠWilcoxon test.

The majority of ESR vent macroconsumers had $\delta^{34}S$ values less than or equal to the 10‰ threshold, indicating chemosynthetic food sources [49]. Species exceeding the 10‰ value occurred mainly at E2 in the anemones *Pacamanactis* sp. and cf *Marianactis* sp, the sponge *Cladorhiza* sp., the pycnogonids *C. elephantis* and the stichasterid seastar along with *Sericosura* spp. and stichasterid seastar at E9S. All had $\delta^{34}S$ values between 10‰ and 16‰. Mixing of epipelagic photosynthetic and hydrothermal vent chemosynthetic production sources at these sites cannot be ruled out.

Determining intra-site differences in food sources and trophic interactions using $\delta^{34}S$ is challenging for macroconsumers with $\delta^{34}S$ values <10‰ because the $\delta^{34}S$ values of inorganic substrates and the net fractionation effect between inorganic substrates and products for primary producers and consumers are uncertain. At E9, $\delta^{34}S$ appeared to increase from macroconsumers living closest to vent openings and within diffuse flow areas (i.e. *Kiwa* sp., Peltospiroidea sp. and *Lepetodrilus* sp.) to those in the periphery (i.e. anemones, stichasterid seastars and *Colossendeis* spp.). It is unclear why an increase in $\delta^{34}S$ occurred from the centre of the vent to the periphery: it may be the result of changes in sulfide speciation [50] or other sulfur sources with increasing distance from the vent opening [33], differences in levels of sulfide exposure [50], incorporation of epipelagic photosynthetic primary production or a combination of the above.

Stichasterid seastars, cf *Actinostola* sp. and *Colossendeis* spp. consistently had the highest $\delta^{15}N$ values of all the ESR vent macroconsumers, which suggested they occupied the highest trophic positions of those predators sampled. Behavioural observations [22] and $\delta^{13}C$ values indicated that *Kiwa* sp. is consumed by stichasterid seastar and cf *Actinostola* sp. 1 but only the stichasterid seastar had $\delta^{15}N$ values indicative of a higher trophic position than *Kiwa* sp. In the case of *Colossendeis* spp., feeding on anemones occurs at the ESR vent sites [22] and at E2 all three stable isotopes indicated a strong predator-prey link. At E9N there was a large difference in $\delta^{13}C$ and $\delta^{34}S$ between cf *Actinostola* sp. 1 and the two species of *Colossendeis* as well as lower $\delta^{15}N$ in these pycnogonids compared to cf *Actinostola* sp. 1. This suggests that at E9N the feeding incidents between cf *Actinostola* sp. 1 and *Colossendeis* spp. are either rare or stable isotope values of *Colossendeis* spp. are strongly affected by isotopic mixing of different energy sources ($\delta^{13}C$ and $\delta^{34}S$) and feeding over multiple trophic positions ($\delta^{15}N$).

It is evident from the ESR hydrothermal vent food webs that predators may have similar or lower $\delta^{15}N$ values than their prey. Calculating trophic position assuming taxon specific nitrogen trophic discrimination factors [23] or applying the more universal value of 3.4‰ [12] was not undertaken within this study because they may have provided erroneous results. Establishing a suitable $\delta^{15}N$ baseline is problematic because: the macroconsumer with the lowest $\delta^{15}N$ differed among locations, is confounded by the use of different tissues (e.g. whole animals, muscle) to construct the food webs [32] and the observed high $\delta^{15}N$ variability in potential food sources. Compound-specific amino acid stable isotope analysis may provide higher resolution information on the organic nitrogen compounds assimilated by vent macroconsumers because the isotopic values of different amino acids record trophic and basal source information [51,52]. Thus it may circumvent some of the limitations of bulk $\delta^{15}N$ analysis and provide a better understanding of nitrogen cycling at hydrothermal vents.

Spatial Patterns in Macroconsumer Trophodynamics

Large spatial differences in $\delta^{13}C$ values for *Kiwa* sp., Stichasteridae sp. and *Sericosura* spp. were attributed primarily to differences in carbon fixation pathways at the base of the food web, which is in turn transferred to higher trophic positions. $\delta^{13}C$ values of *Kiwa* sp. differed by ~9‰ between E2 and E9S as did that of associated *Kiwa* sp. epibionts. Also, epsilon-Proteobacteria dominated the epibiont community at E9 with gamma-Proteobacteria largely absent, compared to a mix of gamma- and epsilon-Proteobacteria at E2 (K. Zwirglmaier unpublished data). All epsilon-Proteobacteria to date use the rTCA cycle to fix carbon while gamma-Proteobacteria predominantly use the CBB cycle [11]. *Riftia pachyptila* has similar differences in $\delta^{13}C$ among vent sites, but this is attributed to its endosymbionts shifting between rTCA and CBB cycles [53] rather than changes in the microbial community it consumes. Alvinoconchid gastropods have $\delta^{13}C$ values that differ by >20‰ among vent fields, which relates to whether epsilon- or gamma-Proteobacteria are the endosymbionts [54]. It is unclear why *Kiwa* sp. epibiont diversity is different between E2 and E9. At other hydrothermal vent locations differences in vent fluid chemical composition influences microbial communities [46,55] and it may be similar at the ESR vent fields. The difference in carbon fixation appeared to be transferred through *Kiwa* sp. to the predatory stichasterid seastar. Such a predator-prey interaction may also explain the large difference in $\delta^{13}C$ values between E2 and E9N in *Sericosura* spp. At E2 *Sericosura* spp. were collected from amongst anemones that had $\delta^{13}C$ values indicative of a mixed carbon source but at E9 they were collected from amongst peltospiroid gastropods dependent on CBB fixed carbon, although *Sericosura* spp. were not observed directly feeding on either anemones or Peltospiroidea sp.

Relatively small differences in stable isotope values were observed among sites in Peltospiroidea sp., *Lepetodrilus* sp. and *Vulcanolepas* sp. To date, Peltospiroidea sp. contains a single strain of gamma-Proteobacteria endosymbiont (K. Zwirglmaier unpublished data), which means spatial differences in $\delta^{13}C$ and $\delta^{34}S$ are unlikely to be the result of differences in the type of endosymbiont [55]. The differences were potentially a result of site-specific variations in the $\delta^{13}C_{DIC}$ and inorganic $\delta^{34}S$ values used by the endosymbionts during chemoautotrophy or physiological temperature-related effects on isotopic discrimination. Small differences among sites for the grazer *Lepetodrilus* sp. and suspension feeder *Vulcanolepas* sp. are harder to explain because of the various factors that are likely to influence their food source. $\delta^{13}C$ values indicated these organisms consume a mixed diet of free-living microbes and particulate material. However, differences in $\delta^{13}C$ values within sites may be related to the organism's distribution within the vent field [56] and in turn the composition of the microbial community [46], the stable isotope values of the inorganic substrate used during chemoautotrophy [57] and temperature effects on trophic discrimination. *Lepetodrilus* sp. and *Vulcanolepas* sp. were collected from single points within each vent site and, therefore, it is not clear whether the difference in stable isotope values among sites is greater or less than that within sites.

Because of the snap-shot nature of this study, it is difficult to identify factors that caused the spatial differences in the *Kiwa* sp. epibiont communities that resulted in a greater range of $\delta^{13}C$ values at the E9 sites compared to E2. Higher concentrations of dissolved sulfides in vent fluids may favour the rTCA pathway resulting in increasing numbers of organisms with $\delta^{13}C$ values greater than $-16‰$ [12]. On the ESR, E9 has higher hydrogen sulfide and lower chloride concentrations than E2 meaning that there are greater concentrations of available gases for microbial primary production due to phase separation [6,21]. Higher concentrations of reduced compounds and gases may be one of the drivers of the differences in trophic structure at the ESR vents. However, hydrothermal vent communities also undergo changes in community composition with age [58] and fluctuating hydrothermal activity [59], which will have an effect on trophic

structure. As data presented here were obtained concurrently with the discovery of the new biogeographical province it is not possible to determine whether the communities at E2 and E9 represent different successional stages, are a product of varying chemistry or a mix of such processes.

Conclusion

Trophic structure differed substantially between the E2 and E9 vents fields, and only slightly between E9N and E9S. $\delta^{13}C_{DIC}$ of the end-member fluid and diffuse flow samples were similar among the sites but large differences in the $\delta^{13}C$ values of some vent macroconsumers indicated spatial variations in the way microbes were fixing carbon at the base of the food chain. $\delta^{13}C$ values $>-13‰$ at the E9N and E9S suggest that the relative contribution to the macroconsumer food web of carbon fixed via the rTCA cycle is likely to be greater than at E2. The greater range of $\delta^{34}S$ values at E2 and E9S indicated a potentially greater influence of epipelagic photosynthetic primary production than at E9N. The greater contribution of rTCA fixed carbon at the E9 vent field may ultimately be related to differences in vent fluid, but more work is required to link vent fluid chemistry with microbial primary production and the related trophic structure at hydrothermal vents.

Acknowledgments

We thank the officers, crew, ROV team & scientists onboard JC42 for assistance in sample collection and sorting. We also thank Veerle Huvenne for producing the bathymetric map.

Author Contributions

Conceived and designed the experiments: WDKR CJS BDW KZ JAH KL NVCP. Performed the experiments: WDKR CJS KZ JAH KL. Analyzed the data: WDKR CJS BDW KZ JAH RARM NVCP. Contributed reagents/materials/analysis tools: WDKR CJS KZ JAH RARM. Wrote the paper: WDKR CJS BDW KZ JAH RARM KL NCVP. Participants on JC042: WDKR CJS JAH KZ KL.

References

1. Kelley DS, Karson JA, Blackman DK, Fruh-Green GL, Butterfield DA, et al. (2001) An off-axis hydrothermal vent field near the Mid-Atlantic Ridge at 30°N Nature 412: 145–149.

2. Staudigel H, Hart SR, Pile A, Bailey SE, Baker ET, et al. (2006) Vailulu'u seamount, Samoa: Life and death on an active submarine volcano. Proc Natl Acad Sci USA 103: 6448–6453.

3. Tunnicliffe V, Juniper SK, Sibuet M (2003) Reducing environments of the deep-sea floor. In: Tyler PA, editor. Ecosystems of the Deep Ocean. Amsterdam: Elsevier Science. 81–110.

4. Cavanaugh CM, Gardiner SL, Jones ML, Jannasch HW, Waterbury JB (1981) Prokaryotic cells in the hydrothermal vent tube worm *Riftia pachyptila* Jones - possible chemoautotrophic symbionts. Science 213: 340–342.

5. German CR, Ramirez-Llodra E, Baker MC, Tyler PA (2011) Deep-water chemosynthetic ecosystem research during the Census of Marine Life decade and beyond: A proposed deep-ocean road map. PLoS ONE 6: 16.

6. German CR, Von Damm KL (2003) Hydrothermal Processes. In: Elderfield H, editor. The ocean and marine geochemistry. Oxford: Elsevier. 181–222.

7. Karl DM (1995) Ecology of free-living, hydrothermal vent microbial communities. In: Karl DM, editor. The microbiology of deep-sea hydrothermal vents. Boca Raton: CRC Press Inc. 35–124.

8. McCollom TM, Shock EL (1997) Geochemical constraints on chemolithoautotrophic metabolism by microorganisms in seafloor hydrothermal systems. Geochim Cosmochim Acta 61: 4375–4391.

9. Campbell BJ, Cary SC (2004) Abundance of reverse tricarboxylic acid cycle genes in free-living microorganisms at deep-sea hydrothermal vents. Appl Environ Microbiol 70: 6282–6289.

10. Desbruyeres D, Alaysedanet AM, Ohta S, Antoine E, Barbier G, et al. (1994) Deep-sea hydrothermal communities in the southwestern Pacific back-arc basins (the North Fiji and Lau Basins)- composition, microdistribution and food web. Mar Geol 116: 227–242.

11. Hugler M, Sievert SM (2011) Beyond the Calvin Cycle: Autotrophic carbon fixation in the ocean. Ann Rev Mar Sci. Palo Alto: Annual Reviews. 261–289.

12. De Busserolles F, Sarrazin J, Gauthier O, Gelinas Y, Fabri MC, et al. (2009) Are spatial variations in the diets of hydrothermal fauna linked to local environmental conditions? Deep Sea Res Part II Top Stud Oceanogr 56: 1649–1664.

13. Riou V, Colaco A, Bouillon S, Khripounoff A, Dando PR, et al. (2010) Mixotrophy in the deep sea: a dual endosymbiotic hydrothermal mytilid assimilates dissolved and particulate organic matter Mar Ecol Prog Ser 405: 187–201.

14. Colaco A, Dehairs F, Desbruyeres D (2002) Nutritional relations of deep-sea hydrothermal fields at the Mid-Atlantic Ridge: a stable isotope approach. Deep Sea Res Part I Oceanogr Res Pap 49: 395–412.

15. Rau GH, Hedges JI (1979) Carbon-13 depletion in a hydrothermal vent mussel - suggestion of a chemosynthetic food source. Science 203: 648–649.

16. Van Dover CL, Fry B (1994) Microorganisms as food resources at deep-sea hydrothermal vents. Limnol Oceanogr 39: 51–57.

17. Levin LA, Mendoza GF, Konotchick T, Lee R (2009) Macrobenthos community structure and trophic relationships within active and inactive Pacific hydrothermal sediments. Deep Sea Res Part II Top Stud Oceanogr 56: 1632–1648.

18. Limen H, Juniper SK (2006) Habitat controls on vent food webs at Eifuku Volcano, Mariana Arc. Cah Biol Mar 47: 449–455.

19. Van Dover CL (2002) Trophic relationships among invertebrates at the Kairei hydrothermal vent field (Central Indian Ridge). Mar Biol 141: 761–772.

20. German CR, Livermore RA, Baker ET, Bruguier NI, Connelly DP, et al. (2000) Hydrothermal plumes above the East Scotia Ridge: an isolated high-latitude back-arc spreading centre. Earth Planet Sci Lett 184: 241–250.

21. Rogers AD, Tyler PA, Connelly DP, Copley JT, James R, et al. (2012) The discovery of new deep-sea hydrothermal vent communities in the Southern Ocean and implications for biogeography. PLoS Biol 10: e1001234.

22. Marsh L, Copley JT, Huvenne VAI, Linse K, Reid WDK, et al. (2012) Microdistribution of faunal assemblages at deep-sea hydrothermal vents in the Southern Ocean. PLoS ONE 7: e48348.

23. Bergquist DC, Eckner JT, Urcuyo IA, Cordes EE, Hourdez S, et al. (2007) Using stable isotopes and quantitative community characteristics to determine a local hydrothermal vent food web. Mar Ecol Prog Ser 330: 49–65.

24. Van Dover CL, Fry B (1989) Stable isotopic compositions of hydrothermal vent organisms. Mar Biol 102: 257–263.

25. Fisher CR, Childress JJ, Macko SA, Brooks JM (1994) Nutritional interactions in Galapagos Rift hydrothermal vent communities - inferences from stable carbon and nitrogen isotope analyses. Mar Ecol Prog Ser 103: 45–55.

26. Guy RD, Fogel ML, Berry JA (1993) Photosynthetic fractionation of the stable isotopes of oxygen and carbon. Plant Physiology 101: 37–47.

27. Robinson JJ, Scott KM, Swanson ST, O'Leary MH, Horken K, et al. (2003) Kinetic isotope effect and characterization of form II RuBisCO from the chemoautotrophic endosymbionts of the hydrothermal vent tubeworm *Riftia pachyptila*. Limnol Oceanogr 48: 48–54.

28. Roeske CA, O'Leary MH (1984) Carbon isotope effects on the enzyme-catalyzed carboxylation of ribulose bisphosphate. Biochemistry 23: 6275–6284.

29. House CH, Schopf JW, Stetter KO (2003) Carbon isotopic fractionation by Archaeans and other thermophilic prokaryotes. Org Geochem 34: 345–356.

30. Suzuki Y, Sasaki T, Suzuki M, Nogi Y, Miwa T, et al. (2005) Novel chemoautotrophic endosymbiosis between a member of the Epsilonproteobacteria and the hydrothermal-vent gastropod *Alviniconcha* aff. *hessleri* (Gastropoda: Provannidae) from the Indian Ocean. Appl Environ Microbiol 71: 5440–5450.

31. Wirsen CO, Sievert SM, Cavanaugh CM, Molyneaux SJ, Ahmad A, et al. (2002) Characterization of an autotrophic sulfide-oxidizing marine *Arcobacter* sp that produces filamentous sulfur. Appl Environ Microbiol 68: 316–325.

32. Michener RH, Kaufman L (2007) Stable isotope ratios as tracers in marine food webs: an update. In: Michener RH, Lajtha K, editors. Stable isotopes in ecology and environmental science. 2nd ed. Singapore: Blackwell Publishing. 238–282.

33. Herzig PM, Hannington MD, Arribas A Jr (1998) Sulfur isotopic composition of hydrothermal precipitates from the Lau back-arc: implications for magmatic contributions to seafloor hydrothermal systems. Mineralium Deposita 33: 226–237.

34. Erickson KL, Macko SA, Van Dover CL (2009) Evidence for a chemoautotrophically based food web at inactive hydrothermal vents (Manus Basin). Deep Sea Res Part II Top Stud Oceanogr 56: 1577–1585.

35. Reid WDK, Wigham BD, McGill RAR, Polunin NVC (2012) Elucidating trophic pathways in benthic deep-sea assemblages of the Mid-Atlantic Ridge north and south of the Charlie-Gibbs Fracture Zone. Mar Ecol Prog Ser 463: 89–103.

36. Layman CA, Arrington DA, Montana CG, Post DM (2007) Can stable isotope ratios provide for community-wide measures of trophic structure? Ecology 88: 42–48.

37. Quinn GP, Keough MJ (2002) Experimental Design and Data Analysis for Biologists. Cambridge: Cambridge University Press. 537 p.

38. R Core Team (2011) R: A language and environment for statistical computing. R Foundation for Statistical Computing, Vienna, Austria. ISBN 3–900051–07–0, URL http://www.R-project.org/.

39. Brooks JM, Kennicutt MC, Fisher CR, Macko SA, Cole K, et al. (1987) Deep-sea hydrocarbon seep communities - evidence for energy and nutritional carbon sources. Science 238: 1138–1142.

40. Childress JJ, Fisher CR (1992) The biology of hydrothermal vent animals-physiology, biochemistry and autotrophic symbioses. Oceanogr Mar Biol 30: 337–441.

41. Campbell BJ, Engel AS, Porter ML, Takai K (2006) The versatile epsilon-Proteobacteria: key players in sulphidic habitats. Nat Rev Microbiol 4: 458–468.

42. Campbell BJ, Stein JL, Cary SC (2003) Evidence of chemolithoautotrophy in the bacterial community associated with *Alvinella pompejana*, a hydrothermal vent polychaete. Appl Environ Microbiol 69: 5070–5078.

43. Levesque C, Juniper SK, Marcus J (2003) Food resource partitioning and competition among alvinellid polychaetes of Juan de Fuca Ridge hydrothermal vents. Mar Ecol Prog Ser 246: 173–182.

44. Suzuki Y, Kojima S, Sasaki T, Suzuki M, Utsumi T, et al. (2006) Host-symbiont relationships in hydrothermal vent gastropods of the genus Alviniconcha from the Southwest Pacific. Appl Environ Microbiol 72: 1388–1393.

45. Thurber AR, Jones WJ, Schnabel K (2011) Dancing for food in the deep sea: bacterial farming by a new species of yeti crab. PLoS ONE 6: e26243.

46. Flores GE, Campbell JH, Kirshtein JD, Meneghin J, Podar M, et al. (2011) Microbial community structure of hydrothermal deposits from geochemically different vent fields along the Mid-Atlantic Ridge. Environ Microbiol 13: 2158–2171.

47. Takai K, Nunoura T, Horikoshi K, Shibuya T, Nakamura K, et al. (2009) Variability in microbial communities in black smoker chimneys at the NW Caldera vent field, Brothers Volcano, Kermadec Arc. Geomicrobiol J 26: 552–569.

48. Nakagawa S, Takai K (2008) Deep-sea vent chemoautotrophs: diversity, biochemistry and ecological significance. FEMS Microbiol Ecol 65: 1–14.

49. Vetter RD, Fry B (1998) Sulfur contents and sulfur-isotope compositions of thiotrophic symbioses in bivalve molluscs and vestimentiferan worms. Mar Biol 132: 453–460.

50. Luther GW, Rozan TF, Taillefert M, Nuzzio DB, Di Meo C, et al. (2001) Chemical speciation drives hydrothermal vent ecology. Nature 410: 813–816.

51. Chikaraishi Y, Ogawa NO, Kashiyama Y, Takano Y, Suga H, et al. (2009) Determination of aquatic food-web structure based on compound-specific nitrogen isotopic composition of amino acids. Limnol Oceanogr Methods 7: 740–750.

52. McClelland JW, Montoya JP (2002) Trophic relationships and the nitrogen isotopic composition of amino acids in plankton. Ecology 83: 2173–2180.

53. Markert S, Arndt C, Felbeck H, Becher D, Sievert SM, et al. (2007) Physiological proteomics of the uncultured endosymbiont of *Riftia pachyptila*. Science 315: 247–250.

54. Suzuki Y, Sasaki T, Suzuki M, Tsuchida S, Nealson KH, et al. (2005) Molecular phylogenetic and isotopic evidence of two lineages of chemoautotrophic endosymbionts distinct at the subdivision level harbored in one host-animal type: The genus *Alviniconcha* (Gastropoda : Provannidae). FEMS Microbiol Lett 249: 105–112.

55. Trask JL, Van Dover CL (1999) Site-specific and ontogenetic variations in nutrition of mussels (*Bathymodiolus* sp.) from the Lucky Strike hydrothermal vent field, Mid-Atlantic Ridge. Limnol Oceanogr 44: 334–343.

56. Levesque C, Juniper SK, Limen H (2006) Spatial organization of food webs along habitat gradients at deep-sea hydrothermal vents on Axial Volcano, Northeast Pacific. Deep Sea Res Part I Oceanogr Res Pap 53: 726–739.

57. Levesque C, Limen H, Juniper SK (2005) Origin, composition and nutritional quality of particulate matter at deep-sea hydrothermal vents on Axial Volcano, NE Pacific. Mar Ecol Prog Ser 289: 43–52.

58. Shank TM, Fornari DJ, Von Damm KL, Lilley MD, Haymon RM, et al. (1998) Temporal and spatial patterns of biological community development at nascent deep-sea hydrothermal vents (9 degrees 50' N, East Pacific Rise). Deep Sea Res Part II Top Stud Oceanogr 45: 465–515.

59. Cuvelier D, Sarrazin J, Colaco A, Copley JT, Glover AG, et al. (2011) Community dynamics over 14 years at the Eiffel Tower hydrothermal edifice on the Mid-Atlantic Ridge. Limnol Oceanogr 56: 1624–1640.

13

Evolutionary Strategies of Viruses, Bacteria and Archaea in Hydrothermal Vent Ecosystems Revealed through Metagenomics

Rika E. Anderson[1]*, Mitchell L. Sogin[2], John A. Baross[1]

1 School of Oceanography and Astrobiology Program, University of Washington, Seattle, Washington, United States of America, **2** Josephine Bay Paul Center, Marine Biological Laboratory, Woods Hole, Massachusetts, United States of America

Abstract

The deep-sea hydrothermal vent habitat hosts a diverse community of archaea and bacteria that withstand extreme fluctuations in environmental conditions. Abundant viruses in these systems, a high proportion of which are lysogenic, must also withstand these environmental extremes. Here, we explore the evolutionary strategies of both microorganisms and viruses in hydrothermal systems through comparative analysis of a cellular and viral metagenome, collected by size fractionation of high temperature fluids from a diffuse flow hydrothermal vent. We detected a high enrichment of mobile elements and proviruses in the cellular fraction relative to microorganisms in other environments. We observed a relatively high abundance of genes related to energy metabolism as well as cofactors and vitamins in the viral fraction compared to the cellular fraction, which suggest encoding of auxiliary metabolic genes on viral genomes. Moreover, the observation of stronger purifying selection in the viral versus cellular gene pool suggests viral strategies that promote prolonged host integration. Our results demonstrate that there is great potential for hydrothermal vent viruses to integrate into hosts, facilitate horizontal gene transfer, and express or transfer genes that manipulate the hosts' functional capabilities.

Editor: Vladimir N. Uversky, University of South Florida College of Medicine, United States of America

Funding: Funding for sequencing of the viral metagenome was provided by the Gordon and Betty Moore Foundation. All other funding was provided by a NASA Astrobiology Institute grant through Cooperative Agreement NNA04CC09A to the Geophysical Laboratory at the Carnegie Institution for Science. R.A. was funded by a NSF Graduate Research Fellowship through NSF grant number DGE-0718124, an NSF IGERT grant to the University of Washington Astrobiology Program, and the ARCS Foundation. The funders had no role in study design, data collection and analysis, decision to publish, or preparation of the manuscript.

Competing Interests: The authors have declared that no competing interests exist.

* Email: rikander@u.washington.edu

Introduction

The deep subsurface below hydrothermal systems hosts a high diversity of archaea, bacteria, and viruses that must tolerate extremely variable environmental conditions. High-temperature, reduced hydrothermal fluids mix with cold, oxidized seawater both above and below the seafloor to establish strong gradients in temperature, pH, and chemical and mineralogical composition [1–5]. Wide variations in environmental parameters can occur over centimeter scales. Constant fluid flux throughout and above the subsurface transports organisms from one region to the next, exposing them to a range of environmental conditions. Gradients that dominate this environment create a highly diverse microbial community consisting of both archaea and bacteria [6]. Physical and chemical parameters vary according to fluid mixing and volcanic activity, leading to niche partitioning in microbial communities across both space [1] and time [7,8]. Moreover, hyperthermophiles are routinely cultured from fluids that exit at low temperatures (5–30°C) [9,10], indicating that organisms in vent systems are frequently flushed from their native habitats, most likely from the deep subsurface. Microbial communities in hydrothermal systems are known to form biofilms that coat mineral surfaces, including within high-temperature chimney structures [5]. Such biofilms, which are likely to occur within the subsurface as well, host high-density communities with potentially high contact rates between organisms.

The dynamic, diverse and dense nature of this habitat should foster frequent exchange of genes within the microbial community. Previous work with vent samples has shown that the genes responsible for this process, including transposases and integrases, were observed to occur at high frequency in cellular metagenomes from hydrothermal systems as compared to other environments [11,12]. Analysis of fully sequenced genomes of thermophiles, including many from vent systems, suggests that gene transfers occur more frequently among thermophiles than mesophiles or psychrophiles [13,14] and that these transfers sometimes cross domains [13–15]. The prevalence of horizontal gene transfer in vent systems may expand the functional repertoire of a given species, expanding the pangenome and providing access to different ecological niches. This expanded metabolic flexibility would provide a strong advantage in hydrothermal vent environments where fluid flux and environmental gradients expose communities to wide extremes in temperature, pH, and chemical composition.

Here, we use comparative metagenomics to elucidate the role that viruses play in facilitating gene flow and manipulating host genetic potential in hydrothermal systems. Viruses are known to play pivotal roles in the transfer of genes and the alteration of host phenotype, particularly in the pelagic oceans (see Breitbart 2012 [16] for review). Bacterial and archaeal viruses introduce foreign genetic material through transduction and expression of virally encoded genes during infection. Transduction, or virally-mediated horizontal gene transfer, occurs on a massive scale in the surface oceans. Up to 10^{14} transduction events can occur per year in Tampa Bay estuary [17], and virus-like particles that serve as gene transfer agents (GTAs) may boost these transduction rates by one million-fold [18]. Viruses are known to encode auxiliary metabolic genes, or AMGs, which play critical roles in facilitating biochemical or metabolic processes [19]. For example, cyanophage transcribe and express photosynthesis genes during lytic infection of their cyanobacterial hosts [20–23], potentially to support the host during infection, or to redirect host metabolism to support phage deoxyribonucleotide biosynthesis [24]. Lysogenic viruses can have similar impacts on their hosts: the expression of genes encoded by integrated viruses (also known as proviruses, or prophage in bacteria) can manipulate host phenotype, such as in the case of the cholera toxin expressed by a prophage integrated in the *Vibrio cholerae* genome [25]. Selection should favor expression of genes within lysogenic viruses that enhance host fitness while the virus is integrated in the genome. For example, it has been hypothesized that proviruses express genes that suppress host metabolism to conserve resources under low-energy or low-nutrient conditions [26].

Despite increasing evidence that viruses play a crucial role in manipulating host genotype and phenotype in the surface oceans, this phenomenon has yet to be explored in the dynamic environment of hydrothermal vents. Viruses are abundant in hydrothermal systems [27] and have the potential to infect many different taxa of bacteria and archaea [3]. It has been suggested that up to 80% of archaea and bacteria in the deep ocean contain proviruses in their genomes [28]. Induction experiments have suggested that proviruses are particularly abundant in the genomes of archaea and bacteria from hydrothermal vent fluids compared to those in water from the deep ocean or the deep chlorophyll maximum [29]. Considering the abundance of viruses in these systems, and lysogenic viruses in particular, several questions arise: do these viruses transfer genes between hosts? Do they express fitness factors while integrated in the host genome? If so, which genes are expressed? Do viruses contribute to host genomic plasticity and facilitate their adaptation to changing conditions? Does selection act differently on viral genes compared to cellular genes?

To address these questions, we used a cultivation-independent approach that provides a community-wide perspective of both the viral gene pool and the bacterial and archaeal gene pool (hereafter referred to as the "cellular" gene pool) in hydrothermal systems. Specifically, we analyzed the unamplified viral and cellular metagenomes of high-temperature diffuse flow hydrothermal fluid from Hulk hydrothermal vent in the Main Endeavour Field on the Juan de Fuca Ridge. We compared the relative content of each of these gene pools and inferred the modes of genetic interaction between viruses and their hosts. This analysis focused on a unique fluid sample that contained organisms native to a wide range of ecological niches within the gradient-dominated hydrothermal environment, all with the potential to come into contact through constant fluid flux. Given the potential for gene and viral exchange across these niches, these metagenomes can provide insights into interactions within the diverse communal gene pool of the hydrothermal vent microbial community.

Comparative analysis of the cellular and viral metagenomes from this sample addressed whether viruses have the potential to manipulate the physiology or metabolism of their hosts. The presence of genes facilitating horizontal gene transfer and lysogenic virus integration described the genetic potential for these processes in the vent environment. We compared the relative abundance of genes in the viral and cellular gene pools in order to determine the types of genes enriched in the viral gene pool relative to the cellular gene pool. Finally, we asked how evolution has shaped the viral and cellular gene pools by examining relative selection pressures on viral and cellular genes. Together, these analyses provide insight into the broader question of how evolution has shaped the genomes of viruses and their hosts in some of the more extreme environments of the planet.

Materials and Methods

Sample collection and DNA extraction

We collected a 170-L hydrothermal vent fluid sample from Hulk vent at the Main Endeavour Field on the Juan de Fuca Ridge (47°57.00′ N, 129°5.81′ W) as described previously [3]. No specific permissions were required for these locations or sampling activities. The vent fluid was obtained using a large barrel sampler equipped with two 100-L sterile bags. We placed the sample collection funnel atop a region of diffuse venting, adjacent to a colony of tube worms on the side of a large sulfide structure. While the tube worms were surrounded by fluid at measured temperatures of 13–30°C, the average temperature of the metagenome fluid sample was calculated from its silica chemistry to be about 125°C [3]. This result indicates that we most likely sampled fluid ranging from cool background seawater to high-temperature hydrothermal fluid (up to 300°C) from the sulfide structure adjacent to the sample site, illustrating the dynamic fluid flux of these systems. The organisms collected in the sample therefore represent a range of habitats in the hydrothermal environment, including psychrophiles, mesophiles and thermophiles, and aerobic, microaerophilic, and anaerobic organisms. Some of these organisms may have been derived from deep subsurface fluids, whereas others from entrained seawater. Having been sampled from the same site, these organisms have the potential to come into contact within the vent environment due to dynamic fluid flux. We included available metadata about this vent sample in Table S1.

We collected the cellular fraction by filtering the 170 L of hydrothermal vent fluid through three 0.22 μl Steripaks (Millipore, USA) while the sample and filtrate were held on ice. The filtrate was retained for subsequent virus sampling. Filters were frozen at −80°C while shipboard and until sample processing. We extracted DNA from one Steripak using a modified DNA extraction procedure described by Anderson et al. [1]. Briefly, DNA extraction buffer (0.1 M Tris-HCl, 0.2 M Na-EDTA, 0.1 M NaH2PO4, 1.5 M NaCl, and 1% cetyltrimethylammonium bromide) was added to each filter, then the filters were capped and freeze-thawed five times. Lysozyme (50 mg/mL solution), proteinase K (1% solution), and SDS (20% solution) were added to each filter and incubated. Lysate was removed from filters and centrifuged; DNA was extracted from the supernatant using a phenol/chloroform/isoamyl extraction method described by Anderson et al. [1].

For virus collection, we concentrated the sample filtrate using tangential flow filtration (30 kDa cutoff) to approximately 400 mL in a 4°C cold room, and froze concentrated filtrate into six aliquots at −80°C until further processing. One aliquot was further concentrated by adding 10% w/v polyethylene glycol 8000 (PEG), incubating overnight at 4°C, and centrifuging at 13

000×g for 50 min. The pellet was resuspended in Tris-EDTA buffer and incubated for 15 min with 0.7 volume of chloroform to lyse any remaining cellular contamination. Free DNA was removed by incubating with 10% DNAse I for 2 h at 37°C, then inactivated by adding EDTA to a final concentration of 0.02 M. The QIAamp MinElute Virus Spin Kit (Qiagen) was used to extract the viral DNA, yielding approximately 90 ng, which was not amplified for downstream sequencing. PCR tests of extracted viral DNA using universal 16S archaeal and bacterial primers showed no amplification from contaminating cellular 16S rRNA, whereas positive controls of DNA extracted from *E. coli* showed successful amplification.

Metagenomic sequencing

The viral metagenome was generated on a Roche Genome Sequencer FLX (GSFLX) with GS FLX Titanium 454 sequencing protocols by the Broad Institute. For the cellular metagenome, libraries were created using the Nexterra transposon-mediated method (Epicentre) at the Josephine Bay Paul Center at the Marine Biological Laboratory, then sequenced using Roche Titanium 454 sequencing protocols on a GSFLX. Both metagenomes are publicly available on the MG-RAST v3 database [30], with accession numbers 4469452.3 for the viral metagenome and 4481541.3 for the cellular metagenome. The viral metagenome was previously described in Anderson et al. [3] and was uploaded under MG-RAST v2, and is still available at 4448187.3.

We used TagCleaner [31] to trim tags from the 5′ end of each sequence in the cellular metagenome. Assembly of both the viral and cellular metagenomes was conducted in Geneious [32] using the "Medium Sensitivity" method, with a word length of 14, a maximum gap size of 2, maximum gaps per read of 15, and maximum mismatches of 2. To classify sequences using di, tri, and tetranucleotide analysis, we created a boutique database of bacterial and archaeal virus sequences (Table S2) as a training set to accompany the existing cellular dataset in PhylopythiaS [33], which was used to identify archaea, bacteria, archaeal viruses and bacterial viruses. Metagenomes were assembled in Geneious prior to classification with PhylopythiaS; only contigs over 1000bp in length were used.

Enrichment of proviruses and mobile genetic elements

To identify the numbers of reads in each metagenome matching lysogenic viruses, metagenomes were compared to a database of sequences from the "Prophages" category in the ACLAME database [34]. To assess abundance of mobile genetic elements, we compared all metagenomes to a dataset of Pfam seed sequences [35] matching transposases, recombinases, resolvases and integrases, listed in Table S3, using tblastn with an e-value cutoff of 10^{-5}. The number of unique reads with a match to a sequence from the query sequence collection was tallied and normalized to the number of reads in the metagenome. Only metagenomes generated with 454 pyrosequencing were used for the analysis, so that all metagenomic reads had a length ranging from approximately 100 to 300 bp.

Relative enrichment of gene categories

We tallied gene categories by adding "abundance" counts for each functional category as defined by the KEGG Orthology database [36], the Clusters of Orthologous Groups (COG) database [37], or the SEED Subsystems database [38] in MG-RAST v3, using an e-value cutoff of 10^{-3}. For the combined analysis of 20 cellular metagenomes and 23 viral metagenomes, all abundance counts for either viral or cellular metagenomes were tallied together. We used Xipe-Totec [39], a nonparametric

method of statistical analysis using a difference of medians analysis, to determine whether abundance differences were statistically significant. For this analysis, we used a confidence level of 95% and a sample size of 5000 to determine significance.

Fragment recruitment

Cellular metagenomic reads were recruited to genomes of hydrothermal vent isolates using NUCmer, part of the MUMmer 3.0 package [40], with the following parameters for the command line: -minmatch 10 -breaklen 1200 -maxgap 1000 -mincluster 50. Fragment recruitments were visualized using mummerplot in the MUMmer package. Coverage plots were created by using the show-coords command in MUMmer, then in-house Python scripts were used to calculate coverage for each base pair position. Coverage plots were created with a convolution function in numpy, using a moving average window size of 50000.

Calculation of dN/dS

Prior to calculation of dN/dS ratios for genes mapped by each metagenome, metagenomes were subjected to stringent error filtering using Prinseq [41] with the following parameters: minimum sequence length of 60bp; minimum mean quality score of 30; maximum number of allowed Ns per sequence of 4; and low-complexity threshold of 70 (using Entropy). The dN/dS ratio measures selection pressures by calculating whether the number of non-synonymous substitutions (dN) in a gene is greater or fewer than the number expected by chance compared to the number of synonymous substitutions (dS). A majority-rule consensus was calculated from the mapped reads; the number of possible synonymous or nonsynonymous substitutions was then tallied and compared to the number of actual synonymous and nonsynonymous substitutions.

We mapped reads from both the viral and cellular metagenomes to the vent isolate genomes using CLC Genomics Workbench with the criteria of 80% identity and 80% coverage, using previously established benchmarks [42]. Mapping results were exported in ACE format; dN/dS was calculated for each gene using the Python scripts described in Tai *et al.* [42]. Polymorphisms were only tallied for positions with a mapping depth of at least 5X; only genes with at least 100 nucleotides at 5X depth were included in the analysis. Redundant genes were deleted from the analysis; only the dN/dS value for the gene with higher coverage was retained. The files used to define gene coordinates were downloaded from JGI IMG, with the exception of *T. kodakarensis*, which was derived from a.gff file obtained from NCBI. The 95% confidence interval was calculated for all genes mapped by the viral and cellular metagenomes with dN/dS less than 1 (subject to purifying selection) using alpha = 0.05.

Results and Discussion

General features of the metagenomes

Metagenomic sequencing of the cellular and viral fractions from the large sample of hydrothermal fluid yielded a total of 808,051 and 231,246 sequence reads, respectively. The cellular metagenome contained reads from a wide variety of bacterial and archaeal taxa, including *Thermococcales*, methanogens, Marine Groups I and II, *Proteobacteria, Bacteroidetes, Firmicutes*, and many others, reflecting the wide range of ecological niches represented in the sample. Figure 1 indicates that approximately 31% of the cellular metagenome and 47% of the viral metagenome (virome) sequences had no matches to the M5NR database (classified here as "Unknown") [43]. An analysis of the viral metagenome alone, including the taxa and diversity of viruses and their general host

A

● Unknown ● Archaea ● Bacteria ● Eukaryota ● Viruses ● Other

Cellular metagenome

B

● Unknown ● Archaea ● Bacteria ● Eukaryota ● Viruses ● Other

Viral metagenome

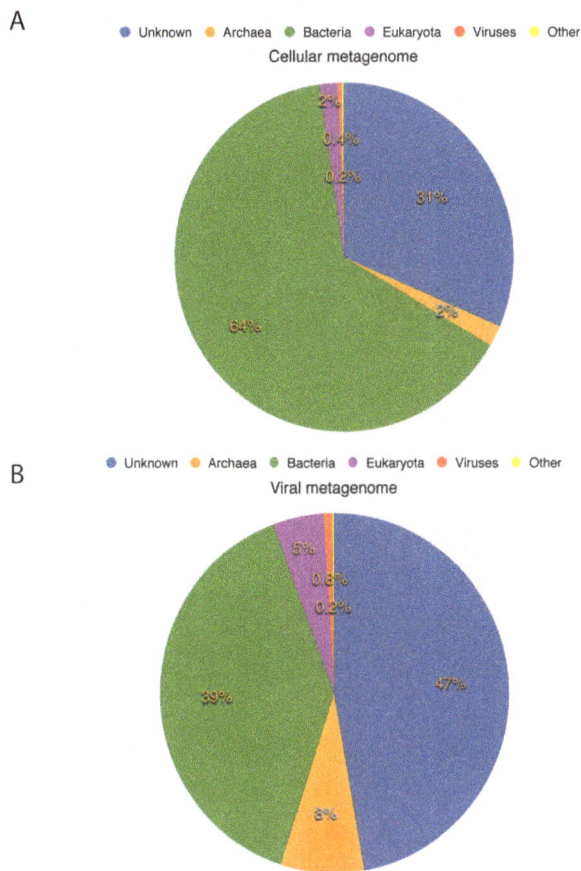

Figure 1. Pie charts showing breakdown of read classification for the cellular metagenome (A) and the viral metagenome (B) according to annotation by the M5NR database. Reads were annotated with a minimum e-value cutoff of 1e-05.

range, has been published previously [3]. Close matches between CRISPR spacers identified in the cellular metagenome and sequences in the viral metagenome suggest an active and relatively recent relationship between the two gene pools (see File S1). Classification of the cellular and viral metagenomes showed that only 2% of the reads from the cellular fraction matched archaeal genes, whereas nearly 8% of the viral metagenome reads matched archaeal genes (Figure 1). Nucleotide signature matching of assembled contigs longer than 1000 bp using PhylopythiaS [33] indicated that a disproportionate percentage of contigs in both metagenomes matched nucleotide compositional patterns of archaeal viruses, given that bacterial reads dominate both metagenomes (Figure S1). The percentage of contigs matching bacterial nucleotide compositional patterns was greater than the percentage of contigs with archaeal patterns by a ratio of 2.8 to 1 in the cellular metagenome. However, contigs matching bacterial virus patterns outnumbered contigs with archaeal virus patterns by only 1.8 to 1 in the viral metagenome. Taken together, these observations suggest that archaeal viruses may be disproportionally abundant in the vent habitat compared to the relative abundance of bacteria to archaea.

Assembly of viral metagenome contigs yielded many short, high-coverage (>20X coverage) contigs shorter than 6 kb (Figure S2). The cellular metagenome did not exhibit contigs with such high coverage and short length. In contrast, several contigs in the viral metagenome had relatively low coverage but were quite long.

We annotated reads based on the best matches to the SEED database, and determined the taxonomy of each contig based on the annotation of the majority of the contained reads [3]. Many of the long, low-coverage contigs were annotated as bacterial or archaeal, whereas many shorter, high-coverage contigs were annotated as unknown (Figure S2).

We considered the long, low-coverage contigs with cellular annotation to be more likely to be derived from cellular contamination, whereas the high-coverage, shorter contigs with unknown annotation are more likely to be derived from viruses. These high-coverage contigs may include certain viral genomes that occurred at high frequency.

While we attempted to eliminate cellular contamination through size fractionation and chloroform and DNAse treatment prior to sequencing, and 16S PCR tests of extracted viral DNA returned no amplification of cellular material, we sought to further reduce cellular contamination *in silico*. We created a "viral subset" of the viral metagenome consisting of reads assembled into contigs with a coverage of 8 or greater, or assembled into contigs annotated as viral or unknown. The aim of producing a subset was to remove low-coverage contigs more likely to be derived from cellular contamination. We chose a coverage cutoff of 8 because most long contigs annotated as bacterial or archaeal had a coverage of approximately 8 or lower (Figure S2). The goal of this subset was not to generate a "pure" viral metagenome but rather to reduce the number of reads that may represent cellular contamination, which is frequently an issue in viral metagenomics [44,45]. While the high percentage of unknown contigs in viral metagenomes makes any attempts at removing contamination difficult, this subset eliminates reads from low-coverage contigs with bacterial or archaeal annotation that are more likely to be derived from cellular contamination. Except in specific cases noted below, the analyses presented here or in the supplementary figures include both the original virome as well as the "viral subset" so that both may be compared.

Assessing the potential for horizontal gene transfer

To assess the degree to which cells and viruses in hydrothermal ecosystems are capable of horizontal gene transfer or integration of proviruses, we determined the relative abundances of provirus genes (Table 1) and genes related to DNA transfer or mobilization (Table 2) in the hydrothermal vent cellular and viral metagenomes. We used the "Prophage" dataset in the ACLAME database [34] to identify provirus-related proteins in 22 pyrosequenced cellular metagenomes. These metagenomes were chosen to represent a range of aquatic and terrestrial environments, while controlling for sequencing method. Table 1 indicates that relative to the 22 cellular metagenomes, the hydrothermal vent cellular metagenome contained a high percentage of reads (approximately 4%) that match provirus-coding regions. This result provides compelling molecular evidence of abundant proviruses in cellular genomes in vents and complements descriptions of high proportions of lysogenic cells in vents and the deep ocean based upon mitomycin C induction experiments [28,29]. We also analyzed the relative abundance of mobile genetic elements in 40 viral and cellular metagenomes, also selected to represent a range of environments, and controlled for sequencing method. We found a relative enrichment of mobile genetic elements in both the viral and cellular gene pools at Hulk hydrothermal vent (Table 2). Thus, not only do the genomes of archaea and bacteria in vents encode a higher abundance of mobile elements, on average, than genomes native to other habitats, but vent viruses encode a high abundance of mobile elements as well. These data suggest that selection has favored enrichment of mobile elements in these

Table 1. Percent of reads in cellular metagenomes matching a protein in the "Prophage" grouping of the ACLAME database.

Metagenome	Reads	ACLAME Prophage hits	Percent reads	Biome	Sampling details	Reference	Accession number
Monterey Bay	192162	9759	5.08	Open ocean	Monterey Bay, California, surface waters, October	-	4443713.3
Hulk hydrothermal vent	**808051**	**32595**	**4.03**	**Hydro-thermal vent**	**Juan de Fuca Ridge, Northeast Pacific Ocean, 2198 m depth**	**This study**	**4481541.3**
Glacial ice	1076539	40695	3.78	Glacial ice	Glacial ice of the Northern Schneeferner, Germany	Simon et al., 2009	CAM_PROJ_IceMetagenome
North Atlantic Spring Bloom	257471	6913	2.68	Open ocean	Bermuda Atlantic Time-Series site	-	4443725.3
Human oral microbiota	339503	5596	1.65	Human	Dental plaque from 25 human volunteers	Belda-Ferre et al., 2012	4447970.3
Healthy fish microbiota	51498	610	1.18	Fish	Healthy aquaculture fish, San Diego, CA	Angly et al., 2009	4440055.3
Guaymas Basin	4970673	58638	1.18	Hydro-thermal vent	Hydrothermal plumes from Guaymas Basin, CA	Baker et al., 2012	CAM_P_0000545
Cow rumen	320471	3678	1.15	Cow	Fiber-adherent microbiome from cow rumen	Brulc et al., 2009	4441681.3
Healthy fish microbiota	60580	541	0.89	Fish	Samples from aquacultured fish gut contents	Angly et al., 2009	4440059.3
Medium salinity saltern	108725	742	0.68	Salt water	Salinity 12–14% from solar salterns, California	Rodriguez-Brito et al., 2010	4440425.3
Salton Sea	161912	992	0.61	Sediments	Sulfidic, anoxic sediments of the Salton Sea	Swan et al., 2010	4440329.3
Tilapia fish pond	344260	1712	0.50	Fresh water	Water samples from aquaculture facility raising striped bass	Rodriguez-Brito et al., 2010	4440440.3
Peru Margin 1mbsf	100093	489	0.489	Deep biosphere-marine sediment	Peru Margin ODP Leg 201 Site 1229, 1 meter below seafloor	Biddle et al., 2008	4440961.3
Peru Margin 50mbsf	63258	288	0.455	Deep biosphere-marine sediment	Peru Margin ODP Leg 201 Site 1229, 50 meters below seafloor	Biddle et al., 2008	4459941.3
Peru Margin 32mbsf	135429	479	0.354	Deep biosphere-marine sediment	Peru Margin ODP Leg 201 Site 1229, 32 meters below seafloor	Biddle et al., 2008	4459940.3
Low salinity saltern	31948	111	0.35	Salt water	Salinity 6–8% from solar salterns, California	Rodriguez-Brito et al., 2010	4440426.3
Microbialites	257573	802	0.311	Micro-bialites	Highborne Cay, Bahamas	Desnues et al. 2008, Dinsdale et al., 2008	4440061.3
High salinity saltern	33356	98	0.29	Salt water	Salinity 27–30% from solar salterns, California	Rodriguez-Brito et al., 2010	4440419.3
Peru Margin 16mbsf	121414	191	0.157	Deep biosphere-marine sediment	Peru Margin ODP Leg 201 Site 1229, 16 meters below seafloor	Biddle et al., 2008	4440973.3
Porites compressa coral	1053275	947	0.0900	Coral	Samples collected at the Hawaii Institute for Marine Biology	Vega Thurber et al., 2009	CAM_PROJ_CoralMetagenome
Soudan Mine	248038	193	0.0778	Deep biosphere-terrestrial mine	Water and sediments in mine, 714 m below surface, Soudan Mine, MN	Edwards et al., 2006	4440282.3
Line Islands	178628	120	0.0672	Seawater	Water sampled near coral reefs, Christmas Island	Dinsdale et al., 2008	4440041.3

Matches were found using tblastn with a minimum e-value of 10^{-5}. All metagenomes listed here were generated with shotgun pyrosequencing.

Table 2. Percent of reads in cellular and viral metagenomes matching a mobile element.

Meta-genomes	Cellular or viral	Reads	Mobile elements	% reads	Biome	Sampling details	Ref	Accession number
Glacial ice	cellular	1076539	5598	0.52	Glacial ice	Glacial ice of the Northern Schneeferner, Germany	Simon et al., 2009	CAM_PROJ_IceMetagenome
Hydro-thermal vent	**viral subset**	**64599**	**252**	**0.39**	**Hydro-thermal vent**	**Juan de Fuca Ridge, Northeast Pacific Ocean, 2198m depth**	**This study**	**4469452.3**
Healthy fish microbiota	cellular	51498	193	0.37	Fish	Healthy aquacultured fish, San Diego, CA	Dinsdale et al., 2008	4440055.3
Healthy fish microbiota	cellular	60580	191	0.32	Fish	Samples from aquacultured fish gut contents	Angly et al., 2009	4440059.3
Hydro-thermal vent	**cellular**	**808051**	**2539**	**0.31**	**Hydro-thermal vent**	**Juan de Fuca Ridge, Northeast Pacific Ocean, 2198m depth**	**This study**	**4481541.3**
Cow rumen	cellular	320471	976	0.30	Cow	Fiber-adherent microbiome from cow rumen	Brulc et al., 2009	4441681.3
Hydro-thermal vent	**viral**	**231246**	**579**	**0.25**	**Hydro-thermal vent**	**Juan de Fuca Ridge, Northeast Pacific Ocean, 2198m depth**	**Anderson et al., 2011a**	**4469452.3**
Antarctic Lake summer	viral	30515	66	0.22	Fresh water	Freshwater oligotrophic lake, Byers Peninsula, Antarctica (summer)	Lopez-Bueno et al., 2009	4441558.3
Reclaimed water	viral	1531954	2330	0.15	Fresh water	Viral fraction of reclaimed water	Rosario et al., 2009	CAM_PROJ_ReclaimedWaterVirues
Monterey Bay	cellular	192162	278	0.14	Seawater	Monterey Bay, California, October 2000, surface waters	-	4443713.3
Human oral microbiota	cellular	339503	450	0.13	Human	Dental plaque from 25 human volunteers	Belda-Ferre et al., 2012	4479970.3
Healthy fish microbiota	viral	55690	66	0.12	Fish	Samples from aquacultured fish gut contents	Angly et al., 2009	4440065.3
Tilapia Pond	cellular	344260	352	0.10	Fresh water	Water samples from aquaculture facility raising striped bass	Rodriguez-Brito et al., 2010	4440440.3
High salinity saltern	cellular	33356	33	0.10	Salt water	Salinity 27-30% from solar salterns, California	Rodriguez-Brito et al., 2010	4440419.3
Guaymas Basin	cellular	4970673	4728	0.10	Hydro-thermal vent	Hydrothermal plumes from Guaymas Basin, CA	Baker et al., 2012	CAM_P_0000545
Arctic Ocean	viral	688590	605	0.09	Sea-water	10-3246m, Fall 2002, Arctic Ocean	Angly et al., 2006	4441621.3
Medium salinity saltern	cellular	108725	95	0.09	Salt water	Salinity 12-14% from solar salterns, California	Rodriguez-Brito et al., 2010	4440425.3
Bay of British Columbia	viral	138347	107	0.08	Sea-water	0-245m, sampled over several dates, Bay of British Columbia	Angly et al., 2006	4441623.3
North Atlantic Spring Bloom	cellular	257471	193	0.07	Seawater	Bermuda Atlantic Time-Series site	-	4443725.3
Peru Margin 1mbsf	cellular	100093	69	0.07	Deep biosphere-marine sediment	Peru Margin ODP Leg 201 Site 1229, 1 meter below seafloor	Biddle et al., 2008	4440961.3
Salton Sea	cellular	161912	98	0.06	Sediments	Sulfidic, anoxic sediments of the Salton Sea	Swan et al., 2010	4440329.3
Gulf of Mexico	viral	263908	153	0.06	Seawater	0-164m, sampled over several dates, Gulf of Mexico	Angly et al., 2006	4441625.3
Micro-bialites	viral	621110	359	0.06	Micro-bialites	Pozas Azules, Mexico; Rio Mesquites, Mexico; Highborne Cay, Bahamas	Desnues et al., 2008	4440320.344403211.344403233
Peru Margin 50mbsf	cellular	63258	28	0.04	Deep biosphere-marine sediment	Peru Margin ODP Leg 201 Site 1229, 50 meters below seafloor	Biddle et al., 2008	4459941.3

Table 2. Cont.

Meta-genomes	Cellular or viral	Reads	Mobile elements	% reads	Biome	Sampling details	Ref	Accession number
Soudan Mine	cellular	248038	105	0.04	Deep biosphere-terrestrial mine	Water and sediments in mine, 714 m below surface, Soudan Mine, MN	Edwards et al., 2006	4440282.3
Antarctic Lake spring	viral	31691	13	0.04	Fresh water	Freshwater oligotrophic lake, Byers Peninsula, Antarctica (spring)	Lopez-Bueno et al., 2009	4441778.3
Coral	viral	36354	14	0.04	Coral	Porites compressa coral samples collected at the Hawaii Institute for Marine Biology	Vega Thurber et al., 2009	4440374.3
Low salinity saltern	viral	56810	21	0.04	Salt water	Salinity 6–8% from solar salterns, California	Rodriguez-Brito et al., 2010	4440420.3
Peru Margin 32mbsf	cellular	135429	49	0.04	Deep biosphere-marine sediment	Peru Margin ODP Leg 201 Site 1229, 32 meters below seafloor	Biddle et al., 2008	4459940.3
Salton Sea	viral	27689	7	0.03	Sediments	Sulfidic, anoxic sediments of the Salton Sea	Swan et al., 2010	4440328.3
Tilapia Pond	viral	231521	48	0.02	Fresh water	Water samples from aquaculture facility raising striped bass	Rodriguez-Brito et al., 2010	4440439.3
Low salinity saltern	cellular	31948	6	0.02	Salt water	Salinity 6–8% from solar salterns, California	Rodriguez-Brito et al., 2010	4440426.3
Medium salinity saltern	viral	33291	6	0.02	Salt water	Salinity 12–14% from solar salterns, California	Rodriguez-Brito et al., 2010	4440427.3
Peru Margin 16mbsf	cellular	121414	18	0.01	Deep biosphere-marine sediment	Peru Margin ODP Leg 201 Site 1229, 16 meters below seafloor	Biddle et al., 2008	4440973.3
Coral	cellular	1053275	144	0.01	Coral	Porites compressa coral samples collected at the Hawaii Institute for Marine Biology	Vega Thurber et al., 2009	CAM_PROJ_CoralMetagenome
Tampa Bay	viral	257075	32	0.01	Fresh water	Prophages induced with mitomycin C from Tampa Bay water samples	McDaniel et al., 2008	4440102.3
High salinity saltern	viral	136564	13	0.01	Salt water	Salinity 27–30% from solar salterns, California	Rodriguez-Brito et al., 2010	4440421.3
Microbialites	cellular	257573	20	0.01	Micro-bialites	Highborne Cay, Bahamas	Breitbart et al., 2009	4440061.3
Sargasso Sea	viral	399343	22	0.01	Open ocean	80 m, sampled June 2005, Sargasso Sea	Angly et al., 2006	4441624.3

These include transposases, integrases, recombinases, and resolvases as defined by a keyword search in Pfam (database file included in supplementary material). Matches found using tblastn with a minimum e-value of 10^{-5}. All metagenomes listed here were generated with shotgun pyrosequencing.

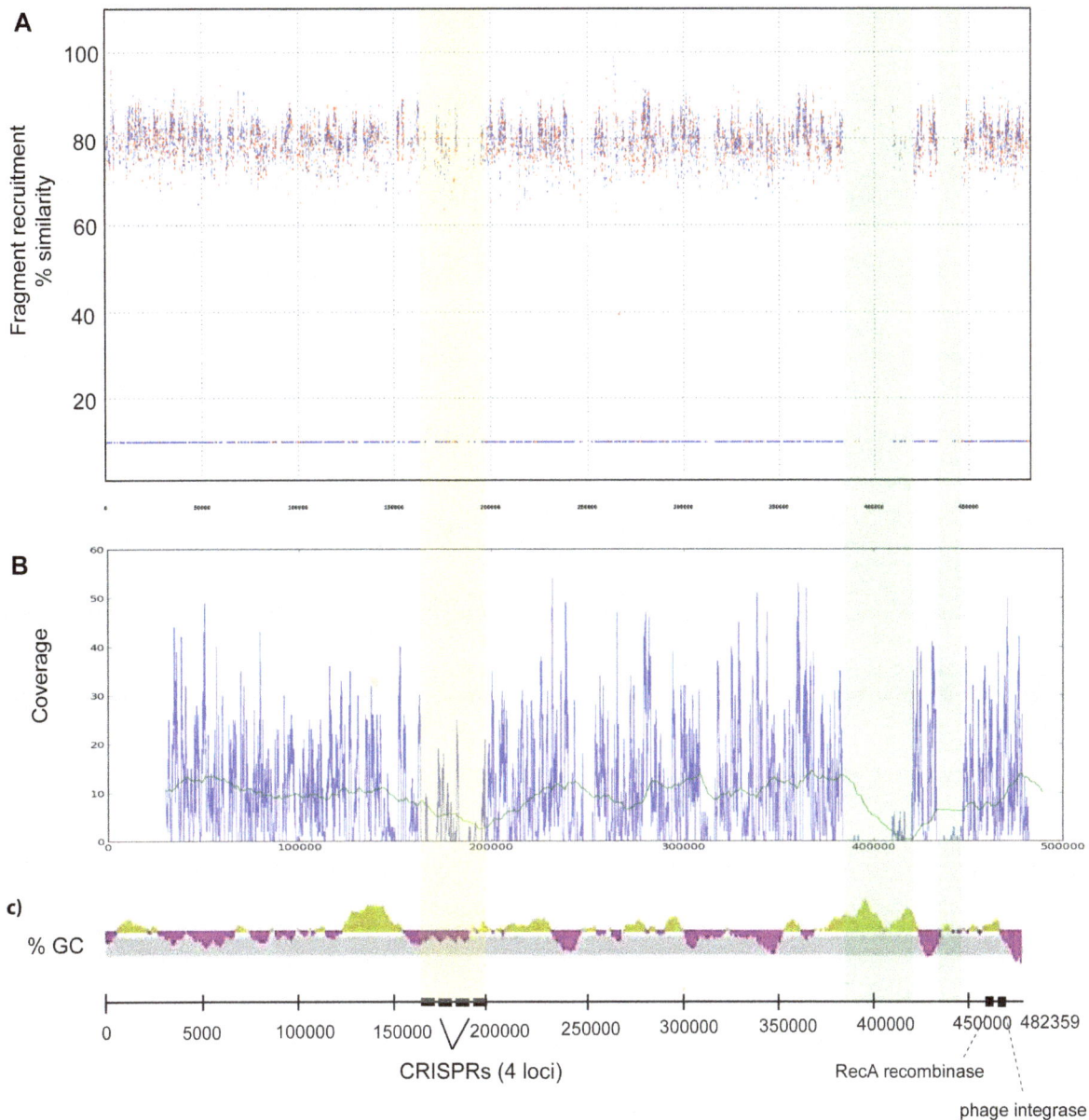

Figure 2. Recruitment plot of metagenomic reads to *Caminibacter mediatlanticus* TB-2. Cellular metagenomic reads were mapped to the longest contig of the draft genome of *C. mediatlanticus* TB-2, with percent similarity on the y-axis and base pair numbers on the x-axis (A). Coverage plot of read recruitment is shown per base pair, with blue line showing actual coverage and green line showing a convolution function of the coverage plot using a weighting of 50000 (B). Percent GC plot for the same contig is shown on the same scale, with base pair numbers marked below (C), and are annotated with CRISPR loci and recombinases or integrases found on the contig. Orange shading shows the location of CRISPR loci on the genome; green shading shows the location of two metagenomic islands.

cellular and viral genomes, potentially leading to increased rates of horizontal gene flow among cells and viruses.

Evidence of genomic plasticity in vent genomes

Having found evidence for gene transfer and viral integration in the genomes of vent cells and viruses, we sought evidence for either lysogenic virus integration or gene transfer events in other hydrothermal vent isolates. Of the 34 available fully sequenced vent genomes, 20 (59%) contain integrated viruses (Table S4). Most of these viruses encode capsid genes or genes such as DNA ligases, which have been identified before as being particularly abundant in the viral fraction of this hydrothermal vent sample

[46]. However, identification of auxiliary metabolic genes (AMGs) in these viral genomes is difficult, partly due to the high abundance of unknown genes and partly because the boundaries between the viral and cellular genome are not always clearly delineated.

To better identify regions that have been transferred in vent genomes, we compared genomes from bacterial or archaeal isolates with sequences sampled directly from the environment. This strategy can identify potential hypervariable regions, or "genomic islands," that display lower coverage than the rest of the genome [47–49]. Previous work with *Haloquadratum walsbyi* DSM 16790 [48] and *Prochlorococcus* genomes [47] used this technique to identify genomic islands that most likely represented regions of virally-mediated lateral gene transfer; in the case of *Prochlorococcus,* these genes are differentially expressed under light and nutrient stress [47].

We performed fragment recruitment of the hydrothermal vent cellular metagenome against the genomes of all isolates of hydrothermal vent bacteria and archaea available in the NCBI database. In most cases, the metagenome did not recruit to the isolate genomes with high enough coverage to yield useful data. *Nautilia profundicola* AmH successfully recruited reads at high coverage, but no metagenomic islands were found. Recruitment of the cellular metagenome to the longest contig (ABCJ01000001) of the draft genome of *Caminibacter mediatlanticus* TB-2, a chemolithotrophic, nitrate-ammonifying *Epsilonproteobacterium* that was isolated from the walls of a hydrothermal vent chimney on the Mid-Atlantic Ridge [50], yielded a number of regions with relatively low coverage (Figure 2). The first, a region with relatively low coverage between 160000 and 200000 bp, contains a series of CRISPR loci. CRISPR regions are dedicated to viral and plasmid immunity by effectively creating a library of previous infection, and therefore are highly specific to a given environment. Therefore recruitment to CRISPR loci should naturally yield lower coverage, particularly for a metagenome sampled in a different geographic location than this isolate. Aside from the CRISPR region, recruitment yielded two distinct genomic islands: one region of approximately 40 kbp, followed by a second low-coverage region of approximately 15 kbp (Figure 2), separated from each other by a 20 kbp region that includes a ribosome. The first genomic island coincides with a region with relatively high GC content, which suggests this region was transferred into the *C. mediatlanticus* genome. A phage integrase gene is located approximately 58 kbp downstream of the 3′ end of the first genomic island, though its presence there is not necessarily conclusive evidence that the region was introduced by a virus. The first genomic island begins near a tRNA gene, a common site for integration of horizontally transferred regions [51]. Many of the genes in both the first and second genomic islands encode proteins that interact with the environment, including sugar and nitrate membrane transporters, and proteins related to energy metabolism, including hydrogenases (Table S5). Possession of a diverse suite of hydrogenases can enhance metabolic flexibility in variable redox conditions, and has been observed in other *Epsilonproteobacteria* isolated from hydrothermal vents [52].

The presence of this hypervariable region indicates that genomic islands from genomes in vent environments encode genes related to environmental interactions and energy metabolism. The genomic island shown here is unique to a vent isolate from the Mid-Atlantic Ridge. The *C. mediatlanticus* strains present in our sample on the Juan de Fuca Ridge most likely have genomic islands of their own, though these cannot be identified without a fully sequenced strain from the Juan de Fuca Ridge. None of the sequenced strains from the Juan de Fuca Ridge had high enough coverage with our metagenome to identify genomic islands. However, the genomic island on *C. mediatlanticus* provides evidence of horizontal transfer of genes that facilitate metabolic flexibility in an environment that is very similar to the Juan de Fuca Ridge. From this we can hypothesize that genes related to environmental interactions and metabolic activity are transferred in our sample site.

Quantification of relative gene enrichment in the viral fraction

We next sought to determine whether there were differences in the relative enrichment of gene types in the cellular and viral fractions. In order to account for biases or omissions in various

databases, we annotated reads in both the viral and cellular metagenomes according to three functional databases: the SEED Subsystems database [38], the Clusters of Orthologous Groups (COG) database [37], and the KEGG Orthology (KO) database [36] (Figures 3 and 4). Among all three databases, certain trends appeared. First, in all three cases the results indicated a high similarity in the overall relative abundances between functional genes in the cellular and viral fractions. A similar study of relative abundances of functional groups in viral and cellular metagenomes, conducted by Kristensen *et al.* [44] based on data collected by Dinsdale *et al.* [53], found a similar trend. While some of this is most likely due to cellular contamination in the viral fraction, as well as the presence of viruses in the cellular fraction, Kristensen *et al.* suggest this trend may also be due in part to the choice of available functional categories, which encompass functions that are generally cellular rather than viral. However, despite the likely presence of contamination in these datasets, direct comparison in this way enables us to compare the relative enrichment of certain types of genes in the viral and cellular gene pools. Comparisons between the viral subset and the cellular metagenome are shown in Figures 3 and 4; comparisons between the cellular metagenome, original virome, and viral subset are shown in Figure S3.

There was a statistically significant enrichment of reads matching phage, prophage, and transposable elements in the virome relative to the cellular metagenome (2.3% vs. 1.5%). These annotations were made according to the SEED Subsystems classification (Figure 3A), and we used Xipe-Totec [39] to assess statistical significance. Most viral gene hits were to distantly related head-tail viruses and cyanophage. The low percentage of matches to viral sequences overall is likely due in part to the relative lack of environmental viral sequences in public databases. We also observed an enrichment of reads matching cofactors, vitamins, prosthetic groups and pigments in the viral fraction (Figure 3A), which falls in line with studies identifying these types of genes on cyanophage genomes. Genes encoding vitamin B12 [53,54], and the pigments psbA/D [55–58] and pebS [23] have previously been identified on cyanophage genomes. These virally encoded genes are thought to supplement host metabolism during infection, and similar processes may occur in this vent ecosystem.

We observed a statistically significant enrichment of reads matching amino acids and derivatives in the cellular fraction relative to the viral fraction, according to the COG (Figure 3B) and KO (Figure 4A) databases. These include genes related to the biosynthesis and metabolism of various amino acids, and are evidently not commonly carried on viral genomes in vent ecosystems, suggesting that viruses rely on their hosts for amino acid biosynthesis.

We also observed enrichment in reads related to replication, recombination and repair in the viral fraction, as annotated by both the COG and KO databases (Figure 3B, Figure 4A). These genes probably serve necessary functions in the synthesis and replication of viral DNA during the course of viral infection, and include genes like DNA and RNA polymerase, which are commonly encoded on viral genomes. The "replication and repair" category includes DNA ligases, which occur at very high abundances in the virome, as previously reported [46].

Finally, we also observed a statistically significant enrichment of genes related to energy metabolism as annotated by the KO database. Among reads annotated as matching the energy metabolism sub-category, 6.5% from the cellular metagenome and 8.4% from the viral subset were annotated as sulfur metabolism, whereas 8.2% of cellular versus 3.7% of viral metagenome reads annotated as belonging to methane metabolism (Figure 4B). The enrichment of energy metabolism genes corre-

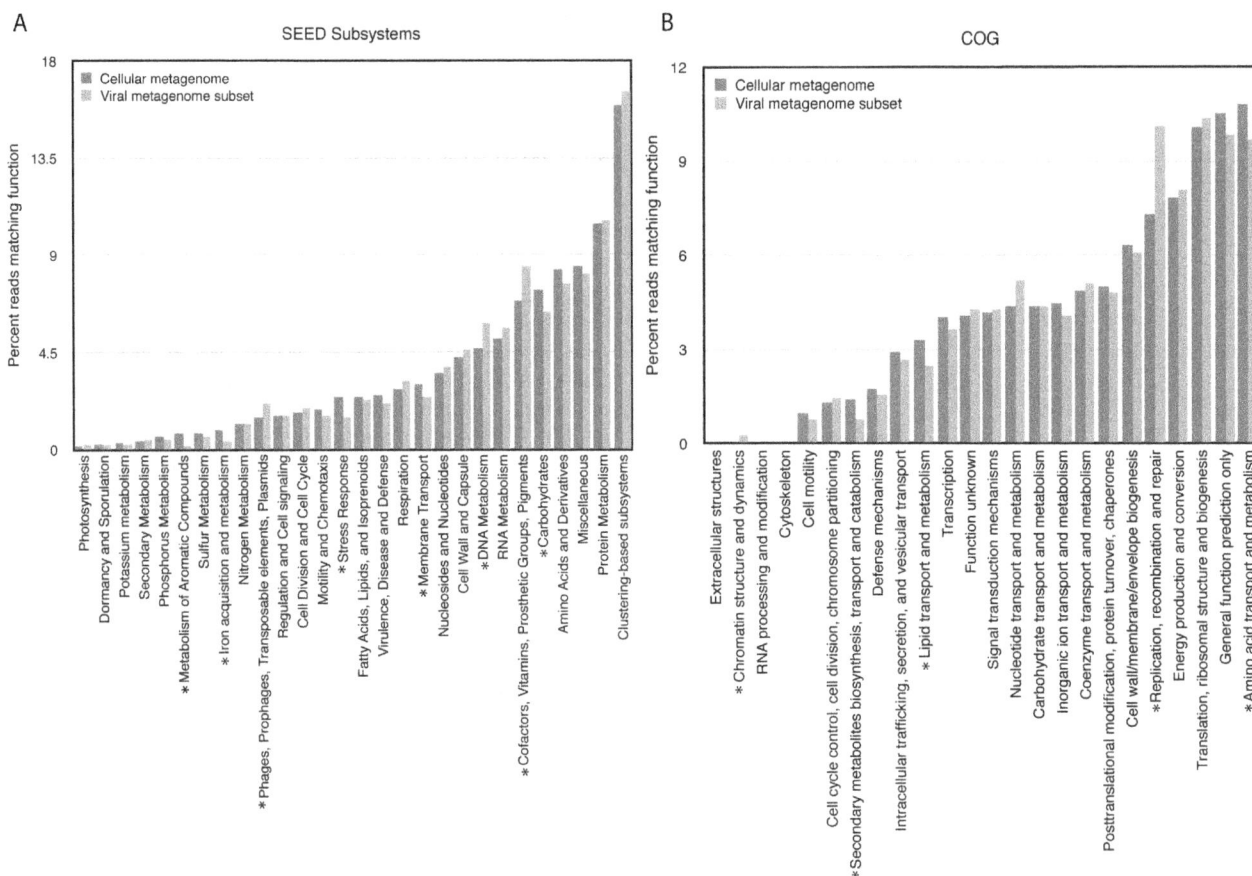

Figure 3. Functional comparisons of the hydrothermal vent cellular and viral subset metagenomes according to the SEED subsystems and Clusters of Orthologous Groups (COG) databases. Metagenomes were annotated in MG-RAST with a minimum e-value of 1e-03 and a minimum identity cutoff of 60%. A single asterisk indicates a significant difference in abundance between the viral subset and the cellular metagenome. A) Matches to the SEED subsystems database; B) matches to the COG database.

sponds with what we might expect given the observation of auxiliary metabolic genes (AMGs) in viral genomes from the surface oceans. A similar enrichment in energy metabolism genes was found in the viral fraction relative to the cellular fraction in samples from the Indian Ocean [59]. Similarly, photosynthesis genes encoded in cyanophage are known to be expressed during viral infection [20,60,61], and modeling work has indicated that these photosynthesis genes can enhance host fitness [62,63]. For lytic viruses, energy metabolism genes may be expressed as a means to supplement host metabolism or to redirect resources for phage particle synthesis. Our analysis of fragment recruitment (Figure 2) indicated that genes related to energy metabolism had been successfully transferred between genomes in the past, and viruses are potential vectors for such gene transfer. One potential explanation for this enrichment, in line with the fragment recruitment results, is that highly abundant proviruses in the vent system express these genes while integrated in the host genome. By providing their hosts with new or supplemental means of surviving a challenging, dynamic environment, these proviruses boost host fitness, and in turn, enhance their own fitness.

To compare these results with other cellular-viral metagenome comparisons, we conducted an analysis of 20 cellular metagenomes and 23 viral metagenomes from other environments, all sequenced with 454 pyrosequencing and annotated with the KO database. Each of the cellular metagenomes had at least one viral counterpart sampled from the same environment. Metadata

regarding these metagenomes are summarized in Table S6; the functional profiling analysis is depicted in Figure S4. The overall patterns of relative gene abundance between the viral and cellular fractions were similar to those observed in our sample. Specifically, in other environments functional annotations of viral and cellular metagenomic reads were strongly correlated, but viral metagenomes were more enriched in reads annotated as nucleotide metabolism and replication and repair. However, the relative enrichment of genes related to energy metabolism in the viral fraction, while apparent in the vent environment, was not a universal characteristic of viruses in other environments.

Does selection operate differentially on viral and cellular genes?

Integral to the study of viral and cellular evolution is understanding how selection shapes the viral and cellular gene pools, and which genes are subject to stronger or weaker selection. Differing life strategies for cells and viruses, as well as disparate roles for functional genes within each of the respective gene pools, should leave different selective signatures on genes within each of these gene pools. To measure differential selection among viral and cellular genes, we calculated dN/dS ratios of genes encoded by the viral and cellular fractions. The challenge of calculating dN/dS ratios with shotgun metagenomic data is that the short sequences make it difficult to align long blocks of sequences to the same region of a gene. To circumvent this problem, we used the

A

B

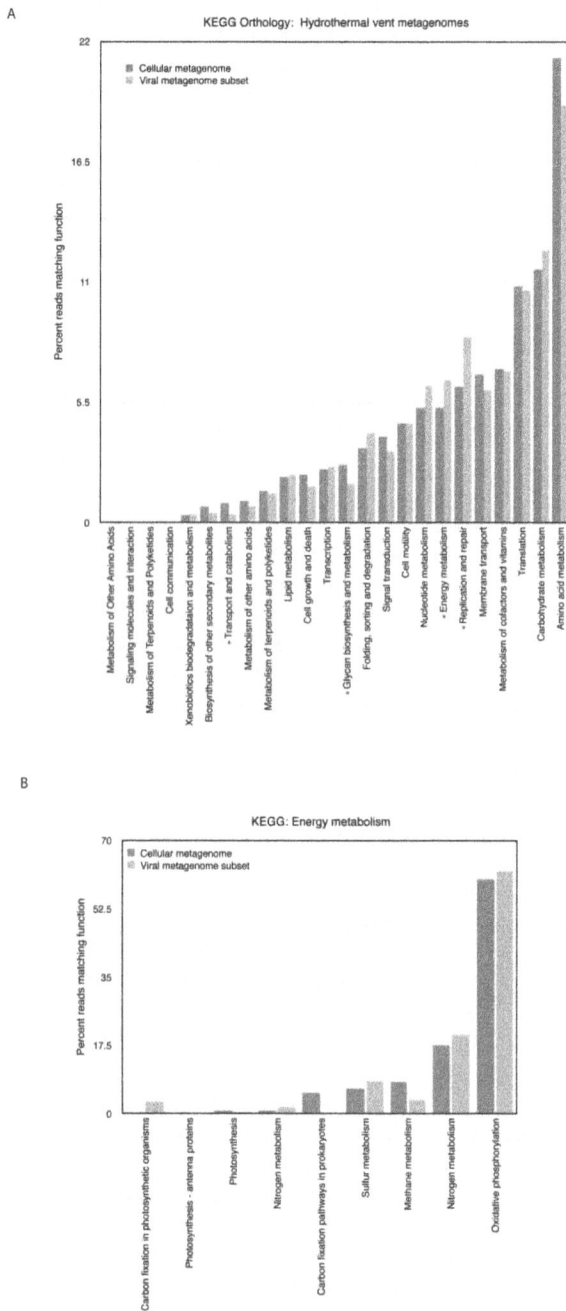

Figure 4. Functional comparisons of the hydrothermal vent cellular and viral subset metagenomes according to the KEGG Orthology annotation system. Metagenomes were annotated in MG-RAST with a minimum e-value of 1e-03 and a minimum identity cutoff of 60%. A single asterisk indicates a significant difference in abundance between the viral subset and the cellular metagenome. A) Matches to the KEGG Orthology database; B) Matches to the energy metabolism category of the KEGG Orthology database.

method developed by Tai *et al.* [42] to calculate dN/dS ratios from metagenomic reads, in which sequencing reads are mapped to the genomes of previously sequenced isolates. We used 80% identity as the threshold for mapping to genomes, a convention established previously [42]. While some work suggests that microbial "species" share an average nucleotide identity of 95% across the genome [64], tiling at that percentage did not yield

enough hits for statistical analysis. Previous work has also indicated that below 80% identity, the number of reads recruited drops drastically, implying a biological threshold at 80% similarity [42]. These sequences therefore define the "population." However, since the sequences used for analysis may be derived from multiple taxa, and we do not know the specific phylogenetic relationship of these sequences to each other, this method cannot determine which polymorphisms have become "fixed" in the population. Instead, this method provides an indicator of diversification within the environmental gene pool.

Both metagenomes were mapped to pre-existing hydrothermal vent isolates as a high-throughput means to align reads to many genes at once. The mapping analyses and calculation of dN/dS ratios are calculated by tiling metagenomic reads to the genomes of *Nautilia profundicola* AmH, *Thermococcus kodakarensis* KOD1, *Thermococcus onnurineus* NA1, *Caminibacter mediatlanticus* TB-2 (contig ABCJ01000001), and *Nitratiruptor* sp. SB155-2, which represent abundant strains in the vent environment. We attempted to map the virome to several existing viral genomes from various environments, but none exhibited sufficient depth of coverage to calculate dN/dS ratios, indicating that the genes encoded by viruses from this hydrothermal system are vastly different from those sequenced previously.

Overall, the cellular metagenome mapped to 831 bacterial and 32 archaeal genes, with an average dN/dS of 0.22. This result indicates that genes encoded by cells in the vent environment are subject to purifying selection. The viral metagenome mapped to 85 bacterial and 106 archaeal genes, with an average dN/dS of 0.15, and the viral metagenome subset mapped to 39 bacterial and 25 archaeal genes, with an average dN/dS of 0.13 (Figure 5). These dN/dS values are significantly lower than the dN/dS of genes matching the cellular metagenome, within a confidence interval of 95%. This pattern was consistent for each of the genomes mapped (Figure S5). The viral and cellular metagenomes mapped to different genes in each of the strains listed above, and so slightly different sets of genes were used to make this calculation. However, the difference in overall dN/dS is not due solely to differences in the types of genes to which each metagenome mapped: when we examined the dN/dS for only the genes to which both metagenomes mapped, the calculated dN/dS was, on average, lower for the viral fraction compared to the cellular fraction (Figure S6). There were no clear trends in dN/dS for different gene categories (Figure S7).

These results indicate that both the viral and cellular fractions in this hydrothermal system are subject to purifying (negative) selection, but the viral gene pool is under stronger purifying selection than the cellular gene pool. One possible explanation for the overall difference in dN/dS between the viral and cellular gene pools is that very little variation in viral genes is permitted. In this scenario, viral genes are under such strong selection that deviations from the consensus protein sequence produce enough of a fitness difference to eliminate the viral mutant. The viral genes included in this analysis mapped to cellular genomes, and therefore are likely to be auxiliary metabolic genes carried on viral genomes, suggesting that AMGs carried by viruses are subject to purifying selection to a greater degree than their cellular counterparts.

This scenario becomes more complicated as a result of the relationship between virus and host. If a virus were primarily lytic, then a selective sweep could act directly on the viral particles, reducing phenotypic variation (and therefore nonsynonymous polymorphisms). If, however, a virus were primarily lysogenic, a larger proportion of time would be spent integrated in the genome of the host. In this situation, the selective sweeps would act on both

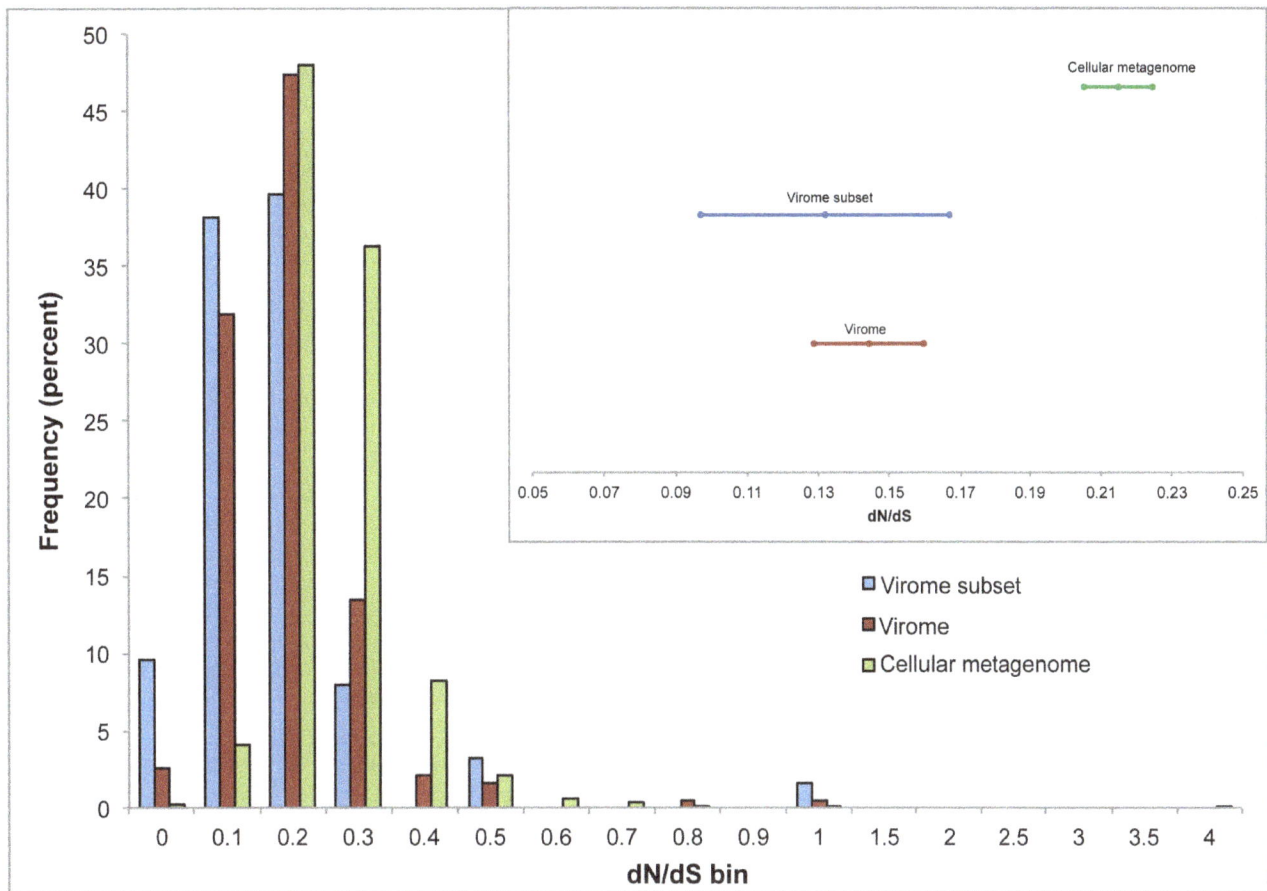

Figure 5. Histogram of dN/dS ratios for each metagenome. A total of 863 genes were included for the cellular metagenome calculation, 191 for the viral metagenome, and 64 for the viral subset. Values are shown only for genes that had a minimum depth coverage of 5 and minimum nucleotide coverage of 100. Frequency values are normalized by percent. Bins are scaled in increments of 0.1 until 1, and then in increments of 0.5. Inset shows mean and 95% confidence intervals for calculated dN/dS for all three data sets, indicating that the average cellular dN/dS is significantly greater than the average dN/dS for the viral metagenome.

host and virus, and we might not expect to see a significant difference in the dN/dS ratios between viral and cellular genes. The difference observed here may indicate that in this system, selection acts most strongly on viruses while they are in the process of replicating or when they are in virion form (free in the environment), and can therefore be selected separately from the host.

Conclusions

The dynamic, recirculating conditions of sulfide-hosted hydrothermal vent systems, combined with the vast diversity of archaea, bacteria, and their accompanying viruses, create an ecosystem with high potential for widespread sharing of the communal gene pool. The unique sample analyzed here, which most likely pulled fluid both from high-temperature subsurface hydrothermal fluid and from cold background seawater, included microorganisms from a wide range of ecological habitats within the vent system. Having been sampled from the same point source, this implies that these diverse microorganisms have the potential to come into close contact in these vent systems. In doing so, there is potential for these microorganisms to exchange genes as well as viruses. Mobile elements were abundant in both the viral and cellular metagenomes we obtained, likely reflecting the abundance of lysogenic

viruses, which require integrases for viral genome insertion into the host, as well as high potential for horizontal gene transfer. In the genomes of vent inhabitants, we found evidence for horizontal transfer of genes for environmental interaction and energy metabolism, suggesting selection for enhanced phenotypic plasticity. Moreover, a slight enrichment of genes related to processes such as cofactor synthesis and energy metabolism suggests that viruses in this hydrothermal system carry auxiliary metabolic genes.

Any genes found in the viral gene pool must have some utility in order to be retained on small viral genomes. The genes encoded on viral genomes may be used by lytic viruses as a means to facilitate the manufacture of viral particles, as has been observed in cyanophage previously. However, the prevalence of lysogenic viruses in vent habitats and the selective signatures observed here suggest that these vent viruses are selected to spend much of their time as integrated proviruses rather than as free virions. In this case, auxiliary metabolism genes may be expressed by integrated proviruses to benefit the host. Selection should favor traits by which proviruses boost host fitness while the fitness of the virus and host are intertwined. In turn, host cells benefiting from provirus-encoded genes may gain an adaptive advantage through enhanced metabolic flexibility, which may favor selection for cells harboring proviruses. This advantage complicates the symbiotic relationship between virus and host. While still capable of wreaking destruction upon the cells they infect, the data described here suggest a viral evolutionary strategy in which the virus-host relationship transcends from a parasitic relationship into a mutualistic one, as both host and virus seek to survive the dynamic, extreme environment in which they coexist.

Supporting Information

Figure S1 Assignment of metagenomic contigs for the cellular metagenome (A) and the viral metagenome (B), based on di-, tri-, and tetranucleotide abundance determined by PhylopythiaS (McHardy et al., 2007). Boutique PhylopythiaS training datasets were created to classify contigs in the cellular and viral metagenomes as archaeal, bacterial, archaeal virus or bacterial virus.

Figure S2 Coverage and length of assembled contigs in the viral metagenome. Average coverage, shown on the x-axis, indicates the average coverage per base pair across the entire contig. Contig length is shown on the y-axis. Contigs are colored according to the assigned taxonomy.

Figure S3 These graphs accompany Figs. 3 and 4 in the main document, but include the original virome for comparison. One asterisk indicates a significant difference between the viral subset and the cellular metagenome; two asterisks indicate a significant difference between the original virome and the cellular metagenome.

Figure S4 Mean percentages of 20 cellular metagenomes and 23 viral metagenomes annotated according to the KEGG Orthology annotation system. Metagenomes were annotated in MG-RAST with a minimum e-value of 1e-03, minimum identity cutoff of 60%, and minimum alignment length of 15, and were derived from studies that directly compared viral and microbial metagenomes. Data from Dinsdale et al. 2008. Error bars indicate standard deviation of the mean.

Figure S5 Histograms of dN/dS for genes in three different genomes mapped by the cellular metagenome, virome, and virome subset. *Caminibacter mediatlanticus* TB-2 and *Nitratiruptor* sp. SB155-2 are not shown because the virome subset mapped only to four and zero genes in each of these genomes, respectively. Number of genes included in each histogram is indicated in parentheses.

Figure S6 Values of dN/dS for genes mapped by the virome versus dN/dS for the same genes mapped by cellular metagenome. The line has a slope of 1.

Figure S7 Box-and-whisker plots of dN/dS values for genes mapped each of the three metagenomic datasets. Genes are categorized according to KO annotation. A dotted line indicates where dN/dS = 1 and selection is neutral. Boxes indicate upper and lower quartiles; whiskers denote 1.5 times the interquartile range. Numbers below gene categories indicate the number of genes included for that category for the cellular and viral metagenomes, respectively.

Table S1 Summary of temperature and bacterial and viral counts from Hulk vent in the Main Endeavour Field. Temperature minimum was measured by temperature probes on a hydrothermal fluid sampler, temperature maximum was extrapolated based on dissolved silica concentrations.

Table S2 List of viruses used to train PhylopythiaS for distinguishing between archaeal viruses and bacterial viruses.

Table S3 Pfam domains included in search for genes associated with mobile genetic elements.

Table S4 Numbers of proviruses identified in hydrothermal vent bacterial and archaeal genomes using Prophage Finder (Bose *et al*, 2006).

Table S5 Annotation and best hit of reads within the low-coverage region of fragment recruitment from the Hulk cellular metagenome to the longest contig in the *Caminibacter mediatlanticus* TB-2 draft genome. See methods for details of fragment recruitment. Annotations are as listed by the draft annotation file released by the JCVI. Organism best hit determined by using blastn against the nr database.

Table S6 List of viral and cellular metagenomes used for functional profiling of viral and cellular metagenomes using the KEGG Orthology database. Metagenomes obtained from the MG-RAST database were first analyzed by Dinsdale et al. (2009).

File S1 Supplementary methods regarding the analysis of CRISPRs in the metagenomes.

Acknowledgments

We gratefully acknowledge Jim Holden, David Butterfield, Mark Spear, Bruce Strickrott, Pat Hickey, Tor Bjorklund, and Sanjoy Som as well as the captain and crew of the R/V *Atlantis* and DSV *Alvin* for assistance with shipboard sample collection. David Butterfield also measured the silica and magnesium chemistry and calculated the average temperature of the fluid sample. We thank Vera Tai for providing Python scripts and invaluable assistance with dN/dS calculations. We would like to thank Hilary Morrison for assistance with metagenomic sequencing of the Hulk cellular metagenome, and William Brazelton for thoughtful discussions. The viral metagenome was sequenced by the Broad Institute with support by the Gordon and Betty Moore Foundation, and the cellular metagenome was sequenced by the Josephine Bay Paul Center at the Marine Biological Laboratory. Funding was provided by a NASA Astrobiology Institute grant through Cooperative Agreement NNA04CC09A to the Geophysical Laboratory at the Carnegie Institution for Science. R.A. was funded by a NSF Graduate Research Fellowship through NSF grant number DGE-0718124, an NSF IGERT grant to the University of Washington Astrobiology Program, and the ARCS Foundation.

Author Contributions

Conceived and designed the experiments: RA JB. Performed the experiments: RA. Analyzed the data: RA JB. Contributed reagents/materials/analysis tools: RA MS JB. Contributed to the writing of the manuscript: RA MS JB.

References

1. Anderson RE, Beltrán MT, Hallam SJ, Baross JA (2013) Microbial community structure across fluid gradients in the Juan de Fuca Ridge hydrothermal system. FEMS Microbiol Ecol 83: 324–339. doi:10.1111/j.1574-6941.2012.01478.x.

2. Anderson RE, Brazelton WJ, Baross JA (2013) The deep viriosphere: Assessing the viral impact on microbial community dynamics in the deep subsurface. Rev Mineral Geochemistry 75: 649–675.

3. Anderson RE, Brazelton WJ, Baross JA (2011) Using CRISPRs as a metagenomic tool to identify microbial hosts of a diffuse flow hydrothermal vent viral assemblage. FEMS Microbiol Ecol 77: 120–133.

4. Baross JA, Hoffman SE (1985) Submarine hydrothermal vents and associated gradient environments as sites for the origin and evolution of life. Orig Life Evol Biosph 15: 327–345.

5. Schrenk MO, Kelley DS, Delaney JR, Baross JA (2003) Incidence and diversity of microorganisms within the walls of an active deep-sea sulfide chimney. Appl Environ Microbiol 69: 3580–3592. doi:10.1128/AEM.69.6.3580-3592.2003.

6. Huber JA, Mark Welch DB, Morrison HG, Huse SM, Neal PR, et al. (2007) Microbial population structures in the deep marine biosphere. Science (80-) 318: 97–100. doi:10.1126/science.1146689.

7. Huber JA, Butterfield DA, Baross JA (2002) Temporal changes in archaeal diversity and chemistry in a mid-ocean ridge subseafloor habitat. Appl Environ Microbiol 68: 1585–1594. doi:10.1128/AEM.68.4.1585-1594.2002.

8. Huber JA, Butterfield DA, Baross JA (2003) Bacterial diversity in a subseafloor habitat following a deep-sea volcanic eruption. FEMS Microbiol Ecol 43: 393–409. doi:10.1111/j.1574-6941.2003.tb01080.x.

9. Holden JF, Summit M, Baross JA (1998) Thermophilic and hyperthermophilic microorganisms in 3–30°C hydrothermal fluids following a deep-sea volcanic eruption. FEMS Microbiol Ecol 25: 33–41. doi:10.1111/j.1574-6941.1998.tb00458.x.

10. Summit M, Baross JA (2001) A novel microbial habitat in the mid-ocean ridge subseafloor. Proc Natl Acad Sci U S A 98: 2158–2163. doi:10.1073/pnas.051516098.

11. Elsaied H, Stokes HW, Nakamura T, Kitamura K, Fuse H, et al. (2007) Novel and diverse integron integrase genes and integron-like gene cassettes are prevalent in deep-sea hydrothermal vents. Environ Microbiol 9: 2298–2312. doi:10.1111/j.1462-2920.2007.01344.x.

12. Brazelton WJ, Baross JA (2009) Abundant transposases encoded by the metagenome of a hydrothermal chimney biofilm. ISME J 3: 1420–1424.

13. Koonin EV, Makarova KS, Aravind L (2001) Horizontal gene transfer in prokaryotes: Quantification and classification. Annu Rev Microbiol 55: 709–742. doi:10.1146/annurev.micro.55.1.709.

14. Beiko RG, Harlow TJ, Ragan MA (2005) Highways of gene sharing in prokaryotes. Proc Natl Acad Sci U S A 102: 14332–14337. doi:10.1073/pnas.0504068102.

15. Nelson KE, Clayton RA, Gill SR, Gwinn ML, Dodson RJ, et al. (1999) Evidence for lateral gene transfer between Archaea and Bacteria from genome sequence of Thermotoga maritima. Nature 399: 323–329.

16. Breitbart M (2012) Marine Viruses: Truth or Dare. Ann Rev Mar Sci 4: 425–448. doi:10.1146/annurev-marine-120709-142805.

17. Jiang SC, Paul JH (1998) Gene transfer by transduction in the marine environment. Appl Environ Microbiol 64: 2780–2787.

18. McDaniel LD, Young E, Delaney J, Ruhnau F, Ritchie KB, et al. (2010) High frequency of horizontal gene transfer in the oceans. Science (80-) 330: 50.

19. Breitbart M, Thompson LR, Suttle CA, Sullivan MB (2007) Exploring the vast diversity of marine viruses. Oceanography 20: 135–139.

20. Lindell D, Jaffe JD, Johnson ZI, Church GM, Chisholm SW (2005) Photosynthesis genes in marine viruses yield proteins during host infection. Nature 438: 86–89.

21. Lindell D, Jaffe JD, Coleman ML, Futschik ME, Axmann IM, et al. (2007) Genome-wide expression dynamics of a marine virus and host reveal features of co-evolution. Nature 449: 83–86. doi:10.1038/nature06130.

22. Clokie MRJ, Shan J, Bailey S, Jia Y, Krisch HM, et al. (2006) Transcription of a "photosynthetic" T4-type phage during infection of a marine cyanobacterium. Environ Microbiol 8: 827–835. doi:10.1111/j.1462-2920.2005.00969.x.

23. Dammeyer T, Bagby SC, Sullivan MB, Chisholm SW, Frankenberg-Dinkel N (2008) Efficient phage-mediated pigment biosynthesis in oceanic cyanobacteria. Curr Biol 18: 442–448.

24. Thompson LR, Zeng Q, Kelly L, Huang KH, Singer AU, et al. (2011) Phage auxiliary metabolic genes and the redirection of cyanobacterial host carbon metabolism. Proc Natl Acad Sci U S A 108: E757-64. doi:10.1073/pnas.1102164108.

25. Waldor MK, Mekalanos JJ (1996) Lysogenic conversion by a filamentous phage encoding cholera toxin. Science (80-) 272: 1910–1914. doi:10.1126/science.272.5270.1910.

26. Paul JH (2008) Prophages in marine bacteria: dangerous molecular time bombs or the key to survival in the seas? ISME J 2: 579–589. doi:10.1038/ismej.2008.35.

27. Ortmann AC, Suttle CA (2005) High abundances of viruses in a deep-sea hydrothermal vent system indicates viral mediated microbial mortality. Deep Sea Res Part I Oceanogr Res Pap 52: 1515–1527.

28. Weinbauer MG, Brettar I, Hölfe MG (2003) Lysogeny and virus-induced mortality of bacterioplankton in surface, deep, and anoxic marine waters. Limnol Oceanogr 48: 1457–1465.

29. Williamson SJ, Cary SC, Williamson KE, Helton RR, Bench SR, et al. (2008) Lysogenic virus-host interactions predominate at deep-sea diffuse-flow hydrothermal vents. ISME J 2: 1112–1121.

30. Meyer F, Paarmann D, D'Souza M, Olson R, Glass EM, et al. (2008) The metagenomics RAST server - a public resource for the automatic phylogenetic and functional analysis of metagenomes. BMC Bioinformatics 9: 386. doi:10.1186/1471-2105-9-386.

31. Schmieder R, Lim YW, Rohwer F, Edwards R (2010) TagCleaner: Identification and removal of tag sequences from genomic and metagenomic datasets. BMC Bioinformatics 11: 341. doi:10.1186/1471-2105-11-341.

32. Drummond A, Ashton B, Cheung M, Heled J, Kearse M, et al. (2009) Geneious v4. 7. Biomatters Ltd.

33. McHardy AC, Martín HG, Tsirigos A, Hugenholtz P, Rigoutsos I (2007) Accurate phylogenetic classification of variable-length DNA fragments. Nat Methods 4: 63–72. doi:10.1038/nmeth976.

34. Leplae R, Lima-Mendez G, Toussaint A (2010) ACLAME: a CLAssification of Mobile genetic Elements, update 2010. Nucleic Acids Res 38: D57-61. doi:10.1093/nar/gkp938.

35. Finn RD, Mistry J, Tate J, Coggill P, Heger A, et al. (2010) The Pfam protein families database. Nucleic Acids Res 38: D211-22. doi:10.1093/nar/gkp985.

36. Kanehisa M, Goto S, Sato Y, Furumichi M, Tanabe M (2012) KEGG for integration and interpretation of large-scale molecular data sets. Nucleic Acids Res 40: D109-14. doi:10.1093/nar/gkr988.

37. Tatusov RL, Fedorova ND, Jackson JD, Jacobs AR, Kiryutin B, et al. (2003) The COG database: an updated version includes eukaryotes. BMC Bioinformatics 4: 41. doi:10.1186/1471-2105-4-41.

38. Overbeek R, Begley T, Butler RM, Choudhuri JV, Chuang, et al. (2005) The subsystems approach to genome annotation and its use in the project to annotate 1000 genomes. Nucleic Acids Res 33: 5691–5702. doi:10.1093/nar/gki866.

39. Rodriguez-Brito B, Rohwer F, Edwards RA (2006) An application of statistics to comparative metagenomics. BMC Bioinformatics 7: 162. doi:10.1186/1471-2105-7-162.

40. Kurtz S, Phillippy A, Delcher AL, Smoot M, Shumway M, et al. (2004) Versatile and open software for comparing large genomes. Genome Biol 5: R12. doi:10.1186/gb-2004-5-2-r12.

41. Schmieder R, Edwards R (2011) Quality control and preprocessing of metagenomic datasets. Bioinformatics 27: 863–864. doi:10.1093/bioinformatics/btr026.

42. Tai V, Poon AFY, Paulsen IT, Palenik B (2011) Selection in coastal Synechococcus (cyanobacteria) populations evaluated from environmental metagenomes. PLoS One 6: e24249. doi:10.1371/journal.pone.0024249.

43. Wilke A, Harrison T, Wilkening J, Field D, Glass EM, et al. (2012) The M5nr: a novel non-redundant database containing protein sequences and annotations from multiple sources and associated tools. BMC Bioinformatics 13: 141. doi:10.1186/1471-2105-13-141.

44. Kristensen DM, Mushegian AR, Dolja VV, Koonin EV (2009) New dimensions of the virus world discovered through metagenomics. Trends Microbiol 18: 11–19.

45. Hurwitz BL, Deng L, Poulos BT, Sullivan MB (2013) Evaluation of methods to concentrate and purify ocean virus communities through comparative, replicated metagenomics. Environ Microbiol 15: 1428–1440. doi:10.1111/j.1462-2920.2012.02836.x.

46. Anderson RE, Brazelton WJ, Baross JA (2011) Is the genetic landscape of the deep subsurface biosphere affected by viruses? Front Extrem Microbiol 2.

47. Coleman ML, Sullivan MB, Martiny AC, Steglich C, Barry K, et al. (2006) Genomic islands and the ecology and evolution of Prochlorococcus. Science (80-) 311: 1768–1770. doi:10.1126/science.1122050.

48. Cuadros-Orellana S, Martin-Cuadrado, Legault B, D'Auria G, Zhaxybayeva O, et al. (2007) Genomic plasticity in prokaryotes: the case of the square haloarchaeon. ISME J 1: 235–245. doi:10.1038/ismej.2007.35.

49. Rodriguez-Valera F, Martin-Cuadrado, Rodriguez-Brito B, Pasic L, Thingstad TF, et al. (2009) Explaining microbial population genomics through phage predation. Nat Rev Microbiol 7: 828–836. doi:10.1038/nrmicro2235.

50. Voordeckers JW, Starovoytov V, Vetriani C (2005) Caminibacter mediatlanticus sp. nov., a thermophilic, chemolithoautotrophic, nitrate-ammonifying bacterium isolated from a deep-sea hydrothermal vent on the Mid-Atlantic Ridge. Int J Syst Evol Microbiol 55: 773–779. doi:10.1099/ijs.0.63430-0.

51. Reiter W-D, Palm P, Yeats S (1989) Transfer RNA genes frequently serve as integration sites for prokaryotic genetic elements. Nucleic Acids Res 17: 1907–1914. doi:10.1093/nar/17.5.1907.

52. Campbell BJ, Smith JL, Hanson TE, Klotz MG, Stein LY, et al. (2009) Adaptations to submarine hydrothermal environments exemplified by the genome of Nautilia profundicola. PLoS Genet 5: e1000362. doi:10.1371/journal.pgen.1000362.

53. Dinsdale EA, Edwards RA, Hall D, Angly F, Breitbart M, et al. (2008) Functional metagenomic profiling of nine biomes. Nature 452: 629–632. doi:10.1038/nature06810.

54. Sullivan MB, Coleman ML, Weigele P, Rohwer F, Chisholm SW (2005) Three Prochlorococcus cyanophage genomes: signature features and ecological interpretations. PLoS Biol 3: e144. doi:10.1371/journal.pbio.0030144.

55. Sullivan MB, Huang KH, Ignacio-Espinoza JC, Berlin AM, Kelly L, et al. (2010) Genomic analysis of oceanic cyanobacterial myoviruses compared with T4-like myoviruses from diverse hosts and environments. Environ Microbiol 12: 3035–3056. doi:10.1111/j.1462-2920.2010.02280.x.

56. Mann NH, Cook A, Millard A, Bailey S, Clokie M (2003) Marine ecosystems: Bacterial photosynthesis genes in a virus. Nature 424: 741. doi:10.1038/424741a.

57. Lindell D, Sullivan MB, Johnson ZI, Tolonen AC, Rohwer F, et al. (2004) Transfer of photosynthesis genes to and from Prochlorococcus viruses. Proc Natl Acad Sci U S A 101: 11013–11018. doi:10.1073/pnas.0401526101.

58. Sullivan MB, Lindell D, Lee JA, Thompson LR, Bielawski JP, et al. (2006) Prevalence and evolution of core photosystem II genes in marine cyanobacterial viruses and their hosts. PLoS Biol 4: e234.

59. Williamson SJ, Allen LZ, Lorenzi HA, Fadrosh DW, Brami D, et al. (2012) Metagenomic exploration of viruses throughout the Indian Ocean. PLoS One 7: e42047. doi:10.1371/journal.pone.0042047.

60. Clokie MRJ, Mann NH (2006) Marine cyanophages and light. Environ Microbiol 8: 2074–2082. doi:10.1111/j.1462-2920.2006.01171.x.

61. Sharon I, Alperovitch A, Rohwer F, Haynes M, Glaser F, et al. (2009) Photosystem I gene cassettes are present in marine virus genomes. Nature 461: 258–262.

62. Bragg JG, Chisholm SW (2008) Modeling the fitness consequences of a cyanophage-encoded photosynthesis gene. PLoS One 3: e3550. doi:10.1371/journal.pone.0003550.

63. Hellweger FL (2009) Carrying photosynthesis genes increases ecological fitness of cyanophage in silico. Environ Microbiol 11: 1386–1394. doi:10.1111/j.1462-2920.2009.01866.x.

64. Konstantinidis KT, Ramette A, Tiedje JM (2006) The bacterial species definition in the genomic era. Philos Trans R Soc Lond B Biol Sci 361: 1929–1940. doi:10.1098/rstb.2006.1920.

14

Probable Ankylosaur Ossicles from the Middle Cenomanian Dunvegan Formation of Northwestern Alberta, Canada

Michael E. Burns[1]*, Matthew J. Vavrek[2]

1 Biological Sciences, University of Alberta, Edmonton, Alberta, Canada, **2** Pipestone Creek Dinosaur Initiative, Clairmont, Alberta, Canada

Abstract

A sample of six probable fragmentary ankylosaur ossicles, collected from Cenomanian deposits of the Dunvegan Formation along the Peace River, represent one of the first dinosaurian skeletal fossils reported from pre-Santonian deposits in Alberta. Specimens were identified as ankylosaur by means of a palaeohistological analysis. The primary tissue is composed of zonal interwoven structural fibre bundles with irregularly-shaped lacunae, unlike the elongate lacunae of the secondary lamellar bone. The locality represents the most northerly Cenomanian occurrence of ankylosaur skeletal remains. Further fieldwork in under-examined areas of the province carries potential for additional finds.

Editor: Peter Dodson, University of Pennsylvania, United States of America

Funding: Funding to MEB was provided by the Department of Biological Sciences. Funding for fieldwork was provided by the Dinosaur Research Institute to MJV. The funders had no role in study design, data collection and analysis, decision to publish, or preparation of the manuscript.

Competing Interests: The authors have declared that no competing interests exist.

* E-mail: il:mburns@ualberta.ca

Introduction

Although Alberta is one of the most intensely studied areas in the world in regards to dinosaur palaeontology [1–2], skeletal fossils of dinosaurs from pre-Santonian rocks are virtually unknown [3]. Other than a recently recovered ankylosaur from marine sediments of the Albian Clearwater Formation of northeastern Alberta [4], and an isolated *Ichthyornis* sp. humerus from the Turonian Kaskapau Formation [5], no other dinosaur skeletal remains have been described from Alberta in rocks older than Santonian, although there does exist a non-descriptive reference to Dunvegan Formation dinosaurs in an encyclopedia article [6]. During the course of research for this paper, the authors were informed of some undescribed, indeterminate bones (possibly ornithischian) from the Blairmore Formation of southwestern Alberta (D. Brinkman, pers. comm.), however this material is highly fragmentary.

Even outside of Alberta, there are virtually no skeletal remains of pre-Santonian dinosaurs from north of the 49th Parallel. There are records of two possible ornithischian vertebrae and several birds from the from the Belle Fourche Member of the Ashville Formation in eastern Saskatchewan [7–8], undescribed dinosaur bones from the Turonian Kaskapau Formation of British Columbia [9], an indeterminate hadrosaur from the Turonian Matanuska Formation of Alaska [10–11], and a possible bone fragment from the Late Jurassic of Alaska [12]. Despite the lack of a recognized fossil record, there are relatively large exposures of lower Upper Cretaceous rocks in Alberta, primarily concentrated along large river channels in the northern portion of the province.

During a survey of the Cenomanian-aged Dunvegan Formation along the Peace River, a handful of vertebrate remains were recovered, including six probable fragmentary ankylosaur ossicles.

Although the fossils, in terms of preservational quality and completeness, are limited relative to some other dinosaur finds in western Canada, they represent an important stratigraphic and biogeographic data point because they come from an age that is poorly known in Canada, and from a location at least a thousand kilometres away from the next nearest contemporaneous skeletal record of dinosaurs.

Geological and Climatic Setting

The Dunvegan Formation represents a middle Cenomanian-aged delta complex that occurs primarily in northeastern British Columbia and northwestern Alberta, with small extensions into the Northwest Territories [13–16]. The formation consists of a repeated succession of alluvial and shallow marine sandstones, siltstones and shales, and ranges between 90 and 270 m in thickness [14–15]. During deposition, the delta complex prograded a maximum of 400 km into the Western Interior Seaway [14–15]; in general, the formation becomes more terrestrial moving towards the west. The Dunvegan Formation is underlain by the marine shales of the Shaftesbury Formation, and is overlain by the marine sandstones and shales of the Kaskapau Formation [13–14].

Locally, the fossils were found weathered out of a siltstone to fine sandstone bed near the top of the Peace River valley, approximately 4 km upstream of the Dunvegan Bridge (Fig. 1). The exact source of the fossils could not be determined. However, because the fossils were found near the top of an exposure, they were likely close to their original level. They were found in association with numerous ironstone pebbles that had likely weathered out from either the same beds or beds that were in close proximity.

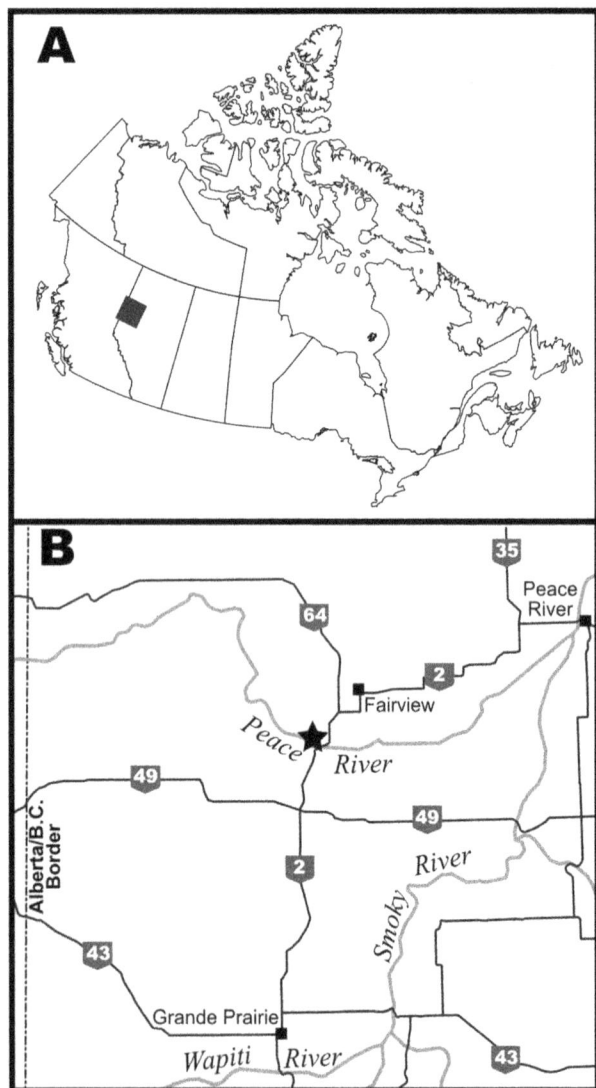

Figure 1. Locality map. A) Overview map showing location of inset map (B). B) Map of the Peace Country with major rivers and towns indicated. The fossil locality is denoted by a black star.

During the Cenomanian, the Dunvegan delta was located near the Arctic Circle, at about 65° N [14–17] Land temperatures during this time were, however, much warmer on average than the present day [18] with higher latitude regions in particular experiencing a greater amount of warming compared to more equatorial areas, leading to a reduced equator to pole thermal gradient [19]. A study of a terrestrial ecosystem from the younger Kaskapau Formation from about the same latitude as the Dunvegan Formation suggests mean annual temperatures were around 14°C, with a cold month mean likely warmer than 5.5 C [9]. These temperatures may be high as global temperatures were on a warming trend through the Cenomanian into the Turonian [20–21], although the difference was likely not considerable.

Materials and Methods

All ankylosaur skeletal specimens in this study were collected under permits obtained from Alberta Culture (Alberta Palaeontological Permit No. 12-029) and are catalogued as TMP (Royal

Tyrrell Museum of Palaeontology, Drumheller, Alberta, Canada) 2012.054.0002. Photographs of ossicles were taken on a Zeiss SteREO Discovery.V8 with a Plan Apo S 0.63× objective and an attached Nikon DXM 1200C camera using NIS-Elements F 2.20 SP3 (Build 244) imaging software. To confirm its identification, one ossicle was selected for paleohistological analysis. It was stabilized via resin impregnation using Buehler EpoThin Low Viscosity Resin and Hardener. A thin section was prepared petrographically to a thickness of 100 μm and polished to a high gloss using CeO_2 powder. The section was examined and photographed on a Nikon Eclipse E600POL trinocular polarizing microscope with an attached Nikon DXM 1200F digital camera. A composite image was constructed in Adobe Photoshop CS6 v. 13.0.1×64.

Descriptive terminology for ossicles follows [14–18]. Palaeohistological terminology for osteoderms follows [22–24]. The definition for "ossicles" adopted here is modified [25]: small (< 70 mm), amorphous mineralized dermal elements often found interstitial to major osteodermal elements. The term interwoven structural fiber bundles (ISFB; sensu [23–26] is used to refer to mineralized metaplastic tissue dominated by large, structural collagen fibers.

Results

The ossicles (Fig. 2) are all irregular in shape and some are fragmentary, so it is unknown exactly how many ossicles are represented (some may be fragments of larger osteoderms). Although their exact in vivo orientation is difficult to discern, the external and basal surfaces are distinguishable. The external surfaces are irregularly rugose and pitted. The basal surfaces are flat and have a distinctive pattern representing the ISFB making up the primary tissue of the ossicles.

The primary tissue of the sectioned ossicle from TMP 2012.054.0002 (Fig. 3) is composed entirely of ISFB. Near the margins, zonation is overprints the pattern of structural fibers (Fig. 4). The lines of zonation, however, are not as distinct enough to be confidently called lines of arrested growth because they do

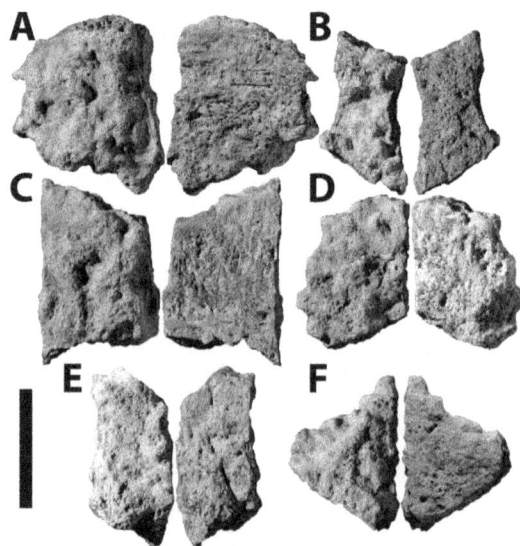

Figure 2. Dunvegan ossicle morphology. Ankylosaur ossicles (TMP.2012.054.0002) from the Cenomanian Dunvegan Formation, near the Peace River, British Columbia, Canada, in external and basal views. Scale bar equals 1.0 mm.

not show a discreet hypermineralized line indicating a cessation of element growth. The ISFB are not as highly organized in terms of arrangement as those reported for the osteoderms of derived nodosaurids or the ossicles of sauropods [22–23], [27–28]. This is similar, however, to the condition reported for the ossicles of *Edmontonia* and *Euoplocephalus* [22] and unlike the radial pattern described for the ossicles of the basal ankylosaur *Antarctopelta* [29]. Secondary tissue comprises almost 50% of the cross sectional area of the ossicle and consists of trabecular bone. Osteocyte lacunae in the primary tissue, unlike the elongate lenticular lacunae of the secondary lamellar bone, are irregularly shaped (Fig. 5). Lacunae in both the primary and secondary tissue have canalliculi.

Discussion

Other dinosaurian groups known to have had ossicles (namely stegosaurs and sauropods) are unknown in Late Cretaceous Canadian sediments. To date, secondary remodeling is not known to occur in sauropod ossicles and their orthogonal pattern of ISFB is stronger than in ankylosaurs [22–27]. Although stegosaurs cannot be ruled out on the basis of macro/microstructure, because their ossicle/gular osteoderm histology is unknown at present, the occurrence of the Dunvegan ossicles in the Upper Cretaceous makes them less parsimonious candidates than ankylosaurs.

The best possible alternative candidate for the taxonomic identification of the Dunvegan is crocodilian. Crocodilian osteoderms are common as fossils from Upper Cretaceous sediments in western Canada. Like some ankylosaur osteoderms/ossicles, the basal surface has a fine cross-hatch pattern corresponding to the connective tissue fascia that separates the element from epaxial musculature. The external surface, however, is characterized by distinctive sculpturing composed of numerous round pits and grooves that radiate from a midline keel [22]. This is a noticeable difference between crocodilian and ankylosaur osteoderms. In addition, crocodilian osteoderms largely develop in the loose, superficial dermis, leading to less dense structural fibers in the primary mineralized tissue [30–34]. Ankylosaur osteoderms/ossicles show dense ISFB networks, suggesting a greater contribution from the dense dermis [22]. Therefore, the external morphology and histology of the Dunvegan ossicles indicate that they are from an ankylosaur.

Unlike larger osteoderms, ossicles show no consistent differences among taxa, at least among derived nodosaurids and ankylosaurids [22]. Therefore, the ossicles sampled here cannot be identified

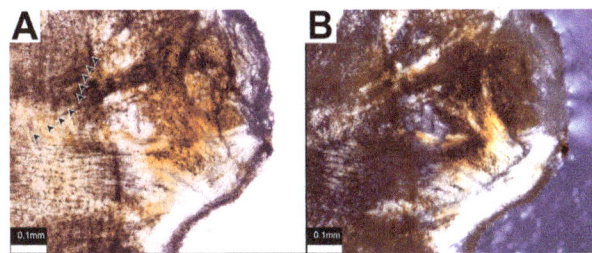

Figure 4. Primary ossicle tissues. Details of primary tissue in ossicle of TMP 2012.054.0002. A) Plane-polarized light showing zonation (growth marks indicated by arrowheads). B) Cross-polarized light showing ISFB.

to any particular ankylosaur group. Zonation in the form of annuli in modern crocodilians has been strongly correlated with age [35]; however, this association has not been tested for ankylosaur osteoderms or ossicles. Ankylosaurs likely had a delayed onset of osteoderm mineralization, more so than modern crocodilians, and similar to the heterochronic condition reported for stegosaur osteoderm mineralization [22],[28],[36]. The zonation observed in TMP 2012.054.0002 may be annual, but the growth marks are not continuous around the entire circumference of the ossicle and, in places, have been reabsorbed by secondary remodeling. The utility of growth marks in ankylosaur ossicles will need to be tested against another form of age determination (i.e., postcranial long bone histology).

Although the Dunvegan Formation has been previously known to contain abundant shark teeth from other localities near the Dunvegan Bridge on the Peace River [37], as well as an articulated, well-preserved fish from a fortuitous subsurface encounter [15], this is the first published record of dinosaur skeletal remains from the formation. Further west in Alberta and into British Columbia, dinosaur ichnites have previously been recorded from the Dunvegan Formation, including trackways of ankylosaurs [14],[38]. This discovery of skeletal fossils opens the possibility that further finds may be able to link some of these trackways with their potential trackmakers more closely.

Finally, the spatial location of the fossils is interesting from a biogeographic perspective. This locality represents the most northerly occurrence of ankylosaur skeletal remains during the Cenomanian, and further solidifies this group, and dinosaurs in general, as persistent residents of high latitude regions [39–41]. The region experienced large fluctuations in solar radiation through the year due to its high latitude, likely leading to seasonal growth patterns in vegetation, however these dinosaurs likely persisted in the area as they would have been physically unable to

Figure 3. Dunvegan ossicle histology. Composite mosaic image of thin section through ossicle of TMP 2012.054.0002 in cross-polarized light. Orientation uncertain. Scale bar equals 1.0 mm.

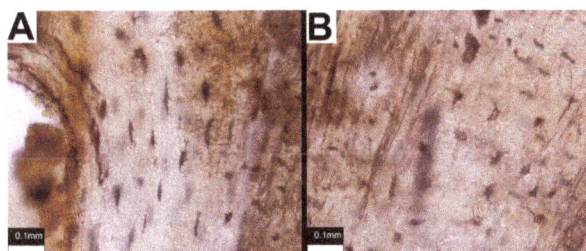

Figure 5. Ossicle osteocyte lacunae. Comparison of osteocyte lacuna morphology in primary (A) and secondary tissue (B) in an ossicle of TMP 2012.054.0002. Orientation uncertain. Scale bar equals 0.10 mm.

migrate long distances [15],[41–42]. Although temperatures would have been warmer and more equable at the time [19], they would have nevertheless experienced at least somewhat cooler winter temperatures.

Conclusions

Although the ankylosaur remains from the Dunvegan Formation are generically indeterminate, their unique geographic and temporal position makes them an important data point for dinosaur biogeography. As well, the presence of dinosaurian remains from pre-Santonian deposits in Alberta suggests that, with further effort in many of these under-examined areas, there is the potential for additional finds in the region. High latitude dinosaur-bearing deposits are more poorly known in general than mid-latitude regions, although this is not due to the animals not being present, but more likely a function of search intensity.

Acknowledgments

P Currie read over an early version of this manuscript. M Caldwell (University of Alberta) provided access to the petrographic microscope and A Murray (University of Alberta) provided access to the Zeiss stereo microscope. We also thank Richard McCrea and an anonymous reviewer for suggestions that greatly improved the quality of this manuscript.

Author Contributions

Conceived and designed the experiments: MEB MJV. Performed the experiments: MEB MJV. Analyzed the data: MEB MJV. Contributed reagents/materials/analysis tools: MEB MJV. Wrote the paper: MEB MJV.

References

1. Ryan MJ, Russell AP (2001) Dinosaurs of Alberta (exclusive of Aves). In: Tanke DH, Carpenter, K, editors. Mesozoic vertebrate life: New research inspired by the paleontology of Philip J. Currie: Bloomington: Indiana University Press. pp. 279–297.
2. Currie PJ (2005) History of research. In: Currie PJ, Koppelhus EB, editors. Dinosaur Provincial Park: A spectacular ancient ecosystem revealed. Bloomington: Indiana University Press. pp. 3–33.
3. McCrea RT, Buckley LG, Plint AG, Currie PJ, Haggart JW, et al. (2014) A review of vertebrate track-bearing formations from the Mesozoic and earliest Cenozoic of western Canada with a description of a new theropod ichnospecies and reassignment of an avian ichnogenus. New Mexico Museum of Natural History and Sciences Bulletin 52: 5–93.
4. Henderson D (2012) Adrift at sea in the Early Cretaceous: The Fort McMurray armoured dinosaur. Alberta Palaeontological Society Bulletin 27: 7.
5. Fox RC (1984) *Ichthyornis* (Aves) from the early Turonian (Late Cretaceous) of Alberta. Can J Earth Sci 21: 258–260.
6. Coy C (1997) Canadian Dinosaurs. In Currie PJ, Padian K, editors. Encyclopedia of Dinosaurs. Academic Press. pp. 90–91.
7. Tokaryk TT, Cumbaa SL, Storer JE (1997) Early Late Cretaceous birds from Saskatchewan, Canada: the oldest diverse avifauna known from North America. Journal of Vertebrate Paleontology 17:172–176.
8. Cumbaa SL, Tokaryk TT (1999) Recent discoveries of Cretaceous marine vertebrates on the eastern margins of the Western Interior Seaway. Summary of investigations 1: 94–99.
9. Rylaarsdam JR, Varban BL, Plint AG, Buckley LG, McCrea RT (2006) Middle Turonian dinosaur paleoenvironments in the Upper Cretaceous Kaskapau Formation, northeast British Columbia. Can J Earth Sci 43: 631–652.
10. Pasch AD, May KC (1997) First occurrence of a hadrosaur (Dinosauria) from the Matanuska Formation (Turonian) in the Talkeetna Mountains of south-central Alaska. In Clough JG, Larson F, editors. Alaska Department of Natural Resources, Professional Report 118. Short notes on Alaska geology. Pp 99–109.
11. Pasch AD, May KC (2001) Taphonomy and paleoenvironment of a hadrosaur (Dinosauria) from the Matanuska Formation (Turonian) in south-central Alaska. In Tanke DH, Carpenter K, editors. Mesozoic vertebrate life: New research inspired by the paleontology of Philip J. Currie. Indiana University Press. pp. 219–236.
12. Fiorillo AR (2006). Review of the dinosaur record of Alaska with comments regarding Korean dinosaurs as comparable high-latitude fossil faunas. Journal of the Paleontological Society of Korea 22: 15.
13. Stott DF (1982) Lower Cretaceous Fort St. John Group and Upper Cretaceous Dunvegan Formation of the foothills and plains of Alberta, British Columbia, District of Mackenzie and Yukon Territory. Geological Survey of Canada Bulletin 328: 1–124.
14. Plint AG (2000) Sequence stratigraphy and paleogeography of a Cenomanian deltaic complex: the Dunvegan and lower Kaskapau formations in subsurface and outcrop, Alberta and British Columbia, Canada. B Can Petrol Geol 48: 43–79.
15. Hay MJ, Cumbaa SL, Murray AM, Plint AG (2007) A new paraclupeid fish (Clupeomorpha, Ellimmichthyiformes) from a muddy marine pro-delta environment: middle Cenomanian Dunvegan Formation, Alberta, Canada. Can J Earth Sci 44: 775–790.
16. McCarthy PJ, Faccini UF, Plint AG (1999) Evolution of an ancient coastal plain: palaeosols, interfluves and alluvial architecture in a sequence stratigraphic framework, Cenomanian Dunvegan Formation, NE British Columbia, Canada. Sedimentology 46: 861–891.
17. Irving E, Wynne PJ, Globerman BR (1993) Cretaceous paleolatitudes and overprints of North American craton. In: Caldwell WGE, Kauffman EG, editors. Evolution of the Western Interior Basin. Geological Association of Canada, Special Paper 39. pp. 91–96.
18. Frakes LA (2002) Estimating the global thermal state from Cretaceous sea surface and continental temperature data. In: Barrerea E, Johnson CC, editors.

Evolution of the Cretaceous ocean–climate system. Geological Society of America, Special Paper 332. pp. 49–57.
19. Upchurch GR, Wolfe JA (1993) Cretaceous Vegetation of the Western Interior and Adjacent Regions of North America. In: Caldwell WGE, Kauffman EG, editors. Evolution of the western interior basin. Geological Association of Canada, Special Paper 39. pp. 243–281.
20. Wolfe JA, Upchurch GR (1987) North American nonmarine climates and vegetation during the Late Cretaceous. Palaeogeogr, Palaeoclimatol, Palaeoecol 61: 33–77.
21. Huber BT, Norris RD, MacLeod KG (2002) Deep-sea paleotemperature record of extreme warmth during the Cretaceous. Geology 30: 123–126.
22. Burns ME, Currie PJ External and internal structure of ankylosaur (Dinosauria; Ornithischia) osteoderms and their systematic relevance. J Vretebr Paleontol 34. In Press.
23. Scheyer TM, Sander PM (2004) Histology of ankylosaur osteoderms: Implications for systematics and function. J Vertebr Paleontol 24: 874–893.
24. Cerda IE, Desojo JB (2011) Dermal armour histology of aetosaurs (Archosauria: Pseudosuchia), from the Upper Triassic of Argentina and Brazil. Lethaia 44: 417–428.
25. Blows WT (2001) Dermal armor of the polacanthine dinosaurs. In: Carpenter K, editor. The Armored Dinosaurs: Bloomington, Indiana University Press. pp. 363–385.
26. Scheyer TM, Sánchez-Villagra MR (2007) Carapace bone histology in the giant pleurodiran turtle *Stupendemys geographicus*: phylogeny and function. Acta Palaeontol Pol 52: 137–154.
27. Cerda IA, Powell JB (2010) Dermal armor histology of *Saltasaurus loricatus*, an Upper Cretaceous sauropod dinosaur from northwest Argentina. Acta Palaeontologica Polonica, 55: 389–398.
28. Hayashi S, Carpenter K, Scheyer TM, Watabe M, Suzuki D (2010) Function and evolution of ankylosaur dermal armor. Acta Palaeontol Pol 5: 213–228.
29. De Ricqlès A, Pereda-Suberbiola X, Gasparini Z, Olivero E (2001) Histology of dermal ossifications in an ankylosaurian dinosaur from the Late Cretaceous of Antarctica. Asociación Paleontológica Argentina, Publicación Especial 7: 171–174.
30. Martill DM, Batten DJ, Loydell DK. (2000) A new specimen of the thyreophoran dinosaur cf. *Scelidosaurus* with soft tissue preservation. Palaeontology 43: 549–559.
31. Salisbury SW, Frey E (2000) A biomechanical transformation model for the evolution of semi-spheroidal articulations between adjoining vertebral bodies in crocodilians. In Grigg GCF, Seebacher F, Franklin CE, editors. Crocodilian Biology and Evolution. Chipping Norton, NSW: Surrey Beatty & Sons. pp. 85–134.
32. Vickaryous MK, Hall BK (2008) Development of the dermal skeleton in *Alligator mississippiensis* (Archosauria, Crocodylia) with comments on the homology of osteoderms. Journal of Morphology 269: 398–422.
33. Vickaryous MK, Sire JY (2009) The integumentary skeleton of tetrapods: origin, evolution, and development. Journal of Anatomy 214: 441–464.
34. Burns ME, Vickaryous MK, Currie PJ (2013) Histological variability in fossil and recent crocodylian osteoderms: systematic and functional implications. Journal of Morphology 274: 676–686.
35. Tucker AD (1997) Validation of skeletochronology to determine age of freshwater crocodiles (*Crocodylus johnstoni*). Mar Freshwater Res 48: 343–351.
36. Hayashi S, Carpenter K, Suzuki D (2009) Different growth patterns between the skeleton and osteoderms of *Stegosaurus* (Ornithischia: Thyreophora). J Vertebr Paleontol 29: 123–131.
37. Cook TD, Wilson MVH, Murray AM (2008) A middle Cenomanian euselachian assemblage from the Dunvegan Formation of northwestern Alberta. Can J earth Sci 45: 1185–1197.
38. McCrea RT, Lockley MG, Meyer CA (2001) Global distribution of purported ankylosaur track occurrences. In: Carpenter K, editor. The armoured dinosaurs. Bloomington: Indiana University Press. pp. 413–454.

39. Chinsamy A, Thomas DB, Tumarkin-Deratzian AR, Fiorillo AR (2012). Hadrosaurs Were Perennial Polar Residents. The Anatomical Record 295: 610–614.

40. Fiorillo AR, Gangloff RA (2001) The Caribou Migration Model for Arctic Hadrosaurs (Dinosauria: Ornithischia): A Reassessment. Historical Biology 15: 323–334.

41. Vavrek MJ, Hills LV, Currie PJ. A hadrosaurid (Dinosauria: Ornithischia) from the Late Cretaceous (Campanian) Kanguk Formation of Axel Heiberg Island, Nunavut, Canada and its ecological and geographical implications. Arctic. In press.

42. Bell PR, Snively E (2008) Polar dinosaurs on parade: a review of dinosaur migration. Alcheringa 32: 271–284.

15

Phenotypic Variation and Fitness in a Metapopulation of Tubeworms (*Ridgeia piscesae* Jones) at Hydrothermal Vents

Verena Tunnicliffe[1,2]*, Candice St. Germain[1], Ana Hilário[3]

1 Department of Biology, University of Victoria, Victoria, British Columbia, Canada, **2** School of Earth & Ocean Sciences, University of Victoria, Victoria, British Columbia, Canada, **3** Departamento de Biologia and Centro de Estudos do Ambiente e do Mar, Universidade de Aveiro, Campus de Santiago, Aveiro, Portugal

Abstract

We examine the nature of variation in a hot vent tubeworm, *Ridgeia piscesae*, to determine how phenotypes are maintained and how reproductive potential is dictated by habitat. This foundation species at northeast Pacific hydrothermal sites occupies a wide habitat range in a highly heterogeneous environment. Where fluids supply high levels of dissolved sulphide for symbionts, the worm grows rapidly in a "short-fat" phenotype characterized by lush gill plumes; when plumes are healthy, sperm package capture is higher. This form can mature within months and has a high fecundity with continuous gamete output and a lifespan of about three years in unstable conditions. Other phenotypes occupy low fluid flux habitats that are more stable and individuals grow very slowly; however, they have low reproductive readiness that is hampered further by small, predator cropped branchiae, thus reducing fertilization and metabolite uptake. Although only the largest worms were measured, only 17% of low flux worms were reproductively competent compared to 91% of high flux worms. A model of reproductive readiness illustrates that tube diameter is a good predictor of reproductive output and that few low flux worms reached critical reproductive size. We postulate that most of the propagules for the vent fields originate from the larger tubeworms that live in small, unstable habitat patches. The large expanses of worms in more stable low flux habitat sustain a small, but long-term, reproductive output. Phenotypic variation is an adaptation that fosters both morphological and physiological responses to differences in chemical milieu and predator pressure. This foundation species forms a metapopulation with variable growth characteristics in a heterogeneous environment where a strategy of phenotypic variation bestows an advantage over specialization.

Editor: Steffen Kiel, Universität Göttingen, Germany

Funding: Natural Sciences and Engineering Research Council of Canada (nserc.ca) funding through the Canadian Healthy Oceans Network (http://www.chone.ca/)(PC-11) and Discovery Grants programme. The funders had no role in study design, data collection and analysis, decision to publish, or preparation of the manuscript.

Competing Interests: The authors have declared that no competing interests exist.

* Email: verenat@uvic.ca

Introduction

In source-sink metapopulation scenarios, optimal habitat supports highly productive adults but the local demographics of sinks result in low contribution to the metapopulation usually because of marginally suitable habitat conditions [1]. Selection will favour the traits of the source population unless there is some reproductive contribution to the metapopulation from the sink [2,3]. A fragmented landscape usually presents variable habitat patch quality for inhabitants in which the proportion of optimal to marginal habitat can be very low. This landscape may encompass enough environmental variability to invoke phenotypic response in individuals of a deme [4]. Thus, extensive marginal patches may constitute a relative sink with some genetic contribution especially when phenotypic adaptations enhance fitness in that sink. The extent to which that response is a genotypic adaptation will depend on factors such as relative habitat frequency and migration among habitats [5]. An alternative adaptation is phenotypic plasticity in which one genotype produces distinct phenotypes under different environmental conditions [6]. A plastic response to marginal conditions can increase individual fitness to support some genetic contribution to the metapopulation [7,8]. In the context of developing conservation strategies, recognizing sources for critical species and the role of sinks in their maintenance is an important aspect [9], the source of the variability notwithstanding.

The habitat at hydrothermal vents is highly variable in space and time. Patches are controlled by subsurface fluid flows that can change abruptly with tectonic movements or volcanic activity [10,11] that may affect an entire field. However, across a vent field, environmental dynamics tend to be independent as flow adjustments are localized. At these small scales, fluid flow in basalt cracks or through a sulphide chimney may shift causing death or invigoration of animal assemblages over the course of weeks [12,13]. Vestimentiferan tubeworms (Polychaeta, Siboglinidae) often dominate faunal biomass at vents. They host bacterial symbionts in the trunk that depend on a sustained supply of dissolved sulphide delivered through the gills and vascular system from venting fluids. *Ridgeia piscesae* is the only tubeworm at vents on the ridges of the northeast Pacific (herein "Juan de Fuca Ridge"). This species dominates biomass and is a foundation species providing habitat for nearly all vent assemblages [14]. Gill filaments on the upper obturaculum emerge above the muscular vestimentum that wedges the animal in the tube. The trunk contains gonad and the trophosome organ housing the symbionts. Sperm masses are captured on female branchiae and fertilization in *R. piscesae* is internal; oocytes are released from the upper ovisacs and begin cleavage after release with ensuing pelagic larval development of several weeks [15–17].

Variation in *Ridgeia piscesae* is well-documented beginning with the initial description of two distinct species by Jones [18]. From morphology, Galkin [19] identified a continuous range of characters encompassing the original descriptions of *R. piscesae* and *R. phaeophiale*. Morphologies ranging from "short-fat", "small-contorted" to "long-skinny" correspond to habitats of different temperatures [20]. When allozymes provided no genetic evidence basis for differentiation between these morphotypes, *Ridgeia* was re-described as a single species [21]. Subsequent targeted tests using a mitochondrial gene, a nuclear gene and fragment length polymorphisms could detect no genetic basis for the phenotypes [22–24]. Studies finding notable differences in levels of dissolved carbon and sulphide-binding amino acids between extreme morphologies invoked differential gene expression in different habitats [25,26]; subsequently, the role of local environment in determining levels of haemoglobin gene expression was demonstrated by Carney et al. [27]. Liao et al [28] document metabolic "flexibility" in the use of nitrate by symbionts in conditions of differing ammonia levels in the fluid environment around different *R. piscesae* morphotypes and suggest both partners in the association are involved in optimizing the fitness of the symbiosis in different environments. Puetz [29] examined the same specimens that we study below ("Hi/Lo" pairs) to find strong support for a single gene pool at our main study site and no consistent differences among phenotypes in a mitochondrial gene region. Vrijenhoek [30] proposes that the morphological variability is a phenotypically plastic response to habitat. However, as transplant experiments to monitor growth change in different conditions are difficult because of handling damage, evidence for phenotypic plasticity rather than local adaptation remains circumstantial.

The objective of this work is to define how the variable hydrothermal fluid environment influences growth, body characteristics and reproductive condition in *Ridgeia piscesae*. We test the hypothesis that, in higher temperature habitats with greater dissolved sulphide flux, *R. piscesae* is fast-growing, short-lived and has a distinctive morphology with high reproductive fitness compared to animals in habitats of low fluid flux. We explore the components of fitness that may be enhanced by phenotypic variation in different fluid settings. We assess the reproductive condition of tubeworms in a variety of habitats to estimate the likely contribution of those habitats to the metapopulation. As metals markets push deep-sea prospecting into hydrothermal vents, designing reserve areas for the unique species inhabiting the deposits is a rising concern [31]. The means and extent to which a foundation species can function over a large habitat range is relevant to resilience in, and restoration of, degraded ecosystems.

Materials and Methods

The hydrothermal vents on Juan de Fuca Ridge (Figure 1) are located at discrete sites on separate segments of the ridge. Axial Volcano is one such site where vents within the caldera at ~1550 m depth are short-lived due to lava eruptions. On Endeavour Segment, venting in the axial spreading valley is concentrated in several "vent fields" where fluids emerge through mineralized chimneys and through cracked seafloor basalts. Extrusive volcanism has not occurred within observation history here. Fields, separated by about 2 km, have multiple black smoker chimneys, clusters of animals around vent openings, and scattered tubeworms in low density. Collections came from Clam Bed, Main Field and Mothra Field at ~2200 m depth (Figure 1). Deep-sea access was provided by the remotely operated vehicle *ROPOS* and the occupied vehicle *Alvin*, both of which used advanced force feed-back manipulators for collecting samples. Collection permits for this work were issued by Fisheries and Oceans Canada.

The terms "short-fat" and "long-skinny" are present in the literature for two morphotypes of *Ridgeia piscesae*, however, we also sampled other forms. To avoid these terms, we use "high flux" and "low flux" to indicate collection habitat following Sarrazin and Juniper [32] and Bates et al. [33] who define visual cues for emerging fluid of relatively high temperature and dissolved sulphide: shimmering, turbulent water. "Low flux" venting fluids are rarely visible but marked by clustered vent animals at low temperatures. Four sets of samples (totalling 731 individuals) were collected during seven sampling opportunities to study growth and condition in different venting settings and address these questions (see Table 1):

1. How does initial *Ridgeia* growth respond to habitat conditions? On January 28 1998, an eruption on Axial Volcano extruded new lavas where a field of *R. piscesae* had existed the previous year [11]. Vents formed on the new lavas while some *R. piscesae* survived on the adjacent old lavas. In August 1998, we sampled newly settled worms at four new vents (Nascent, N41, M113, T&S), and survivors at two old vents (Oldwrms, Lrgwrms) nearby and one more distal (Bob) (Figure 1; Table S1). At each vent, fluid was assessed for maximum temperature and dissolved sulphide [34]. Markers were placed to pinpoint the site and in summers of 1999 and 2000, Nascent and N41 were re-sampled.

2. How do gonad and trophosome development relate to size? Five samples were collected in 1998 and 1999 from Main Endeavour Field and Clam Bed (Figure 1), preserved in 10% seawater formalin and transferred to 70% ethanol. Vestimentum width was used as a standard size measure following Thiébaut et al. [35]. Using a random stratified sampling design, the trunk of each individual was divided in 10 equal sections from which a 1 mm segment was chosen. The 10 segments were dehydrated in 90% propan-2-ol overnight followed by nine hours in 100% propan-2-ol with solution changed every 3 hours, then cleared with 100% xylene and impregnated in wax at 70°C. Subsequently, 5 μm sections were stained with Mayer's hematoxylin and eosin. One section from each segment was digitised using a binocular microscope to measure gonad and trophosome areas (Jandel Scientific's SigmaScan-Pro v4.01).

3. How do body and reproductive condition compare in different habitats? In June 2008, we systematically sampled vents in three vent fields of Endeavour Segment where a range of morphological types of *R. piscesae* lives in close proximity. We paired the samples, "Hi and Lo" (Table 1), such that collections in a pair were within 4 m and located on the same structure or seafloor crack; thus, the fluid source was the same but delivery rate differed. Temperatures were measured at the branchial plumes of

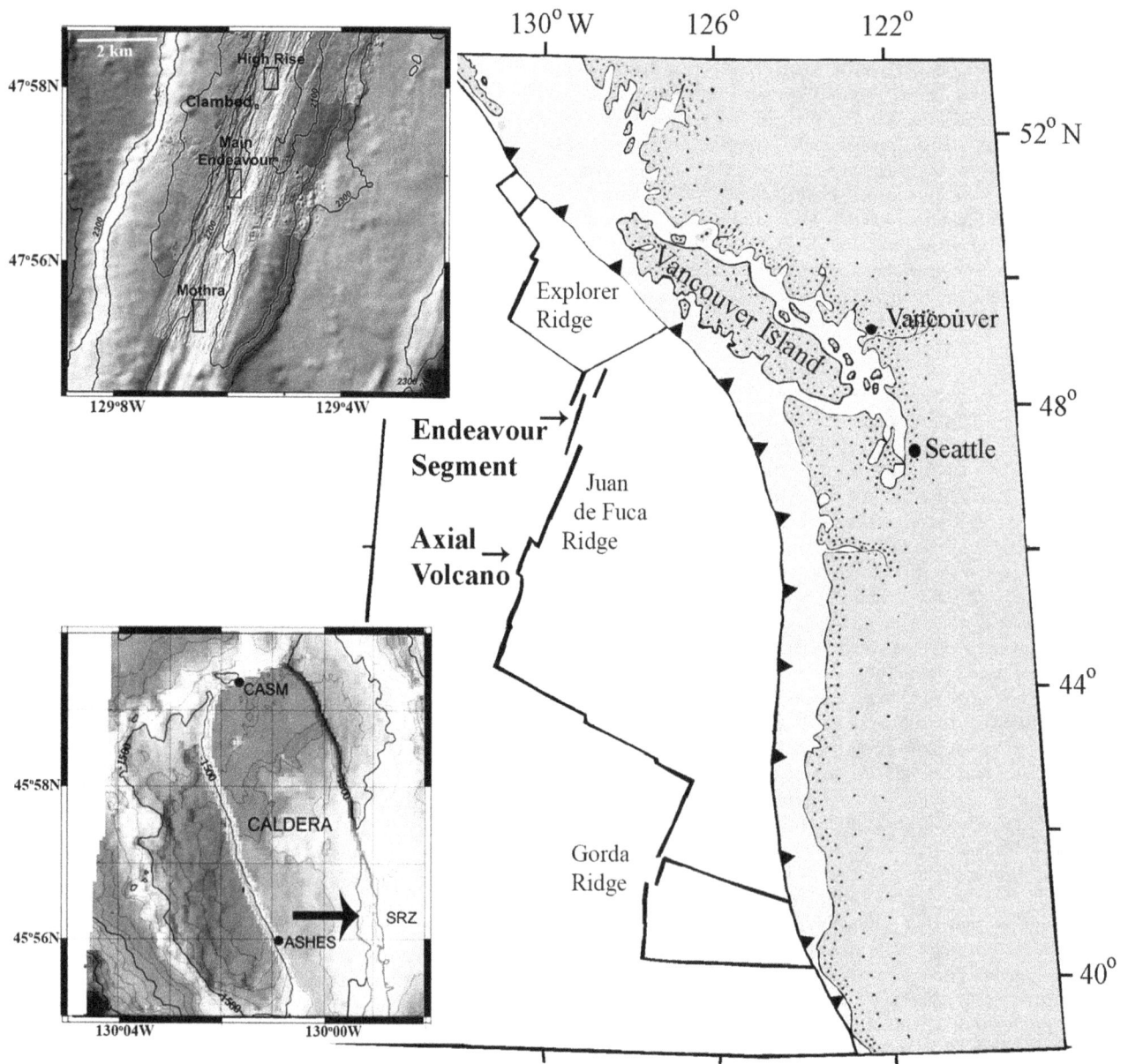

Figure 1. Locations of sampling sites on Juan de Fuca Ridge. Endeavour inset shows the three sampled vent fields: Clam Bed, Main Endeavour and Mothra. Axial inset shows locations of samples from three fields and the grey line near SRZ (arrow) delineates the extent of the 1998 lava flow.

Ridgeia when the probe found the highest consistent measurement for over a minute. Tubes were grasped by manipulator at the base and moved to closable boxes for recovery. The 25 largest worms were selected, the tubes split and preserved in 95% ethanol. In these Hi/Lo samples, the largest worms are most likely to be reproducing individuals that represent the high end of output from that location; this conservative approach reduced the differences between sample pairs as most worms in Lo samples were not reproductive. The following characters were assessed first: upper tube diameter, vestimentum diameter, obturaculum-vestimentum length, trunk length, wet weight of total body and of trunk alone, branchial plume condition and presence of sperm bundles in the vestimental groove. In some low flux specimens, the base of the trunk was missing so length is a conservative estimate. Plume condition was rated as: still developing, or little to extreme predation (0 to 3). Non-parametric tests (Mann-Whitney) in 'R'

software examined differences between Hi and Lo characters. We generated the logistic curves in 'R' then optimized the logit model with the Excel Solver add-in. A Principal Components Analysis used normalized data with a maximum of five axes to determine common groupings.

The trunk was opened to determine gender and gonad development. An animal was deemed "non-reproducing" if gametes were absent in the anterior ovisacs or sperm sacs (but gametes might be seen in the gonad). To assess the relevance of blotted wet weight as a measure of trunk condition, 36 worms were processed for ash-free dry weight by first drying then burning; ash-free dry weight was highly correlated to dry weight ($r^2 = 0.996$).

4. <u>Do lipid content and gamete development reflect habitat?</u> In summer 2009, additional specimens were collected from four sites at Axial and Endeavour (Table 1). The largest 4 males and 4 females from each site were frozen at $-80°C$. Lipids (a measure of

gonad investment; [36]) were extracted by homogenizing bodies in chloroform and methanol then diluting to 2:2:1.8 (c:m:water) [37]. The lipid-containing layer was washed several times with a stock inorganic layer solution and dried under nitrogen at 50°C. Three formalin fixed females from each of high, moderate and low flux sites were sectioned at 0%, 30% and 60% of trunk length. We tested whether oocyte size varied among worms by measuring cross sectional area (Image J© software) of oocytes sectioned through the nucleus within the ascending oviduct. Total gonoduct length was measured in the dissected females with a "packing distance" of 60 μm assigned to each egg along the gonoduct. The section counts were applied in three portions to derive an overall estimate of egg number as a measure of fecundity.

Results

Ridgeia growth at post-eruption vents on Axial

New vents created by the eruption were emitting fluids over 20°C with substantial levels of dissolved sulphide (Table S1) while surviving vents were cooler. In general, there is more sulphide at higher temperatures, a relationship evident across a much wider suite of samples at Axial [34]. In the following two years, heat and sulphide levels dropped as the volcano cooled. In the first year, tubeworms at these new vents were fatter and shorter than the older animals in less vigorous flow (Figure 2A). Settling larvae recruited and grew to 23 cm length within 280 days (Figure 3A) at which time, of 59 animals, 60% were already producing gametes. In the following two years as sulphide flux diminished somewhat, growth remained vigorous (at a minimum of 50 cm/yr) but the tube form elongate greatly (Figure 2B). As tubes lengthened, so did the trunk ($r^2 = 0.96$, n = 57). One collection at Nascent Vent in 2001 revealed that all long worms had died and empty tube surfaces hosted a second generation of recruits over 30 cm long. The first generation lasted three years.

Maturation of Ridgeia piscesae

We examined trophosome, gonad extent and gamete formation (Table 1) with onset of reproductive maturity in smaller animals. In 19 of 43 females and in 16 of 51 males, no gametes were seen in the sections; not until over 4 mm diameter did a majority of the females have oocytes when gonad began to enlarge. Only four individuals were over 7.5 mm diameter in this sample set but comparison with R. piscesae from the 2009 collection shows at larger sizes (over 10 mm vestimentum diameter) there is a marked increase in gonad extent, most notable in males (Figure 4). Trophosome area increased variably with vestimentum size (Spearman r = 0.58, p<0.01) to a maximum of 50% of the body area; there was no significant correlation with gonad size.

Morphological variation between high flux and low flux habitats

Short, fat R. piscesae in vigorous flows at Endeavour occurred mostly as small clumps on sulphide chimneys with black smokers nearby (Video S1). Temperatures for these samples ranged up to 30°C (Table 2); in 1995, we had scanned the site of the HiC sample to measure sulphide levels around 90 μM sulphide at 18°C. Tubes were white and parchment-like with bright red branchial plumes that extended from nearly all individuals (Figure 3B, D). In contrast, small worms in lower temperatures with little visible flow were much more abundant on these chimneys while the longest worms grew on basalt, forming extensive "fields" (Figure 3E). Here, the obturactula had pale branchiae with very short lamellar filaments (Figure 3C) and were capped with chitin-like plates.

Tubes from the Endeavour collections (total N = 379) showed width/length characteristics that were distinctive between high and low flux with the former resembling the Axial post-eruption worms (Figure 5A). Specimens from high flux samples were significantly wider and shorter than low flux in all but one sample (Figure 5B). The muscular vestimentum constructs the tube thus there is a strong correlation between anterior tube diameter and vestimentum width for both high flux ($r^2 = 0.91$) and low flux ($r^2 = 0.96$) tubeworms. Vestimentum width in high flux individuals was significantly larger than low flux (p<0.05, Mann-Whitney) for all pairs except one site. Similarly, most characters measured between pairs differed significantly with site G showing smaller differences (Table 2; Table S2). The longest worm was 60.5 cm in a 125 cm tube from LoA weighing 1.5 gm wet wgt; the heaviest worm was 6.0 gm with a body length of 19.6 cm in a 47 cm tube from HiG. Overall, the measurements defined wider, heavier but shorter worms from the high flux collections. Wet weight was a good predictor of trunk dry weight (dry = 0.142*wet +0.022; Pearson $r^2 = 0.91$, n = 36).

Not all samples from high flux habitats were similar. In general, sample HiA had the largest worms measured by all characters while HiB, HiD and HiH were significantly smaller than other high flux samples. For most characters, the smallest individuals were from LoC, LoD and LoH (Table 2; Table S2). The

Table 1. Sample collection and preservation data for the four questions addressed.

Question	Year	Site	# samples	# worms	Preservation	Comments
Tube growth post settlement	1998	Axial	7	74	BF to 70% EtOH	Sampled 8 months post eruption.
	1999	Axial	2	52	BF to 70% EtOH	Resampled recruits at two new vents
	2000	Axial	2	80	BF to 70% EtOH	And again the following year
Gonad and trophosome development	1998–99	Endeavour	5	94	BF to 70% EtOH	Combined samples from chimney walls
Condition in high and low flux	2008	Endeavour	16	378	95% EtOH	8 pairs of "high flux" and "low flux" samples collected in close proximity
Lipid content, gamete staging	2009	Axial	3	35	−80°C, BF	Equal sex numbers
	2009	Endeavour	1	18	−80°C, BF	Equal sex numbers

BF = 10% buffered formalin; EtOH = ethanol.

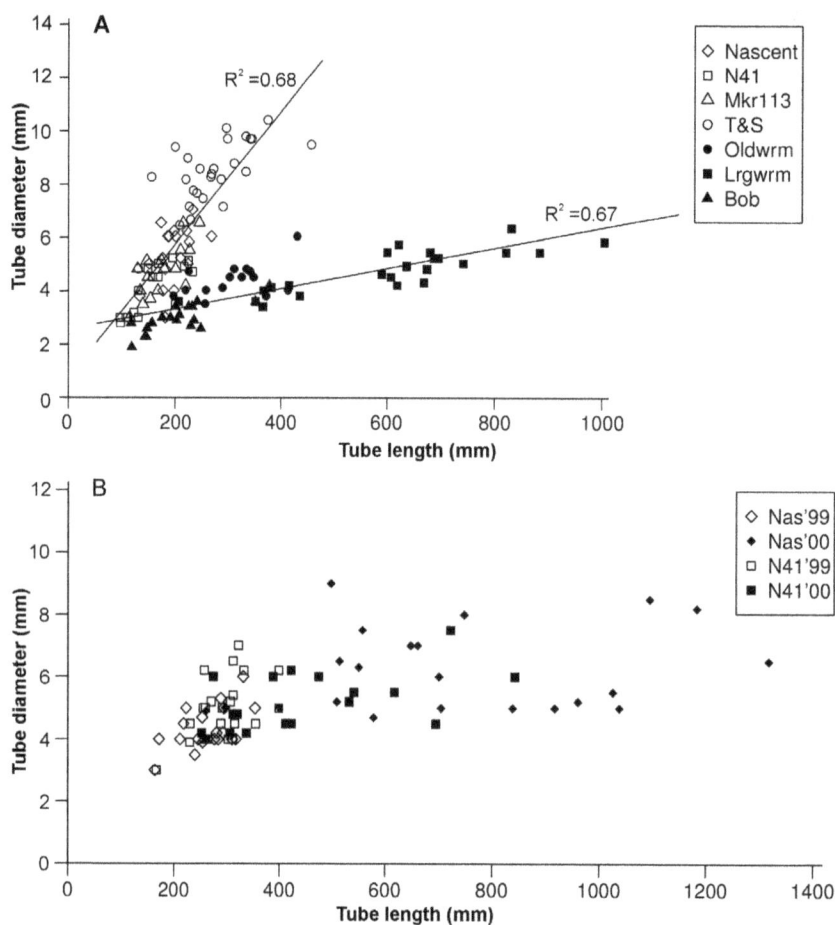

Figure 2. Tube characteristics of *Ridgeia piscesae* **from Axial Volcano 1998 to 2000.** The eruption in late January 1998 formed new vents rapidly colonized by tubeworm larvae. Growth in high flux is rapid and tube form is short and fat. As measured sulphide levels diminish through year 2000, tube form changes to resemble that of surviving worms on old lavas. A. Comparing *Ridgeia* samples from 8-month old vents on new lavas (open symbols) to *Ridgeia* from established vents (filled symbols). Lines are linear regressions on the new and old samples with Pearson correlation values shown. B. Largest worms at Nascent and N41 sampled at 18 and 30 months post eruption for comparison with 1998. In 1999, the tubes remain stocky; in 2000, a greatly elongated growth occurs in diminishing sulphide with little further increase in diameter.

remaining low flux samples were characterized by long, very thin trunks that left most of the tube space empty. A PCA using four characters (obturaculum-vestimentum length, vestimentum width, trunk length and trunk wet weight) identified four groups: Hi samples, Lo samples, HiA and the Lo samples that were immature (Figure 6). The first two axes accounted for 96% of the variability. The HiB, HiD and HiH trio, in which less than half the animals were reproductive, fall near the immature Lo group.

Predation and Branchial Condition

Overall, 44% percent of the 379 worms had excellent branchial condition with well-developed lamellar filaments on the obturaculum and no evidence of predation. Within high flux samples, nearly all animals had the highest condition score (Table S2) reflecting the bushy red plumes in Figure 3 (B, D). The exception was HiB in which many animals had grazed lamellae. In comparison, animals from the low flux sites had evidence of predation including lamellae completely shaved on parts or all of the obturaculum and chunks of tissue missing. For worms in samples LoC, LoD and LoH, and also in half of the LoB sample, we determined that nearly all branchiae had the distinctive features of a developing juvenile [18]. Most Lo worms had many chitinous plates on the obturaculum apex to plug the tube as an

anti-predator device; these plates were rare in Hi worms. In one collection pair, we assessed the presence of predators: the HiC and LoC collections yielded 562 and 720 other macrofauna, respectively. Polynoid polychaetes, mostly *Lepidonotopodium piscesae*, were the predators: in HiC they constituted 4% of the wet weight of this additional fauna while in LoC, they were 22% of the wet weight (26 individuals) thus reflecting field observations of higher abundance in lower flux conditions.

Reproductive Features and Body Condition

We observed tethered sperm bundles streamed from worms in high fluid flux several times. Ovoid bundles attached to strands up to 15 cm long bounced vigorously in the turbulence over the branchial plumes (Video S1), each traceable to an individual worm. Bundles showed no swimming or mobile behaviour, thus fertilization appears to occur between individuals within strand range. Among the sampled worms, 37% of high flux females and 36% of high flux males held sperm bundles compared with only 5% of low flux females and 0% of low flux males. All high flux samples had at least some females with sperm masses in the vestimental fold while only three low flux sites had females with masses (Table 2). In most high flux tubeworms, the gonad was full of gametes and occupied a large part of the trunk whereas in low

Figure 3. *Ridgeia piscesae* **on Juan de Fuca Ridge.** Images taken using the vehicle *ROPOS* (Canadian Scientific Submersible Facility).A. Eight months post-eruption at Nascent Vent, South Rift Zone, Axial; scale 10 cm. B. Branchial plume with white obturaculum of high flux *R. piscesae*; scale 1.5 cm. C. Sparse branchial plumes in low flux grazed by polynoid polychaetes (top centre); scale 1.5 cm. D. Clump of short-fat *R. piscesae* on the side of a smoker chimney; fluids emerging from ledge below. Orange polychaetes are *Paralvinella palmiformis*, a microbial grazer; scale 5 cm. E. Extensive tubeworm clumps on basalts in weak fluid flow between chimneys; image about 2 m across.

flux animals most had little gonad and, in some low flux tubeworms, only the top of the gonad held gametes. All males and over 80% of the females from high flux sites had gametes in the upper storage sacs and were deemed to be reproductively active with the exception of sample HiB from Clam Bed (Table 2).

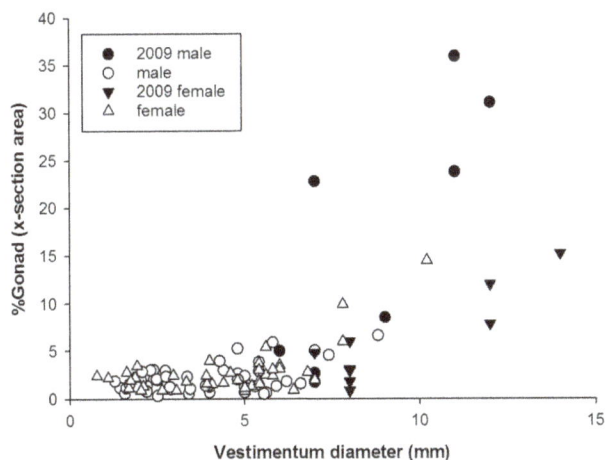

Figure 4. Estimate of gonad extent with increasing size. The majority of animals in our initial samples were small with little gonad development in both sexes (open symbols). Each measure of gonad area is the mean of 10 cross-sections equidistant along the trunk. Several worms from the 2009 study (filled symbols) augment numbers of larger individuals to illustrate gonad extent in full maturity.

From low flux sites, no worms were reproductive in samples LoC, LoD or LoH (about 20% had some gametes). High flux samples had a significantly higher proportion of reproductive individuals than low flux samples (Mann-Whitney test, p<0.05).

The largest lipid values occurred in females from a high temperature setting (26°C) (Figure 7A). Overall gonad volume was much larger in high flux individuals than moderate (10°C) or low (4°C) flux (Mann-Whitney, p<0.01) and oocytes in the oviducts more abundant in high flux (Figure 7B, C). The overall estimate of fecundity was 56,000 oocytes in high flux versus 27,000 in moderate or low when oocytes were present; however, only high flux worms had oocytes in the ovisacs ready for release. There were no significant differences among oocyte sizes from the three sites. Similarly, in males, the high flux animals displayed much greater amounts of gonad and trophosome as illustrated in the trunk cross-sections in Figure S1.

Of the 379 worms assessed for developmental state in the paired samples, 171 were considered non-reproducing with empty ovisacs or seminal vesicles; all these animals, except two, were under 6.7 mm tube diameter. All other tubeworms over 6 mm diameter were reproductive with abundant gametes. Figure 8 illustrates that an individual is likely to become reproductive between 5 and 6.5 mm diameter based on a logistic model that fits the observations well (Hosmer-Lemeshow test of goodness of fit, p< 0.01). The model combines all individuals, however, the likelihood of individuals being reproductive between the size of 5 and 6.5 mm was significantly lower in low flux samples (Chi-Sq; p< 0.01). The proportion of reproductive individuals within a sample follows the logistic curve, $p = 1/(1+EXP(-1*(-11.3+2.1*ATD)))$ where ATD is anterior tube diameter (Figure 8).

Table 2. *Ridgeia* sample location and body data for 2008 paired samples from Endeavour Segment.

Site	Sample	Temp °C	N	Vest Width mm	Sig	Trunk Weight gm	Sig	Trunk/OV Weight ratio	Sig	Sperm Bundles ♀, ♂	Repro-ductive %
Clam Bed	HiA	27.0	25	8.1 (0.2)		3.29 (0.31)	**	2.42 (0.17)		7, 3	96
	LoA	2.4	18	5.5 (0.2)		1.30 (0.17)		2.56 (0.23)		2, 0	67
	HiB	n/a	25	4.8 (0.1)	**	0.47 (0.05)	**	1.35 (0.07)	**	2, 0	44
	LoB	n/a	20	4.5 (0.1)		0.76 (0.08)		2.34 (0.18)		1, 0	30
Main Field	HiC	10.0	25	8.0 (1.0)	**	2.15 (0.16)	**	2.17 (0.13)	**	8, 9	96
	LoC	5.0	23	4.3 (0.6)		0.05 (0.01)		0.88 (0.06)		0, 0	0
	HiD	30.0	25	5.4 (0.2)	**	0.55 (0.05)	**	1.37 (0.08)	**	2, 0	92
	LoD	3.6	24	2.5 (0.1)		0.03 (.003)		0.69 (0.07)		0, 0	0
	HiE	30.0	20	7.6 (0.2)	**	2.02 (0.17)	**	2.00 (0.15)	**	4, 2	100
	LoE	11.4	26	4.3 (0.1)		0.34 (0.03)		1.31 (0.11)		2, 0	23
	HiF	n/a	25	7.8 (0.2)	**	2.52 (0.05)	**	2.43 (0.16)	**	9, 11	100
	LoF	n/a	25	4.3 (0.1)		0.38 (0.05)		1.32 (0.09)		0, 0	12
	HiG	n/a	24	7.4 (0.2)	**	1.79 (0.19)	**	2.02 (0.17)	**	4, 3	100
	LoG	n/a	24	4.9 (0.1)		0.87 (0.05)		2.07 (0.11)		0, 0	20
Mothra	HiH	21.3	25	6.0 (0.3)	**	0.90 (0.12)	**	1.68 (0.09)	**	5, 0	100
	LoH	5	25	3.4 (0.1)		0.14 (0.02)		1.25 (0.13)		0, 0	0

Temperature is the highest measured near the branchiae of the sampled individuals (n/a: probe not available). Body measurements are mean (st. err.) values for the sample. Vest = vestimentum; OV = obturaculum + vestimentum.
** sig p<0.01 Wilcoxon rank (vertical asterisks: Lo value is greater than Hi). The number of sperm bundles within the vestimentum fold is shown for females and males. An animal is considered reproductive if there were gametes in the anterior ovisacs or sperm sacs. The largest animals in each collection were used to maximize reproductive condition.
n/a = not available.
** sig p<0.01 Wilcoxon rank that Hi-Lo pairs differ; * p<0.05.

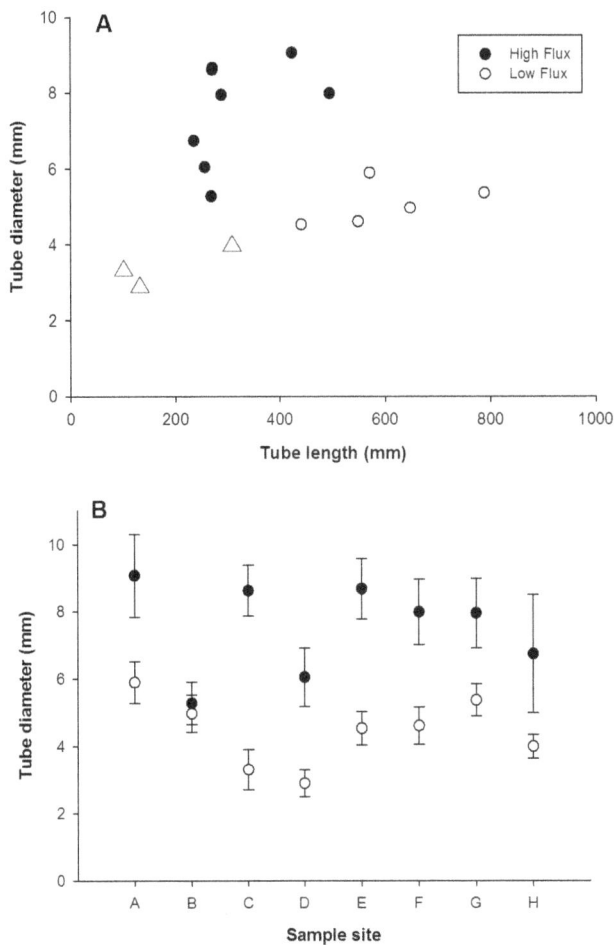

Figure 5. Comparison of tube characteristics between high fluid flow and low flow samples. Each sample represents 20 to 25 of the largest tubeworms to capture maximum growth extent. Tube diameter is highly correlated with vestimentum diameter and usually about 400 μm larger. A. Sample means plotted against tube length for comparison with Figure 2 illustrate that low flux animals are smaller diameter at length. Triangles are low flux samples in which all individuals were non-reproductive and deemed immature: LoC, LoD and LoH. B. In paired samples at each site members of each pair are within metres on the same structure. Bars are standard error. All Mann-Whitney paired tests show significant difference (p<0.01) between high and low flux samples except Site B.

Discussion

A major feature of the *Ridgeia piscesae* population is morphological variation that corresponds to habitat: fast-growing, large animals have high reproductive success and fecundity in higher temperatures (=dissolved sulphide, as in Butterfield et al. [34] and Urcuyo et al. [38]). Long, thin worms with low fertilization and reproductive readiness occur in low temperatures. Abundant small intermediate forms are mostly immature. Morphological differences relate to relative fitness in different habitats. While the primary environmental driver of variation is the flux of dissolved sulphide, a secondary driver is habitat stability as locations (or times) of high flux may be short-lived or subject to substratum disruption. Recruitment, growth and maturation of a short-fat tubeworm phenotype happens within months after volcanic eruption when temperatures were relatively high. On Axial, and also at an adjacent eruption site in 1994 [39], *R.*

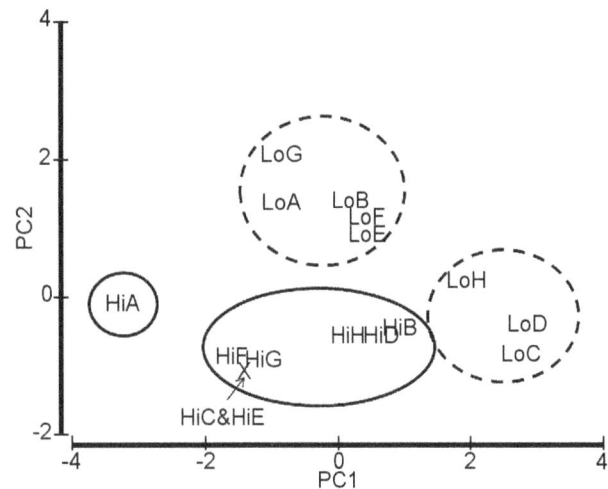

Figure 6. PCA plots paired samples based on morphological features. Using five principal components and four variables representing obturaculum-vestimentum and trunk size four groups of samples separate. PC1 explains 69% of the variability which increases to 96% with the addition of PC2. LoC, LoD and LoH were identified as immature juveniles; HiB also has many immature individuals.

piscesae grew at rates over 50 cm per year and were dead after three years. The tube character changed as sulphide decreased, becoming long and recumbant, likely to bring the branchiae closer to the emerging fluid. On sulphide chimneys, fluid redirection opens new vigorous outlets inducing colonization and high growth rates or causing abrupt death due to high heat or substratum instability [20]. Where low flow is sustained for long periods, tubeworms can grow over 2 m in length as at Clam Bed. Here, growth of marked individuals is very slow; using the upper 10% of measured rates, Urcuyo et al. [40] give conservative estimates of life-spans between 10 and 40 years stating larger animals are likely older.

The branchial plume of the vestimentiferan is the uptake site for oxygen and metabolites that support the symbiotic bacteria in tubeworms [25]. Predation, however, takes a large toll on low flux *R. piscesae* plumes; Urcuyo et al. [38] note predation on *R. piscesae* from Clam Bed at ~95% frequency. They also determine that this phenotype can take up sulphide through a thin-walled posterior extension of the tube that accesses this critical nutrient where hydrothermal fluid exits basalt cracks. In large low flux worms, we see a long trunk that would increase potential uptake area especially when branchiae are reduced. This functional adaptation that reduces reliance on the branchiae is facilitated by backward growth of the tube and trunk.

Our novel observation of the sperm mass trailing from males in turbulent flows completes our understanding of the fertilization process begun by Southward and Coates [15] and MacDonald et al. [17] and underscores the importance of branchial capture. We find no evidence of periodicity in gametogensis so assume that fertilization occurs when sperm are available. The presence of oocytes in ovisacs indicates reproductive readiness in females, however, not all released oocytes develop to embryos, especially from small females [16]. Many traits of fitness indicate that the high flux, short-fat phenotype has a high contribution to the next generation compared to other phenotypes (long, thin or short) (Table 3). "Live long and prosper" is not an option for this species in a time-and-space variable environment. Three scenarios appear to exist: i) early maturation, reproduction and death in high nutrient habitat, ii) slow growth and long life with low reproductive

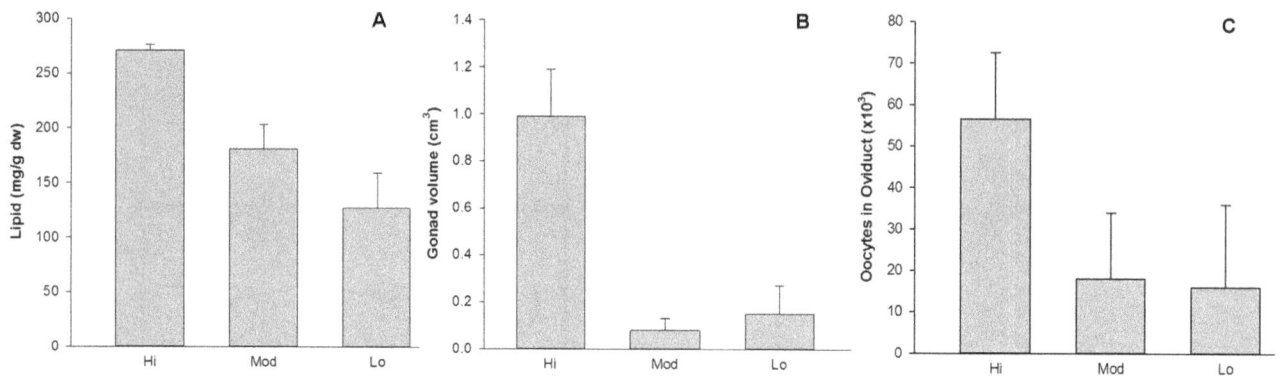

Figure 7. Female reproductive condition in *Ridgeia piscesae* in high, moderate and low flux habitats. Indicators reflect that females from the high flux samples have greater reproductive output; bars are sd. A. Total lipid content in trunks of 12 individuals. B. Gonad volume estimated from trunk sections. C. Total oocytes estimated from section counts and gonad volume. Neither moderate nor low flux worms had oocytes in ovisacs ready for release.

output sustained by regular low sulphide supply, or iii) remaining a small form that never reaches maturity due to inadequate nutrient flux, predation and/or habitat instability. While it is possible that, over time, oocyte output from a long-lived tubeworm could approach that of a short-fat phenotype, our data suggest that fertilization success is likely very low in worms with predator-cropped branchiae. Thus, low flux habitat is occupied by worms with minimal reproductive output while the high flux habitat patches support high generational turnover with constant contribution of progeny.

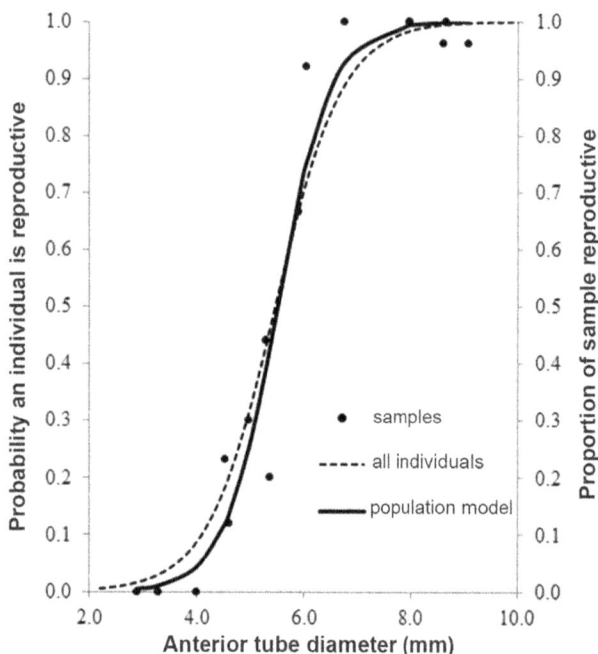

Figure 8. Model of reproductive readiness based on tube diameter. Onset of reproduction with size follows a logistic curve. The likelihood that an individual is reproductive at a given tube diameter is shown in the dotted line (left axis) using all individuals in our 2008 study. Tube diameter is highly correlated body characters but is an easier trait to measure. The right axis represents the proportion of the Endeavour samples that were reproductive (dots) and the full logistic curve (solid line) is the best fit for any sample taken.

The demographic potential of a patch in a metapopulation depends on the quality of the underlying habitat [41]. The sulphide-rich habitat of robust, reproductive *Ridgeia piscesae* is both limited and unstable. It occurs mostly in small patches on black smoker chimneys where new heat bursts and smoker growth sponsor rare, short-lived clumps of worms or in bursts of heat/sulphide through basalt after tectonic or volcanic events. On large chimneys, small tubeworms occupy much of the surface. Sarrazin and Juniper [32] mapped the structure that we sampled a decade later (Vent "C"); low temperature *R. piscesae* dominated biomass but 95% of those they measured were under 7.5 cm body length, thus very small diameter. Only our LoD worms were this small – animals that were non-reproductive. Even in trying to find large low flux *R. piscesae*, we still collected immature samples from chimneys. The evidence that the *Ridgeia piscesae* metapopulation at Endeavour has a source-sink structure is strong. The source is optimal high sulphide habitat that is limited but has a high throughput of very productive animals, thus the rate of offspring production per m^2 is far higher than low flux habitat. Habitat surveys on seven chimneys document about 2% as high flux tubeworm habitat compared to over 80% of the surface with low flux worms [42]. Thus, while the sheer abundance of low flux worms on chimneys is substantial, the animals rarely mature before flow ceases; repeated mapping illustrates these surface changes [12,42].

On basalts, we estimate 2% of the tubeworm coverage as high flux phenotypes from sporadic overflight imagery. Suitable high flux habitat is rare although Clam Bed is an interesting exception where the longest low flow worms also occur in a stable vent setting. Here, we find the highest reproductive condition for low flow among the selected large worms although predation was high and likely to impede fertilization. Urcuyo et al. [38] find over 70% of 500 worms from their Clam Bed grab were under 4.0 mm diameter and, therefore, not reproducing (Figure 8); the largest was 6.5 mm tube diameter, thus, overall reproductive output is limited to few individuals. Suitable high flux habitat is rare enough that contribution from low flow, long-lived animals is probable despite the finding that most low flux habitat animals are non-reproductive. Overall metapopulation fitness may be higher when most recruits are generated regularly from high fecundity, short-lived individuals in small habitat patches, but long-lived individuals in poor quality, stable habitat will store genetic diversity generated from successful recruits over many years. Indeed, the marginal habitats may perform a rescue function through a few

Table 3. Traits in *Ridgeia piscesae* that influence individual fitness.

Functional Trait	Attribute	Fitness Component	Notes
Threshold reproductive size	smaller in Hi	Reproductive probability	diameter significantly different between phenotypes
Maturation rate	sooner in Hi	Reproductive probability	months versus years between phenotypes
Growth rate	faster in Hi	Reproductive probability	varies over two orders of magnitude among phenotypes
Gametic tissue allocation	greater in Hi	Embryo production	increases with size; high flux phenotype grows larger
Physiological condition	variable	Embryo production; Survivorship	lipids greater in high flux but see adaptation of haemoglobin levels and trunk dimensions in low flux phenotype
Branchial condition	better in Hi	Embryo production; Survivorship	healthy branchiae enhance metabolite uptake and fertilization; growth of defense structures in low flux phenotype
Aging rate	slower in Lo	Survivorship; Recruitment	low turnover (decades) in low flux enhances individual fitness and provides long-term settlement surface for recruits.

The main environmental driver is the level of dissolved sulphide flux to sustain the bacterial symbionts. Habitat stability, which is low in high flux habitat, is likely an additional factor. This study did not assess factors that influence recruitment success (a fitness component). Most attributes are assessed in this study but additional literature information includes growth rates [38,39], hemoglobin levels [27] and sulphide uptake [38].

offspring repopulating high flux habitat after times of high regional habitat instability.

Despite the high proportion of marginal habitat, we do not see fixation of the low flux phenotype as might be predicted if the phenotypes are a result of genetic specialization (e.g. Dias [43]). Given this point, the observations presented here, plus past studies, it is probable that variation is a plastic response to habitat. Phenotypic plasticity is adaptive when the appropriate phenotype in each environment has a higher fitness than the alternative phenotype [44]. If plasticity exists in *Ridgeia piscesae*, the morphological response is expressed in at least two ways: i) alteration of the trunk and tube to optimize the main site of sulphide uptake (branchiae versus tube base); and ii) formation of stiff caps on the obturaculum that plug the tube when predators are more abundant: an inducible defense trait. Additional physiological plasticity in levels of haemoglobin gene expression [27] and of certain amino acids [26] would relate to the mobilization of dissolved sulphide within the body. The plasticity cost to the low flux habitat worms lies in the resources needed to construct additional structural parts but the long, thin morphology confers greater fitness in low sulphide supply. An overall fitness is likely balanced by relative habitat frequency and the rate at which propagules are delivered to the gene pool. Plasticity is common in sessile organisms occupying a variable environment in which individual fitness can increase by modification of a morphological or physiological trait [45]. Given the nature of unpredictability of high quality habitat, it is adaptive to maintain a reaction norm that includes fitness in low quality habitat so that part of the population will survive a major disturbance in this volatile setting. However, further work through reciprocal transplants is necessary to prove that a tubeworm can alter its form under different fluid flow conditions.

Consequences of phenotypic variation (plastic or otherwise) on ecological processes can be profound [46]. *R. piscesae* is a foundation species that provides structured habitat for many species in which tube surface area, tube form and surface character shape the associated communities as does tube longevity. The architecture of the tubeworm bush is dictated by phenotype; bushes with small twisted worms of several recruit generations host the greatest number of species while the long, thin phenotype hosts the least [14]. Thus, the indirect effects of a range of morphotypes on the vent community are notable.

Consideration of metapopulation structure including the disposition of sources and sinks should be a factor in conservation [43,47]. This work was conducted with a Marine Protected Area with zoned uses under Canadian jurisdiction. We recommend that a management strategy include monitoring of human impacts on high flow habitat in all zones. As mining scenarios approach reality for the sulphide chimneys that form the habitat for many hydrothermal vent animals [31], application of appropriate strategies for biological reserves that support diversity protection and restoration potential is critical. Dynamic, changeable landscapes present further challenges for metapopulations in which suitable habitat persistence may be highly variable [48]. The cumulative effects of anthropogenic and natural disturbance bears further consideration in vent ecosystems.

Supporting Information

Table S1 Maximum temperature and dissolved sulphide concentrations from vents on Axial Volcano before and after the eruption of January 1998. Measurements were taken in July to September of each year. *Vents with asterisk remained from pre-eruption; Sonne Vent was paved over by the eruptive lavas. n/a = not available.

Table S2 Additional measurements on *Ridgeia piscesae* from Endeavour Hi/Lo samples. Body measurements are mean (st. err.) values for the sample. Obt-Vest = Obturaculum + Vestimentum. Predation is a qualitative scale from 0 (none) to 3 (major damage).

Video S1 Tubeworm sperm bundles released on long filaments in turbulent flow. A large cluster of short-fat *Ridgeia piscesae* on a sulphide chimney at Endeavour Segment, Juan de Fuca Ridge (2190 m depth). The small black spire to the left is emitting hydrothermal fluid over 350°C. Initial image is about 1.5 m across. Zoom at 15 seconds shows tubeworms, alvinellid polychaetes and small white limpets. To the left, four sperm packets from *R. piscesae* are moving in the turbulent flow tethered to the males. Sperm packets are also present entangled in the red branchiae.

Figure S1 Cross-sections through upper trunk of male *Ridgeia piscesae* comparing long-skinny to short-fat. Scale bar is 1 mm. A. Long-skinny worm from Mod flux Site K (Axial). Trophosome is reduced and gonad is barely visible. (Vestimentum diameter, 7 mm.). B. Short-fat worm from high flux Site I (Endeavour). Trophosome and gonad occupy most of the trunk area. (Vestimentum diameter, 14 mm.). c – coelom; cf – coelomic fluid; dv – dorsal blood vessel; fm – feather muscle; g – gonad; t – trophosome.

Acknowledgments

We thank ship and submersible vehicle crews from Coast Guard Canada, UNOLS fleet, WHOI (*Alvin*) and CSSF (*ROPOS*) and the following Chief Scientists who facilitated sample collection: R.W. Embley, C. Fisher, J. Holden, S.K. Juniper and R. Lee. J. Rose assisted in laboratory analyses. AH thanks P.A. Tyler for his insights. We acknowledge discussion opportunities through the NSERC Canadian Healthy Oceans Network (CHONe) meetings. We particularly thank T. Vines, J. Shurin and two reviewers at Axios (axiosreview.org) for commentary that improved the manuscript.

Author Contributions

Conceived and designed the experiments: VT CSG AH. Performed the experiments: CSG VT AH. Analyzed the data: VT CSG AH. Contributed reagents/materials/analysis tools: VT AH. Wrote the paper: VT CSG. Edited manuscript: AH.

References

1. Hanski I (1999) Metapopulation Ecology; May RMaHPH, editor. Oxford: Oxford University Press. 313 p.
2. Kawecki T (1995) Demography of source—sink populations and the evolution of ecological niches. Evolutionary Ecology 9: 38–44.
3. Sultan SE, Spencer HG (2002) Metapopulation structure favors plasticity over local adaptation. American Naturalist 160: 271–283.
4. Hanski I, Mononen T, Ovaskainen O (2011) Eco-Evolutionary Metapopulation Dynamics and the Spatial Scale of Adaptation. The American Naturalist 177: 29–43.
5. Tienderen PHV (1997) Generalists, specialists, and the evolution of phenotypic plasticity in sympatric populations of distinct species. Evolution 51: 1372–1380.
6. Pigliucci M (2005) Evolution of phenotypic plasticity: where are we going now? Trends in Ecology & Evolution 20: 481–486.
7. Kawecki TJ (2008) Adaptation to marginal habitats. Annual Review of Ecology, Evolution, and Systematics 39: 321–342.
8. Chevin LM, Lande R (2011) Adaptation to marginal habitats by evolution of increased phenotypic plasticity. Journal of Evolutionary Biology 24: 1462–1476.
9. Wiens JA, Van Horne B (2011) Sources and sinks: what is the reality? In: Liu J, V . Hull, AT . Morzillo and JA . Wiens, editor. Sources, Sinks and Sustainability. Cambridge: Cambridge University Press.
10. Johnson HP, Hutnak M, Dzlak RP, Fox CG, Urcuyo I, et al. (2000) Earthquake-induced changes in a hydrothermal system on the Juan de Fuca mid-ocean ridge. Nature 407: 174–177.
11. Embley RW, Chadwick WW Jr, Clague D, Stakes D (1999) 1998 Eruption of Axial Volcano: Multibeam anomalies and seafloor observations. Geophysical Research Letters 26: 3425–3428.
12. Sarrazin J, Robigou V, Juniper SK, Delaney J (1997) Biological and geological dynamics over four years on a high-temperature sulfide structure at the Juan de Fuca Ridge hydrothermal observatory. Marine Ecology Progress Series 153: 5–24.
13. Tunnicliffe V, Juniper SK (1990) Dynamic character of the hydrothermal vent habitat and the nature of sulphide chimney fauna. Progress in Oceanography 24: 1–14.
14. Tsurumi M, Tunnicliffe V (2003) Tubeworm-associated communities at hydrothermal vents on the Juan de Fuca Ridge, northeast Pacific. Deep-Sea Research Part I Oceanographic Research Papers 50: 611–629.
15. Southward EC, Coates KA (1989) Sperm masses and sperm transfer in a vestimentiferan, *Ridgeia piscesae* Jones, 1985 (Pogonophora: Obturata). Canadian Journal of Zoology 67: 2776–2781.
16. Hilário A, Young CM, Tyler PA (2005) Sperm storage, internal fertilization, and embryonic dispersal in vent and seep tubeworms (Polychaeta: Siboglinidae: vestimentifera). Biological Bulletin (Woods Hole) 208: 20–28.
17. MacDonald I, Tunnicliffe V, Southward E (2002) Sperm transfer in the vestimentiferan *Ridgeia piscesae* Jones: an event observed at Endeavour Segment, Juan de Fuca Ridge. Cahiers de Biologie Marine 43: 395–399.
18. Jones ML (1985) On the Vestimentifera, new phylum: six new species, and other taxa, from hydrothermal vents and elsewhere. Bulletin of the Biological Society of Washington 6: 117–158.
19. Galkin SV (1998) Morphological variability and taxonomic position of Vestimentiferans of the genus *Ridgeia* from the hydrothermal community of Axial Seamount, Juan de Fuca Ridge, Pacific Ocean. Russian Journal of Marine Biology 24: 313–319.
20. Tunnicliffe V, Garrett JF, Johnson HP (1990) Physical and biological factors affecting the behaviour and mortality of hydrothermal vent tubeworms (vestimentiferans). Deep-Sea Research Part A Oceanographic Research Papers 37: 103–125.
21. Southward EC, Tunnicliffe V, Black M (1995) Revision of the species of *Ridgeia* from northeast Pacific hydrothermal vents, with a redescription of *Ridgeia piscesae* Jones (Pogonophora: Obturata = Vestimentifera). Canadian Journal of Zoology 73: 282–295.
22. Southward EC, Tunnicliffe V, Black MB, Dixon DR, Dixon LRJ (1996) Ocean-ridge segmentation and vent tubeworms (Vestimentifera) in the NE Pacific. In: MacLeod CJ, Tyler PA, Walker CL, editors. Tectonic, Magmatic, Hydrothermal and Biological Segmentation of Mid-Ocean Ridges: Geological Society Special Publication. pp. 211–224.
23. Carney S, Peoples J, Fisher C, Schaeffer S (2002) AFLP analyses of genomic DNA reveal no differentiation between two phenotypes of the vestimentiferan tubeworm *Ridgeia piscesae*. Cah Biol Mar 43: 363–366.
24. Black MB, Trivedi A, Maas PAY, Lutz RA, Vrijenhoek RC (1998) Population genetics and biogeography of vestimentiferan tube worms. Deep-Sea Research II 45: 365–382.
25. Scott KM, Bright M, Macko SA, Fisher CR (1999) Carbon dioxide use by chemoautotrophic endosymbionts of hydrothermal vent vestimentiferans: affinities for carbon dioxide, absence of carboxysomes, and d13C values. Marine Biology (Berlin) 135: 25–34.
26. Brand GL, Horak RV, Bris NL, Goffredi SK, Carney SL, et al. (2007) Hypotaurine and thiotaurine as indicators of sulfide exposure in bivalves and vestimentiferans from hydrothermal vents and cold seeps. Marine Ecology 28: 208–218.
27. Carney SL, Flores JF, Orobona KM, Butterfield DA, Fisher CR, et al. (2007) Environmental differences in hemoglobin gene expression in the hydrothermal vent tubeworm, *Ridgeia piscesae*. Comparative Biochemistry and Physiology B Biochemistry & Molecular Biology 146: 326–337.
28. Liao L, Wankel SD, Wu M, Cavanaugh CM, Girguis PR (2014) Characterizing the plasticity of nitrogen metabolism by the host and symbionts of the hydrothermal vent chemoautotrophic symbioses Ridgeia piscesae. Molecular Ecology 23: 1544–1557.
29. Puetz L (2014) Connectivity within a metapopulation of the ecologically significant species, *Ridgeia piscesae* Jones (Annelida, Siboglinidae), from the Endeavour Hydrothermal Vents Marine Protected Area on Juan de Fuca Ridge. Victoria: University of Victoria.
30. Vrijenhoek RC (2009) Cryptic species, phenotypic plasticity, and complex life histories: Assessing deep-sea faunal diversity with molecular markers. Deep-Sea Research Part II Topical Studies in Oceanography 56: 1713–1723.
31. Boschen RE, Rowden AA, Clark MR, Gardner JPA (2013) Mining of deep-sea seafloor massive sulfides: A review of the deposits, their benthic communities, impacts from mining, regulatory frameworks and management strategies. Ocean & Coastal Management 84: 54–67.
32. Sarrazin J, Juniper SK (1999) Biological characteristics of a hydrothermal edifice mosaic community. Marine Ecology Progress Series 185: 1–19.
33. Bates AE, Tunnicliffe V, Lee RW (2005) Role of thermal conditions in habitat selection by hydrothermal vent gastropods. Marine Ecology Progress Series 305: 1–15.
34. Butterfield DA, Massoth GJ, McDuff RE, Lupton JE, Lilley MD (1990) Geochemistry of hydrothermal fluids from Axial Seamount Hydrothermal Emissions Study vent field, Juan de Fuca Ridge: subseafloor boiling and subsequent fluid-rock interaction. Journal of Geophysical Research 95: 12895–12921.
35. Thiébaut E, Huther X, Shillito B, Jollivet D, Gaill F (2002) Spatial and temporal variations of recruitment in the tube worm *Riftia pachyptila* on the East Pacific Rise (9°50'N and 13°N). Marine Ecology Progress Series 234: 147–157.
36. Hilário A, Tyler PA, Pond DW (2008) A new method to determine the reproductive condition in female tubeworms tested in Seepiophila jonesi (Polychaeta: Siboglinidae: Vestimentifera). Journal of the Marine Biological Association of the United Kingdom 88: 909–912.
37. Bligh E, Dyer W (1959) A rapid method of total lipid extraction and purification. Canadian Journal of Biochemistry and Physiology 37: 911–917.
38. Urcuyo IA, Massoth GJ, Julian D, Fisher CR (2003) Habitat, growth and physiological ecology of a basaltic community of *Ridgeia piscesae* from the Juan de Fuca Ridge. Deep-Sea Research Part I Oceanographic Research Papers 50: 763–780.
39. Tunnicliffe V, Embley RW, Holden JF, Butterfield DA, Massoth GJ, et al. (1997) Biological colonization of new hydrothermal vents following an eruption on Juan

de Fuca Ridge. Deep-Sea Research Part I Oceanographic Research Papers 44: 1627–1644.

40. Urcuyo I, Bergquist D, MacDonald I, VanHorn M, Fisher C (2007) Growth and longevity of the tubeworm *Ridgeia piscesae* in the variable diffuse flow habitats of the Juan de Fuca Ridge. Marine Ecology Progress Series 344: 143–157.

41. Figueira W, Crowder L (2006) Defining patch contribution in source-sink metapopulations: the importance of including dispersal and its relevance to marine systems. Population Ecology 48: 215–224.

42. Dancette R, Juniper S (2007) Biodiversity and research impacts at Main Endeavour and Mothra Fields. F1103-060079. Nanaimo: Fisheries and Oceans Canada

43. Dias PC (1996) Sources and sinks in population biology. Trends in Ecology & Evolution 11: 326–330.

44. Thompson JD (1991) Phenotypic plasticity as a component of evolutionary change. Trends in Ecology & Evolution 6: 246–249.

45. Dudley S (2004) The functional ecology of phenotypic plasticity in plants. In: DeWitt TJ, Scheiner S, editors. Phenotypic Plasticity: Functional and Conceptual Approaches. Oxford: Oxford University Press. pp. 151–172.

46. Miner BG, Sultan SE, Morgan SG, Padilla DK, Relyea RA (2005) Ecological consequences of phenotypic plasticity. Trends in Ecology & Evolution 20: 685–692.

47. Gaines SD, White C, Carr MH, Palumbi SR (2010) Designing marine reserve networks for both conservation and fisheries management. Proceedings of the National Academy of Sciences 107: 18286–18293.

48. Wilcox C, Cairns BJ, Possingham HP (2006) The role of habitat disturbance and recovery in metapopulation persistence Ecology 87: 855–863.

16

Microbial Ecology of Thailand Tsunami and Non-Tsunami Affected Terrestrials

Naraporn Somboonna[1]*, **Alisa Wilantho**[2], **Kruawun Jankaew**[3], **Anunchai Assawamakin**[4], **Duangjai Sangsrakru**[2], **Sithichoke Tangphatsornruang**[2], **Sissades Tongsima**[2]

1 Department of Microbiology, Faculty of Science, Chulalongkorn University, Bangkok, Thailand, **2** Genome Institute, National Center for Genetic Engineering and Biotechnology, Pathumthani, Thailand, **3** Department of Geology, Faculty of Science, Chulalongkorn University, Bangkok, Thailand, **4** Department of Pharmacology, Faculty of Pharmacy, Mahidol University, Bangkok, Thailand

Abstract

The effects of tsunamis on microbial ecologies have been ill-defined, especially in Phang Nga province, Thailand. This ecosystem was catastrophically impacted by the 2004 Indian Ocean tsunami as well as the 600 year-old tsunami in Phra Thong island, Phang Nga province. No study has been conducted to elucidate their effects on microbial ecology. This study represents the first to elucidate their effects on microbial ecology. We utilized metagenomics with 16S and 18S rDNA-barcoded pyrosequencing to obtain prokaryotic and eukaryotic profiles for this terrestrial site, tsunami affected (S_1), as well as a parallel unaffected terrestrial site, non-tsunami affected (S_2). S_1 demonstrated unique microbial community patterns than S_2. The dendrogram constructed using the prokaryotic profiles supported the unique S_1 microbial communities. S_1 contained more proportions of archaea and bacteria domains, specifically species belonging to Bacteroidetes became more frequent, in replacing of the other typical floras like Proteobacteria, Acidobacteria and Basidiomycota. Pathogenic microbes, including *Acinetobacter haemolyticus*, *Flavobacterium* spp. and *Photobacterium* spp., were also found frequently in S_1. Furthermore, different metabolic potentials highlighted this microbial community change could impact the functional ecology of the site. Moreover, the habitat prediction based on percent of species indicators for marine, brackish, freshwater and terrestrial niches pointed the S_1 to largely comprise marine habitat indicating-species.

Editor: Vasu D. Appanna, Laurentian University, Canada

Funding: The research was supported by Research Funds from the Faculty of Science, Chulalongkorn University, Under the A1B1-NS (RES-A1B1-NS-01), and Thai Aviation Refuelling Co., Ltd. The funders had no role in study design, data collection and analysis, decision to publish, or preparation of the manuscript.

Competing Interests: The authors received funding from Thai Aviation Refuelling Co., Ltd.

* E-mail: Naraporn.S@chula.ac.th

Introduction

Phra Thong island, Phang Nga province of southern Thailand (Figure 1), represents a location for comparative studies of tsunami (S_1) and non-tsunami (S_2) affected terrestrial ecosystems. The S_1 and S_2 shared nearby geographies separated by a hill, whereby S_1 terrain was inundated by the Indian Ocean tsunami on 26 December 2004 and S_2 unaffected; otherwise both were comparable based on geological characteristics [1,2]. The tsunami left an Andaman Sea-facing, S_1, distinguished terrestrial layer that was classified by geologist as a sand layer of 5–20 cm thick (layer A in Figure 1; [1]). Interestingly, geological evidence indicated three historic tsunamis also occurred prior to the 2004 tsunami at S_1, and none to S_2. The youngest recorded historic tsunami predating the 2004 tsunami was approximately 600 years ago (600yo) (layer B in Figure 1; [1]).

Each tsunami occurrence could affect the S_1 terrestrial characteristics due to the massive impact of seawater with marine organisms and garbage [1,3,4]. Studies comparing the 2004 tsunami affected versus non-affected (or pre-affected) terrestrials and terrestrial water reported the greater salinity, acidity, conductivity, turbidity and organic contents following the tsunami occurrence [5–7]. Studies also reported widespread disease-carrying vectors, such as mosquitoes, trematodes and snails, after

the 2004 tsunami [3,4]. Several bacterial and fungal infections involved skin and respiratory disorders were documented among repatriated tourists [8] and people working in the tsunami affected area [9]. In addition, the 2004 tsunami sediments consisted of higher concentrations of Mercury and Thallium [10,11]. Together, this chance of terrestrial characteristics could affect the microbial biodiversity and functional ecology.

Nonetheless, the impact of tsunamis on microbial diversity and ecology function remains ill-defined. The present study thereby analyzed the microbial biodiversity and their potential functional composites in the tsunami impacted S_1 terrain, in comparison to the non-affected S_2 site, using 16S and 18S rRNA genes pyrosequencing derived metagenomic DNA approach. For each site, the data included the prokaryotic and eukaryotic diversity profiles categorized into different depth levels corresponding to the terrestrial ages: 2004 tsunami, 1–300yo (pre-dating the 2004), 300–600yo, 600yo tsunami, and >600yo, respectively (starting from the top layer to a deeper layer), and also the amalgamated profiles for each site. Geologists determined the terrestrial age period from its depths below the land surface [1]. The overall results represent for the first time the use of metagenomics in analysing the prokaryotic and eukaryotic microbial biodiversity of the 2004 tsunami and non-tsunami affected terrestrials. Unlike

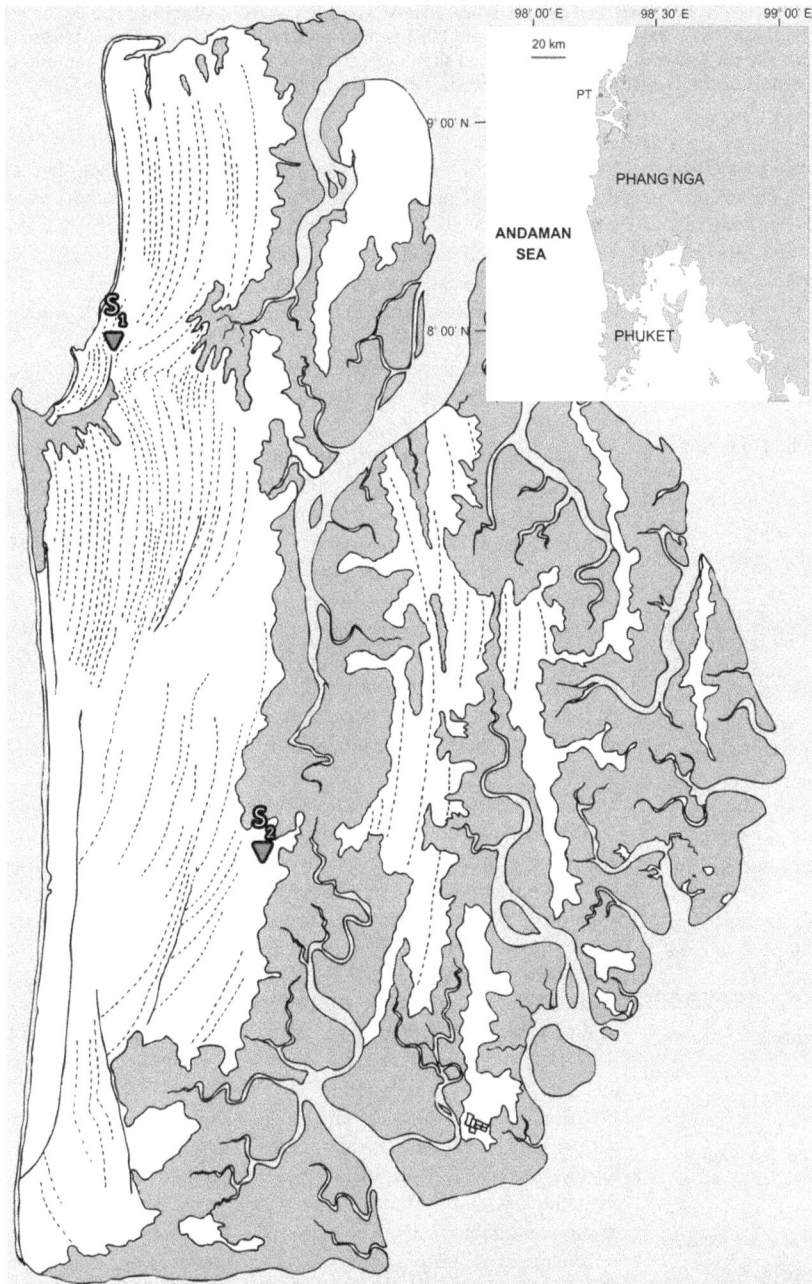

Figure 1. Index map of Phra Thong island relative to Phuket and terrestrial sites where samples were collected. The lower left photograph shows the pit wall of tsunami affected site (S₁). Light color sheets **A** and **B** represent 2004 tsunami and 600yo tsunami affected terrestrial layers, respectively [1]. The lower right photograph shows the pit wall of non-tsunami affected site (S₂) of the parallel geography, and samples of equivalent depths to those of S₁ were collected. Time period of the terrestrial is determined via sample depth [1].

traditional biodiversity study that is conducted via cultivation technique and could reveal merely less than 1% of the true microbiota, metagenomics is a culture-independent technique that has been proved worldwide a robust, reliable and comprehensive tool for obtainment of entire microbiota from diverse environmental and clinical samples [12–17].

Materials and Methods

Sample collection

The owners of the lands gave permission to conduct the study on these sites. We confirm that the study did not involve endangered or protected species.

Phra Thong island provides a location for comparative tsunami (S₁: N9.13194 E98.26250) and non-tsunami (S₂: N9.07250 E98.27222) affected terrestrial studies based on geological evidences (Jankaew, personal communication) [1,2]. S₁ and S₂ are 6.73 km apart. S₁ is 0.40 km from the sea, and S₂ is 2.26 km from the sea (Figure 1). Approximately 1 kg samples were collected, each in sterile containers, between 11:00–15:00 hours during 23–24 March 2011. S₁ samples comprised: 2004 tsunami (14.5 cm), 1–300yo (22 cm), 300–600yo (29 cm), 600yo tsunami (38 cm), and >600yo (46 cm); S₂ samples comprised: S21 (14.5 cm), S22 (22 cm), S23 (29 cm), S24 (38 cm), and S25 (46 cm). The number in parenthesis represents the depth level where the sample was collected, and that entailed the approximate age of the sample relative to the year 2004. On-site records for color, texture and pH were done. All samples were transported in ice chest, stored in 4°C and processed for the next steps within 14 days.

Metagenomic DNA extraction and DNA quality examination

Each sample was mixed with a sterile spatula, and 15 g each was used for metagenomic DNA extraction [18]. Two independent metagenomic DNA extractions were performed per sample. The samples were dissolved in an extraction buffer (Epicentre, Wisconsin, USA) with Tween 20, low-speed centrifuged to remove large debris, and poured through four-layered sterile cheesecloth to remove particles and organisms of >30 μm in size. Microorganisms between 0.22 and 30 μm were collected by filtering over a sterile 0.22 μm filter membrane (Merck Millipore, Massachusetts, USA) [15]. Total nucleic acid from each sample was extracted using Meta-G-Nome DNA Isolation Kit (Epicentre) following the manufacturer's protocols. Metagenomic DNA quality was assessed using agarose gel electrophoresis. The DNA concentration and purity was further analysed by A_{260} and the ratio of A_{260}/A_{280} spectrophotometry, respectively.

PCR generation of pyrotagged 16S and 18S rDNA libraries

Table 1 lists forward and reverse pyrotagged 16S and 18S rRNA gene primers. For broad-range 16S and 18S rRNA genes amplification, universal prokaryotic 338F (forward) and 803R (reverse) primers [19–21], and universal eukaryotic 1A (forward) and 516R (reverse) primers [15,22,23] were used. Italics denote the eight nucleotides pyrotag sequences, functioning to specify sample names [24]. A 50-μl PCR reaction comprised 1×

EmeraldAmp GT PCR Master Mix (TaKaRa, Shiga, Japan), 0.3 μM of each primer, and 100 ng of the metagenome. PCR conditions were 95°C for 4 min, and 30–35 cycles of 94°C for 45 s, 50°C for 55 s and 72°C for 1 min 30 s, followed by 72°C for 10 min. To generate the pyrotagged 16S or 18S rDNA libraries with minimized stochastic PCR biases, two to three independent PCRs were performed per extracted metagenomes, and two extracted metagenomes per sample, resulting in a minimum of four PCR products to be pooled for pyrosequencing per sample.

Gel purification and pyrosequencing

PCR products (~473 bp for 16S rDNAs; ~577 bp 18S rDNAs) were excised from agarose gels, and purified using PureLink Quick Gel Extraction Kit (Invitrogen, New York, USA). The 454-sequencing adaptors were ligated to all 16S and 18S rDNA fragments, the reactions were purified by MinElute PCR Purification Kit (Qiagen), and the samples were pooled for pyrosequencing on an eight-lane Roche picotiter plate, 454 GS FLX system (Roche, Branford, CT) at the in-house facility of the National Center for Genetic Engineering and Biotechnology, according to the recommendations of the supplier.

Sequence annotation and bioinformatic analyses

After removal of unreliable sequences, including sequences that failed the pyrosequencing quality cut-off and sequences shorter than 50 nucleotides, the sequences were categorized based on the appended pyrotag sequences. Sequences corresponding to the same sample category were inspected for domain and taxon compositions using mg-RAST [25,26] with default parameters. Species were identified by BLASTN [27] with E-value $\leq 10^{-5}$ against 16S rDNA databases including NCBI non-redundant [28], RDP [29] and Greengenes [30], and for 18S rDNAs the databases included NCBI non-redundant [28], EMBL [31,32] and SILVA [33]. Evolutionary distances and phylogenetic tree were computed with default thresholds (E-value $\leq 10^{-8}$, similarity score $\geq 80\%$). Species (or phylum) prevalence was determined by dividing the frequency of reads in the species (or phylum) by the total number of the identifiable reads. The differences in community structures were compared using Yue & Clayton theta similarity coefficients (Thetayc) and Morisita-Horn dissimilarity index, in mothur [34–36]. Low Thetayc and Morista-Horn inferred high community similarity. An unweighted pair group method with arithmetic mean (UPGMA) clustering was constructed using Thetayc, in mothur [36]. Furthermore, functional subsystems and functional groups of the prokaryotic profiles were determined using SEED-based assignments in mg-RAST server [25,26,37]. For habitat classification, the data were compared against a World Register of Marine Species (WoRMS) database [38].

Results

Metagenome abundances and compositions at domain and kingdom levels

On-site records for physical characteristics of S₁ and S₂ were as shown in Figure 1. Alternating layers of black soil-like and grey sand-like comprised most of S₁, whereby more homogeneous layers of grey sand-like predominated in S₂. Differences in pH

Table 1. Pyrotagged16S and 18S rRNA genesuniversal primers.

Sample names	Forward primers (5'-3')	Reverse primers (5'-3')
16S rRNA gene universal primers		
2004 tsunami	*TAGTAGCG*ACTCCTACGGGAGGCAGCAG	*TAGTAGCG*CTACCAGGGTATCTAATC
1–300yo	*AGACGACG*ACTCCTACGGGAGGCAGCAG	*AGACGACG*CTACCAGGGTATCTAATC
300–600yo	*ACTCGTAG*ACTCCTACGGGAGGCAGCAG	*ACTCGTAG*CTACCAGGGTATCTAATC
600yo tsunami	*ACATCGAG*ACTCCTACGGGAGGCAGCAG	*ACATCGAG*CTACCAGGGTATCTAATC
>600yo	*ACGCTATC*ACTCCTACGGGAGGCAGCAG	*ACGCTATC*CTACCAGGGTATCTAATC
S21	*TACTACGC*ACTCCTACGGGAGGCAGCAG	*TACTACGC*CTACCAGGGTATCTAATC
S22	*AGCAGAGC*ACTCCTACGGGAGGCAGCAG	*AGCAGAGC*CTACCAGGGTATCTAATC
S23	*TCAGCTAC*ACTCCTACGGGAGGCAGCAG	*TCAGCTAC*CTACCAGGGTATCTAATC
S24	*AGAGCGAC*ACTCCTACGGGAGGCAGCAG	*AGAGCGAC*CTACCAGGGTATCTAATC
S25	*ATGCTCAC*ACTCCTACGGGAGGCAGCAG	*ATGCTCAC*CTACCAGGGTATCTAATC
18S rRNA gene universal primers		
2004 tsunami	*AGATAGCG*CTGGTTGATCCTGCCAGT	*AGATAGCG*ACCAGACTTGCCCTCC
1–300yo	*TGTAGACG*CTGGTTGATCCTGCCAGT	*TGTAGACG*ACCAGACTTGCCCTCC
300–600yo	*TGCAGTAG*CTGGTTGATCCTGCCAGT	*TGCAGTAG*ACCAGACTTGCCCTCC
600yo tsunami	*TCTGCGAG*CTGGTTGATCCTGCCAGT	*TCTGCGAG*ACCAGACTTGCCCTCC
>600yo	*ATCAGCAG*CTGGTTGATCCTGCCAGT	*ATCAGCAG*ACCAGACTTGCCCTCC
S21	*ATACAGTC*CTGGTTGATCCTGCCAGT	*ATACAGTC*ACCAGACTTGCCCTCC
S22	*ATCATATC*CTGGTTGATCCTGCCAGT	*ATCATATC*ACCAGACTTGCCCTCC
S23	*TGCGATGC*CTGGTTGATCCTGCCAGT	*TGCGATGC*ACCAGACTTGCCCTCC
S24	*ATCGCAGC*CTGGTTGATCCTGCCAGT	*ATCGCAGC*ACCAGACTTGCCCTCC
S25	*TATACTAC*CTGGTTGATCCTGCCAGT	*TATACTAC*ACCAGACTTGCCCTCC

Italic sequence denotes the 8 nt-pyrotagged sequence.

were also evident between the two ecosystems where S_1 ranged from 6–7 (more acidity), while S_2 ranged from 7–7.5.

Following total nucleic acids extraction of 0.22–30 µm sizes, S_1 and S_2 had average metagenomic concentration of 23.16 ng and 27.02 ng per gram of soil, respectively. Libraries of pyrotagged 16S and 18S rRNA gene fragments were constructed, and pyrosequenced to obtain the culture-independent prokaryotic and eukaryotic profiles of the sites. After removal of unreliable sequences, 21,592reads for S_1 and 33,308 reads for S_2 remained for BLASTN species identification. Significant E-values ($\leq 10^{-5}$) were identified for 20,555 reads for S_1 (95.20%) and 31,946 reads for S_2 (95.91%). Reads with non-significant E-values ($>10^{-5}$) were omitted from the analyses.

The domain compositions of S_1 indicated a lower proportion of eukaryotes and higher proportion of prokaryotes than S_2 (Figure 2). The proportion of eukaryotic species was reduced by almost half in S_1. Among prokaryotes, the greatest divergence between S_1 and S_2 was evident among archaea compared to bacteria where found increased in S_1 (Figure 3: 1.15-fold for bacteria, 4.07-fold archaea). These greater representations somewhat caused the reduced diversity of the other 4 kingdoms of lives.

Diversity of prokaryotic phyla and species

Major prokaryotic phyla for S_1 included Proteobacteria, Bacteroidetes, and Actinobacteria; and S_2 included Proteobac-

teria, Acidobacteria, and Actinobacteria, in diminishing order (Figure 4A). Consistent with Figure 3, S_1 demonstrated greater prokaryotic biodiversity than S_2. S_1 contained many new species belonging to uncultured species, i.e. OP3, GN04 and SC3, whereas S_2 still contained high proportion of common environmental phyla, including Proteobacteria and Acidobacteria (Figure 4A). Different sample periods showed slight variation of prokaryotic phyla profiles in S_1, whereas in S_2 more variation among the phyla distributions was evident (Figure 4B). For instances, Proteobacteria comprised 62.23% in the S_1 layer of 2004 tsunami, 58.56% 1–300yo, 59.18% 300–600yo, 60.69% 600yo tsunami, and 62.54% >600yo; while S_2 comprised 46.07% in the S21 layer, 65.72% S22, 77.64% S23, 75.05% S24, and 67.41% S25. Actinobacteria comprised 14.48% in S_1 layer of 2004 tsunami, 13.22% 1–300yo, 13.53% 300–600yo, 13.82% 600yo tsunami, and 13.57% >600yo; while S_2 layers demonstrated 14.32% S21, 7.59% S22, 6.52% S23, 8.92% S24, and 10.47% S25 (Figure 4B). Further distinguished differences were diagnosed upon analysis based on species distributions: no parallel-age pairs of S_1 and S_2 showed similar species distribution pattern with the >600 yo and S25 pair showing the least differences. Examples of predominated species in S_1 were *Acinetobacter haemolyticus*, *Polynucleobacter* sp., *Polynucleobacter necessaries* and *Flavobacterium denitrificans*; and S_2 were *Burkholderia* sp. and *Silvimonas terrae* (Figure 5). Subsequently, the computed Thetayc and Morisita-Horn dissimilarity indices revealed dissimilar community structures between S_1

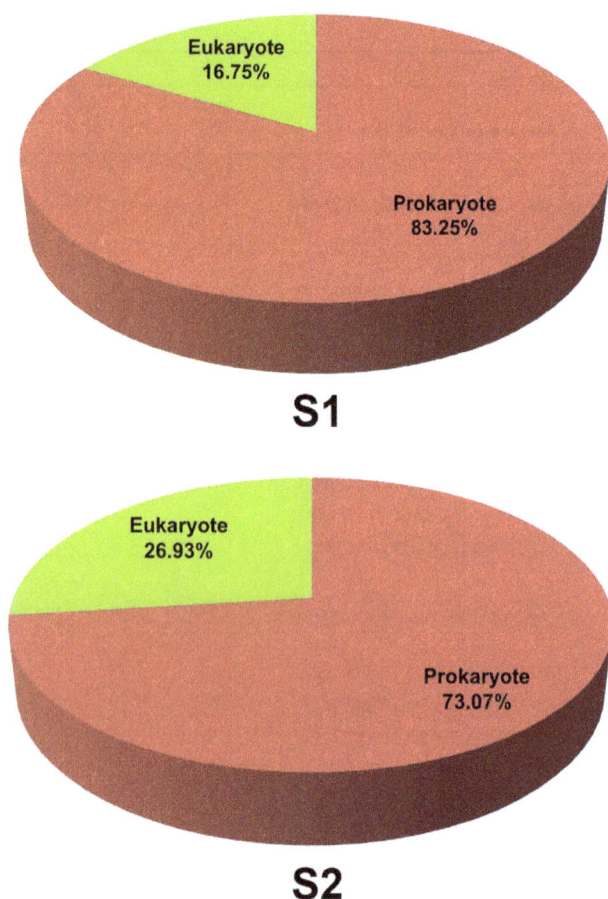

Figure 2. Percentages of prokaryotic and eukaryotic domains in S$_1$ and S$_2$.

and S$_2$. As shown in UPGMA dendrogram in Figure 6A: terrestrial layers corresponding to S$_1$ site were clustered together separated from the S$_2$ layers. Although S21 was grouped with the S$_1$ layers, it was an outer related branch to the S$_1$ group.

Potential metabolic system analysis of prokaryotic communities

From a total of 28 possible metabolic subsystems by mg-RAST [25,26,37], the prokaryotic communities of S$_1$ contained 19 subsystems, and S$_2$ contained 11 subsystems. The predominant subsystems in the S$_1$ included: regulation and cell signalling (14.35%), cell wall and capsule (10.05%), protein metabolism (8.13%), and sulfur metabolism (3.83%) (Figure 7). Overall, S$_1$ inhabited prokaryotic communities with high metabolic potentials for cell wall and capsule (i.e. gram-negative cell wall components by TIdE/PmbA protein), protein metabolism (i.e. protein degradation by prolyl endopeptidase), regulation and cell signaling (i.e. regulation of virulence by two-component system response regulator), and carbohydrates (i.e. sugar alcohols and central carbohydrate metabolism by HPr kinase/phosphorylase). S$_2$ inhabited prokaryotic communities with high metabolic potentials for respiration (i.e. electron donating and accepting reactions by Type cbb3 cytochrome oxidase biogenesis protein), clustering based subsystems (i.e. copper-translocating P-type ATPase), and virulence, disease and defence (resistance to antibiotics and toxic compounds by acriflavin resistance protein).

Diversity of eukaryotic phyla and species

Dominant eukaryotic phyla for S$_1$ were in kingdom Animalia: Brachiopoda (47.82% in 2004 tsunami, 32.53% 1–300yo, 37.32% 300–600yo, 49.15% 600yo tsunami, 25.00% >600yo), and Mollusca (28.88% 2004 tsunami, 29.39% 1–300yo, 31.34% 300–600yo, 30.51% 600yo tsunami, 7.14% >600yo); and kingdom Protozoa: Dinophyta for particularly the >600yo layer (51.99%) (Figures 8A and 8B). For S$_2$, although Brachiopoda and Mollusca were dominant, fungal phylum Basidiomycota (0.08% S21, 66.51% S22, 7.47% S23, 12.82% S24, 12.86% S25) and animal phylum Arthropoda (3.05% S21, 3.42% S22, 28.33% S23, 7.70% S24, 2.86% S25) were the most prevalent. When analyzing the data into individual sample periods, similar finding to Figure 4B were found. Different sample periods of S$_1$ demonstrated less phyla pattern variation than those of S$_2$ (Figure 8B). Distinguished phyla pattern of S$_2$ from S$_1$ were displayed apparently in S22, S23 and S24 layers (Figure 8B), resulting in their divergence from S$_1$ and the other S$_2$ communities by the UPGMA dendrogram constructed using Thetayc dissimilarity indices (Figure 6B). Similar to prokaryotes (Figure 6A), the eukaryotic communities corresponding to the S$_1$ site were relatively clustered together (Figure 6B). Analysis at the species level identified a more diverse fungal and animal species among S$_2$ layers (Figure 9).

Habitat classification

The S$_1$ and S$_2$ prokaryotic and eukaryotic profiles were matched against WoRMS database [38] to further characterize their microbial ecology: how each is related to marine, brackish water, freshwater, and terrestrial species communities. Figure 10 exhibited a substantially higher abundance of marine species habitat with S$_1$ (S$_1$ = 24.11%, S$_2$ = 13.33%), and terrestrial species with S$_2$ (S$_1$ = 0.11%, S$_2$ = 1.42%). Examples of abundant marine prokaryotes in S$_1$ were: *Lutaonella thermophilus*, *Shewanella aquimarina*, *Erythrobacter ishigakiensis* and *Thalassobacter stenotrophicus*. Abundant marine eukaryotes in S$_1$ included: *Dinophysis acuminata*, *Ctenodrilidae sp.*, *Remanella sp.*, *Nemertinoides elongatus*, *Skeletonema grethae*, *Crassostrea gigas*, *Hymenocotta mulli*, *Diplodasys ankeli*, *Pinna muricata*, *Arenicola marina*, *Limopsis marionensis* and *Haliplanella lucia*.

Discussion

On-site records indicated the greater turbidity (Figure 1) and acidity of S$_1$ were in agreement with previous reports [1,2,5–7]. Together with many other tsunami studies, these different soil types suggested terrestrial component changes following tsunami inundation, which could affect the microbial ecology of the site. The terrestrial microbiome representing the 2004 tsunami-affected site has never been studied. Our findings represent the first to utilize metagenomics in gaining databases of these entire terrestrial microbiomes, including prokaryotes and eukaryotes, in tsunami-affected (S$_1$) and non-tsunami affected (S$_2$) sites of Phra Thong island, as of March 2011. The data helped characterize the microbial biodiversity and its impact by tsunami occurrence. This knowledge is essential for scientists and engineers involved with land management and environmental bio-improvement.

Diminished total nucleic acids from S$_1$ suggested a less populated microbial community. Although some fossil DNA and fragments of DNA from live animals (known as extracellular "dirt" DNA) could be included in the extracted metagenomes, and might partly complicate the analysis. Andersen et al. [39] found extracellular "dirt" DNA from the terrestrial surface could reflect an overall taxonomic richness and relative abundance of species of a site at the time of investigation. Hence, some extracellular "dirt"

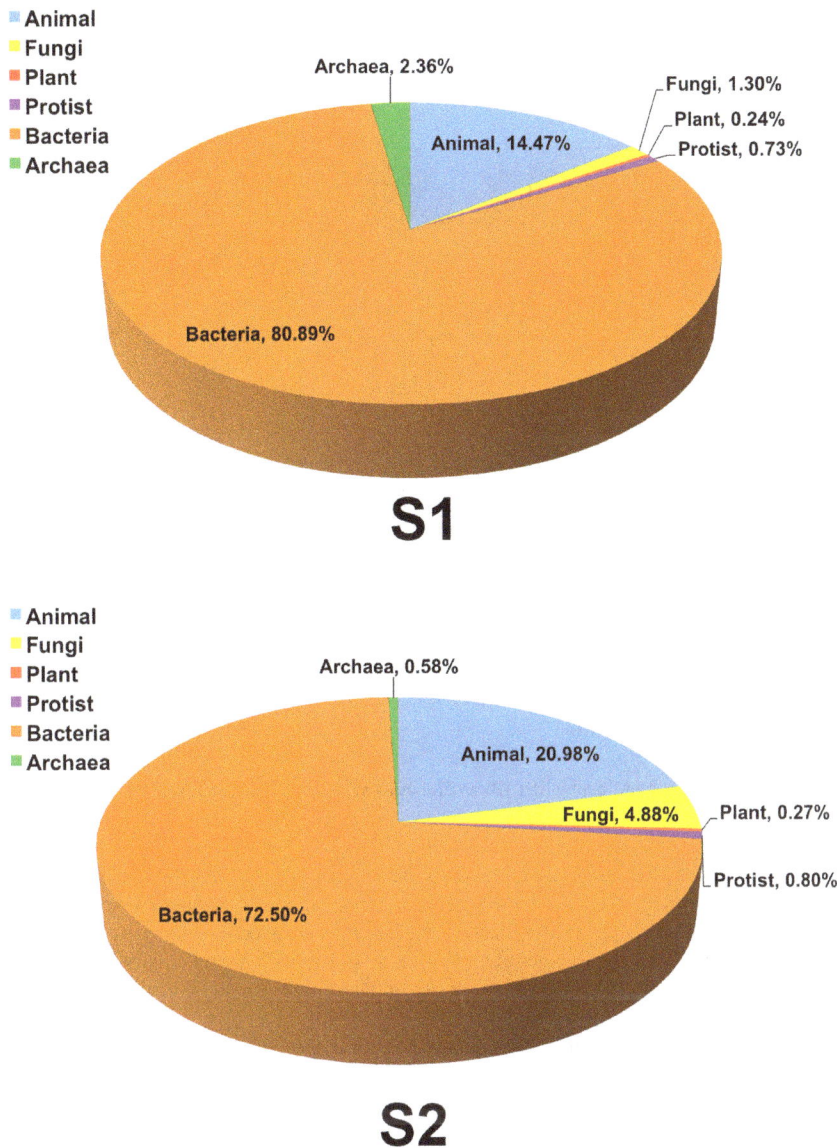

Figure 3. Percentages of 6 kingdoms of lives in S$_1$ and S$_2$.

DNA in our extracted metagenome should also reflect an overall biodiversity.

Libraries of pyrotagged 16S and 18S rRNA gene fragments were successfully constructed and pyrosequenced: 21,592 reads for S$_1$ and 33,308 reads for S$_2$ were retrieved after removal of unreliable sequences. For BLASTN species identification, greater than 95% of the S$_1$ and S$_2$ reads were identified with $\leq 10^{-5}$ E-values. The amount of reads should be sufficient to recapture the relationships among the samples, as Caporaso et al. [40] reported 2,000 reads could recapture the same relationships among samples as did with the full dataset. Additionally, many studies discovered variable regions 3 and 4 of 16S rRNA gene analyses were more effective than random sequence reads analyses in estimating the biodiversity and relationships among the samples [41–43].

In S$_1$, domain of prokaryotes became highly present (Figure 2). In particular, S$_1$ had a richer archaeal population (4.07-fold increase), meanwhile fungi, animals, plants, and protists were decreased (Figure 3). This finding was consistent with the fact that

tsunami inundation might leave a terrestrial site inhospitable, causing archaea and bacteria to be more common due to their flexible life activities and requirements [44–46]. The greater biodiversity of kingdoms in S$_2$ supported the more hospitable terrestrial habitat than S$_1$.

Phyla and species distribution patterns between S$_1$ and S$_2$ were different. Changing the prokaryotic pattern of major phyla, precisely S$_1$ was predominantly comprised of Bacteroidetes with a lower prevalence of Proteobacteria, Actinobaceria and Acidobacteria (Figure 4A), highlighted the modified microbial ecology. Wada et al. [47] reported the similar change of bacterial floras in the sludge brought ashore by the 2011 East Japan earthquake. Bacteroidetes was more evident than Proteobacteria in the affected coastal water area. Additionally, numerous sulfate-reducing bacteria were evident in the sludge, which corresponded with high concentrations of sulfate ions in the sludge and the affected water area. The latter report was consistent with our finding of the higher sulfur metabolism in S$_1$ (Figure 7). Further, among the Bacteroidetes, flavobacteria predominanted which is generally

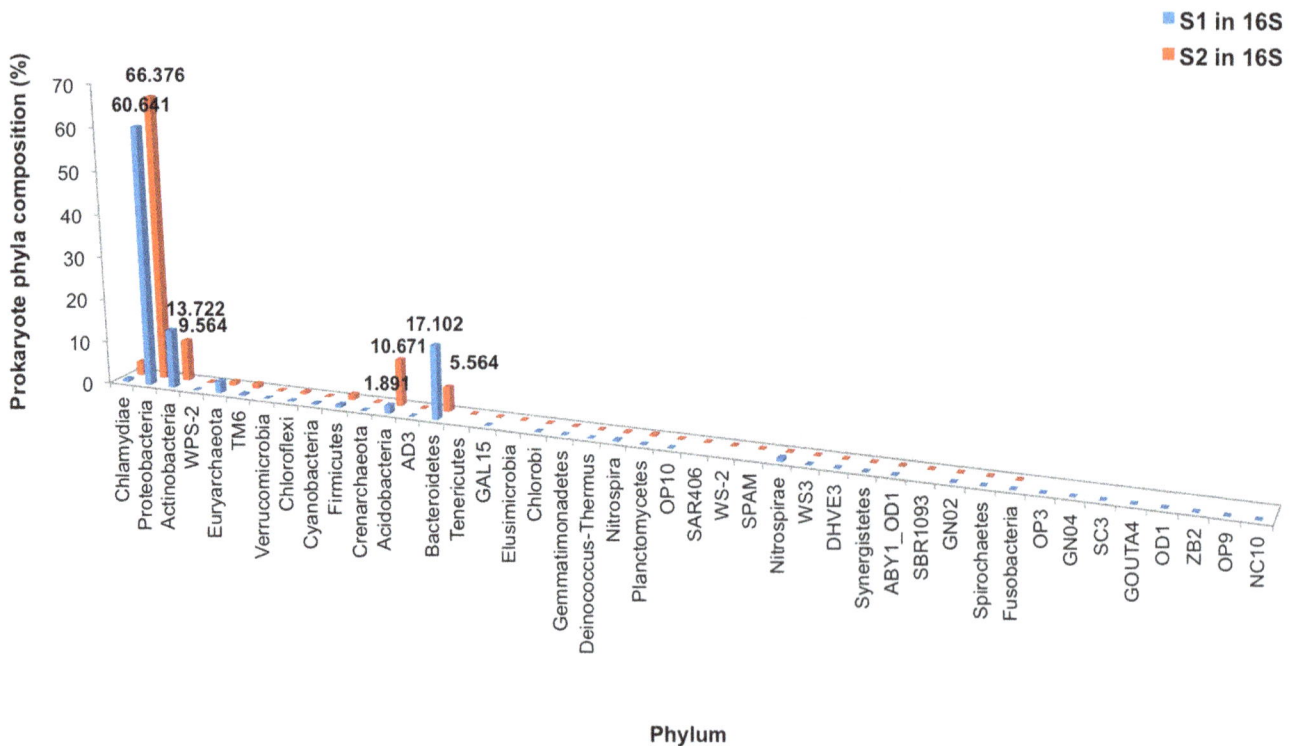

Figure 4. Distribution of prokaryotic phyla in S₁ and S₂, without (A) and with (B) individual sample ages categorization.

classified as an environmental bacterium with both commensal and pathogen species of marine animals and humans. Banning et al. [48] discovered several flavobacteria strains in various marine environments could function as predators on other bacteria. These bacteria have minimal growth requirements, only sea salt and the utilization of the lysed bacteria. The marine Flavobacteria thus could have critical consequences on microbial ecology as they could eradicate certain microbial communities [48]. Additionally, *Polynucleobacter* sp. are bacterioplankton that could survive broad ecological niches due to their ability to obtain energy by consuming organic materials from other organisms through nitrogen fixation, nitrification, remineralisation and methanogenesis. *Photobacterium* sp. is also a genus with metabolic versatility, which can degrade chitin and cellulose for carbohydrates [49]. Consequently, the flavobacteria, bacterioplankton and photobacteria activities could partly support the high metabolic subsystems of carbohydrates and protein metabolism in S₁ (Figure 7), albeit the overall poor living condition. Note the increased regulation and cell signalling, and cell wall and capsule subsystems (Figure 7) could in part symbolize the growth activities of these bacteria, given capsule lies outside the bacterial cell wall and considered a virulent factor. Bacterial capsule protects the bacteria against some hostile environment, such as desiccation, and prevents phagocytosis by host immune cells [50]. For examples, *Acinetobacter haemolyticus* in contaminated seafood produces shiga toxin that causes bloody diarrhea [51], and *Flavobacterium* sp. cause cold water disease in salmon and other fish species [52]. *Photobacterium*, a genus in family *Vibrionaceae*, is primarily marine microorganisms that evolved to become pathogenic to marine animals, causing mortality in crabs and fish, and indirect pathogens of humans through contact or consumption [49,53]. Hence residents and workers in these areas were recommended to minimize direct contact with the affected soil,

sludge and water, to prevent their risk of infection, and frequent hand wash [3,8,9,47]. Note the many new uncultured species in S₁ (Figure 4A) further emphasized its environmental change resulting in new identified species. Our 16S rRNA gene analyses supported the high dissimilarity indices between S₁ and S₂ prokaryotic community structures (Figure 6A).

The metabolic potentials in Figure 7 supported the prior results, showing advanced metabolic subsystems of regulation and cell signalling, cell wall and capsule, protein metabolism, sulfur metabolism, and carbohydrates in S₁. In contrast, S₂ microbial communities carried high metabolic potentials for pathways of respiration, photosynthesis, and drug and bioactive compound production. This finding supported the diversified biodiversity in the non-affected terrestrials, and highlighted the more abundant pharmaceutical related microbial producers in the naturally undisturbed environments [15], like S₂.

For eukaryotic phyla and species distribution patterns, while both mollusks and brachiopods predominated in both terrestrials, given both animals were marine animals and were more prominent in tsunami-inundated S₁ site, fungi Basidiomycota and animal Arthropoda were only highly proportionate in the S₂ area (Figure 8A). Like prokaryotic phyla distribution patterns among various terrestrial depths (Figure 4B), the more similar eukaryotic phyla distribution patterns among various terrestrial depths were evident in S₁ (Figure 8B) highlighted the factor that a massive tsunami hit could destroy the biodiversity within microbial ecosystems. Basidiomycota, which were found more evident in S₂, are higher fungi that play important roles as carbon recycler and nutrient decomposer, and posed the chief source of bioactive natural products [54–56].

Since the 2004 tsunami up to our study period, an effect of microbial population mixing and microbial change due to human or animal activity on S₁ and S₂ sites in Phra Thong island should

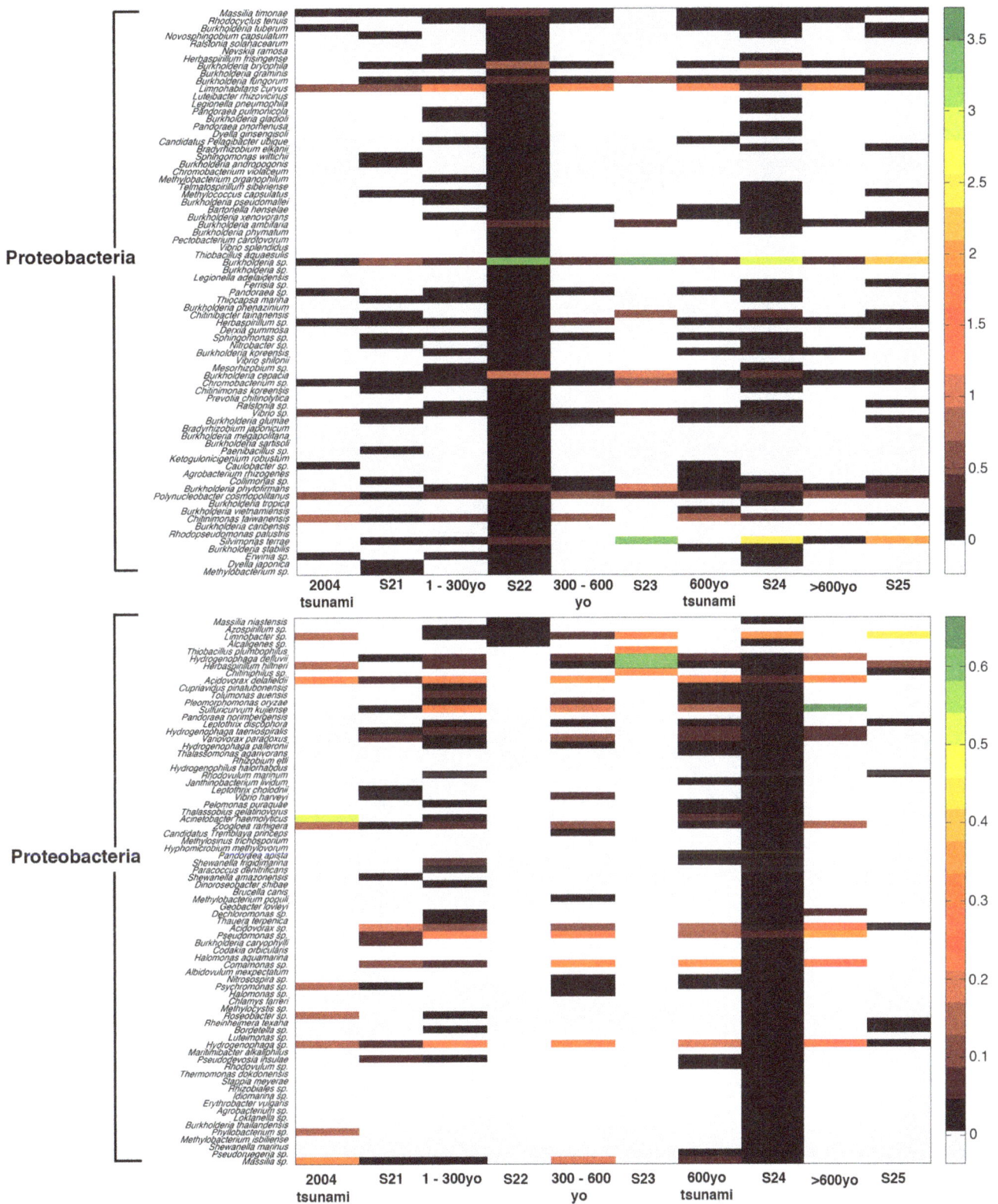

Figure 5. Distribution of prokaryotic species in S$_1$ and S$_2$, categorized by individual sample ages. Different color on the diagram represents a different relative abundance, based on the percent frequency chart on the right.

be minimal. The reasons are because, after the 2004 tsunami, Phra Thong island remains almost no human inhabited and no human activity. The place becomes part of the wildlife sanctuary.

Microbial population mixing through time could only be by rainfall and plant root penetration (mostly grass at both sites); hence it should be minimal and in vertical direction only.

A

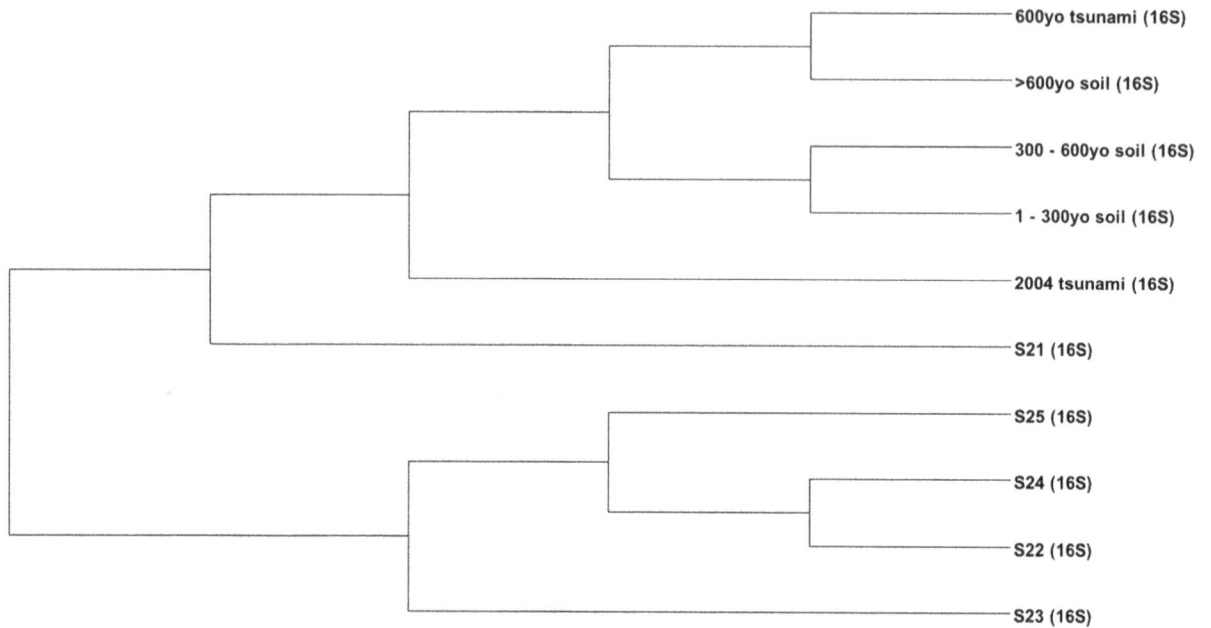

B

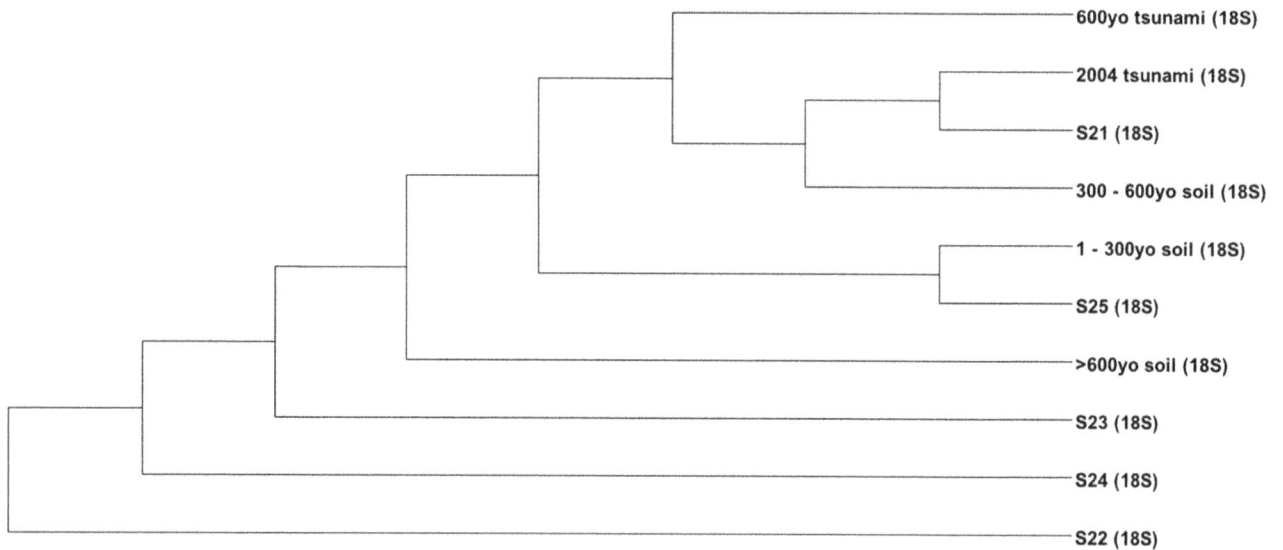

Figure 6. UPGMA clustering comparing relatedness among S₁ and S₂ prokaryotic (A) and eukaryotic (B) profiles.

Together, the 16S and 18S rRNA gene profiles indicated both terrestrials marine habitat, corresponding to the fact that the two sites were located on a small island in Andaman Sea of Thailand. Nevertheless, twice the greater prediction for marine habitat for S_1 (Figure 10) highlighted its much higher number of prokaryotic and eukaryotic species that represent marine habitat species-indicators, and tsunami inundation.

Conclusion

During the past decades, tsunamis have occurred more frequently though the correlation between tsunami disturbance and change of terrestrial microbial ecology remain poorly defined. The present study provided a culture-independent prokaryotic and eukaryotic analyses representing 0.22–30 μm metagenomes belonging to Thailand tsunami and non-tsunami affected terrestrials.

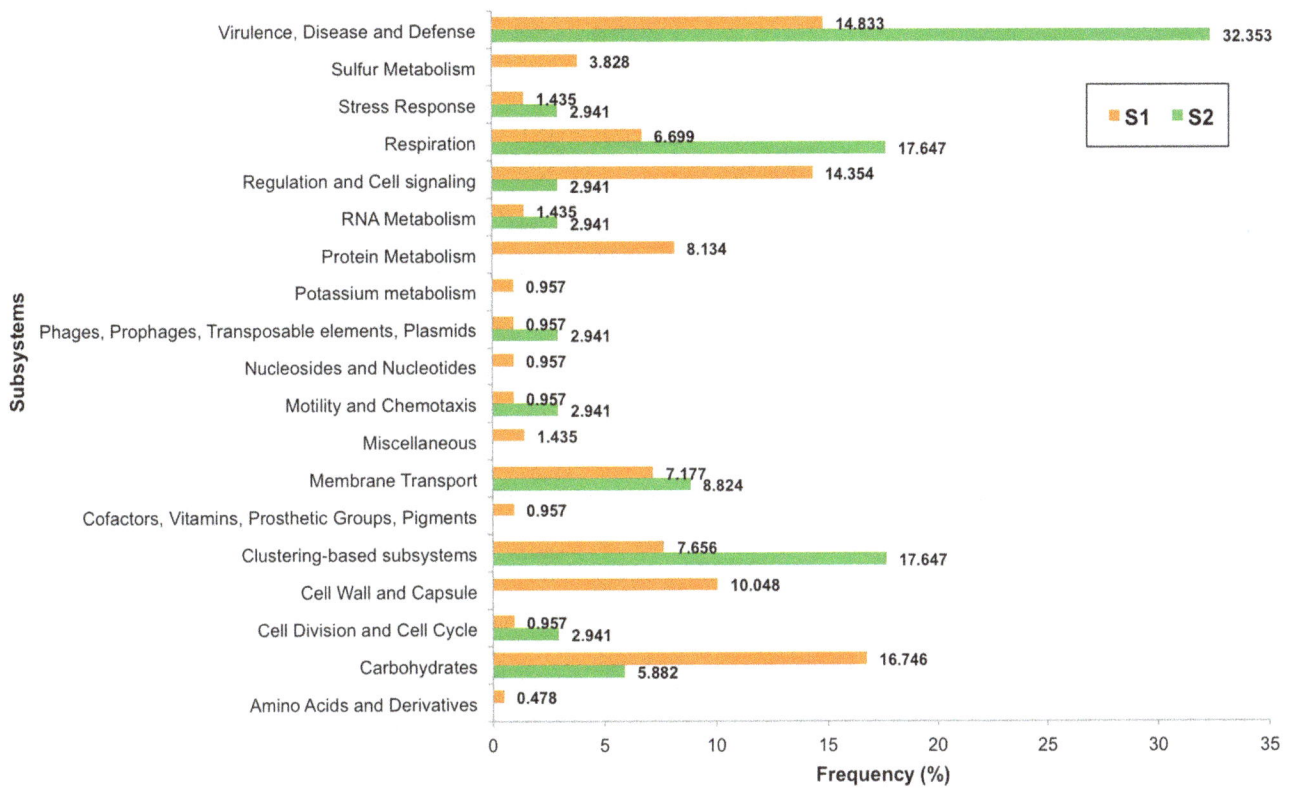

Figure 7. Metabolic subsystems of prokaryotic communities in S₁ and S₂.

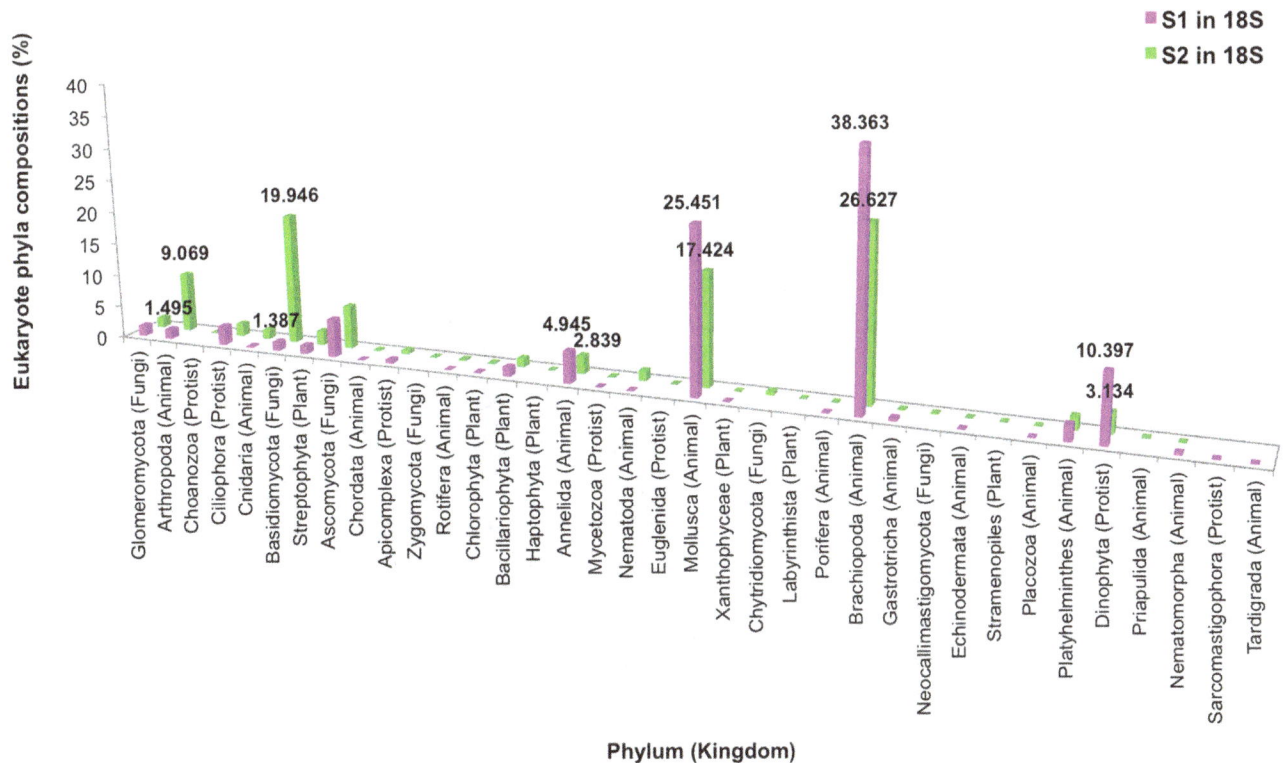

Figure 8. Distribution of eukaryotic phyla in S₁ and S₂, without (A) and with (B) individual sample ages categorization.

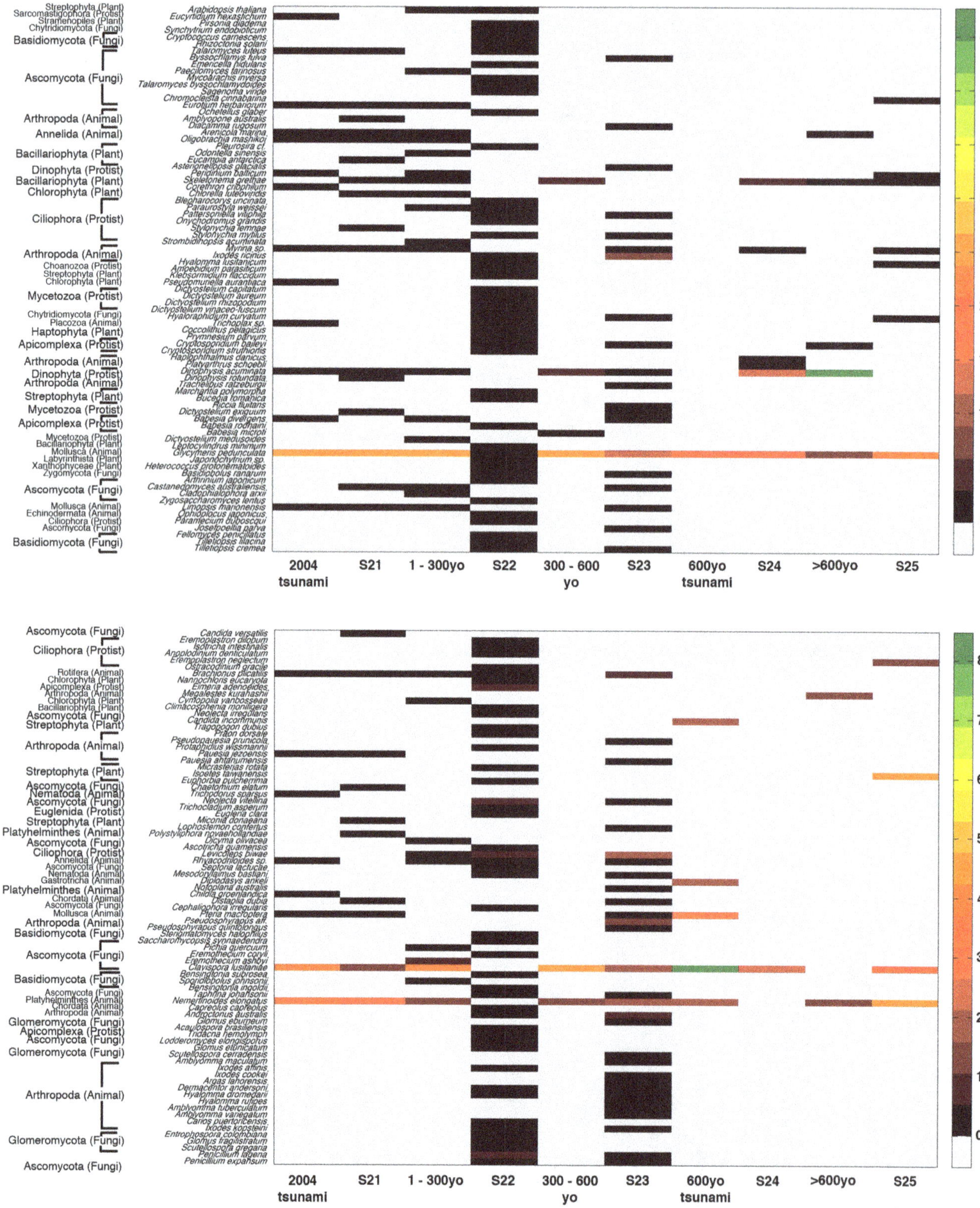

Figure 9. Distribution of eukaryotic species in S₁ and S₂, categorized by individual sample ages. Different color on the diagram represents a different relative abundance, based on the percent frequency chart on the right.

The different prokaryotic and eukaryotic profiles highlighted the differences due to tsunami, and helped fulfill our knowledge of diverse terrestrial microbial ecologies. The biodiverse species of S₁ distinguished its microbial communities and metabolic potentials.

For instances, the finding of predator and prey bacterial relationship, and cell wall and capsule subsystem were common for S₁, whereas bioactive compound producers were more common in S₂.

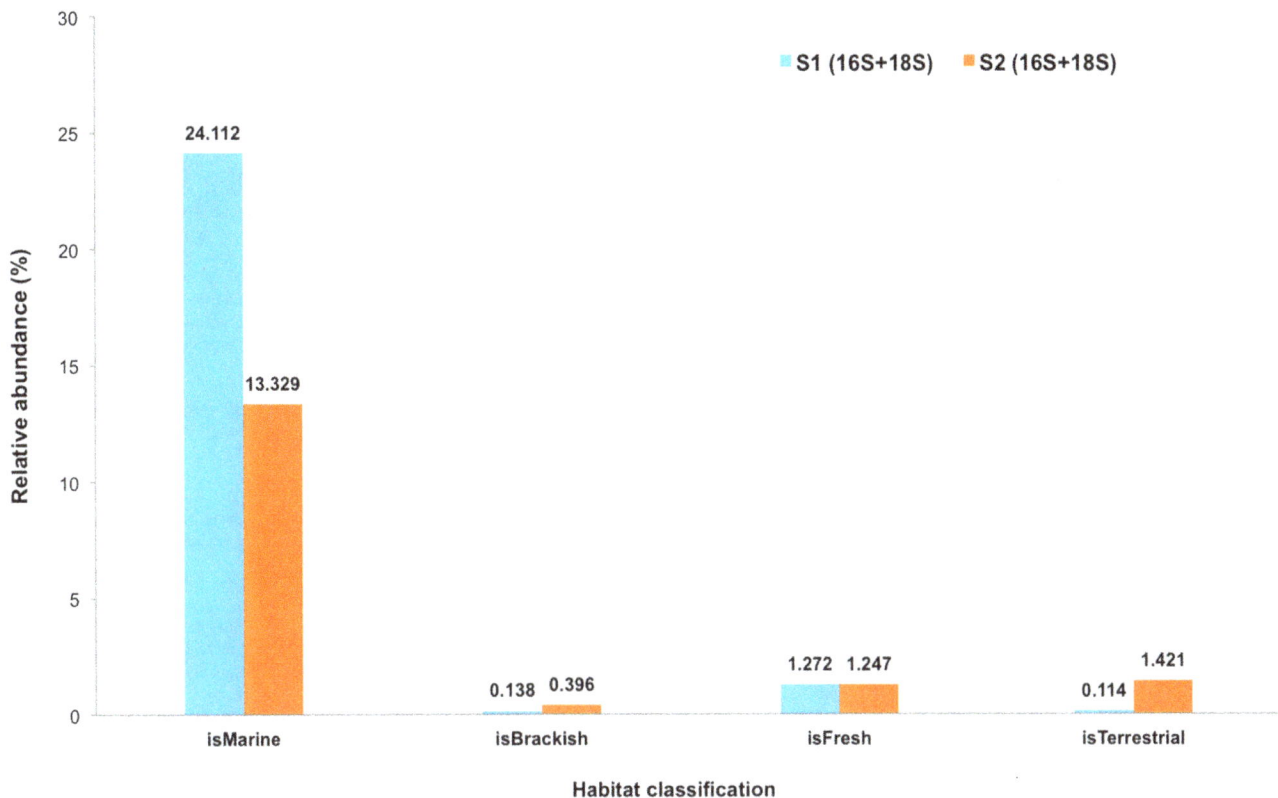

Figure 10. Habitat classification for S₁ and S₂. Prokaryotic and eukaryotic species profiles of S1 and S2 were matched against WoRMS database for habitat classification.

Further, the marine habitat analysis demonstrating the greater percent of marine prokaryotic and eukaryotic species in S_1 could perhaps help serve as another biomarker for the geological history of a terrestrial site. Nonetheless, more researches are required to utilize these as biomarkers for estimating times and presence of any geological incidences. Presently, identification of historic tsunami was restricted to examination of marine and brackish diatoms while silica shell of diatom could be dissolved through time, especially in hot weather of a tropical country [1,2,57,58].

Acknowledgments

The authors thank Sarah Hof and Khunaluck Kidmoa for general help, Dominik Brill for helping with field work, and Troy Skwor for manuscript editing.

Author Contributions

Conceived and designed the experiments: NS AA. Performed the experiments: NS DS S. Tangphatsornruang. Analyzed the data: NS AW. Contributed reagents/materials/analysis tools: NS KJ S. Tongsima. Wrote the paper: NS. Revised the manuscript: NS KJ S. Tongsima.

References

1. Jankaew K, Atwater BF, Sawai Y, Choowong M, Charoentitirat T, et al. (2008) Medieval forewarning of the 2004 Indian ocean tsunami in Thailand. Nature 455:1228–1231.
2. Sawai Y, Jankaew K, Martin ME, Prendergast A, Choowong M, et al. (2009) Diatom assemblages in tsunami deposits associated with the 2004 Indian Ocean tsunami at Phra Thong Island, Thailand. Mar Micropaleontol 73:70–79.
3. Sri-Aroon P, Chusongsang P, Chusongsang Y, Pornpimol S, Butraporn P, et al. (2010) Snails and trematode infection after Indian ocean tsunami in Phang-Nga province, southern Thailand. Southeast Asian J Trop Med Public Health 41: 48–60.
4. Prummongkol S, Panasoponkul C, Apiwathnasorn C, Lek-Uthai U (2012) Biology of *Culex sitiens*, a predominant mosquito in Phang Nga, Thailand after a tsunami. J Insect Sci 12: 11. doi:10.1673/031.021.1101
5. Tharnpoophasiam P, Suthisarnsuntorn U, Worakhunpiset S, Charoenjai P, Tunyong W, et al. (2006) Preliminary post-tsunami water quality survey in Phang-Nga province, southern Thailand. Southeast Asian J Trop Med Public Health 37: 216–220.
6. Collivignarelli C, Tharnpoophasiam P, Vaccari M, De Felice V, Di Bella V, et al. (2008) Evaluation of drinking water treatment and quality in Takua Pa, Thailand. Environ Monit Assess 142: 345–358.
7. Collivignarelli C, Tharnpoophasiam P, Vaccari M, De Felice V, Di Bella V, et al. (2008) Water monitoring and treatment for drinking purposes in 2004 tsunami affected area-Ban Nam Khem, Phang Nga, Thailand. Environ Monit Assess 147: 191–198.
8. Chastel C (2007) Assessing epidemiological consequences two years after the tsunami of 26 December 2004. Bull Soc Pathol Exot 100: 139–142.
9. Huusom AJ, Agner T, Backer V, Ebbehoj N, Jacobsen P (2012) Skin and respiratory disorders following the identification of disaster victims in Thailand. Forensic Sci Med Pathol 8: 114–117.
10. Boszke L, Astel A (2007) Fractionation of mercury in sediments from coastal zone inundated by tsunami and in freshwater sediments from the rivers. J Environ Sci Health A Tox Hazard Subst Environ Eng 42: 847–858.
11. Lukaszewski Z, Karbowska B, Zembrzuski W, Siepak M (2012) Thallium in fractions of sediments formed during the 2004 tsunami in Thailand. Ecotoxicol Environ Saf 80: 184–190.
12. Rusch DB, Halpern AL, Sutton G, Heidelberg KB, Williamson S, et al. (2007) The Sorcerer II Global Ocean Sampling expedition: northwest Atlantic through eastern tropical Pacific. PLoS Biol 5:e77.
13. Yooseph S, Nealson KH, Rusch DB, McCrow JP, Dupont CL, et al. (2010) Genomic and functional adaptation in surface ocean planktonic prokaryotes. Nature 468:60–66.
14. Zinger L, Ammaral-Zettler LA, Fuhman JA, Horner-Devine MC, Huse SM, et al. (2011) Global patterns of bacterial beta-diversity in seafloor and seawater ecosystems. PLoS ONE 6: e24570. doi:10.1371/journal.pone.0024570

15. Somboonna N, Assawamakin A, Wilantho A, Tangphatsornruang S, Tongsima S (2012) Metagenomic profiles of free-living archaea, bacteria and small eukaryotes in coastal areas of Sichang island, Thailand. BMC Genomics 13:S29. doi:10.1186/1471-2164-13-S7-S29

16. Auld RR, Myre M, Mykytczuk NCS, Leduc LG, Merritt TJS (2013) Characterization of the microbial acid mine drainage microbial community using culturing and direct sequencing techniques. J Microbiol Methods 93: 108–115.

17. Redel H, Gao Z, Li H, Alekseyenko AV, Zhou Y, et al. (2013) Quantitation and composition of cutaneous microbiota in diabetic and nondiabetic men. JID 207: 1105–1114.

18. Taberlet P, Prud'Homme SM, Campione E, Roy J, Miquel C, et al. (2012) Soil sampling and isolation of extracellular DNA from large amount of starting material suitable for metabarcoding studies. Mol Ecol 21:1816–1820.

19. Baker GC, Smith JJ, Cowan DA (2003)Review and re-analysis of domain-specific 16S primers. J Microbiol Methods 55:541–555.

20. Humblot C, Guyoet J-P (2009) Pyrosequencing of tagged 16S rRNA gene amplicons for rapid diciphering of the microbiomes of fermented foods such as pearl millet slurries. Appl Environ Microbiol 75: 4354–4361.

21. Nossa CW, Oberdorf WE, Yang L, Aas JA, Paster BJ, et al. (2010) Design of 16S rRNA gene primers for 454 pyrosequencing of the human foregut microbiome. WJG 16:4135–4144.

22. Grant S, Grant WD, Cowan DA, Jones BE, Ma Y, et al. (2006) Identification of eukaryotic open reading frames in metagenomic cDNA libraries made from environmental samples. Appl Environ Microbiol 72:135–143.

23. Bailly J, Fraissinet-Tachet L, Verner M-C, Debaud J-C, Lemaire M, et al. (2007) Soil eukaryotic functional diversity, a metatranscriptomic approach. ISME J 1:632–642.

24. Meyer M, Stenzel U, Hofreiter M (2008) Parallel tagged sequencing on the 454 platform. Nat Prot 3:267–278.

25. Overbeek R, Begley T, Butler RM, Choudhuri JV, Chuang HY, et al. (2005) The subsystems approach to genome annotation and its use in the project to annotate 1000 genomes. Nucleic Acids Res 33: 5691–5702.

26. Meyer F, Paarmann D, D'Souza M, Olson R, Glass EM, et al. (2008) The metagenomics RAST server - a public resource for the automatic phylogenetic and functional analysis of metagenomes. BMC Bioinformatics 9: 386. doi:10.1186/1471-2105-9-386

27. Altschul SF, Madden TL, Schäffer AA, Zhang J, Zhang Z, et al. (1997) Gapped BLAST and PSI-BLAST: a new generation of protein database search programs. Nucleic Acids Res 25: 3389–3402.

28. Sayers EW, Barrett T, Benson DA, Bolton E, Bryant SH, et al. (2010) Database resources of the National Center for Biotechnology Information. Nucleic Acids Res 38: D5–D16.

29. Maidak BL, Cole JR, Liburn TG, Parker CT Jr, Saxman PR, et al. (2001) The RDP-II (ribosomal database project). Nucleic Acids Res 29: 173–174.

30. McDonald D, Price MN, Goodrich J, Nawrocki EP, DeSantis TZ, et al. (2012)An improved Greengenes taxonomy with explicit ranks for ecological and evolutionary analyses of bacteria and archaea. ISME J 6:610–618.

31. Brunak S, Danchin A, Hattori M, Nakamura H, Shinozaki K, et al. (2002) Nucleotide sequence database policies. Science 298: 1333.

32. Leinonen R, Akhtar R, Birney E, Bower L, Cerdeno-Tárraga A, et al. (2011) The European nucleotide archive. Nucleic Acids Res 39: D28–D31.

33. Pruesse E, Quast C, Knittel K, Fuchs BM, Ludwig W, et al. (2007) SILVA: a comprehensive online resource for quality checked and aligned ribosomal RNA sequence data compatible with ARB. Nucleic Acids Res 35: 7188–7196.

34. Yue JC, Clayton MK (2005) A similarity measure based on species proportions. Commun Stat Theor Methods 34: 2123–2131.

35. Chao A, Chazdon RL, Colwell RK, Shen T-J (2006) Abundance-based similarity indices and their estimation when there are unseen species in samples. Biometrics 62: 361–371.

36. Schloss PD, Westcott SL, Ryabin T, Hall JR, Hartmann M, et al. (2009) Introducing mothur: open-source, platform-independent, community-supported software for describing and comparing microbial communities. Appl Environ Microbiol 75: 7537–7541.

37. Aziz RK, Bartels D, Best AA, Dejongh M, Disz T, et al. (2008) The RAST server: rapid annotations using subsystems technology. BMC Genomics 9: 75. doi:10.1186/1471-2164-9-75

38. Appeltans W, Bouchet P, Boxshall GA, De Broyer C, de Voogd NJ, et al. (eds.) (2012) World Register of Marine Species [http://www.marinespecies.org]

39. Andersen K, Bird KL, Rasmussen M, Haile J, Breuning-Madsen H, et al. (2012) Meta-barcoding of 'dirt' DNA from soil reflects vertebrate biodiversity. Mol Ecol 21:1966–1979.

40. Caporaso JG, Lauber CL, Walters WA, Berg-Lyons D, Lozupone CA, et al. (2010) Global patterns of 16S rRNA diversity at a depth of millions of sequences per sample. PNAS: doi:10.1073/pnas.1000080107

41. Manichanh C, Chapple CE, Frangeul L, Gloux K, Guigo R, et al. (2008) A comparison of random sequence reads versus 16S rDNA sequences for estimating the biodiversity of a metagenomic library. Nucleic Acids Res 36: 5180–5188.

42. Maurice CF, Haiser HJ, Turnbaugh PJ (2012) Xenobiotics shape the physiology and gene expression of the active human gut microbiome. Cell 152: 39–50.

43. Ridaura VK, Faith JJ, Rey FE, Cheng J, Duncan AE, et al. (2013) Gut microbiota from twins discordant for obesity modulate metabolism in mice. Science 341. doi:10.1126/science.1241214

44. Henry EA, Devereux R, Maki JS, Gilmour CC, Woese CR, et al. (1994) Characterization of a new thermophilic sulfate-reducing bacterium Thermodesulfovibrio yellowstonii, gen.nov. and sp.nov.: its phylogenetic relationship to Thermodesulfobacterium commune and their origins deep within the bacterial domain. Arch Microbiol 161: 62–69.

45. Huang LN, Zhou H, Chen YQ, Luo S, Lan CY, et al. (2002) Diversity and structure of the archaeal community in the leachate of a full-scale recirculating landfill as examined by direct 16S rRNA gene sequence retrieval. FEMS Microbiol Lett 214:235–240.

46. Kendall MM, Liu Y, Sieprawska-Lupa M, Stetter KO, Whitman WB, et al. (2006) Methanococcus aeolicus sp. Nov., a mesophilic, methanogenic aechaeon from shallow and deep marine sediments. Intl J Syst Evol Microbiol 56: 1525–1529.

47. Wada K, Fukuda K, Yoshikawa T, Hirose T, Ikeno T, et al. (2012) Bacterial hazards of sludge brought ashore by the tsunami after the great East Japan earthquake of 2011. J Occup Health 54: 255–262.

48. Banning EC, Casciotti KL, Kujawinski EB (2010) Novel strains isolated from a coastal aquifer suggest a predatory role for flavobacteria. FEMS Microbiol Ecol 73: 254–270.

49. Vezzi A, Campanaro S, D'Angelo M, Simonato F, Vitulo N, et al. (2005) Life at depth: Photobacterium profundum genome sequence and expression analysis. Science 307: 1459–1461.

50. Yoshida K, Matsumoto T, Tateda K, Uchida K, Tsujimoto S, et al. (2000) Role of bacterial capsule in local and systemic inflammatory responses of mice during pulmonary infection with Klebsiella pneumoniae. J Med Microbiol 49: 1003–1010.

51. Grotiuz G, Sirok A, Gadea P, Varela G, Schelotto F (2006) Shiga toxin 2-producing Acinetobacter haemolyticus associated with a case of bloody diarrhea. J Clin Microbiol 44: 3838–3841.

52. Starliper CE (2011) Bacterial coldwater disease of fishes caused by Flavobacterium psychrophilum. J Adv Res 2: 97–108.

53. Osorio CR, Toranzo AE, Romalde JL, Barja JL (2000) Multiplex PCR assay for ureC and 16S rRNA genes clearly discriminates between both subspecies of Photobacterium damselae. Dis Aquat Org 40: 177–183.

54. Rosa LH, Machado KM, Rabello AL, Souza-Fagundes EM, Correa-Oliverira R, et al. (2009) Cytotoxic, immunosuppressive, trypanocidal and antileishmanial activities of Basidiomycota fungi present in Atlantic rainforest in Brazil. Antonie Van Leeuwenhoek 95: 227–237.

55. Qadri M, Johri S, Shah BA, Khajuria A, Sidiq T, et al. (2013) Identification and bioactive potential of endophytic fungi isolated from selected plants of the Western Himalayas. SpringerPlus 2: 8.

56. Jaszek M, Osińska-Jaroszuk M, Janusz G, Matuszewska A, Stefaniuk D, et al. (2013) New bioactive fungal molecules with high antioxidant and antimicrobial capacity isolated from Cerrena unicolor idiophasic cultures. BioMed Res Intl 2013. doi:10.1155/2013/497492

57. Hemphill-Haley E (1995) Diatoms evidence for earthquake-induced subsidence and tsunami 300 yr ago in southern coastal Washington.GSA Bulletin 107: 367–378.

58. Hemphill-Haley E (1996) Diatoms as an aid in identifying late-Holocene tsunami deposits. Holocene6: 439–448.

17

Distribution of Hydrothermal Alvinocaridid Shrimps: Effect of Geomorphology and Specialization to Extreme Biotopes

Anastasia A. Lunina*, Alexandr L. Vereshchaka

Laboratory of structure and dynamics of plankton communities, P.P. Shirshov Institute of Oceanology of Russian Academy of Sciences, Moscow, Russia

Abstract

The aim of this study is to review of our knowledge about distribution of recently known species of vent shrimps and to analyze factors influencing distribution patterns. Analyses are based upon (1) original material taken during eight cruises in the Atlantic Ocean (a total of 5861 individuals) and (2) available literature data from the Atlantic, Pacific, and Indian Oceans. Vent shrimps have two patterns of the species ranges: local (single vent site) and regional (three - six vent sites). Pacific species ranges are mainly of the local type and the Atlantic species ranges are of the regional type. The regional type of species ranges may be associated with channels providing easy larval dispersal (rift valleys, trenches), while the local type is characteristic for other areas. Specialization of a shrimp genus to extreme vent habitats leads to two effects: (1) an increase in the number of vent fields inhabited by the genus and (2) a decrease of species number within the genus.

Editor: Brock Fenton, University of Western Ontario, Canada

Funding: The studies were supported by the Ministry of Education and Science of the Russian Federation, by the Russian Foundation for Basic Research, and by the Program for a Basic Research of the Presidium of the Russian Academy of Sciences. The funders had no role in study design, data collection and analysis, decision to publish, or preparation of the manuscript.

Competing Interests: The authors have declared that no competing interests exist.

* E-mail: yalunina@list.ru

Introduction

The discovery of hydrothermal vents along the Galápagos Ridge in 1977 [1] has stimulated an increasing research effort examining the diversity, ecology, physiology, and biogeography of vent organisms, as well as new avenues of research into the origins of life on Earth [2]. Unusual characteristics of deep-sea vents compared with other deep-sea habitats, coupled with the ephemeral nature of hydrothermal circulation, may have important implications for deep-sea biology [3]. Decades of exploration have revealed numerous vent sites and faunal assemblages at many mid-ocean ridges and back-arc basins. As the global biogeography of vent organisms has been elucidated, separate biogeographic provinces have been erected for the shallow and deep Atlantic, the East Pacific, the North East Pacific, West Pacific back-arc basins, and the Indian Ocean [4]. Recent studies have modified this general scheme, proving existence of a single province for the Atlantic, a single province for the North West Pacific, a single province for the South West Pacific and Indian Ocean, and a separation of the North East Pacific, North East Pacific Rise, and South East Pacific Rise [5]. However, there are some shortcomings in the methodology of Bachraty et al. [5], some of which have been addressed by Rogers et al. [3].

At the same time, several attempts have been made to understand which factors may affect the observed distributional patterns of vent biota. The effects of spreading rate and geomorphology of the mid-ocean ridge axis have been proposed to be among these factors [6]. We assume that some of patterns will become clearer and much more visible if we revise and analyze the global distribution of a selected taxonomic group,

including species with similar morphological, physiological, biochemical, and reproductive features. This group should be widely distributed and have numerous species to provide statistically significant conclusions. One potential group for such study is the shrimp family Alvinocarididae.

Alvinocaridid shrimps (Caridea: Alvinocarididae) represent the key elements of hydrothermal communities of several vent fields on the Mid-Atlantic Ridge (MAR), and they are members of the hydrothermal communities in other areas of the oceans [7]. We know that numerous species occur in the Atlantic, Pacific, and Indian Oceans, with new species described during the last 10 years [8–18]. Despite the importance and visibility of the group, no recent review of its composition, distribution, and spatial/ecological biogeography is available.

Even among the severe conditions of the deep sea (elevated pressure, complete absence of light), the environments of hydrothermal vents may be considered extreme, with unique physical and chemical properties such as high and rapidly changing temperature (from 2–4 °C to 400 °C), acidic pH, toxic heavy metals, an hydrogen sulfide [19–20]. Vent shrimp genera show numerous adaptations to vent habitats. These adaptations are fewer and less conspicuous in the genus *Alvinocaris*, which is similar to usual deep-sea shrimps, and numerous and prominent in the genus *Rimicaris* [21–23]. Specialization to vent habitats implies morphological adaptations (highly specialized mouthparts covered with soft setae, enlarged branchial chambers formed by the carapace, appearance of dorsal organ – [21–23] and adaptations reflected in life cycles [24]. Specialization to hydrothermal conditions increases in the string *Alvinocaris* - *Opaepele* - *Chorocaris*

- *Mirocaris* - *Rimicaris* [23–24]. Recently the family is ripe for review of its internal phylogeny.

In this paper we summarize original data about composition and distribution of Alvinocaridid vent shrimps and put them in the context of relevant general literature information. Further we try to (1) reveal general patterns of vent shrimp distribution, (2) understand factors determining the species distribution and (3) estimate role of specialization to extreme biotopes in the composition and distribution of vent shrimp genera.

Material and Methods

Original material was taken along the Mid-Atlantic Ridge during eight cruises of R/V "Akademik Mstislav Keldysh" with use of two deep-sea manned submersibles "Mir–1" and "Mir–2" (Fig. 1, Table 1). Seven vent fields were investigated during 1994-2005, including Menez Gwen (37.8417 N, 31.525 W), Lucky Strike (37.2933 N, 32.2733 W), Rainbow (36.23 N, 33.902 W), Broken Spur (29.17 N, 43.1717 W), TAG (26.1367 N, 44.8267 W), Snake Pit (23.3683 N, 44.95 W) and Logatchev (14.752 N, 44.9785 W). No specific permissions were required for field studies for all locations. The field studies did not involve endangered or protected species.

Faunal composition, micro-scale distribution, behavior and population structure of shrimps in various vent microbiotopes were thoroughly investigated. Shrimps were collected using baited traps and submersible suction samplers. Immediately after

retrieval all specimens were sorted, measured, and fixed in 80% alcohol. Measurements follow established methods for shrimp morphological description [25]. Type material is deposited in the Zoological Museum, Moscow, and the Oxford University Museum of Natural History (OUMNH). A total of 5861 individuals of vent shrimps were analyzed.

Analysis of shrimp distribution was made with use of original and all available literature data (see references in Table 2) including original descriptions. Detailed description of material and discussion of the species status may be found in [17], [26–27].

As differences in sampling efforts may affect current records of species' distributions, we analyzed the InterRidge database including all recently recorded active vent sites (http://irvents-new3.whoi.edu). Table 3 illustrates the number of explored active vent sites within major geographic regions and the number of sites within those areas where Alvinocaridid shrimps have been recorded. To examine the possible correlation between these two numbers, we used the Spearman correlation coefficient. For a sample of size n and difference d coefficient ρ is computed from these: $\rho = 1 - (6\Sigma d^2 / n(n^2 - 1))$. All maps were created with use of the CorelDraw (styled CorelDRAW) vector graphics editor, version X6, graphics were made with use of Microsoft Excel spreadsheet application, version 14.0.

Results and Discussion

A list of all recently known species of Alvinocaridid vent shrimps is presented in Table 2. The shrimps represent eight genera and 26 species. Most of the Alvinocaridid species inhabit the Pacific (54% species) and Atlantic (38%) Oceans. Only two species (8%) are recorded from the Indian Ocean, which may be a result of less exploration of this area.

Most species listed in Table 2 (22 of 26) were reported exclusively from hydrothermal vents. A single species, *A. longirostris*, was found both in hot vents (Kermadec fault, [9]) and cold seeps (Sagami Bay, Japan, [28–29]. Three species (*Alvinocaris methanophyla*, *A. muricola*, and *A. stactophyla*) were reported from cold seeps only. These species may be found in hot vents in the future (as occurred with *A. longirostris*), but for now they are excluded from our analyses.

While analyzing the species ranges for the obligate vent fauna, Mironov et al. [30] recognized three types of species ranges: (1) local, (2) regional, and (3) transoceanic ranges. A range is local type if the species has been recorded so far from a single hydrothermal field. A regional type of distribution represents cases where a species has been reported from more than one hydrothermal vent field within a large geographic region (e.g. Eastern Pacific, Western Pacific and the Mid-Atlantic Ridge). Species inhabiting at least two large geographic regions are classified as transoceanic in range type.

For vent shrimps we observe two types of species ranges. Most species of the genera *Alvinocaris*, *Opaepele*, and *Chorocaris* along with the monotypic genera *Nautilocaris*, *Shinkaicaris*, and *Alvinocaridinides* exhibit local ranges (Fig. 2–4).

Regional-type species ranges (Fig. 2–4) are typical for Atlantic vent shrimps and for *Alvinocaris longirostris*. Ranges in this category include 3–6 hydrothermal sites mostly along the Mid-Atlantic Ridge.

As difference in sampling efforts may contribute to species distributions being classified as "local", we tried to compare sampling efforts using data in the Atlantic, Pacific, and Indian Oceans. Table 3 shows that Alvinocaridid vent shrimps occur in most geographic areas with active vents. The proportion of active vents in each region inhabited by Alvinocaridid shrimps varies

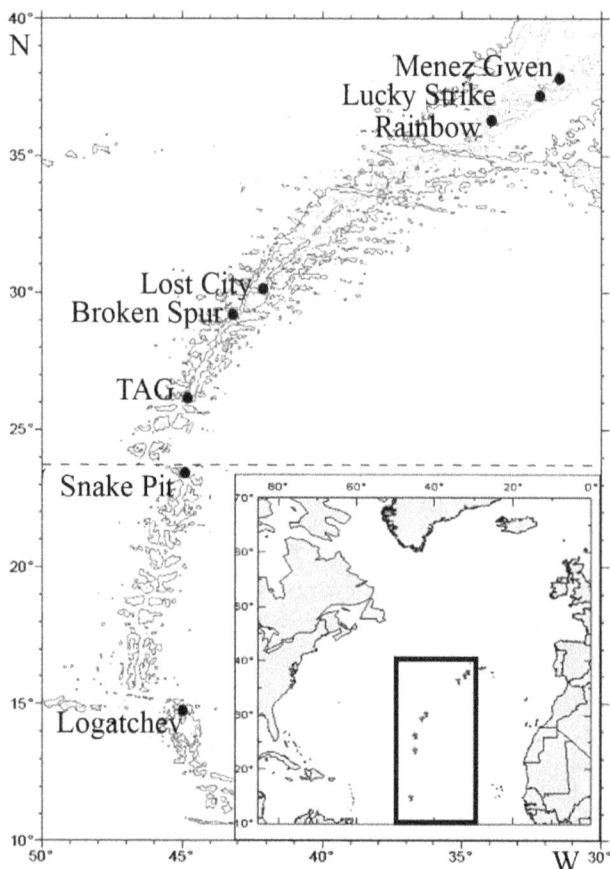

Figure 1. Map of the Atlantic hydrothermal vent fields visited during collection of the original material. Isobaths 500 m, 1000 m, 2000 m, and 3000 m are shown.

Table 1. Vent shrimps, original collections during 34–41 Cruises of R/V "Akademik Mstislav Keldysh", submersibles "Mir", Atlantic Ocean.

Hydrothermal field	Rimicaris exoculata		Alvinocaris markensis	
Hydrothermal field	Date	No of individuals	Date	No of individuals
Menez Gwen	-		01.03.1997	1
Lucky Strike	-		11.06.2002	4
Rainbow	25.10.1998	138	25.10.1998	21
	17.07.2002	40	18.07.2002	12
	17.07.2002	27	03.09.2005	110
	18.07.2002	47	-	-
	19.07.2002	61	-	-
	02.09.2005	79	-	-
	03.09.2005	68	-	-
Broken Spur	03.09.1996	75	6–8.09.1996	4
	04.09.1996	527	01.06.2002	8
	08.09.1996	49	25.08.2005	98
	08.09.1996	299	-	-
	01.07.2002	330	-	-
	01.07.2002	35	-	-
	29.09–01.10.1994	13	-	-
TAG	24.09.1994	330	26–27.06.2002	1
	24.09.1994	199	-	-
	25.06.2002	401	-	-
	26.06.2002	45	-	-
	27.06.2002	474	-	-
	17.09–22.09.1994	21	-	-
Snake Pit	20.06.2002	500	22.06.2002	22.06.2002
	20.06.2002	190	12.08.2003	12.08.2003
	21.06.02	100	-	-
	22.06.2002	82	-	-
	12.08.2003	147	-	-
	12.08.2003	211	-	-
	12.08.2003	232	-	-
	13.08.2003	65	-	-
Logachev	18.11.1998	461	01.02.1995	1
Logachev	-	-	27.07.1998	8

from 0 (South Ocean, Juan de Fuca Ridge, Manus Basin, Tonga Arc) to 0.4–0.5 (Mid-Atlantic Ridge, Central Pacific, Okinawa Trough) and even to 0.8 (Indian Ocean). The average value is 0.22 ± 0.06 (n = 15).

The correlation between the number of sites explored and the number of sites with Alvinocaridid shrimp records in the same area were analyzed. The Spearman correlation coefficient is 0,25 and P-level is 0,36, indicating a low relation between these parameters. We therefore suggest that differences in sampling efforts within various geographic regions do not significantly affect our conclusions.

Table 3 also indicates that all regions with vent shrimp records include vent sites with similar depth ranges, from shelf to ca. 3–5 thousand meters (except Kermadec Arc and Okinawa Trough where maximal depths are slightly less than 2 thousand meters).

That means that the depth factor may not significantly control the types of ranges exhibited by species.

This is remarkable that the local type of species ranges is characteristic for the Pacific Ocean, while the regional-type species ranges are usual for the Atlantic. Most of the Atlantic vent sites occur within a long narrow rift valleys, which is absent in the eastern Pacific Ocean, and we suggest that the global biogeographic difference between eastern Pacific and Atlantic Oceans might be influenced by such geomorphological differences. Indeed, the presence of a large number of species with regional species ranges is characteristic for the rift valleys in the Atlantic Ocean and the Okinawa Trench in the western Pacific. Both areas are similar in having long, narrow, and deep bottom channels. Regional species ranges were also reported for other Atlantic vent animals: on average, the local type of distribution is reported for

Table 2. Recently known hydrothermal vent and seep shrimp species.

Species	Region	Habitat	Coordinates	Depth	Authors
Alvinocaridinides formosa	**Taiwan**	Gueishandao, Yilan County	24°51.231'N, 121°59.204'E	252–275 m	[16]
Alvinocaris alexander	**New Zealand, Kermadec Ridge**	Rumble V Seamount	36°08.27–35 S 178°11.74 E;	730–415 m	[14]
		Brothers Caldera	34°52.89 S 179°03.76 E	1346–1196 m	
Alvinocaris brevitelsonis	**Okinawa Trough**	Minami-Ensei Knoll	28°23.35'N, 127°38.38'E	705 m	[50]
Alvinocaris chelis	**Taiwan**	Gueishandao, Yilan County	24°49–51N 121°59–122°0'E	300–252 m	[16]
Alvinocaris dissimilis	**Okinawa Trough**	Minami-Ensei Knoll	28°23.35'N, 127°38.38'E	705 m	[10]
Alvinocaris komaii	**South-West Pacific: Lau**	Kilo Moana	20°9'S, 176°12'E	2620 m	[15]
		Tow Cam	20°19'S, 176°8'E	2700 m	
		ABE	20°45'S, 176°11'E	2145 m	
Alvinocaris longirostris	**Sagami Bay**	Off Hatsuchima site (cold seep)	35°00' N; 139°14'E	~1100 m	[28–29]
	Okinawa Trough	Iheya Ridge	27°32.70'N, 126°58.20'E	1360 m	[51]
		Hatoma Knoll	24°51'N; 123°50.4'E	~1950 m	[52]
		Miname-Ensei Knoll	28°23.4' N; 127°38.4'E	~700 m	
	New Zealand, Kermadec Ridge	Brothers Caldera	34° 51-53'S 179° 3-4'E	1850–1196 m	[9]
Alvinocaris lusca	**Galapagos Rift:**	Rose Garden area	00°48.15'N, 86°13.29'W	2450 m	[53]
	East Pacific Ridge	9°N site,	09°50.3'N, 104°17.4'W	2520 m	[54–55]
Alvinocaris markensis	**Mid-Atlantic Ridge**	Lucky Strike	37° 17.598'N 32° 16.398'W	1600–1740 m	[56]
		Rainbow	36° 13.800'N33° 54.120'W	2270–2320 m	
		Broken Spur	29° 10.200'N 43° 10.302'W	3100 m	[57]
		TAG	26° 8.202'N 44° 49.602'W	3436–3670 m	[21]
		Snake Pit	23° 22.098'N 44° 57.000'W	3450–3500 m	[55]
		Logatchev	14° 45.120'N 44° 58.710'W	2925–3050 m	[54]
Alvinocaris niwa	**New Zealand: Kermadec Ridge**	Brothers Caldera	34°52.89–52.87'S 179°3.76–3.21'E;	1346–1196 m	[9]
		Rumble V Seamount	36°8.63–8.57'S, 178°11.77–11.50'E;	877–655 m	
Alvinocaris williamsi	**Mid-Atlantic Ridge**	Menez Gwen	37° 50.502'N 31° 31.500'W	840–865 m	[58]
Chorocaris chacei	**Mid-Atlantic Ridge**	Moytirra	45° 28.998'N 27° 51.000'W	3000 m	[59]
		Lucky Strike	37° 17.598'N 32° 16.398'W	1600–1740 m	[60]
		Broken Spur	29° 10.200'N 43° 10.302'W	3100 m	
		TAG	26° 8.202'N 44° 49.602'W	3436–3670 m	
		Snake Pit	23° 22.098'N 44° 57.000'W	3450–3500 m	
		Logatchev	14° 45.120'N 44° 58.710'W	2925–3050 m	
		Ashadze	12° 58.398'N 44° 51.798' W	4080 m	[61]
Chorocaris paulexa	**East Pacific Rise**	17 37'S, EPR, Homer Vent Site	17°37.220'S, 113°15.123'W	2596 m	[11]
		Rapa Nui vent field, Brandon vents	21°33.7'S 114°17.9'W	3640 m	
Chorocaris vandoverae	**Western Pacific Ocean,Mariana Back-Arc Spreading Center**	Alice Springs vent field;	18°12.599'N, 144°42.431'E;	3660 m	[55–56], [62]
Nautilocaris saintlaurentae	**North Fiji Basin**	White Lady site	16°59.50'S, 173°55.47'E,	2000 m	[63]
	Lau Basin	Vaï-Lili site	22°13'S, 176°38'E	1750 m	
Opaepele loihi	**Pacific Ocean, Hawaii**	Loihi Seamount;	18°55'N, 155°16'W	980 m	[64]
	Mariana Arc	NW Rota-1 Volcano	14°36.0'N, 144°46.5'E	530 m	[65]
	Philippine Sea Plate	Nikko Seamount	23° 4.856' N 142° 19.512' E	456 m	[66]

Table 2. Cont.

Species	Region	Habitat	Coordinates	Depth	Authors
Opaepele susannae	**South Mid-Atlantic Ridge**	Semenov	13° 30.822'N 44° 57.780'W	2440 m	[67]
		Mephisto	04°47.834S, 12°22.593W	3045 m	
		Turtle Pits	04°48.57 S, 12°22.41 W	2998 m	[13]
		Sisters Peak	04°48.188S, 12°22.301W	2986 m	
		Lilliput	9° 33.000' S 13° 10.800'W	1500 m	
Opaepele vavilovi	**Mid-Atlantic Ridge**	Broken Spur	29.1700 N 43.1717 E	3100 m	[17]
Rimicaris exoculata	**Mid-Atlantic Ridge**	Moytirra	45° 28.998'N 27° 51.000'W	3000 m	[59]
		Lucky Strike (very low abundance)	37° 17.598'N32° 16.398'W	1600–1740 m	[10]
		Rainbow	36° 13.800'N 33° 54.120'W	2270–2320 m	[60]
		Broken Spur	29° 10.200'N 43° 10.302'W	3100 m	
		TAG	26° 8.202'N 44° 49.602'W	3436–3670 m	
		Snake Pit	23° 22.098'N 44° 57.000'W	3450–3500 m	
		Logatchev	14° 45.120'N 44° 58.710'W	2925–3050 m	
		Ashadze	12° 58.398'N 44° 51.798' W	4080 m	[61]
	South MAR	Mephisto	04°47.834S, 12°22.593W	3045 m	
Rimicaris kairei	**Central Indian Ridge, Rodriguez Triple Junction**	Kairei Field;	25°19.16'S, 70°02.40'E;	2454 m	[8]
		Edmond vent field	23°52.68'S, 69°35.80'E	3290–3320 m	[68]
		Dodo hydrothermal field	18°20.19S, 65°17.99E;	2745 m	[69]
		Solitaire hydrothermal field	19°33.413S, 65°50.888E	2606 m	
	SW Indian Ridge	Dragon	37° 46.998'S 49° 39.000'W	2785 m	Copley J., pers. comm.
Rimicaris hybisae	**Mid-Cayman Spreading Centre**	Beebe	18° 32.688'N 81° 43.170'W	4960 m	[18]
		Von Damm	18° 22.596'N 81° 47.832'W	2300 m	
Shinkaicaris leurokolos	**Okinawa Trough**	Minami-Ensei Knoll	28°23.35'N, 127°38.38'E	705 m	[50]
Mirocaris fortunata	**Mid-Atlantic Ridge**	Moytirra	45° 28.998'N 27° 51.000'W	3000 m	[59]
		Menez Gwen	37° 50.502'N 31° 31.500'W	840–865 m	[70]
		Lucky Strike	37° 17.598'N32° 16.398'W	1600–1740 m	
		Rainbow	36° 13.800'N 33° 54.120'W	2270–2320 m	
		Broken Spur	29° 10.200'N 43° 10.302'W	3100 m	
		TAG	26° 8.202'N 44° 49.602'W	3436–3670 m	
		Snake Pit	23° 22.098'N 44° 57.000'W	3450–3500 m	
		Logatchev	14° 45.120'N 44° 58.710'W	2925–3050 m	
		Ashadze	12° 58.398'N 44° 51.798' W	4080 m	[61]
	South MAR	Turtle Pits	04°48.57 S, 12°22.41 W	2998 m	
Mirocaris indica	**Central Indian Ridge, Rodriguez Triple Junction:**	Kairei Field	25°19.2'S, 70°02.4'E	2422 m	[12]
		Edmond Field	23°52.7'S, 69°35.8'E,	3300 m	
		Solitaire hydrothermal field	19°33.413S, 65°50.888E	2606 m	[69]
	SW Indian Ridge	Dragon	37° 46.998'S 49° 39.000'W	2785 m	Copley J., pers. comm.

Table 3.

Region	Number of sites explored	Number of sites with shrimp records	Share of active vents inhabited by shrimps	Depth range, m	Sites explored
Southern Ocean	3	0	0.00	45–270	ESR; E2, Adventure Caldera, Kemp Caldera
Northwest Atlantic	7	2	0.29	1–4960	Champagne Hot Springs, Kick'em Jenny submarine volcano, Montserrat Volcano, Beebe, Europa, Von Damm, Don Joao de Castro Bank
Mid-Atlantic Ridge	22	11	0.50	350–4200	Ashadze, Ashadze 2, Broken Spur, Bubbylon, Evan, Logatchev, Logatchev 2, Lost City, Lucky Strike, Menez Gwen, Menez Hom, Moytirra, Rainbow, Saldanha, Semyenov, Snake Pit, TAG, Steinaholl Vent Field, Baily's Beads, Lilliput, MAR; 4 48'S, Nibelungen
Indian Ocean	6	5	0.83	1600–3320	Aden, Dodo Field, Edmond Field, Kairei Field, Solitaire Field, SWIR Area A
Central Pacific	5	2	0.40	150–4800	Loihi Seamount, Bounty Seamount, Macdonald Seamount, Teahitia vents, Vailulu'u Seamount
Galapagos Rift	6	1	0.17	1640–2700	Calyfield, Galapagos Mounds, Iguanas-Pinguinos, Navidad, Precious Stone Mountain, Rose Garden
Juan de Fuca Ridge	20	0	0.00	1540–3460	Axial Volcano; ASHES, Axial Volcano; CASM, Axial Volcano; South Rift Zone, Baby Bare Seamount, Central Cleft; off-axis, East Blanco Depression, Floc, Flow, High-Rise Field, Main Endeavour Field, Middle Valley; Dead Dog Vent Field, Middle Valley; ODP Mound, Mothra Field, North Cleft; high temperature, North Cleft; low temperature, Not Dead Yet, Salty Dawg Field, Sasquatch Field, Source, South Cleft
Kermadec Arc	9	2	0.22	130–1800	Brothers volcano, Clark volcano, Giggenbach volcano, Healy volcano, Macauley Caldera, Rumble III volcano, Rumble V volcano, Vulkanolog, Wright volcanic center
Lau Basin	17	4	0.24	1200–2700	ABE, CDE, CLSC; A3, Hine Hina, Kilo Moana, Kulo Lasi, Maka, Mariner, Misiteli, Si'iSi'i, Tahi Moana 2, TELVE, Tow Cam, Tu'i Malila, Vai Lili, Volcano O, White Church
Manus Basin	8	0	0.00	535–2500	DESMOS Cauldron, PACMANUS field, Solwara 11, Solwara 13, Solwara 17, SuSu Knolls, Vienna Woods, Vienna Woods; Hydrothermal Field 4
Mariana Arc and Trough	21	2	0.10	55–3676	Daikoku volcano, East Diamante volcano, Esmeralda Bank, Forecast, Kasuga 2 Seamount, Kasuga 3 Seamount, Maug Caldera, Minami-Hiyoshi submarine volcano, Nikko volcano, Northwest Eifuku, Northwest Rota-1 volcano, Ruby, Seamount X, TOTO Caldera, West Rota volcano, 13 N Ridge Site, Alice Springs Field, Mariana Mounds, Mariana Trough; unnamed, Pika, Snail
North East-Pacific Rise	22	1	0.05	2000–2950	AHA Field, EPR; 10 02'N, EPR; 10 44.6'N, EPR; 11 17'N, EPR; 11 24'N, EPR; 11 42'N, EPR; 13 N, EPR; 13 N; Marginal High, EPR; 21 N, EPR; 3.9 N offset, EPR; 8 38'N, EPR; 9 17'N, EPR; 9 30'N, EPR; 9 33'N, EPR; 9 40'N, EPR; 9 47'N, EPR; 9 50'N, Feather Duster, Medusa, Mounds and Microbes, Red Seamount, Teotihuacan

Table 3. Cont.

Region	Number of sites explored	Number of sites with shrimp records	Share of active vents inhabited by shrimps	Depth range, m	Sites explored
Okinawa Trough	11	5	0.45	30–1850	Iheya Ridge, Irabu Knoll, Izena Cauldron, Kueishan Island, Kueishan Island; offshore, Minami-Ensei Knoll, Natsushima 84-1 Knoll, North Knoll; Iheya Ridge, SPOT; Hatoma Knoll, SPOT; Yonaguni Knoll IV, Yoron Hole
South East-Pacific Rise	27	2	0.07	2064–3050	Animal Farm, EPR; 1.4 S; off-axis, EPR; 11 18'S, EPR; 14 S, EPR; 17 12'S, EPR; 17 34'S, EPR; 17 44'S, EPR; 18 10'S, EPR; 18 15'S, EPR; 18 26'S, EPR; 18 32'S, EPR; 2 S, EPR; 20 06'S, EPR; 21 25'S, EPR; 23 30'S, EPR; 23 50'S, EPR; 26 10'S, EPR; 26.5 S, EPR; 7 25'S, EPR; Ridge 1; 20 40'S propaging rift, EPR; Ridge 3; 20 40'S propaging rift, Nolan's Nook, Pito Seamount, Rapa Nui, Rehu-Marka, Saguaro Field, Stealth
Tonga Arc	7	0	0.00	210–2600	Mata Fitu, Mata Tolu, Monowai Caldera, Tonga Arc; Volcano 1, Tonga Arc; Volcano 19, Tonga Arc; Volcano P, West Mata submarine volcano

75% of the obligate vent species in all oceans [19], while for the Atlantic Ocean this value is 43% [31].

The Mid-Atlantic Ridge has a low spreading rate (ca 5 cm per year). The central part of rift valley is characterized by tectonic

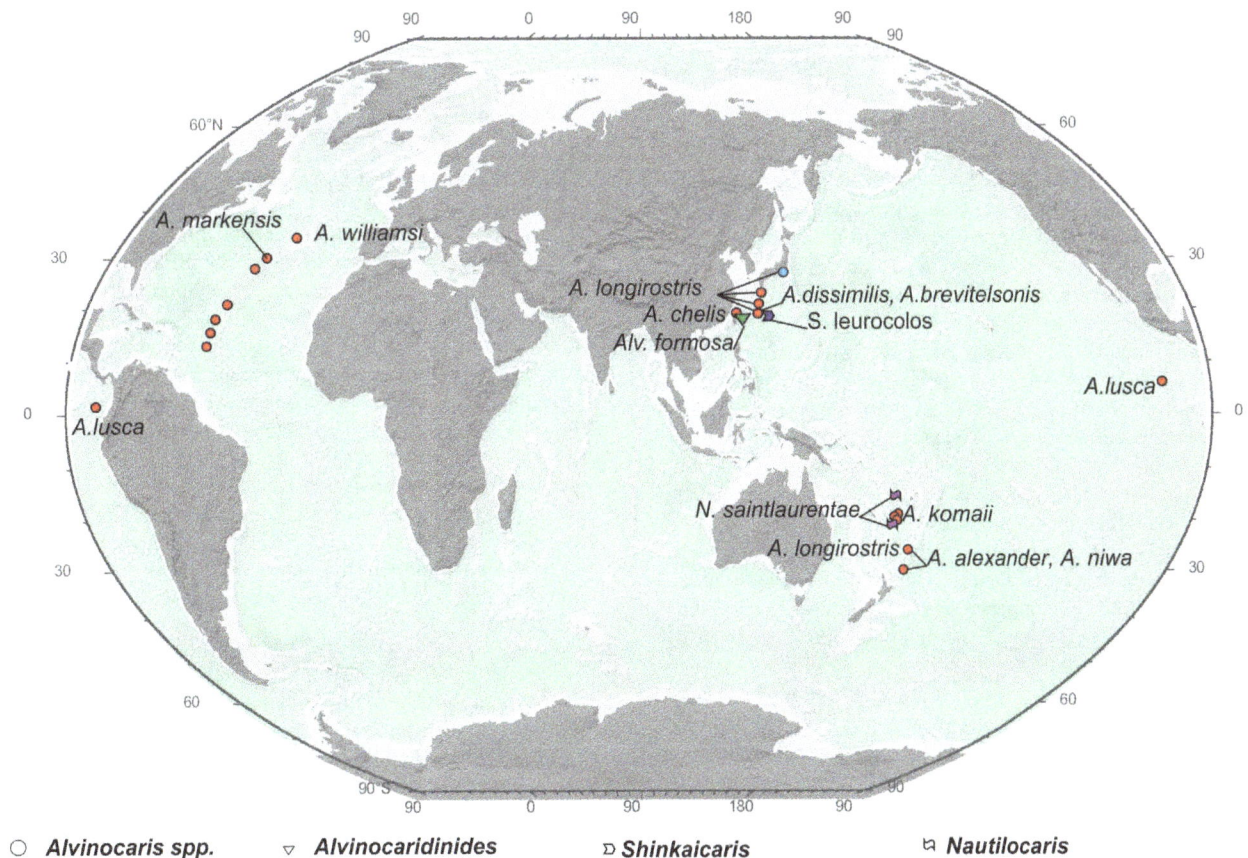

Figure 2. Distribution of the genera *Alvinocaris, Alvinocaridinides, Shinkaicaris,* **and** *Nautilocaris.* The same symbol shape indicates the same genus, and the same symbol color indicates the same species.

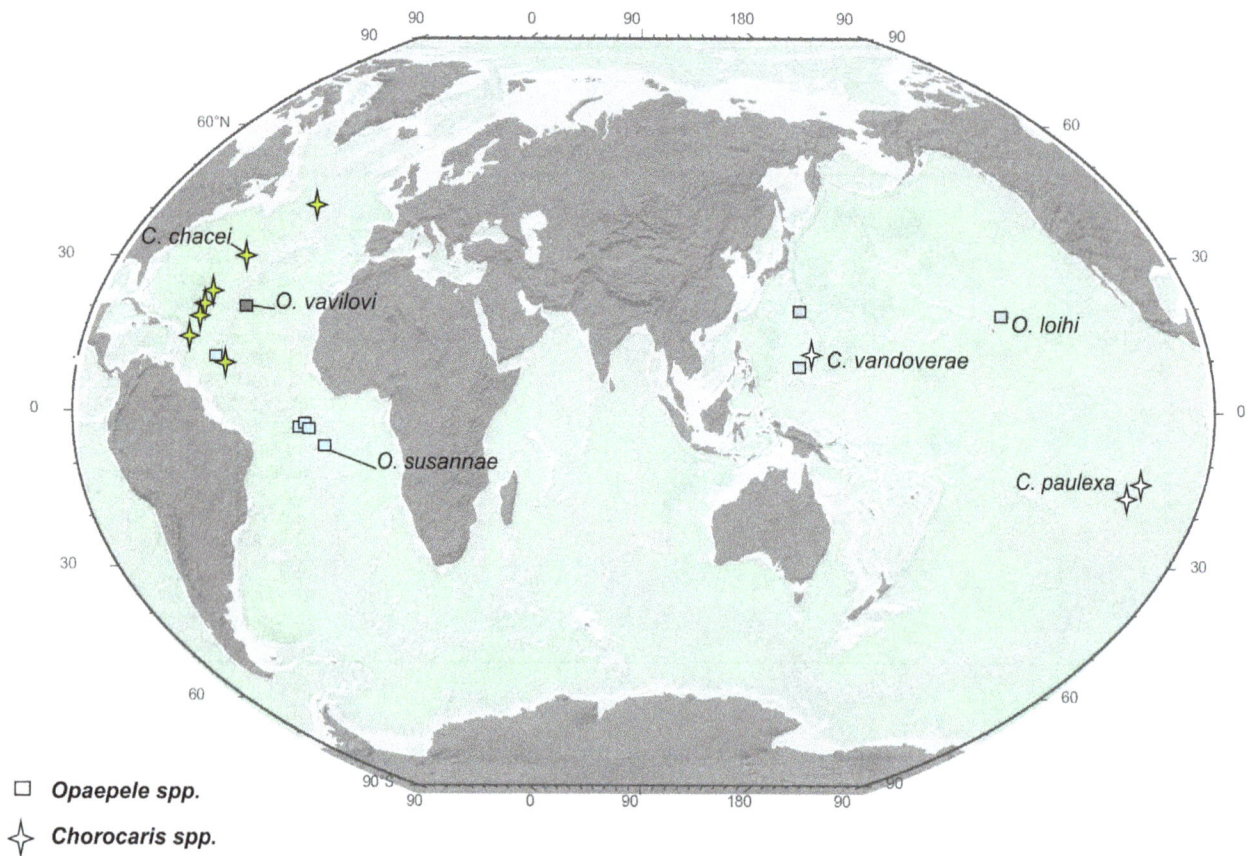

Figure 3. Distribution of the genera *Rimicaris* and *Mirocaris*. The same symbol shape indicates the same genus, and the same symbol color indicates the same species.

and volcanic activity (Fig. 5 A, B); it is the area where the vent communities develop. On average, the rift valley is about 12.5 km wide and 2.8 km deep [32]. The buoyant plume of hydrothermal vent fields cannot go beyond the internal rift and spread along the ridge axis [33]. The buoyant plume carries dispersal stages of shrimps from vent sites to the water column at the level of buoyant plume, approximately 200–300 m above the hydrothermal source [34]. Thus, species spread along the MAR and may colonize a number of hydrothermal fields. The rift valley may therefore serve as a corridor channeling the dispersal of a vent animal's larvae along the valley (Fig. 5A) without considerable loss of individuals.

The spreading rate of the East Pacific Rise is much greater (15 cm per year) than that of the Mid-Atlantic Ridge. The valley is the apical part of the ridge and the hydrothermal plumes rise above the flanking edges [35–36]. Under these circumstances, deep water flows take the dispersal stages out of the rift valley, thus reducing the chance to settle at the neighboring vent fields (Fig. 6B).

The type of recruitment is reflected in the type of the species ranges. The presence of the dispersal corridor in the Atlantic Ocean may provide extensive gene exchange between populations inhabiting neighboring vent and prevent geographic isolation and speciation. In this case, we observe elongated species ranges of regional type along the Mid-Atlantic Ridge. Conversely, the absence of such a dispersal corridor in the eastern Pacific Ocean may lead to the significant loss larvae, considerable geographic isolation and higher speciation. In this case, we observe local species ranges.

Here we are faced with 2 problems:

(1) There are vent taxa such as the siboglinid polychaetes that show a regional distribution along the EPR, in contrast to Alvinocaridid shrimp. A possible explanation for this pattern may be related to differences in the fecundity of these taxa and their levels of gene flow between neighbouring vent fields. At present, we do not have estimates of lifetime fecundity for any siboglinid [37]. Observations on spawning by a group of 155 females showed that a large numbers of oocytes were being spawned each day, but it was impossible to state whether any particular female spawned each day or how much. However, the average spawning rate of oocytes per female per day (over a 7-day period) was 335 ($\pm$130) [38]. Given that spawning lasts at least hundreds/thousands of days, lifetime fecundity for siboglinid could be of order of magnitude 10^{4-5} propagules per female.

Total fecundity of *Alvinocaris muricola* is related to female size and ranges between 1432 to 5798 embryos [39], e.g. 1–2 orders of magnitude less than that suggested for siboglinid polychaetes. Total abundances of alvinocaridid shrimp and siboglinid poly-chaetes at vent fields are difficult to estimate, but abundances of siboglinid species may be at least as high as that of alvinocaridid species. If so, the gene flow between neighbouring fields for siboglinid species could be 1–2 orders of magnitude higher than that of alvinocaridid species, if we assume comparable larval duration and mortality. This difference in potential gene flow could prevent isolation and maintain the regional type of species

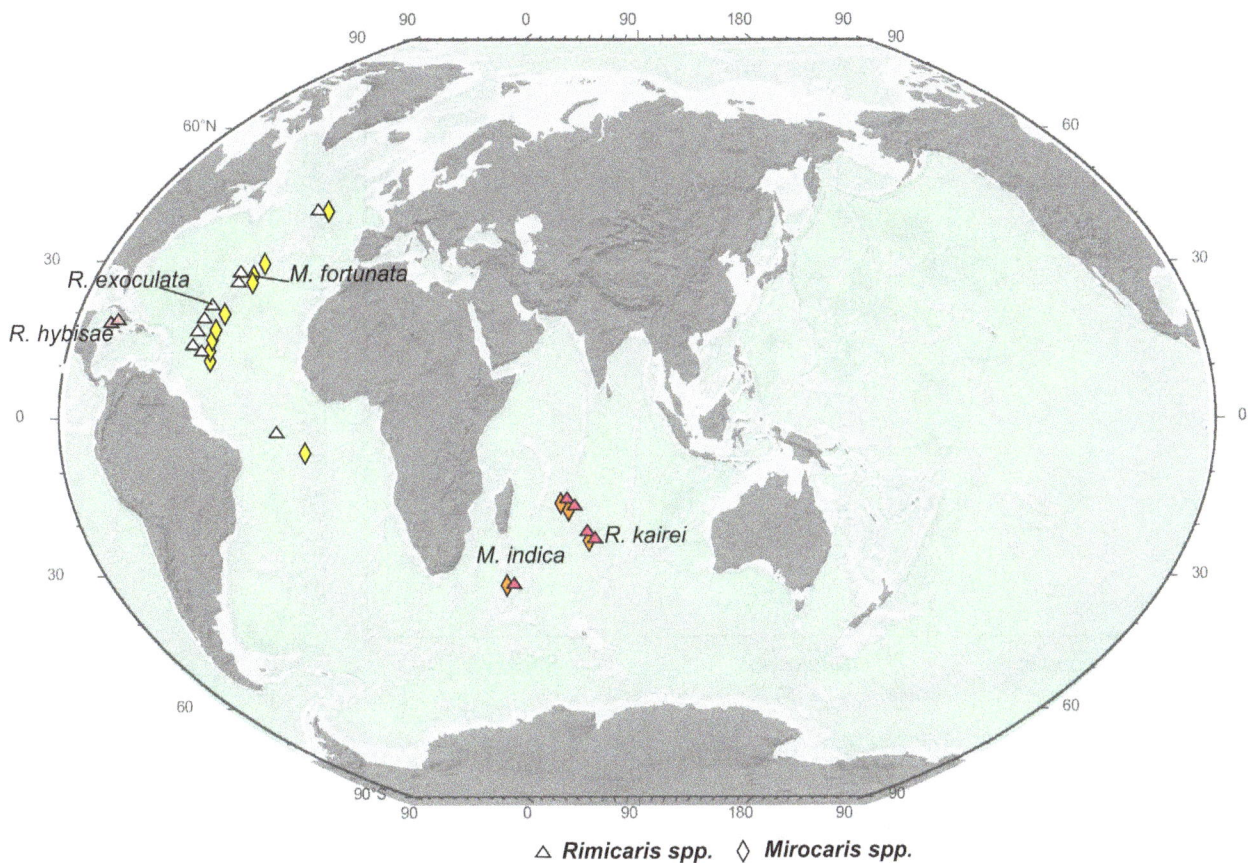

Figure 4. Distribution of the genera *Chorocaris* and *Opaepele*. The same symbol shape indicates the same genus, and the same symbol color indicates the same species.

range for the siboglinids in the eastern Pacific. New data about the fecundity, population densities, larval mortality, and gene flow, for example from the use of molecular methods along with modeling of water mass advection, may test some of the assumptions in this overall hypothesis.

(1) Fundamentally, can a "local" distribution be "real": how can a species endemic to an ephemeral environment such as hydrothermal vents only be present at only one vent field? When venting inevitably ceases at a vent field, offspring of that population must have colonised neighbouring vent fields for the species to persist. Along the MAR during monitoring of vent fields between 1994 and 2005 [26], we recorded different morphs of *Alvinocaris* (identified as species 1–6). However, final examination of hundreds of specimens (based on morphology) and statistical analysis have proven that all individuals represented a single highly diverse species [26]. At any time we observed different populations moving along the MAR. The channeling effect of the MAR rift valley may have promoted panmixia of the population through the region.

Conversely, at the EPR, lack of the channeling effect does not prevent speciation and may lead to appearance of similar related species each inhabiting 1–2 neighboring vent fields. We also acknowledge the additional possibility that a perceived "local" pattern may just be a result of uneven sampling effort so far.

With new information about composition and distribution of vent shrimps, we can consider whether these parameters are related to specialization to vent habitats demonstrated by different

Alvinocaridid genera. Specialization to the vent environment in adult forms increases in the string *Alvinocaris* - *Opaepele* - *Chorocaris* - *Mirocaris* – *Rimicaris* [23]. This appears to lead to two effects: (1) increase of average number of vent fields inhabited by the one species of the genus and (2) decrease of species number within the genus (Fig. 6). The less specialized genus (*Alvinocaris*), with least number of adaptations to vent environments in adult form, was found to be much more speciose than the specialized genera including 2–3 species each (*Mirocaris*, *Rimicaris*). The leap in specialization occurs between *Alvinocaris* and all other genera of Alvinocarididae both morphologically (significantly modified characters) and ecologically (harbouring exosymbionts). It is here that we find difference in number of species within the genus (>10 in *Alvinocaris* and 2–3 each in the other genera). Each species within less specialized genera (*Opaepele*, *Chorocaris*) inhabits one or two vent fields, whilst each species of the most specialized species occurs in numerous vent sites. Genera with intermediate specialization demonstrate intermediate patterns.

The analyses of palaeontological data and living marine mollusks indicate that proportion of monotypic genera may provide an index of the genus origination rate [40]. Since most of vent shrimp genera are nearly monotypic and including two or three very similar species, we may suggests that the genus origination rate (and thus the speciation rate) within the group is high.

Analysis of the global biogeography of vent shrimps provides the clues to understand what factors drive and shape the distribution of animals under extreme environmental conditions. One possible

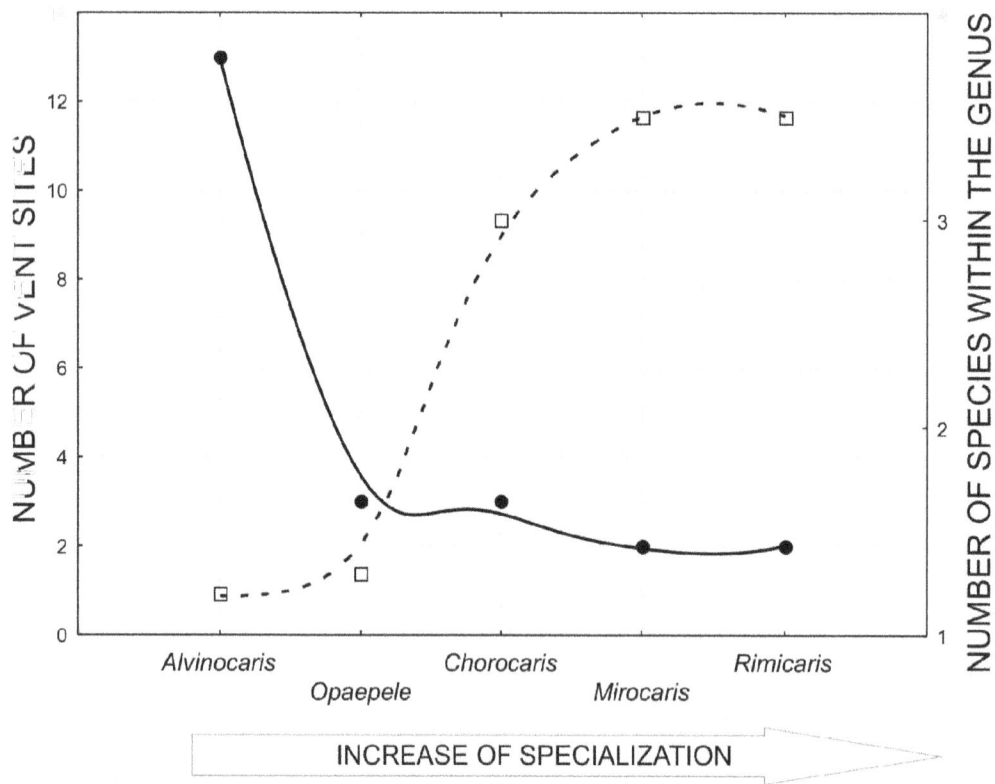

Figure 5. A - Mid-Atlantic Ridge, spreading rate 2.5 cm/year; B – East-Pacific Rise, full spreading rate 15 cm/year.

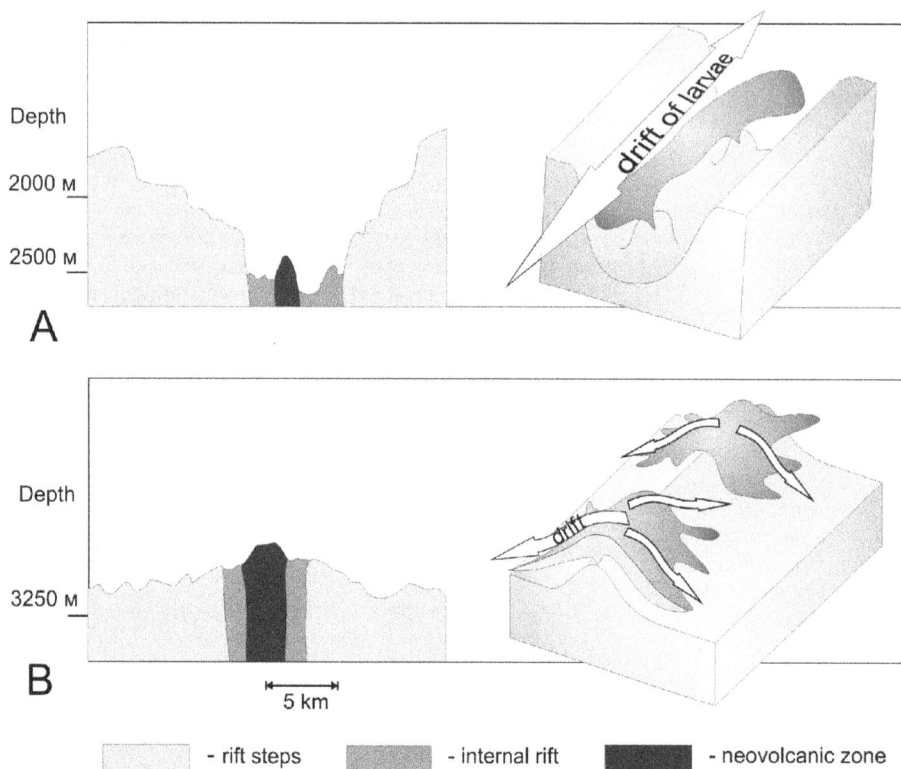

Figure 6. Relation between specialization, speciation and average number of sites inhabited by the species within the genus. Trend lines are made with use of distance weighted least squares analyses. These are just trends, not statistically significant effects. Specialization grows from left to right. Solid line and closed circles correspond to left y-axis. Dotted line and open squares correspond to right y-axis.

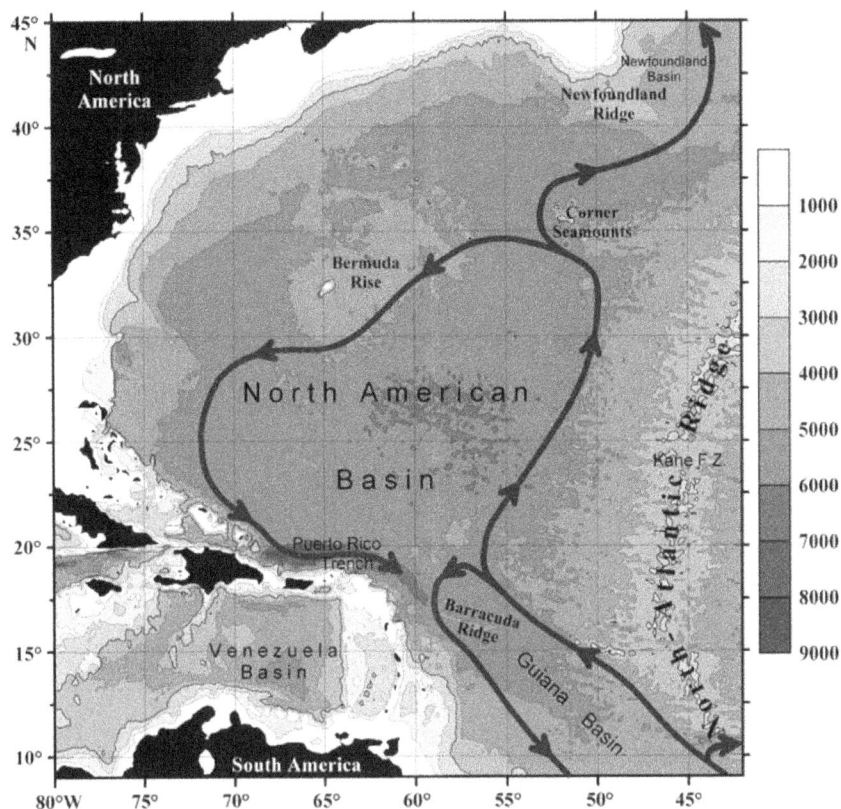

Figure 7. Circulation of Antarctic Bottom Water (Lower Circumpolar Water) in the Central and North Atlantic [41].

factor, as explored here, is the influence of ridge axis geomorphology on larval dispersal. A second possible factor is the specialization to extreme biotopes that may leads to (1) extension of the species range and (2) reducing of species number within the genus, to the limit of monophyly.

Finally, a third possible factor may be global circulation patterns acting over long periods of time: for example, near-bottom circulation in the deep Atlantic. Information about this circulation is scant, but available data indicate that it is dominated by Antarctic Bottom Water, flowing to the North American Basin after passing the Equatorial Channel and Guiana Basin [41]. Antarctic Bottom Water propagates mainly near the western slope of the Mid-Atlantic Ridge [42–43] and the circulation in the basin is cyclonic [43–45] (Fig. 7). Near-bottom meridional transport along with channeling effect is a unique feature of the Atlantic ocean that could also contribute to long regional species ranges within this area. Morphological analyses [26], [46] revealed fast population changes during 1994–2005, gene flow (measured by morphology) being apparently directed northward, coaxially with the main stream of Antarctic Bottom Water.

We may expect that species ranges from active vents in the South Atlantic are also of regional type and that any further discovered sites will appear to be populated by shrimp fauna similar to those of the North and Central Atlantic.

The Antarctic Circumpolar Current may be another factor influencing global shrimp distribution and preventing their occurrence in the Southern Ocean. According to mitochondrial cytochrome oxidase subunit analyses, vent shrimps radiated in the Miocene (less than ~20 Myr; [47]) and since then have been distributed worldwide except the Southern Ocean. Indeed, recent description of fauna associated with high-temperature hydrothermal vents on the East Scotia Ridge in the Southern Ocean indicate an absence of Alvinocaridid shrimps [3]. But it is not just Alvinocaridid shrimp that are excluded from the Southern Ocean vents seen so far: also Bathymodiolid mussels, and indeed any vent taxa with planktotrophic larvae. This is more likely to be a result of "Thorsen's Rule", given the productivity regime of polar regions, than it is to be a result of ACC as a hydrographic barrier. Other non-planktotrophic vent taxa have managed to colonise the region arguably within the period that the ACC has been active (e.g. Kiwidae crabs; [48]). In addition, planktotrophic larvae appear to be rare among the taxa present at Arctic vents (e.g. [49]), also consistent with "Thorsen's Rule".

Acknowledgments

The authors are grateful to A.V. Gebruk, A.N. Mironov, and J. Copley for help and fruitful discussions.

Author Contributions

Analyzed the data: AAL ALV. Contributed reagents/materials/analysis tools: ALV. Wrote the paper: AAL ALV.

References

1. Corliss JB, Dymond J, Gordon LI, Edmond JM, von Herzen RP, et al. (1979) Submarine thermal springs on the Galapagos Rift. Science 203: 1073–1083.

2. Martin W, Baross J, Kelley D, Russell MJ (2008) Hydrothermal vents and the origin of life. Nat. Rev. Microbiol 6: 805–814.

3. Rogers AD, Tyler PA, Connelly DP, Copley JT, James R, et al. (2012) The Discovery of New Deep-Sea Hydrothermal Vent Communities in the Southern Ocean and Implications for Biogeography. PLoS Biol 10(1): e1001234. doi:10.1371/journal.pbio.1001234

4. Van Dover CL, German CR, Speer KL, Parson LM, Vrijenhoek RC (2002) Evolution and biogeography of deep-sea vent and seep invertebrates. Science 295: 1253–1257.

5. Bachraty C, Legrende P, Desbruyeres D (2009) Biogeographic relationships among hydrothermal vent faunas on a global scale. Deep Sea Res Part 1 Oceanogr Res Pap 56: 1371–1378.

6. Ramirez-Llodra E, Shank TM, German CR (2007) Biodiversity and biogeography of hydrothermal vent species: Thirty years of discovery and investigations. Oceanography 20(1): 30–41.

7. Martin JW, Haney TA (2005) Decapod crustaceans from hydrothermal vents and cold seeps: a review through 2005. Zool J Linn Soc 145: 445–522.

8. Watabe H, Hashimoto J (2002) A new species of the genus *Rimicaris* (Alvinocarididae: Caridea: Decapoda) from the active hydrothermal vent field, 'Kairei field', on the Central Indian Ridge, the Indian Ocean. Zoolog Sci 19: 1167–1174.

9. Webber R (2004) A new species of *Alvinocaris* (Crustacea:Decapoda: Alvinocarididae) and new records of alvinocaridids from hydrothermal vents north of New Zealand. Zootaxa 444: 1–26.

10. Komai T, Segonzac M (2006) Decapoda, Caridea. In: Desbruyères DI, Segonzac M, Bright M, editors. Handbook of Deep-Sea Hydrothermal Vent Fauna. Denisia 18: pp. 410–454.

11. Martin J, Shank T (2005) A new species of the shrimp genus *Chorocaris* (Decapoda, Caridea, Alvinocarididae) from hydrothermal vents in the eastern Pacific. Proc. Biol. Soc. Wash. 118: 183–198.

12. Komai T, Martin J, Zala K, Tsuchida S, Hashimoto J (2006) A new species of *Mirocaris* (Crustacea, Decapoda, Caridea, Alvinocarididae) associated with hydrothermal vents on the Central Indian Ridge, Indian Ocean. Si Mar 70: 109–119.

13. Komai T, Giere O, Segonzac M (2007) New record of alvinocaridid shrimps (Crustacea: Decapoda: Caridea) from hydrothermal vent fields on the Southern Mid-Atlantic Ridge, including a new species of the genus *Opaepele*. Species Diversity 12: 237–253.

14. Ahyong S (2009) New Species and New Records of Hydrothermal Vent Shrimps from New Zealand (Caridea: Alvinocarididae, Hippolytidae). Crustaceana, 82: 775–794.

15. Zelnio K, Hourdes S (2009) A new species of *Alvinocaris* (rustacea: Decapoda: Caridea: Alvinocarididae) from hydrothermal vents at the Lau Basin, southwest Pacific, and a key to the species of Alvinocarididae. Proc. Entomol. Soc. Wash. 122: 52–71.

16. Komai T, Chan T-Y (2010) A new genus and two new species of alvinocaridid shrimps (Crustacea:Decapoda: Caridea) from a hydrothermal vent field off northeastern Taiwan. Zootaxa 2372: 15–32.

17. Lunina A, Vereshchaka A (2010) A new vent shrimp (Crustacea: Decapoda: Alvinocarididae) from the Mid-Atlantic Ridge. Zootaxa 2372: 69–74.

18. Nye V, Copley J, Plouviez S (2011) A new species of *Rimicaris* (Crustacea: Decapoda: Caridea: Alvinocarididae) from hydrothermal vent fields on the Mid-Cayman Spreading Centre, Caribbean. J Mar Biol Assoc U.K. 92: 1–16.

19. Tunnicliffe V, McArthur G, McHugh D (1998) A biogeographical perspective of the deep-sea hydrothermal vent fauna. Adv Mar Biol 34: 355–442.

20. Zierenberg RA, Adams MW, Arp AJ (2000) Life in extreme environments: Hydrothermal vents. Proc Natl Acad Sci USA, 97: 12961–12962.

21. Gebruk A, Galkin S, Vereshchaka A, Moskalev L, Southward AJ (1997) Ecology and biogeography of the hydrothermal vent fauna of the Mid-Atlantic Ridge. Adv Mar Biol 32: 93–144.

22. Gebruk A, Southward E, Kennedy H, Southward A (2000) Food sources, behavior, and distribution of hydrothermal vent shrimps at the Mid-Atlantic Ridge. J Mar Biol Assoc U.K. 80: 485–499.

23. Vereshchaka A, Gebruk A (2002) Shrimps (Decapoda Macrura Natantia) In: Gebruk A, editor. Biology of hydrothermal systems. Moscow, KMK. pp. 185–197.

24. Vereshchaka A, Vinogradov G, Ivanenko V (1998) Common features of reproductive biology of some hydrothermal crustaceans (amphipods, copepods, shrimps). Dokl Biol Sci 360: 269–270.

25. Vereshchaka A (2000) Revision of the genus *Sergia* (Decapoda: Dendrobranchiata: Sergestidae): Taxonomy and distribution. Galathea Report 18, 69–207.

26. Lunina A, Vereshchaka A (2008) Hydrothermal vent shrimps *Alvinocaris markensis*: interpopulation variation. Dokl Biol Sci 421: 266–268.

27. Lunina A (2011) Vent shrimps of the Mid-Atlantic Ridge. PhD Thesis, P.P. Shirshov Institute of Oceanology of RAS, Russia, Moscow [in Russian].

28. Fujikura K, Hashimoto J, Fujiwara Y, Okutani T (1995) Community ecology of the chemosynthetic community at Off Hatsushima site, Sagami Bay, Japan. JMSTC Journal of Deep Sea Research 11: 227–241 [in Japanese with English summary].

29. Fujikura K, Hashimoto J, Fujiwara Y, Okutani T (1996) Community ecology of the chemosynthetic community at Off Hatsushima site, Sagami Bay, Japan-II: comparisons of faunal similarity. JMSTC Journal of Deep Sea Research 12: 133–153 [in Japanese with English summary].

30. Mironov A, Gebruk A, Moskalev L (2002) In: Gebruk A, editor. Biology of hydrothermal systems. Moscow, KMK. pp. 410–455.

31. Gebruk A, Mironov A (2006) Biogeography of the Atlantic hydrothermal vents. In: Vinogradov ME, Vereshchaka AL, editors. Ecosystems of the Atlantic hydrothermal vents. Moscow, Nauka. pp. 119–162. [in Russian].

32. Smith D, Cann J (1993) Building the crust at the Mid-Atlantic Ridge. Nature, 365: 707–715.

33. Bogdanov Y, Sagalevich A (2002) Geological deep-sea studies by manned submersibles "Mir". Moscow, Nauchny mir [in Russian].

34. Gurvich E (1998) Metalliferous sediments of World Ocean. Nauchny Mir, Moscow.

35. Tyler P, Young C (2003) Dispersal at hydrothermal vents: a summary of recent progress. Hydrobiologia 503: 9–19.

36. Zonnenshine L, Kuzmin M, Baranov B, Shilovskii P, Poroshina I (1992) Hydrothermal formation of the Mid-Atlantic Ridge, pp.12–44. Moscow, Nauka [in Russian].

37. Hilário A, Capa M, Dahlgren TG, Halanych KM, Little CTS, et al. (2011) New perspectives on the ecology and evolution of siboglinid tubeworms. PLoS ONE 6(2): e16309.

38. Rouse GW, Wilson NG, Goffredi SK, Johnson SB, Smart T, et al. (2009) Spawning and development in *Osedax* boneworms (Siboglinidae, Annelida). Mar. Biol. 156: 395–405.

39. Ramirez-Llodra E, Segonzac M (2006) Reproductive biology of *Alvinocaris muricola* (Decapoda: Caridea: Alvinocarididae) from cold seeps in the Congo Basin. Journal J Mar Biol Assoc U.K. 86: 1347–1356.

40. Foote M (2012) Evolutionary dynamics of taxonomic structure. Biol Lett 8: 135–138.

41. Morozov E, Demidov A, Tarakanov R (2010) Abyssal channels in the Atlantic Ocean: Water structure and flows. Springer. 266 p.

42. Wunsch C (1984) An eclectic Atlantic Ocean circulation model. Part I: The meridional flux of heat. J Phys Oceanogr 14(11): 1712–1733.

43. Stephens JC, Marshall DP (2000) Dynamical pathways of Antarctic Bottom Water in the Atlantic. J Phys Oceanogr 30(3): 622–640.

44. Weatherly GL, Kelley EA (1982) 'Too cold' bottom layers at the base of the Scotian Rise. J Mar Res 40(4): 985–1012.

45. Lavin AM, Bryden HL, Parrilla G (2003) Mechanisms of heat, freshwater, oxygen and nutrient transports and budgets at 24.5° N in the subtropical North Atlantic. Deep Sea Res I 50: 1099–1128.

46. Vereshchaka A (1997) Comparative morphological studies on four populations of the shrimp Rimicaris exoculata from the Mid-Atlantic ridge. Deep-Sea Research I, V. 44 (11), p. 1905–1921.

47. Shank TM, Black MB, Halanych KM, Luts RA, Vrijenhoek RC (1999) Miocene radiation of deep-sea hydrothermal vent shrimp (Caridea: Bresiliidae): evidence from mitochondrial cytochrome oxidase subunit I Mol. Phylogenet. Evol 13: 244–254.

48. Roterman CN, Copley JT, Linse KT, Tyler PA, Rogers AD (2013) The biogeography of the yeti crabs (Kiwaidae) with notes on the phylogeny of the Chirostyloidea (Decapoda: Anomura). Proc Biol Sci 280 (1764).

49. Pedersen RB, Rapp HT, Thorseth IH, Lilley MD, Barriga FJ, et al. (2010) Discovery of a black smoker vent field and vent fauna at the Arctic Mid-Ocean Ridge. Nat Commun 1: 126.

50. Kikuchi T, Hashimoto J (2000) Two new caridean shrimps of the family Alvinocarididae (Crustacea, Decapoda) from a hydrothermal vent field at the Minami-Ensei Knoll in the Mid-Okinawa Trough, Japan. Species Diversity 5: 135–148.

51. Watabe H, Miyake H (2000) Decapod fauna of the hydrothermally active and adjacent fields on the Hatoma Knoll, southern Japan. JAMSTEC Journal of Deep Sea Research 17: 29–34 [in Japanese with English summary].

52. Kikuchi T, Ohta S (1995) Two caridean shrimps of the families Bresiliidae and Hippolytidae from a hydrothermal field on the Iheya Ridge, off the Ryukyu Islands, Japan. Journal of Crustacean Biology 15: 771–785.

53. Williams A, Chace F Jr (1982) A new caridean shrimp of the family Bresiliidae from thermal vents of the Galapagos Rift. Journal of Crustacean Biology 2: 136–147.

54. Shank TM, Black MB, Halanych KM, Lutz RA, Vrijenhoek RC (1999) Miocene radiation of deep-sea hydrothermal vent shrimp (Caridea: Bresiliidae): Evidence from mitochondrial cytochrome oxidase subunit I. Mol Phylogenet Evol 13: 244–254.

55. Shank TM (1997) *Alvinocaris lusca* Williams, Chace Jr., 1982, *Alvinocaris markensis* Williams, 1988, *Chorocaris chacei* (Williams and Rona, 1986), *Chorocaris vandoverae* Martin, Hessler, 1990. In: Desbruyères D, Segonzac M, editors. Handbook of deep-sea hydrothermal vent fauna. Brest: IFREMER. pp. 191–194.

56. Williams A (1988) New marine decapod crustaceans from waters influenced by hydrothermal discharge, brine, and hydrocarbon seepage. Fish. Bull. (Wash.D. C.) 86: 263–287.

57. Segonzac M, de Saint Laurent M, Casanova B (1993) L'énigme du comportement trophique des crevettes Alvinocarididae des sites hydrothermaux de la dorsale médioatlantique. Cah. Biol. Mar. 34: 535–571.

58. Shank T, Martin J (2003). A new caridean shrimp of the family Alvinocarididae from thermal vents at Menez Gwen on the Mid-Atlantic Ridge. Proc. Biol. Soc. Wash. 116: 158–167.

59. Wheeler AJ, Murton B, Copley J, Lim A, Carlsson J, et al. (2013) Moytirra: Discovery of the first known deep-sea hydrothermal vent field on the slow-spreading Mid-Atlantic Ridge north of the Azores. Geochemistry, Geophysics, Geosystems, 14(00), n/a–n/a. doi:10.1002/ggge.20243.

60. Williams A, Rona P (1986) Two new caridean shrimps (Bresiliidae) from a hydrothermal field on the Mid-Atlantic Ridge. J Crustacean Biol 6: 446–462.

61. Fabri M-C, Bargain A, Briand P, Gebruk A, Fouquet Y, et al (2010) The hydrothermal vent community of a new deep-sea field, Ashadze-1, 12°58′N on the Mid-Atlantic Ridge. J Mar Biol Assoc U.K. 91: 1–13.

62. Martin J, Hessler R (1990) *Chorocaris vandoverae*, a new genus and species of hydrothermal vent shrimp (Crustacea, Decapoda, Bresiliidae) from the western Pacific. Contributions in Science, Natural History Museum of Los Angeles 417: 1–11.

63. Komai T, Segonzac M (2004) A new genus and species of alvinocarid shrimp (Crustacea: Decapoda: Caridea) from hydrothermal vents on the North Fiji and Lau Basins, southwestern Pacific. J Mar Biol Assoc U.K. 84: 1179–1188.

64. Williams A, Dobbs F (1995) A new genus and species of caridean shrimp (Crustacea, Decapoda, Bresiliidae) from hydrothermal vents on Loihi Seamount, Hawaii. Proc. Entomol. Soc. Wash. 108: 228–237.

65. Limen H, Juniper SK, Tunnicliffe V, Clement M (2006) Benthic community structure on two peaks of an erupting seamount: Northwest Rota-1 Volcano, Mariana Arc, western Pacific. Cah. Biol. Mar. 47: 457–463.

66. Yang J-S, Lu B, Chen D-F, Yu Y-Q, Yang F, et al. (2013) When did decapods invade hydrothermal vents? Clues from the Western Pacific and Indian Oceans. Mol Biol Evol 30: 305–9.

67. Beltenev V, Ivanov V, Rozhdestvenskaya I, Cherkashov G, Stepanova T, et al. (2009) New data about hydrothermal fields on the Mid-Atlantic Ridge between 11°–14°N: 32nd Cruise of R/V Professor Logatchev. InterRidgeNews 18: 13–17.

68. Van Dover CL, Humphris SE, Fornari D, Cavanaugh C, Collier R, et al. (2001) Biogeography and ecological setting of Indian Ocean hydrothermal vents. Science 294: 818–823.

69. Nakamura K, Watanabe H, Miyazaki J, Takai K, Kawagucci S, et al. (2012) Discovery of new hydrothermal activity and chemosynthetic fauna on the Central Indian Ridge at 18°–20° S. PloS One 7: 1–11.

70. Martin J, Christiansen J (1995) A new species of the shrimp genus *Chorocaris* Martin and Hessler, 1990 (Crustacea, Decapoda, Bresiliidae) from hydrothermal vent fields along the Mid-Atlantic Ridge. Proc. Biol. Soc. Wash. 108: 220–227.

18

Cold Seep Epifaunal Communities on the Hikurangi Margin, New Zealand: Composition, Succession, and Vulnerability to Human Activities

David A. Bowden[1]*, Ashley A. Rowden[1], Andrew R. Thurber[2], Amy R. Baco[3], Lisa A. Levin[4], Craig R. Smith[5]

1 Coasts and Oceans Centre, National Institute of Water and Atmospheric Research, Wellington, New Zealand, **2** College of Earth, Ocean, and Atmospheric Sciences, Oregon State University, Corvallis, Oregon, United States of America, **3** Department of Earth, Ocean and Atmospheric Sciences, Florida State University, Tallahassee, Florida, United States of America, **4** Center for Marine Biodiversity and Conservation, Integrative Oceanography Division, Scripps Institution of Oceanography, La Jolla, California, United States of America, **5** Department of Oceanography, School of Ocean and Earth Science and Technology, University of Hawaii at Manoa, Honolulu, Hawaii, United States of America

Abstract

Cold seep communities with distinctive chemoautotrophic fauna occur where hydrocarbon-rich fluids escape from the seabed. We describe community composition, population densities, spatial extent, and within-region variability of epifaunal communities at methane-rich cold seep sites on the Hikurangi Margin, New Zealand. Using data from towed camera transects, we match observations to information about the probable life-history characteristics of the principal fauna to develop a hypothetical succession sequence for the Hikurangi seep communities, from the onset of fluid flux to senescence. New Zealand seep communities exhibit taxa characteristic of seeps in other regions, including predominance of large siboglinid tubeworms, vesicomyid clams, and bathymodiolin mussels. Some aspects appear to be novel; however, particularly the association of dense populations of ampharetid polychaetes with high-sulphide, high-methane flux, soft-sediment microhabitats. The common occurrence of these ampharetids suggests they play a role in conditioning sulphide-rich sediments at the sediment-water interface, thus facilitating settlement of clam and tubeworm taxa which dominate space during later successional stages. The seep sites are subject to disturbance from bottom trawling at present and potentially from gas hydrate extraction in future. The likely life-history characteristics of the dominant megafauna suggest that while ampharetids, clams, and mussels exploit ephemeral resources through rapid growth and reproduction, lamellibrachid tubeworm populations may persist potentially for centuries. The potential consequences of gas hydrate extraction cannot be fully assessed until extraction methods and target localities are defined but any long-term modification of fluid flow to seep sites would have consequences for all chemoautotrophic fauna.

Editor: Philippe Archambault, Université du Québec à Rimouski, Canada

Funding: Data and samples were collected during R/V Tangaroa voyage TAN0616 "New Zeeps" and R/V Sonne voyages SO191 "New Vents" and SO214 "NEMESYS." TAN0616 was funded by National Oceanic and Atmospheric Administration Ocean Exploration Grants NA05OAR4171076 and NA17RJ1231/58, and National Institute of Water and Atmospheric Research (NIWA) Capability Fund project CRFH073. DAB was funded on TAN0616 and SO191 by NIWA Capability Fund project CPDU073 and on SO214 by NIWA project GNS11304 under subcontract to GNS project CO5X0908 'Gas Hydrate Resources'. Support during analysis for DAB and AAR came from NIWA projects GNS11304 and DSCA113 Vulnerable Deep-Sea Communities. The funders had no role in study design, data collection and analysis, decision to publish, or preparation of the manuscript. The funders had no role in study design, data collection and analysis, decision to publish, or preparation of the manuscript.

Competing Interests: The authors have declared that no competing interests exist.

* E-mail: david.bowden@niwa.co.nz

Introduction

Cold seeps are sites where fluids enriched with hydrocarbons, primarily methane, emerge from the seabed. They are known from both active and passive continental margins, typically occurring on the continental slope but also much shallower [1–3]. The flux of methane-rich fluids at seeps supports distinctive chemoautotrophic food webs in which primary production takes place via microbial oxidation of methane and hydrogen sulphide. Metazoan communities at seeps are characterised by high densities and relatively low diversity of invertebrate taxa which have evolved symbioses with methane- or sulphide-oxidising bacteria, enabling them to exploit the energy potential of the emerging fluids [4].

Although cold seeps are now known to occur on continental margins throughout the global ocean, including the Arctic and Antarctic [5,6] variations in the composition of seep communities in different parts of the world are not well resolved [7]. Recent research initiatives, particularly those within the Chemosynthetic Ecosystem Science (ChEsS) project of the Census of Marine Life [5] have sought to improve understanding of the biogeography of seep faunas by focusing attention on areas of the oceans that have been relatively under-sampled. Concern has also been raised about the threats that seep sites and their communities face from trawling and future mining for gas hydrates [8,9], and the need to provide relevant ecological information to inform management strategies for protection of potentially vulnerable seep communities [10].

New Zealand was one of four regions targeted by the ChEsS programme. The existence of cold seeps around New Zealand was initially deduced from fisheries by-catch and research trawl and dredge samples [11]. The first dedicated biological sampling of seep sites on the Hikurangi Margin off the east coast of the North Island took place in 2006 with the "New Zeeps" voyage (TAN0616) [8] and was continued during geophysically-focused voyages in 2007 (SO191) and 2011 (SO214). To date, at least thirty-two sites of active seepage have been identified on the Hikurangi Margin at depths of 600 to 2000 m [12]. All known seep sites on the Hikurangi Margin are methane-derived authigenic carbonate mounds (sensu [13]). Mud volcanoes are apparently absent from the margin and no evidence of chemo-autotrophic fauna has been recovered from the few areas of pockmarks that have been detected to date in the region [12]. Megafaunal communities associated with the seep sites are dominated by vesicomyid clams, large siboglinid tubeworms, and bathymodiolin mussels [8,14,15]. White bacterial mats and dark sulphide-rich patches which host dense populations of ampharetid polychaetes are characteristic on soft sediments around carbonate concretions [16–18], and frenulate siboglinid worms have been found in high densities on the peripheries of some seep sites [17].

Baco et al. [8] presented initial descriptions of community composition and habitats at eight Hikurangi Margin seep sites based on data from towed camera, epibenthic sled, grab, and multicorer samples collected during TAN0616. These authors speculated that the seep communities on the Hikurangi margin could represent a distinct biogeographical province but this contention remains untested until the taxonomic and genetic identities of the species sampled have been established. Baco et al. [8] also observed that bottom trawls had impacted several seep sites and highlighted the need for further research on seep ecology to inform environmental management of this and potential future threats to seep communities. The Hikurangi Margin has strong potential for large-scale exploitation of gas hydrates as an energy resource [19]. Although seabed exploration of the Hikurangi Margin gas hydrate province has been focused around sites of active seepage, as identified by the presence of acoustic water column flares and chemoautotrophic fauna [12], geophysical surveys using seismic methods have covered much wider areas of the margin [20]. Published analyses from these surveys provide broader spatial and temporal perspectives on the geophysical context in which the highly localised biological communities exist, providing insight into the processes and pathways controlling the availability of methane-rich fluids at the seabed.

In this study, we review the megafaunal ecology of the known seep sites and, by placing this in the contexts of the likely timescales of ecological processes and present and possible future human impacts, we evaluate the potential vulnerability of the fauna to disturbance. Analysis of all fifty-six camera transects from the three research voyages to the Hikurangi Margin cold seeps enables us to describe and compare seep faunal communities across a total of eighteen sites, each of which has been traversed by at least one camera transect. We do not consider here other reported sites from which seep-associated fauna have been recovered but for which seabed imagery is not available (see [11,12]). The primary aims of the study are to: (1) describe the principal seep-associated communities and habitats present at seep sites on the Hikurangi margin; (2) assess variability among seep sites in terms of their communities and physical habitats; (3) use observations from the New Zealand seeps and life-history information from the literature to develop a hypothetical succession model for the Hikurangi Margin seeps, and (4) assess

the vulnerability of seep communities to current and potential future human activities on this margin.

Methods

Study area and sites

The Hikurangi Margin is an active subduction zone off the east coast of North Island, New Zealand, representing the southern extension of the Tonga-Kermadec subduction system [20]. Cold seep exploration to date has been concentrated on the seabed from Ritchie Ridge in the north (39°30'S) to Opouawe Bank in the south (41°47'S), and active seeps have been discovered at water depths from ca. 600 to 1200 m. Five principal regions of seep activity have been studied: Ritchie Ridge, Rock Garden, Omakere Ridge, Uruti Ridge, and Opouawe Bank (Figure 1) across which a total of thirty-two active seeps have been confirmed [12]. Evidence of chemoautotrophic faunal communities has been found at eighteen of these sites (Table 1 – note, of the individual sites listed by Greinert et al. [12], we treat the following groups as single named sites: Kea and Kaka; Weka A, B, and C; Faure Sites A and B).

Multibeam sonar bathymetry maps have been generated for all regions, and for Omakere Ridge and Opouawe Bank extensive high-resolution side-scan sonar (SSS) images of the seabed have also been developed [14,15]. Areas of high backscatter in these images show the locations of exposed and shallow sub-surface carbonate concretions associated with cold seeps and are an accurate indication of seep extent. Some estimation of the vertical relief of the carbonates can also be made from the intensity of shadow areas in the SSS image [14]. Here, we used SSS maps in a geographic information system to measure the areal extent of seep sites. Where SSS coverage was not available, area was estimated from the length of camera transect in which seep fauna or habitats

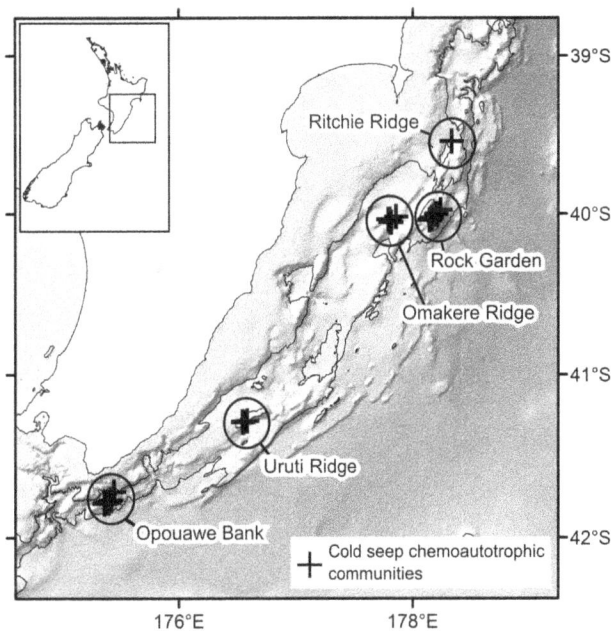

Figure 1. Cold seep sites on the Hikurangi Margin of New Zealand. Labelled circles indicate regions referred to in the text, crosses show individual seep sites at which live chemoautotrophic communities have been sampled (see Table 1 and Greinert et al. [12] for site details).

Table 1. Cold seep sites on the Hikurangi Margin sampled by towed camera transects.

Region	Site name	Site code	Latitude (S)	Longitude (E)	Depth (m)	Area (ha)	Camera station numbers			Analysed	
							TAN0616	SO191	SO214	Stills	Video
Builder's Pencil	Builder's Pencil	BPL	39°32.636	178°19.944	797	7.0	19, 20, 26, 27, 28, 29			142	00:55:31
Rock Garden	LM-3	LM3	39°58.607	178°14.221	863	0.1		42 [2], 176 [19]		26	01:14:00
	Weka (A to C)	WEK	40°00.286	178°11.905	655	12.0		41 [1a]			01:51:00
	Faure Site (A)	FAU	40°01.960	178°09.401	654	na		41 [1]			00:55:00
	Rock Garden knoll	RGK	40°02.379	178°08.599	767	na	5, 14, 15				01:35:02
Uruti Ridge	Hihi	HIH	41°17.687	176°33.548	744	1.5	66			85	00:37:16
	Kereru	KER	41°17.161	176°35.469	727	1.5	64, 69				00:20:08
	LM-10	LM10	39°24.993	178°24.268	729	1.5	65				00:42:31
Omakere Ridge	Bear's Paw	BPW	40°03.187	177°49.252	1100	2.2		52-3 [5], 81 [7]		25	00:38:52
	Kea & Kaka	KAK	40°02.240	177°47.714	1168	8.1		52-1 [3], 80 [6], 167 [17]			04:59:33
	LM-9	LM9	40°00.603	177°52.358	1150	1.5	44, 55, 56				01:27:05
	Moa reef	MOA	40°03.235	177°48.802	1118	11.0		52-3[5], 81[7], 166(16)			04:21:47
	SW Moa	SWM	40°03.336	177°48.247	1120	12.2		81 [7]	94 [10]	57	00:54:00
Opouawe Bank	North Tower	NTR	41°46.911	175°24.083	1052	6.1	75, 76, 85, 114, 115	106 [9]	41 [1]	163	04:59:43
	South Tower	STR	41°47.300	175°24.521	1056	4.5	75, 77, 87, 117, 119, 120, 124				01:26:46
	Piwakawaka	PIW	41°47.664	175°22.348	1095	2.5		292 [22]	70 [5], 71 [6]	57	00:57:20
	Pukeko	PUK	41°47.153	175°23.465	1060	4.3		105 [8], 155 [15], 272 [21]			01:04:04
	Takahe	TAK	41°46.368	175°25.651	1058	6.5	128	107 [10]	64 [2], 65 [3]	23	00:51:04
	Tui	TUI	41°43.288	175°27.091	815	5.2	129	108 [11], 129 [12], 154 [14]		21	03:24:56

Sampling took place during voyages of *RV Tangaroa* (TAN0616, November 2006) and *RV Sonne* (SO191, February 2007; SO214, April 2011). Table shows; region; seep site names and abbreviations; latitude and longitude (of flare position or seep centre); seabed depth; approximate seabed area; station numbers of all camera transects at each site (numbers in square brackets are sequential deployment numbers for *RV Sonne's* OFOS camera system), and the numbers of still images and hours of video (hh:mm:ss) analysed from each site.

were recorded and the local topography of the site shown in multibeam sonar maps.

Sampling

The most intensive sampling of the Hikurangi Margin seep fauna, in terms of both the number of deployments and the number of seep sites sampled, has been by towed camera systems. Cameras were deployed at potential seep sites during each of the three research voyages (TAN0616, SO191, and SO214), yielding a total of fifty-seven transects covering approximately 61,000 linear metres of seabed (Table 1). The intensity of sampling within each region was influenced primarily by the number of active seeps detected. Thus, overall effort was greatest at Opouawe Bank and Omakere Ridge because these regions contained the highest densities of active seeps as indicated by broad-scale acoustic detection of water-column flares [12]. All sampling was conducted under the authority of New Zealand Ministry of Fisheries Special Permits 318 (TAN0616 and SO191) and 421 (SO214) issued to the National Institute of Water and Atmospheric Research (NIWA) for the purposes of investigative research and education.

Two camera systems were used; NIWA's deep towed imaging system (DTIS [21]) on TAN0616, and *RV Sonne*'s Ocean Floor Observation System (OFOS) on SO191 and SO214. Both systems recorded continuous digital colour video with a separate digital still image camera capturing higher-resolution images ("stills") automatically at either 20 s (TAN0616) or 15 s (SO191 and SO214) intervals. The combination of continuous video and intermittent stills afforded two perspectives on each transect. First, video provided a full record of habitat transitions and megafaunal population densities, including infrequently-occurring fauna and microhabitats (i.e. characteristic combinations of fauna and substrata) likely to be missed in intermittent stills. Second, still images enabled accurate density counts for high-density fauna and quantitative characterisation of habitat structure at smaller spatial scales ($<$1 m^2). All cameras were oriented directly downwards, to facilitate quantitative analyses, and pairs of red lasers at 0.2 m spacing, parallel to the optical axis of each camera, were projected onto the seabed for image scaling. All transects were tracked using ultra-short baseline acoustic systems (Simrad HPR on *RV Tangaroa* and Ixsea Posidonia on *RV Sonne*), yielding seabed positions with an accuracy of ca. ±20 m at ca. 2–5 s intervals depending on depth. On TAN0616, digital video was recorded in 1080 50i high definition (HD) format with 8 megapixel (mp) digital stills. For SO191 video was recorded in 720×576 standard definition (SD) format with 4 mp stills. For SO214, both HD and SD video cameras were used simultaneously and 10 mp stills taken but HD and useable stills were only recorded successfully on half of the transects. All transects were run at target camera altitude of 2.5 m and speed of 0.25 to 0.5 ms^{-1}. Sections of transects which were outside the target altitude and speed ranges, or where quality was poor for technical reasons, were not used in analyses. These criteria resulted in video frame widths ranging from ca. 1 to 3 m, and still image areas from 1.8 m^2 to 7.0 m^2. In total, 59.5 h of video and 10,560 still images were collected across the three voyages.

Analyses

Camera transect analysis. Transects were initially cropped to the section bounded by the first and the last occurrence of seep-associated fauna or substrata (typically clam shells, sulphidic patches, or bacterial mats). Thirty-six video transects were analysed using methods described in Jones et al. [14] and Klaucke et al. [15]. Briefly, cropped transects were reviewed using Ocean Floor Observation Protocol software (OFOP, http://ofop.texel.

com) to record the occurrence of all seabed megafauna ($>$ca. 50 mm) and substratum types. Substrata were recorded as coarse level categories (e.g. "muddy sediment", "carbonate rock", see [14]) with modifiers indicating overlying biogenic substrata (e.g. "*Calyptogena* sp. shell"). As described in Jones et al [14], substratum categories were treated as continuous, each recorded observation propagating onward through the video transect record at 1 s intervals until the next substratum observation was made. Fauna were recorded as counts of individuals but densities of *Lamellibrachia* sp. exceeded the analyst's ability to discriminate between individuals in places. At these points the OFOP output file becomes 'saturated', representing high population density rather than absolute numbers. For transects where this occurred, the rank order of densities between transects was compared with the more precise counts from the still images and video counts were adjusted where necessary to match the stills rank order. The distributions of seep-associated megafauna and habitats were then plotted against SSS maps, where available, to visualise spatial relationships.

Still images were analysed from eleven transects chosen to be representative of each of the major seep regions. For smaller seep sites ($<$ca. 200 m across) all useable images along the transect within the site were analysed. For larger sites, because individual seep habitats were highly localised at metre scales, subsets of still images were analysed with the aim of representing all microhabitats present and capturing the full range of population densities, including the highest. Using ImageJ (http://rsbweb.nih.gov/ij/) analysis software, images were first scaled by reference to the laser points projected on the seabed. Principal substratum types were then measured as areas and converted to percentages of the whole image. Fauna were identified to the finest practicable taxonomic resolution and counted. Counts were then standardised to numbers of individuals m^{-2} based on imaged seabed area.

Comparisons between seep sites. For sites where still images were analysed in detail, population densities of the principal seep-associated taxa (as individuals m^{-2}) were plotted together with substratum type (as % of total seabed area of the image) against distance across the seep site. These profiles illustrate differences in the spatial scales of individual seep sites, variations in the principal fauna present and their population densities, and the spatial relationships between habitats and fauna.

Data from video analyses were used in multivariate analyses to explore variability between seep sites and regions. Counts of seep-associated fauna were standardised to unit area by dividing by the seabed area of the transect (cropped transect distance multiplied by image frame width, which was approximated as 2 m for all transects). Count data were log (x+1) transformed to reduce the influence of variations in the high population density regions of the video transects, where absolute counts were less precise (see above). Because substrata counts were effectively continuous (i.e. with a categorical value recorded at every 1 s increment along the transect), they were expressed as proportions of the total transect by dividing the sum of counts recorded for each substratum type by the total number of seconds in the analysed portion of the transect. Thus, faunal data were rendered as counts of individuals m^{-2} and substrata as percentages of the total transect.

Data for seven seep-associated fauna characteristics from the video analyses: *Lamellibrachia* sp. tubeworms; *Calyptogena* sp. live clams; Bathymodiolin mussels; *Calyptogena* sp. shell valves; *Stelletta* sp. sponges; dark sediment patches (a proxy for ampharetid polychaete beds), and bacterial mats, together with observations of thicket-forming scleractinian corals, were used to generate a matrix of Hellinger distances [22] between each pair of transects. Non-metric multidimensional scaling (NMDS) was then used to produce a two-dimensional ordination of relationships between the thirty-six analysed video transects. Hellinger distance was chosen because the data included both abundance and proportion values, but trials using Bray-Curtis similarity showed very similar results. The relative influence of each of the measured faunal variables in the final ordination was visualised by superimposing vectors proportional to their Pearson correlations with the 2-dimensional ordination. The same procedure was used to illustrate correlations with three physical environmental factors: depth; percentage of carbonate rock substrata recorded in video, and the number of trawls per site for the period 1998 to 2005 (see below for details). Formal statistical tests for differences between survey regions were not appropriate because of the limited number of variables and differences in numbers of sites between regions (e.g. only one site at Richie Ridge). Multivariate analyses were run in PRIMER [23].

Assessing impacts of human activities

Trawling. The Hikurangi Margin seep sites have been subject to disturbance from bottom trawling, mainly for orange roughy (*Hoplostethus atlanticus*) and oreo (*Pseudocyttus maculates*), since at least 1989 [8]. Using trawl data compiled for the period 1989 to 2005 [24]. We extended the analysis of trawl frequency at the seep sites conducted by Baco et al [8] to include all sites in the study area from which evidence of chemoautotrophic fauna has been collected. Trawling intensity was calculated as the number of trawl tracks intersecting a 250 m radius circle centred on each of the seep locations detailed by Greinert et al. [12]. In most instances, these locations are the coordinates of the seabed origin of acoustic flares [12] but some are identified as the centre of high backscatter areas in SSS images (e.g. Bear's Paw, at which no water column flares were detected during initial surveys). Analyses were run using XTools Pro in ArcGIS v.10.

Gas hydrate extraction. Techniques for seabed gas hydrate extraction have yet to be fully developed, and the optimal locations for exploitation on the Hikurangi Margin remain a subject of debate (e.g. [25]), with current opinion erring towards sandy strata in the back limb areas of thrust ridges (see Barnes et al. [20] for geophysical background). Our assessment here, therefore, was restricted to (1) consideration of relationships between seep sites and patterns of fluid flow through the geological structures underlying them, as understood from recent published interpretations of seismic survey data [20,25–28] and (2) the proximity of known seep communities to potential drill sites currently under discussion.

Results

Fauna & habitats

With the exception of Takahe, all seep sites were associated with authigenic carbonate rock structures visible at the sediment surface. Photographs and physical samples indicated two broad categories of carbonate formation: light-brown blocks embedded in soft sediments, usually associated with high densities of chemoautotrophic fauna, and referred to as "chemoherm" [14,29] or "fresh" [30]; and darker, more extensive, "weathered-looking" [30] rock associated with larger geomorphologic structures, sessile heterotrophic taxa and few, if any, chemoautotrophic fauna.

At the optical resolution of towed camera transects, seep sites were characterised by communities dominated by five conspicuous metazoan epifaunal taxa: vesicomyid clams in the genus *Calyptogena*; siboglinid tubeworms in the genus *Lamellibrachia*; mussels in the sub-family *Bathymodiolinae*, and sponges in the genera *Pseudosuberites* and *Stelletta* (Figure 2). In addition to these, white bacterial mats and dark sediment patches colonised by

ampharetid polychaete worms were conspicuous and locally common (Figure 2 A and B). Individual ampharetids were not visible in images but their populations are known to be strongly associated with these dark, sulphide-rich sediments, in which they form a characteristic 'rain drop' patterning of small pits visible at larger scales (Figure 2 B) [8,16,17,31]. These taxa were present in varying combinations across the study area, forming a limited spectrum of characteristic combinations of physical habitat and faunal community within and around active seep sites. Although other chemoautotrophic taxa, including frenulate siboglinid worms and *Acharax* sp. bivalves, were certainly present at many, if not all, sites [8,17] these were not reliably recorded in camera transects because of their small size and infaunal habit. Four mobile heterotrophic invertebrate taxa characteristic of surrounding slope habitats were also commonly observed within and around seep sites: lithodid crabs were seen at the edges of sulphidic patches containing *Calyptogena* clams (Figure 3 D); pagurid crabs were common on carbonate and sediment substrata at Hihi, Builder's Pencil and North Tower; predatory gastropod molluscs (mostly *Fusitriton* sp.) were common, particularly at Piwakawaka, North Tower, Bear's Paw, and Tui, and the regular echinoid *Gracilechinus multidentatus* occurred in high densities around the periphery of Tui (Figure 3 C).

Dark sediment patches with the distinctive pitting caused by ampharetid polychaetes were observed at all seep sites where soft sediments were present. The pitting was observed in all areas of sulphide-rich sediments, both inside and outside the authigenic carbonate chemoherm of the seeps, and whether colonised by clams or not. Isolated ampharetid patches on open sediments occurred most frequently to the southeast of North Tower, at Takahe, and between Bear's Paw and Moa, with recorded patch size up to ca. 2 m². Bacterial mats were also most common in these areas and were often, but not always, associated with the dark sediment patches. Mats were generally smaller than the dark sediment patches, typically ca. 0.2 m² to the southeast of North Tower but up to ca. 1 m² at Takahe.

Past or present occurrence of *Calyptogena* sp. clams at all sites was indicated by disarticulated shell valves on the seabed (Figures 2 & 3), ranging from small, highly localised patches (<1 m²) on muddy sediments at Takahe to ~70,000 m² of mostly carbonate rock covered by shell valves and fragments at Builder's Pencil. Live clams were identifiable as they were intact (i.e. with valves joined and tightly closed) and in characteristic living position, with only their posterior end protruding from the sediment (see Figure 2 C in [8]). Clam shells were typically 40 to 50 mm in length but individuals reached 80 mm. Shell morphologies indicate that *Calyptogena tuerkayi* is the predominant vesicomyid clam species at all sites (Elena Krylova personal communication) but two valves recovered from LM-9 (station TAN0616-045) suggest that a species of *Laubericoncha* may also be present. Live clams occurred only in highly localised patches (<1 m²) of closely packed individuals and were always associated with sulphide-rich sediments with ampharetid pits and surrounded by larger areas of disarticulated valves (Figure 2C). The maximum recorded density of live *Calyptogena* sp. in a single image was 24 individuals m⁻² at Piwakawaka on Opouawe Bank but because patches of live clams were always somewhat smaller than the total image area, actual densities within patches will be considerably greater than this.

Lamellibrachid tubeworms occurred only in areas where carbonate rocks were present, and appeared to be the same species at all sites. Initial molecular studies suggest the species is *Lamellibrachia columna* (AB, unpublished data), first described from the Lau Basin at the northern end of the Tonga-Kermadec

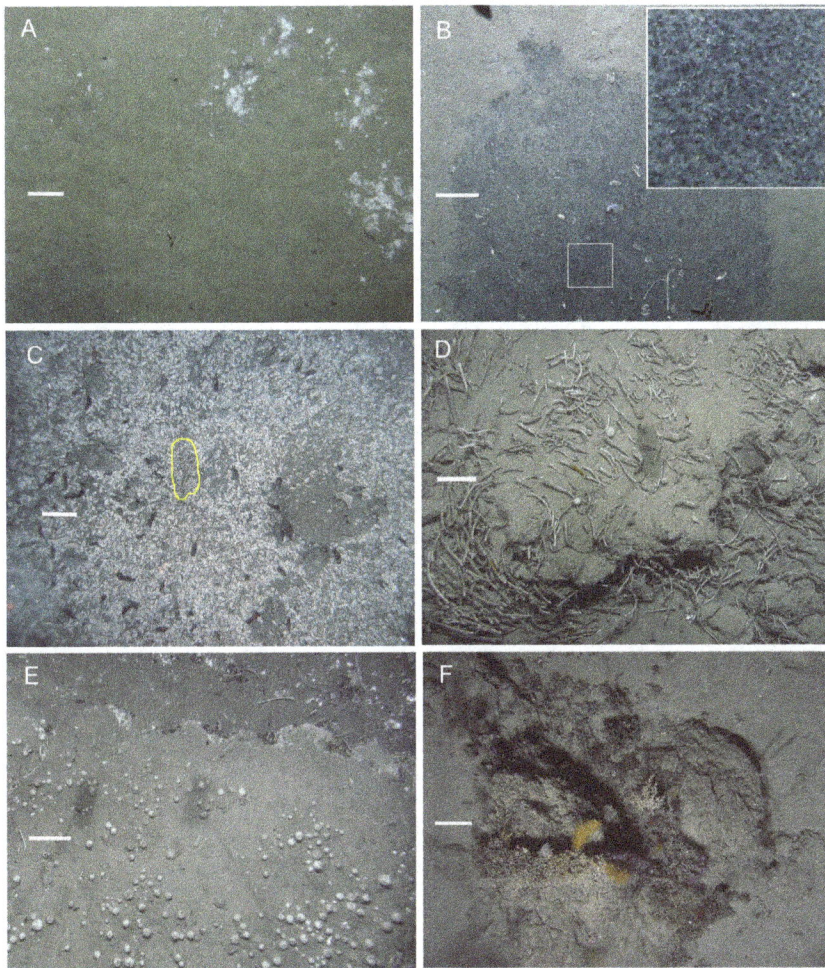

Figure 2. Hikurangi Margin cold seep habitats and fauna (1). (A) bacterial mats on soft sediments at North Tower; (B) sulphide-rich sediment patch with high density ampharetid polychaete population and (inset) characteristic 'raindrop' patterning at North Tower; (C) extensive *Calyptogena* sp. shell valves surrounding a population of live clams (yellow outline) with live and dead *Bathymodiolus* sp. mussels at Tui; (D) high density *Lamellibrachia* sp. tubeworm population at Southwest Moa showing tubes emerging from upper surface of carbonate concretions; (E) high densities of *Stelletta* n. sp. sponges and *Bathymodiolus* sp. mussels (along ledge at top) at Southwest Moa, and (F) scleractinian and gorgonian corals on weathered carbonates at Moa. Scale bars show 0.2 m.

subduction system [32]. Tubes were smooth with indistinct annular rings in places, typically 10 to 15 mm diameter at the anterior opening, and with up to ~1 m but more usually 0.2–0.5 m visible above the substratum. The anterior portion of the tube was generally straight or slightly curved, lying parallel to the substratum or rising at a shallow angle, and becoming more convoluted at the seabed and within carbonates (Figure 2 D). An incomplete tube extracted from a carbonate block sampled by grab at North Tower (SO191 station 138 TV-G5) had a 16 mm anterior tube diameter and was 1.32 m long, two thirds of this length being within the carbonate block (Figure 3 B). Anterior sections of tubes were often a pale off-white colour in images, presumably representing recent growth, but older parts were usually overgrown with epifauna, including hydroids and sponges (see Figure 2 A in [8]). Tubeworms occurred in loose aggregations, often with their tubes aligned with each other, but not forming the dense 'thickets' documented for *L. luymesi* in the Gulf of Mexico [33]. At most sites, tubes were seen only in crevices or under ledges, with none projecting above the upper surface of carbonate blocks or pavements. At Omakere Ridge sites, however, particularly Southwest Moa, tubes were seen emerging directly from the

upper surfaces of carbonates (Figure 2 D). Occurrence was patchy within seep sites, with maximum recorded density in a single image of 51 individuals m^{-2} at Southwest Moa.

Live bathymodiolin mussels were recorded at eight of the eighteen sites from which video transects were analysed, and were observed in all regions except Uruti Ridge. Mussels ranged from ca. 50 to 100 mm in length and most appeared to be *Bathymodiolus tangaroa*, based on shell morphology [8 and references therein]. However, at least three species of bathymodiolin have been sampled from the Hikurangi seeps [8], so diversity in this group is likely to be higher than represented here. All live mussels observed were attached to carbonate rocks; highest densities occurring at Southwest Moa (31 individuals m^{-2}, Figure 2 E) and in a single small patch within extensive carbonate platforms at LM-3 (77 individuals m^{-2}). At Builder's Pencil and Tui, mussels occurred in isolated clusters of a few individuals, often amongst extensive areas of disarticulated *Calyptogena* sp. shells.

The sponges *Pseudosuberites* sp. and *Stelletta* n. sp. (Michelle Kelly, NIWA, unpublished data) were found only on 'fresh' carbonate chemoherm within the central areas of seep sites and both taxa reached highest densities in areas where tubeworms and mussels

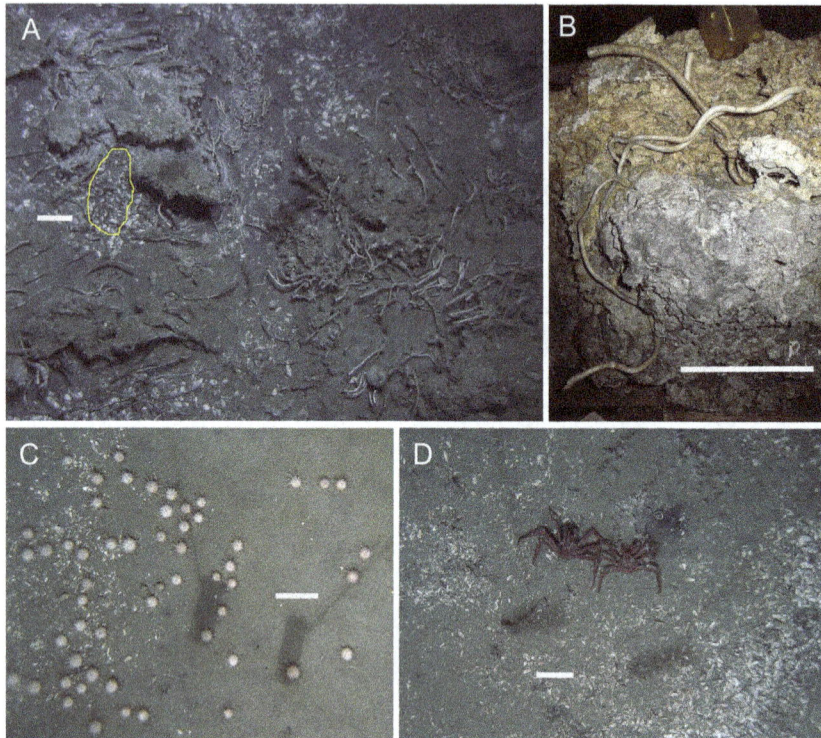

Figure 3. Hikurangi Margin cold seep habitats and fauna (2). (A) 'Chemoherm' habitat at Southwest Moa showing *Lamellibrachia* sp. tubeworms in association with authigenic carbonates, live *Calyptogena* sp. clams (yellow outline) and disarticulated shell valves, two spherical *Stelletta* n. sp. sponges, grey bacterial mats on carbonates (at upper left), and sulphidic sediments; (B) Underside of a carbonate block recovered by grab from chemoherm at North Tower, showing *Lamellibrachia* sp. tubes rooted in soft, grey, sulphide-rich, anoxic material at the base of the block; (C) *Gracilechinus multidentatus* sea urchins with *Calyptogena* sp. shell debris on soft sediments at the periphery of Tui seep; (D) *Lithodes aotearoa* crabs amongst *Calyptogena* sp. shell debris at Tui seep. Scale bars show 0.2 m.

were also abundant. Whereas the encrusting species *Pseudosuberites* sp. was observed at most sites where fresh carbonates and chemoautotrophic megafauna were present, the spherical sponge *Stelletta* n. sp. was recorded only at Omakere Ridge seep sites, notably Southwest Moa, reaching maximum densities of 50 individuals m^{-2}.

Site descriptions

Opouawe Bank. *North Tower* is the most intensively sampled of the Hikurangi sites [8,12,15,17] and illustrates most of the seep-associated habitats and fauna characteristic of the region (Figure 4). The first signs of seep-associated habitats at the periphery of the seep were scattered *Calyptogena* sp. shells and, on the southeast approaches especially, small patches of white bacterial mat (cm scale) and cm to m scale ampharetid patches. Authigenic carbonates occurred more frequently approaching the outskirts of the high backscatter SSS region and were often associated with low densities of *Lamellibrachia* sp. tubeworms, sometimes protruding directly from the sediment but always in close proximity to carbonate rocks. At the boundary of the high-backscatter region, there was an abrupt transition from predominantly muddy sediments to authigenic carbonate blocks interspersed with dark sediments populated by ampharetids. Tubeworms were strongly associated with carbonates within the seep site (Figure 5). Vesicomyid clam shells covered the seabed in localised patches (m^2 scale) but live *Calyptogena* sp. were restricted to infrequent, small (<1 m^2 scale) populations occurring in patches of dark sediments.

South Tower. No additional biological sampling has been undertaken at South Tower since TAN0616, findings from which are described by Baco et al. [8] and Klaucke et al [15]. Review of video and stills from the site in the present study confirmed habitats and fauna similar to those recorded at North Tower.

Piwakawaka showed in SSS images as a region of stronger backscatter covering a seabed area of ca. 2.45 ha. The site is similar to North Tower in terms of both substrata and fauna (Figure 5), but with low densities of bathymodiolin mussels in places.

Tui forms the summit of a rise in the northern part of Opouawe Bank. The site was similar to North Tower in that the central area consisted of exposed authigenic carbonate blocks interspersed by muddy sediments, with tubeworms, clams, and ampharetid patches patchily distributed among them. It differed, however, in having a greater proportion of the substratum covered by *Calyptogena* sp. shells, more live *Calyptogena* sp. patches, and localised populations of bathymodiolin mussels. On muddy sediments surrounding the site, areas of disarticulated clam shells were common and often associated with populations of *G. multidentatus* (Figure 4 C).

Takahe showed in SSS images as a roughly circular area of elevated backscatter ca. 250 m in diameter but no carbonates were visible at the seabed. Patches of dark sediment with high densities of ampharetid polychaete pits and white bacterial mats were common (Figure 5), particularly on the north-eastern circumference of the site. Infrequent patches of *Calyptogena* sp. shell were recorded but live clams were seen in only one image. No other chemoautotrophic megafauna were observed. On the last occasion

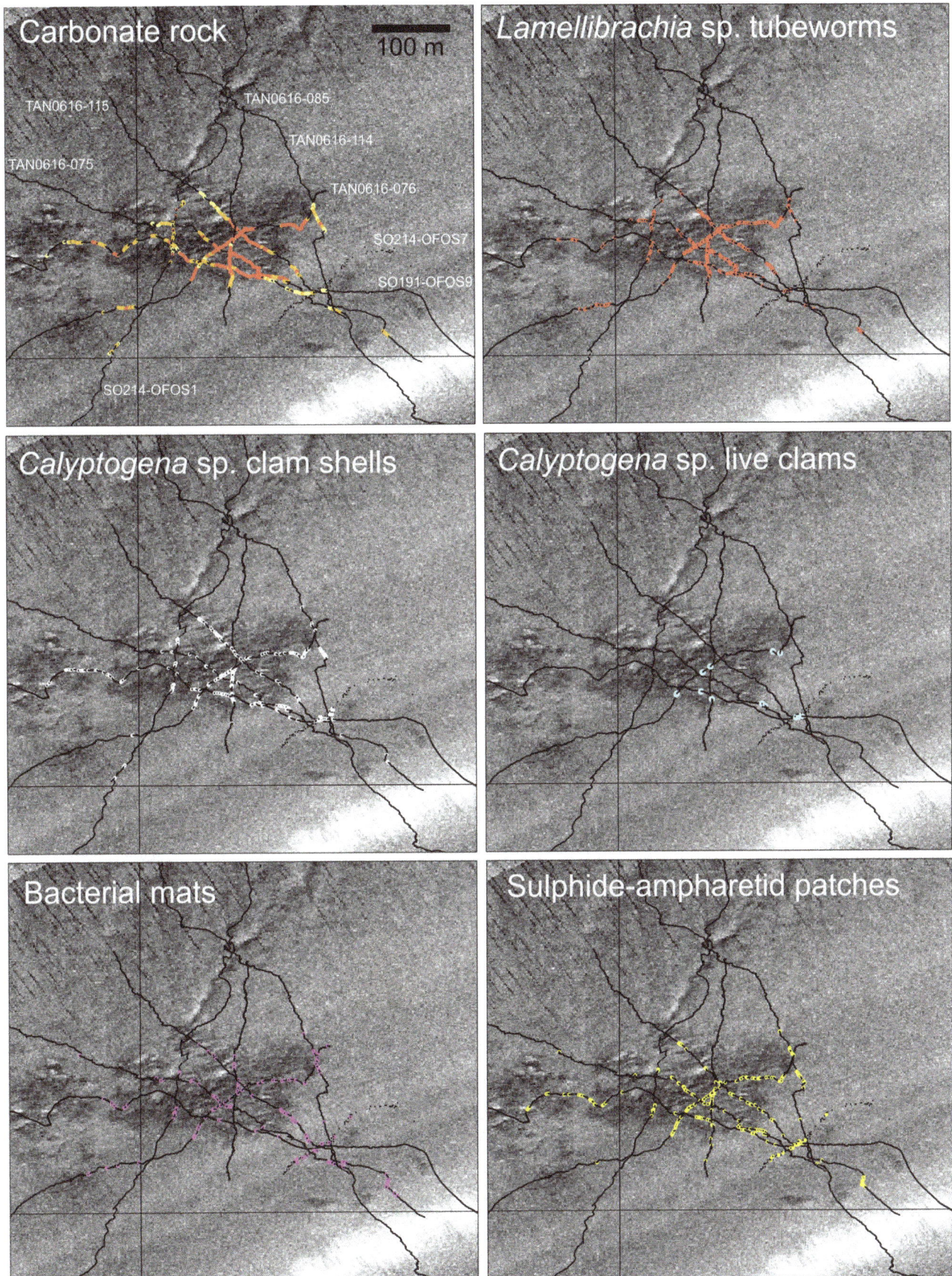

Figure 4. North Tower seep, Opouawe Bank. Observations from towed video camera transects showing distributions of: authigenic carbonate

rocks (red, "chemoherm"; orange, "large blocks (>20 cm)"; yellow, "small blocks (<20 cm)"); *Lamellibrachia* sp. siboglinid tubeworms; disarticulated shell valves of *Calyptogena* sp. vesicomyid clams; live *Calyptogena* sp. clams; bacterial mats, and dark, sulphide-rich sediment patches colonised by ampharetid polychaetes. Transects are plotted against a side-scan sonar image of the seabed generated during SO191-2 (darker pixels indicate stronger acoustic backscatter). Labels in the top left panel show voyage and deployment number for each transect: TAN0616, *RV Tangaroa* 2006; SO191, *RV Sonne* 2007; SO214, *RV Sonne* 2011.

this site was sampled (SO214, April 2011), the dark, sulphide-rich sediments were less distinct in camera transects than in earlier years (2006, 2007). Subsequent core samples showed that the sulphide-rich patches were overlain by deposits of fresh planktonic detrital matter, which obscured the dark sediments below.

Uruti Ridge. *Hihi*, *LM-10*, and *Kereru*. No additional sampling has been undertaken at these sites since TAN0616. Habitats and fauna have been described by Baco et al. [8]. All three sites consist primarily of extensive weathered carbonate rock platforms with sparse *Lamellibrachia* sp. populations close to the flare co-ordinates, scattered *Calyptogena* sp. shell fragments, and some patches of bacteria-covered rock (Figure 6). Scleractinian coral rubble and evidence of trawling (abandoned gear and scour marks) were widespread.

Omakere Ridge. *LM-9*, *Kea and Kaka*, *Bear's Paw*, and *Moa*. Other than a single multicorer deployment at Bear's Paw during SO214 (SO214 station 095), no additional biological sampling has been undertaken at these sites since SO191. Substratum and faunal characteristics have been described by Jones et al. [14] but re-analysis of images here showed that *Stelletta* n. sp. sponges were present on carbonates at all sites, and that bathymodiolin mussels were present in low densities at all sites except LM-9.

Moa (reef). As described by Jones et al. [14], this is an area of massive, weathered, carbonate rock colonised by sessile heterotrophic fauna including scleractinian and antipatherian corals, and sponges. No live seep-associated fauna were recorded.

Southwest Moa. Jones et al. [14] noted that the south-western extension of Moa had very different physical and faunal

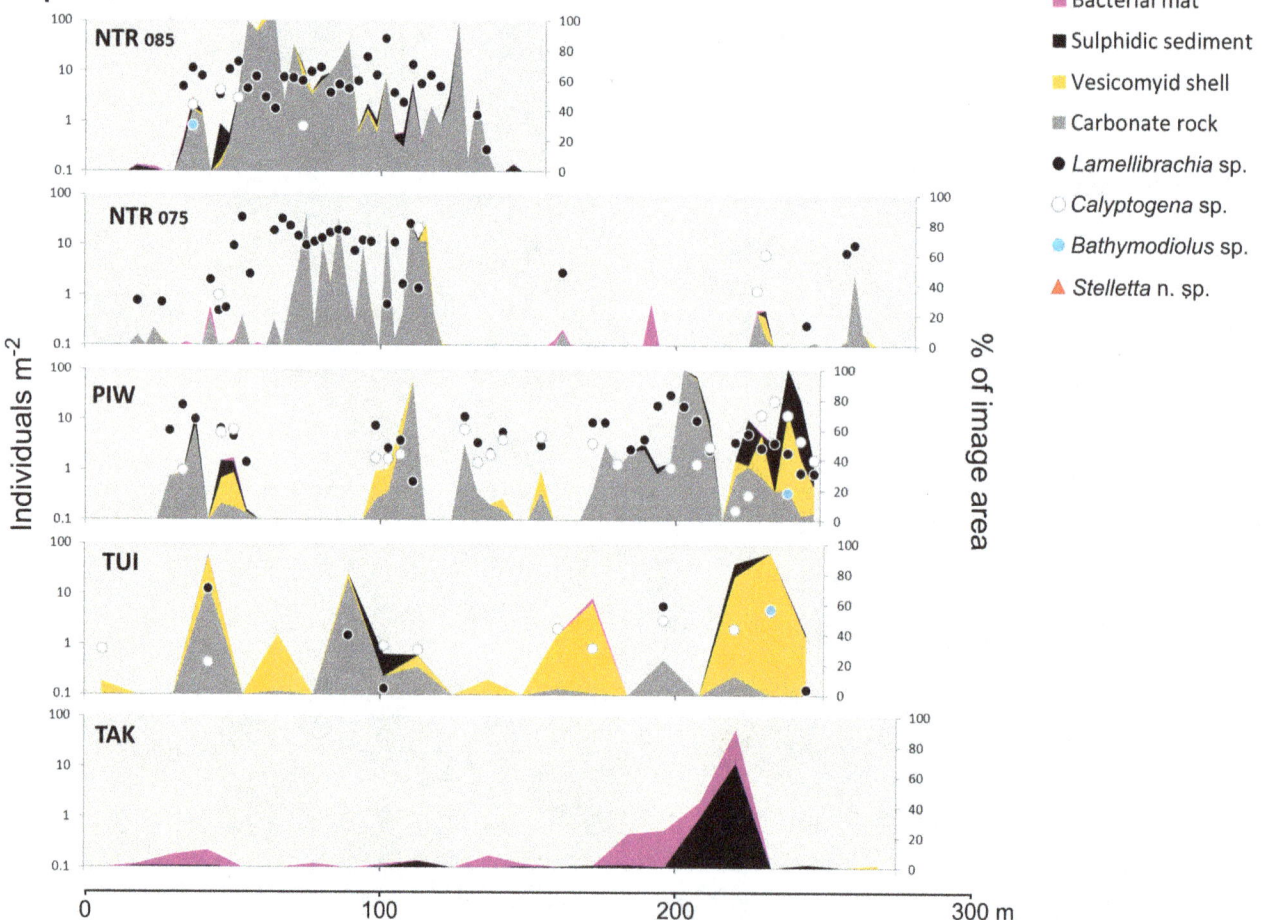

Figure 5. Representative profiles across selected seep sites at Opouawe Bank. Site name abbreviations: NTR, North Tower; PIW, Piwakawaka; TUI, Tui; TAK, Takahe (numbers on North Tower profiles show station numbers; see Figure 4. Station numbers for other sites: PIW, SO214-070; TUI, SO191-129; TAK, SO214-065). Data are from analyses of still images taken during towed camera transects across each site. Point markers show population densities of seep-associated fauna (individuals m^{-2}, log scale, left y-axis), area fills show substratum type as proportion of the full image area (% of image area, right y-axis). Profiles are scaled to horizontal distance travelled (x axis).

Uruti Ridge

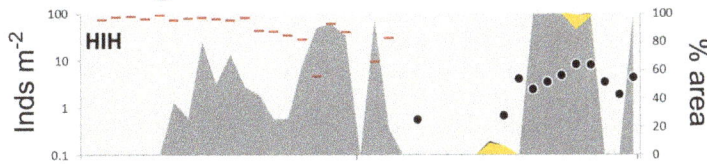

Omakere Ridge

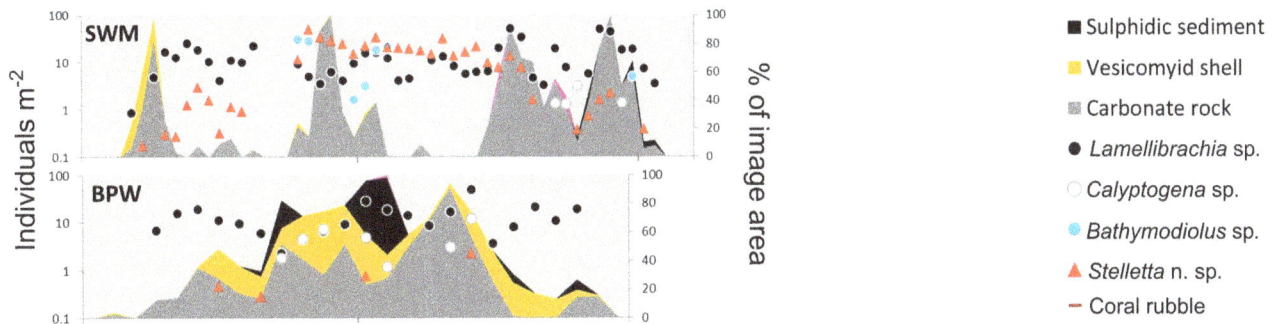

- ■ Bacterial mat
- ■ Sulphidic sediment
- ■ Vesicomyid shell
- ▦ Carbonate rock
- ● *Lamellibrachia* sp.
- ○ *Calyptogena* sp.
- ● *Bathymodiolus* sp.
- ▲ *Stelletta* n. sp.
- — Coral rubble

Rock Garden

Ritchie Ridge

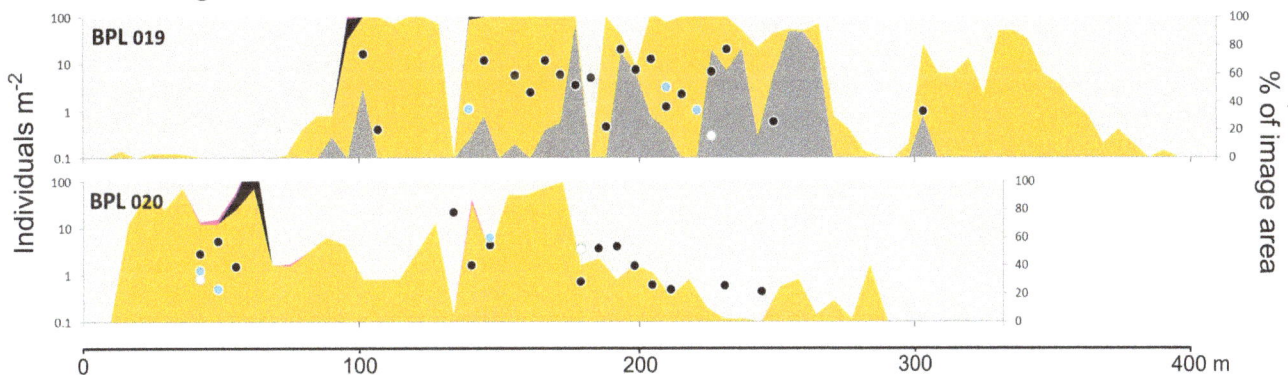

Figure 6. Representative profiles across selected seep sites at Uruti Ridge, Omakere Ridge, Rock Garden, and Ritchie Ridge. Site name abbreviations: HIH, Hihi; SWM, Southwest Moa; BPW, Bear's Paw; LM3, LM-3; BPL, Builder's Pencil (figures on Builder's Pencil profiles show station numbers. Station numbers for other sites: HIH, TAN0616-066; SO214-094; BPW, SO191-52-3; LM3, SO191-042). Data, symbols, and scale as for Figure 5.

characteristics to the main part of Moa. Additional SSS coverage collected during SO214 together with a single camera transect across the western extremity of the site confirmed that Southwest Moa is an area of active seepage supporting the highest densities of chemoautotrophic taxa recorded at any site. (Figure 2 E, Figure 6). *Stelletta* n. sp. sponges were particularly abundant and, together with *Lamellibrachia* sp. tubeworms commonly occurred on the upper surfaces of carbonate blocks and pavements, in contrast to most other sites where fauna rarely projected above the upper surfaces of the carbonates.

Rock Garden. *Rock Garden Knoll.* Extensive weathered carbonate substratum was recorded but with no seep fauna recorded other than a single live *Lamellibrachia* sp. tubeworm.

Faure Site. As described by Naudts et al. [34], transects showed primarily muddy sediments with *Calyptogena* sp. clam shells in some areas, low densities of *Lamellibrachia* sp. tubes, and some patches of dark sediments. Still images were intermittent and of low quality at this site but video showed apparently live Bathymodiolin mussels associated with one of the areas of clam shell.

Weka. Image quality here was also poor but video revealed extensive areas of weathered carbonate platform with widespread *Calyptogena* sp. shells and, at the Weka *a* site, occasional *Lamellibrachia* sp. tubes. No live seep fauna were recorded.

LM-3. As described by Naudts et al [34], the only camera transect across this site showed extensive weathered carbonates with a single small patch (ca. 25 m across, Figure 6) with very high densities of chemoautotrophic fauna including Bathymodiolin mussels, *Lamellibrachia* sp., and *Pseudosuberites* sp. (Figure 6).

Ritchie Ridge. *Builder's Pencil* is on a ridge of weathered carbonates and is remarkable for extensive areas of seabed overlain by accumulations of *Calyptogena* sp. clam shells. Video analyses confirmed a seabed area of ca. 70,000 m^2 covered by clam shells (*cf* [8]), many of which were worn and had brown Fe–Mn oxide discolouration characteristic of long-term presence on the seabed. Live *Lamellibrachia* sp. tubeworms were widespread but sparse, reaching maximum recorded densities of 21 individuals m^{-2} and were always either in close contact with the substratum or in crevices and under ledges. Bathymodiolin mussels occurred frequently but at low densities, and only one patch of live *Calyptogena* sp. clams was recorded. On the flanks of the shell-covered ridge, carbonate rocks were colonised by sponges, crinoids, and antipatherian, stylasterid, and gorgonian corals.

Comparisons among sites

Profiles across seep sites in each of the regions, using data from analysis of still images illustrate differences in substratum and community composition within and among regions (Figures 5 and 6). Conspicuous distinctions among sites are the absence of carbonate rocks at Takahe, the extensive areas covered by *Calyptogena* sp. shells at Builder's Pencil, and the high faunal densities at Omakere Ridge sites, including *Stelletta* n. sp. which was recorded only in this region.

The NMDS ordination based on observations of seep-associated fauna and corals in video transects (Figure 7) indicated that communities range between three extreme states represented by transects at Takahe, Southwest Moa, and Moa Reef, respectively. The Takahe extreme was characterised primarily by the absence of all mega-epifauna other than highly localised populations of *Calyptogena* sp. clams associated with small areas (m^2 scale) of disarticulated shell, and the presence of bacterial mats and ampharetid patches. The Southwest Moa extreme was characterised by the highest densities of all seep-associated taxa, and moderate areas of *Calyptogena* sp. shell. The third extreme, represented by Moa, was characterised by scleractinian corals and low densities, or absence of, chemoautotrophic taxa, including ampharetid patches and bacteria. Vectors of variable contributions (Figure 7 B) show that variability among sites along an axis from top left (Takahe) to bottom right (Moa) in the ordination was driven primarily by changes in the relative occurrence of bacterial mats and ampharetid patches (increasing towards top left), and scleractinian corals (increasing towards bottom right). However, there was also marked variability along an axis orthogonal to this which was associated primarily with changes in the occurrences of live populations of the four megafaunal chemoautotrophic taxa; occurrence of clams, tube worms, mussels, and *Stelletta* n. sp. sponges decreasing from upper right to bottom left. The horizontal dimension of the ordination was correlated with the percentage of carbonate rock substratum (Pearson correlation, 0.52); carbonates increasing from the Takahe to the Moa extremes (Figure 7 C). The vertical dimension was correlated with both depth (-0.57) and trawling (0.41); deepest sites being at the top of the ordination, and number of trawls per site increasing from the South-west Moa extreme towards the lower left of the ordination.

The spread of data points between the three extreme states in the ordination indicated considerable variability at both within-region and within-site scales. For instance, Opouawe Bank sites span a large proportion of the total ordination space, ranging from the extreme, no-carbonate, no-tubeworm, no-mussel state of Takahe to massive carbonates with large areas of *Calyptogena* sp. shells at Tui, and moderate to high densities of tubeworms and carbonates at North Tower. Similarly, in the Omakere Ridge region, Moa and Southwest Moa are adjoining sites yet contain strongly contrasting habitats and fauna; transects at these sites representing two of the three extreme states in the ordination.

The ordination also indicated some separation between regions, particularly between Opouawe Bank and the other regions. This was driven in part by higher incidence of bacterial mats and ampharetid patches at Opouawe Bank sites, but also by the presence of *Stelletta* n. sp. sponges at most Omakere Ridge sites, and corals at Moa and the Uruti Ridge sites. However, there was convergence between sites independent of spatial separation. For instance, Tui and Builder's Pencil were similar, despite being located on Opouawe Bank and Ritchie Ridge, respectively; the similarity in this case being driven largely by high proportions of seabed covered by *Calyptogena* shells at both sites.

Assessing impacts of human activities

Trawling. Sites in the Rock Garden region (Rock Garden Knoll, Faure Sites, Weka, and LM-3) were the most heavily trawled, with cumulative trawl numbers for the period 1998–2005 ranging from 43 to 54 trawls at Faure Site up to 150 trawls at LM-3 (Figure 8). In the Opouawe Bank region, the number of trawls ranged from eight at South Tower to 43 at Tui, with 28 at North Tower. Sites on Uruti Ridge (Hihi, LM-10, and Kereru) were each trawled ca. 20 times. Omakere Ridge sites were least impacted by trawling, with seven trawls at LM-9, one at Kea and Kaka, one at Southwest Moa, and three at Moa. Comparing these values to the distribution of sites in the NMDS ordination (Figure 7 C) showed that increasing trawl intensity was associated with the gradient of decreasing occurrence of live chemoautotrophic fauna identified above, but was also inversely correlated with depth; shallower sites having fewer live fauna and greater trawl intensity.

Gas hydrate extraction. Seismic surveys of the Hikurangi Margin [20,25–28] have shown that bottom simulating reflectors (BSRs), which are indicative of free gas underlying gas hydrates, are widespread across much of the margin (Figure 9). Recent interpretations suggest that most BSRs on the margin represent gas hydrates in fine-grained sediments and mudstones [25] which are unlikely to be suitable for commercial hydrate extraction [35]. The seep sites are associated with stronger BSRs in areas where methane-rich fluids pass through the gas hydrate stability zone (GHSZ; where pressure and temperature conditions favour hydrate formation) and reach the seabed either via direct vertical 'chimneys' or by migration along stratigraphic pathways and emergence through networks of extensional faults [26,36]. Current resource interest centres on hydrates and free gas reserves in sand reservoir strata, which are not necessarily associated with BSRs [35]. These strata have not yet been well-defined across the Hikurangi Margin [37] and thus the likely locations of any future extraction of hydrates in relation to the seep communities are not yet known. A number of sites have been proposed for research drilling, however, including Opouawe Bank and sand channel systems north of Uruti Ridge [38], both of which would be in close proximity to known chemoautotrophic communities (Figure 9).

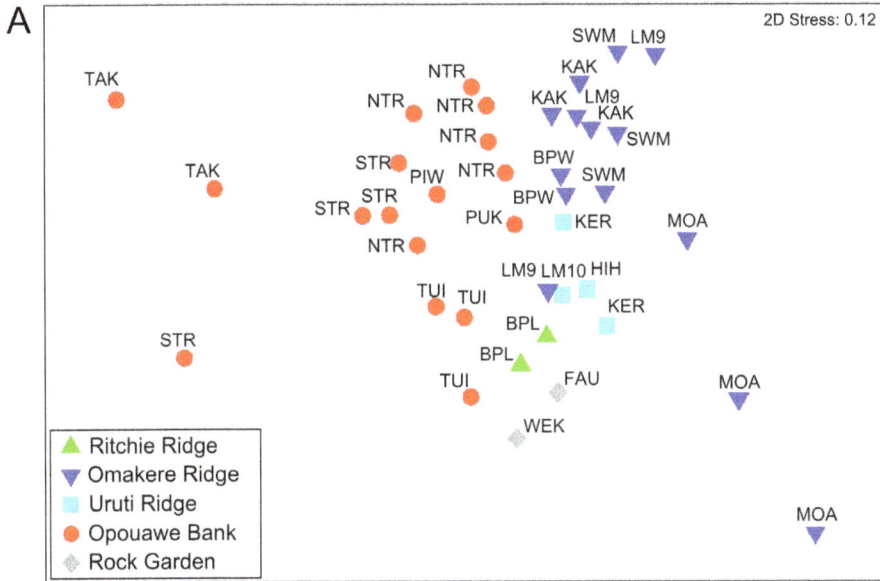

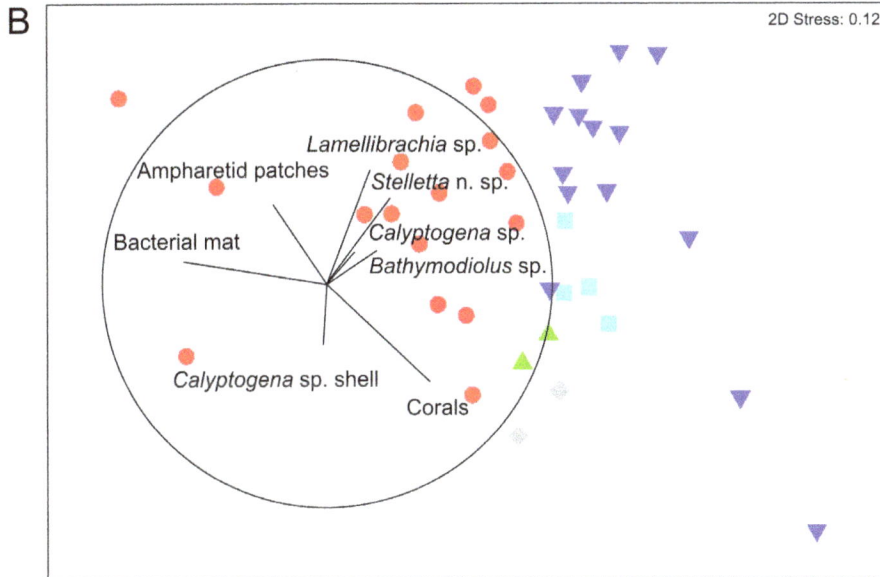

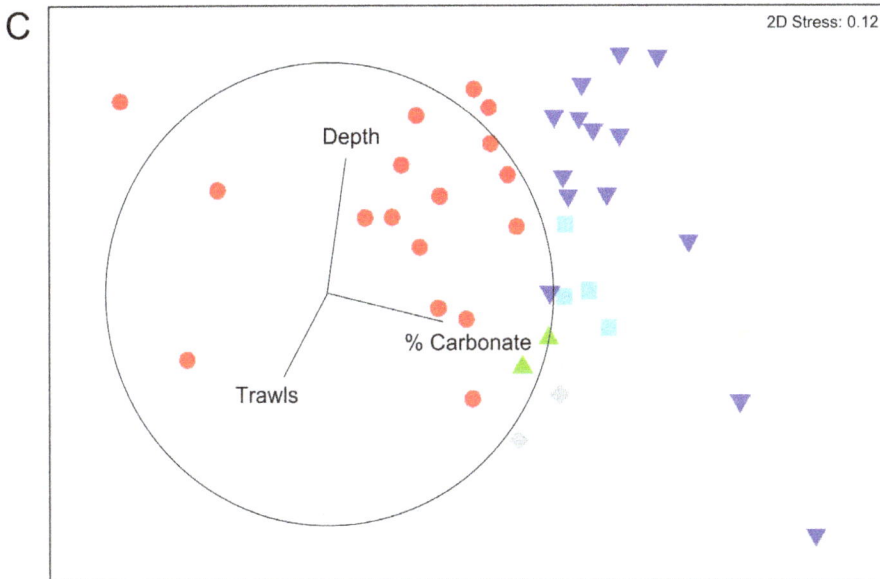

Figure 7. Non-metric multivariate scaling (NMDS) ordination of Hellinger distances between seep communities observed in video transects. Underlying data are log-transformed occurrence records of chemoautotrophic megafauna (*Lamellibrachia* sp. tubeworms, *Calyptogena* sp. clams, *Bathymodiolus* sp. mussels, *Stelletta* n. sp. sponges), seep-associated substrata (bacterial mats, ampharetid patches and *Calyptogena* sp. clam shells), and intact scleractinian coral matrix. A – Transects distinguished by region (symbols; see Figure 1 for context) and seep site (labels; see Table 1 for full names). B – Contributions (Pearson correlations) of faunal variables to between-sample distances. C – Relationships (Pearson correlations) between three environmental variables and the distribution of transects in the ordination: depth (−0.57); % carbonate rock substratum (0.52), and number of trawls per site from 1998 to 2005 (0.41).

Discussion

Habitat and community composition and structure

Study of the Hikurangi Margin cold seep sites, primarily using towed camera systems but augmented with information from corer, grab and epibenthic sled sampling [8,12,14–18], has enabled description of the principal chemoautotrophic megafauna present, their population densities, the physical habitats they are associated with, and how these parameters vary across the study area. The small spatial scale of seep microhabitats in relation to the intensity of sampling to date makes it likely that isolated patches of live fauna will have been missed at any given site, but broad conclusions concerning distributions of fauna and substrata are well-supported. For instance, intensive sampling with a range of methods at sites on Opouawe Bank did not record the distinctive spherical sponge *Stelletta* n. sp., which was abundant at sites in the Omakere Ridge region, nor the scleractinian corals that were recorded at Moa and at the Uruti Ridge sites. Similarly, the relative extents of different seep-associated substrata described here, including authigenic carbonates, sulphide-rich sediments, and clam shells, are likely to be reliable, particularly across the

Opouawe Bank and Omakere Ridge regions where full SSS coverage provides accurate measures of seep site extent.

The seep communities on the Hikurangi Margin share some characteristics with those from both seeps and vents in other parts of the world [3,7,39–45], including dominance of mega-epifaunal communities by large siboglinid tubeworms, vesicomyid clams, and bathymodiolin mussels, but there are also differences in the combinations of fauna in different regions. For instance, while the well-oxygenated New Zealand seep sites [16] are characteristically populated by mixed communities of tubeworms, clams, mussels, and sponges, those on the Californian and Oregon margins are dominated by bacterial mats and vesicomyid clams within the oxygen minimum zone [44] but by tubeworms and clams in oxygenated areas [45]. In the Atlantic, seeps in the Gulf of Mexico and on the equatorial West African margin are dominated by tubeworms and mussels, although clams may also be present [39,42]. Perhaps more significantly, the distinctive ampharetid beds appear to be ubiquitous at the New Zealand sites. Although these beds are not unique to the New Zealand region, similar tube-building ampharetids having been observed around vesicomyid clam beds at Hydrate Ridge [18], their density, biomass, and frequency of occurrence are apparently greater on the Hikurangi

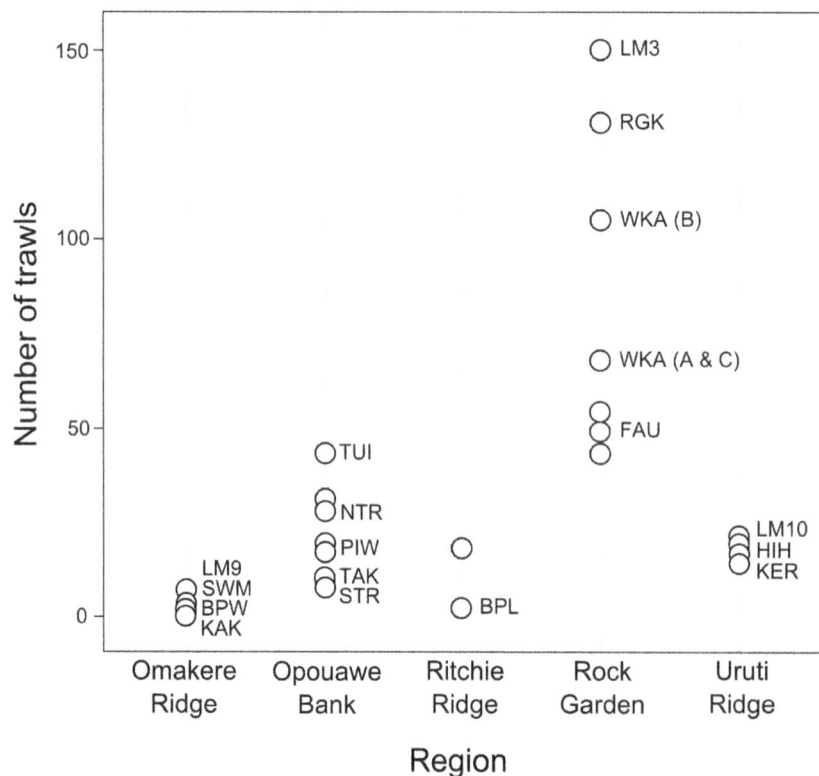

Figure 8. Trawl intensity at seep sites by region. Selected seep sites are identified by name (see Table 1 for full names). Trawl data are cumulative totals for the period 1989–90 to 2004–5 [24] calculated as the number of trawl tracks intersecting a 250 m radius circle around seep site positions given in Greinert et al. [12].

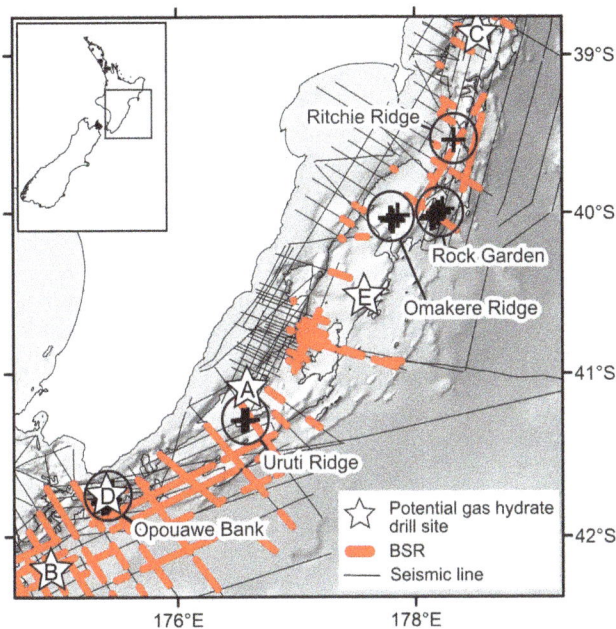

Figure 9. Cold seep communities and gas hydrates. Distribution of cold seep faunal communities (black crosses, regions as in Figure 1) on the Hikurangi Margin in relation to occurrence of bottom simulating reflectors (BSR) in seismic survey data and sites of potential interest for gas hydrate exploration (white stars): A, channel system north of Uruti Ridge; B, Pegasus Basin; C, Tuaheni Basin; D, Opouawe Bank; E, Porangahau Ridge.

Margin ([18], AT and LL unpublished data). Thus there appear to be structural differences between the seep communities of the Hikurangi Margin and those elsewhere in the world.

Inter-site differences. Differences among sites were associated with variations in the relative abundances and presence or absence of taxa, and the types and proportions of seep-associated substrata present. While there was some distinction by region, most obviously with the presence of *Stelletta* n. sp. only at Omakere Ridge sites, most taxa and most habitats were observed in most regions. There was, however, wide variability within regions in terms of seep size, dominant substrata, and community composition. We suggest that this variability is driven (1) by the ages of the seeps (i.e. the history of fluid flow at each site) and thus the successional stage of the faunal communities inhabiting them, and (2) by the history of disturbance from trawling. We expand on these points below, starting with evaluations of the ecology of the principal chemoautotrophic fauna and their likely roles in succession.

Succession. Although Levin [4] highlighted the need for *in situ* measurements over extended periods for developing full understanding of successional processes at cold seeps, a growing literature on the trophic and reproductive ecology of the principal seep megafaunal taxa provides a logical framework around which to develop a successional model for the Hikurangi Margin seeps (as has been done for whale falls [46] and later stages of Gulf of Mexico seeps [33]). This is an important goal because understanding of the sequence and rate of succession is essential for evaluating the vulnerability of these sites to present and potential future disturbances. We do not yet have life-history information for any of the New Zealand megafaunal taxa but detailed studies of congeneric species of *Lamellibrachia* [47–51] and *Calyptogena* [52–56] in other seep provinces, when combined with inferences from

in situ observations here, enable us to construct a first hypothetical model of succession for the Hikurangi seeps.

From their ubiquitous presence in Hikurangi Margin seep environments and what is known of their chemistry and ecology [16–18,31,57], it is likely that the ampharetid populations play a part in the early stages of seep community development. Sommer et al. [16,31] have speculated that the ampharetids play an important role in early successional stages by increasing the concentration of sulphides in surficial sediments. Their hypothesis suggests that tube building by the ampharetids actively promotes growth of the bacteria they feed on by providing conduits that simultaneously increase rates of methane flux through the sediment-water interface and concentrations of sulphate-rich seawater in near-surface sediments. Together, these processes would also promote increased rates of microbial anaerobic oxidation of methane (AOM, [58]) beneath the ampharetid patch and thus explain the higher sulphide concentrations measured in ampharetid patches than in surrounding sediments [16]. Regardless of the actual pathways involved, measurements of much higher rates of oxygen consumption, methane flux, and sulphide concentrations within ampharetid patches [16] support a hypothesis that the ampharetid populations modify habitats in a way that increases the availability of sulphide and methane at the seabed, thus facilitating seep community development.

Sommer et al. [16] and Dale et al. [57] have argued that ampharetid beds are a transitional stage in seep development between the onset of fluid flow and the establishment of bacterial mats. Examination of seabed images and consideration of the trophic ecology of the ampharetids, however, suggests that the sequence might be the reverse of this; bacterial mats preceding establishment of ampharetid populations. The morphology and stable isotope signatures of the ampharetids show that they are heterotrophic and that most of their carbon intake is derived from aerobic methanotrophy; probably by direct consumption of methane-oxidising bacteria [17,18]. This suggests that the ampharetids recruit to patches in which their food, bacteria, is abundant. Furthermore, seabed images show that bacterial mats often occurred on sediments where there was no sign of associated ampharetid pits or colonisation by any other seep-associated metazoans.

The *Calyptogena* sp. clams have thiotrophic symbionts [17] and in this study live clams were only observed embedded in dark, sulphide-rich, sediments colonised by ampharetid polychaetes. We suggest, therefore, that the clams recruit preferentially to patches that have first been 'conditioned' by the activities of ampharetid polychaetes to have high sulphide concentrations. While this might not be an obligatory relationship, our observations suggest a hypothesis that the presence of ampharetid populations facilitates clam settlement by enhancing sulphide availability. From this study we know that adult *Calyptogena* spp. clams are ~40 to 80 mm long, live partially buried in the sulphide-rich sediment patches and, based on the ratio of disarticulated shells to live populations, apparently experience high rates of mortality. Congeneric vesicomyids also have thiotrophic symbionts [1,59], can have relatively rapid growth (*C. kilmeri*; 80% of asymptotic length at 6.6 y, [52]), and can exhibit near-continuous reproduction [54]; characteristics that are associated with taxa in ephermeral habitats [60,61]. Although both slower growth [56] and indirect evidence for seasonal reproduction [55] have been reported, these observations set bounds on habitats that *Calyptogena* spp. can colonise and suggest a minimum necessary habitat persistence time: they require soft sediments with high concentrations of sulphide accessible within ~80 mm of the sediment surface, which persist for at least 5 to 10 years (i.e. the shortest generation time reported for a congeneric [52]). Thus, in combination with our observations that live clam populations are small and highly-localised, and that areas of disarticulated shells are widespread, we might characterise *Calyptogena* spp. clams on the Hikurangi Margin as early colonisers of potentially short-lived (years to decades) sulphide-rich surficial sediment patches. Over short time scales (years) cessation of fluid flux in a given patch will result in extinction of the local (m^2 scale) population as sulphides are depleted. Over much longer time scales (100 s–1000 s y) capping of fluid flow by accumulation of carbonate precipitates will have the same effect over larger spatial scales (ha scale). This expectation matches the common observation of disarticulated shells, often in very large numbers, in the present study but leaves us with a poorly-constrained idea of the actual timescales involved. Thus, while it is possible that the quantity of accumulated shell material at a site is correlated with the local history of fluid flow, and thus seep age, we cannot yet put bounds on this on the basis of the clams alone.

By introducing seawater, and thus sulphate, into sediments, the clams also enhance AOM [62–64]. Over long time scales, if fluid flux to the patch persists, locally-enhanced rates of AOM result in the build-up of authigenic carbonate particles in surficial sediments [65], thus generating a local environment which is rich in sulphide and methane but which also has hard substrata. These are presumably the conditions required for recruitment by *Lamellibrachia* tubeworms, bathymodiolin mussels, and the seep-associated sponges, all of which need hard substrata for settlement.

The lamellibrachid tubeworms have thiotrophic endosymbionts (but see [17]) and their tubes attach to, and extend deep within, carbonate rock concretions. *Lamellibrachia luymesi* tubeworms at cold seeps in the Gulf of Mexico are of comparable size to the New Zealand species [51] and are extremely long-lived (100 s y) [46–49]. There is also persuasive evidence that *L. luymesi* actively promotes AOM, and thus the continued generation of sulphides, around its 'root' by releasing sulphate into sediments beneath carbonate concretions [66–68]. If the Hikurangi Margin species has similar characteristics, the primary constraints on colonisation for the tubeworms will be the availability of sulphides for energy, hard substrata for settlement, and established methane-rich fluid flow over extended periods (10 s–100 s y). Thus, while the tubeworms exploit the same energy resource (sulphide) as the clams, their attachment to hard substrata, longevity, and ability to access, and promote, generation of sulphides deeper in the sediments indicate recruitment at a later successional stage and very different life-history characteristics. They colonise only when fluid flow has been established long enough for carbonate precipitates to form, providing substrata for settlement, but may persist for as long as methane flux to sediments beneath the accumulating carbonate continues. That they might actively promote AOM in the sediments further suggests an important role in modifying physical habitats within the seep. Once established, increases in the rate and duration of AOM caused by the worms themselves will result in increased generation of carbonates, thus accelerating the development of carbonate chemoherm habitat.

The bathymodiolin mussels at the Hikurangi seeps rely primarily on methanotrophic symbionts [17] and thus exploit a different energy resource to the clams and tubeworms. Because they attach to hard substrata, however, they are dependent on availability of carbonates and thus will be later successional species than *Calyptogena* sp., recruiting to mature seep sites with carbonate substrata and methane flux to the water-column. Furthermore, because methane flux at seeps is variable across a range of spatial and temporal scales [15,29,34], rapid colonisation and growth,

coupled with short generation times, are also likely characteristics [61].

Little is known about the ecology of the two sponge species common at the seep sites; *Pseudosuberites* sp. and *Stelletta* n. sp. but the former, at least, can support a diverse commensal macrofaunal community characterised by high levels of methane-derived carbon [17]. Thurber et al. [17] hypothesise that the sponge may be chemoautotrophic and play a significant role in facilitating transfer of methane into the metazoan food web in hard substrata habitats at the seep sites.

Finally, we also know that all chemoautotrophic taxa are ultimately dependent on the flux of methane-rich fluids into seabed sediments. Consequently, when flow to a site ceases completely, all chemoautotrophic populations will die out and conditions will become suitable for colonisation by heterotrophic fauna. On carbonates, these are likely to include sponges, corals, and other sessile heterotrophic epifauna. The transition to heterotrophic communities is a gradual process, however, as many 'background' slope taxa can be tolerant of sulphide and may colonise seep sites before fluid flow ceases [69].

Liebetrau et al. [29] used uranium–thorium (U–Th) dating to estimate the age of carbonate rocks at North Tower, LM-10, Bear's Paw, and Moa. Bear's Paw was the youngest site, with a maximum recorded age of 2,360±70 years before present, North Tower was older (4,950±650 ybp), while LM-10 was considerably older (12,400±160 ybp). At Moa, only surficial carbonates were sampled, yielding dates for the most recent carbonates of 4,390±130 ybp. The times of both the onset and the most recent seepage are likely to be relevant to the structure of the seep communities and habitats observed in the present. The time since onset will influence the extent and thickness of carbonates, and thus the types of habitat available for colonisation, while the time of most recent activity will influence whether present habitats are more suitable for chemoautotrophic or heterotrophic fauna. The youngest carbonates measured at each of these sites ranged from 2,090 to 4,390 ybp, with the sites ranked from oldest to youngest as Moa (4,390±130 ybp), LM-10 (4,120±40 ybp), North Tower (3,960±50 ybp), Bear's Paw (2,090±850 ybp). Obvious matches between this chronology and our observations are that 'weathered' carbonates colonised by cold water corals were present at Moa and LM-10, whereas 'fresh' carbonates colonised by high densities of chemoautotrophic fauna were present at North Tower and Bear's Paw. Furthermore, while the youngest carbonates at LM-10 are of similar age to those at North Tower, the oldest carbonates at LM-10 are much older, indicating a considerably longer history of fluid flow.

While the U–Th data provide useful indicative ages, it is also important to note that these dates relate to the consolidated upper layers of the carbonates. Beneath the carbonates, at the interface between carbonate blocks and the underlying sediments, the process of carbonate build-up via AOM continues in the present. Evidence for this comes from observations of calcite particles in anoxic sediments beneath a large block retrieved from North Tower during SO191 (A. Eisenhauer, IFM-Geomar and DB, personal observation – Figure 3 B), as well as the presence of live chemoautotrophic fauna, which are reliable indications of continuing seepage in the present.

Much of this reasoning about seep succession is necessarily speculative but many of the basic ecological constraints (e.g. soft vs. hard substrata, thiotrophy vs. methanotrophy vs. heterotrophy), as well as likely time scales (e.g. years to decades for establishment of ampharetid beds and *Calyptogena* populations vs. centuries for establishment of *Lamellibrachia* populations vs. millennia for development of carbonate reefs) are likely to be reliable. Putting these bounds and processes together, we propose a hypothetical sequence of succession at Hikurangi Margin cold seep sites consisting of ten principal steps which result in five successional stages identifiable from seabed observations (Figure 10). Dates are poorly constrained in this scheme but indicative ranges are given as years since the onset of fluid flux in a patch, based on literature values for geochemical and ecological processes as discussed above:

1. Onset of **localised flux of methane-rich fluid** from deeper strata to the sediment-water interface [20,36].

2. **Anaerobic oxidation of methane** by microbial consortia within sediments generates patches of sulphide close to the sediment surface [58], but most methane continues to reach the sediment-seawater boundary layer [57].

3. Availability of methane and sulphide within the patch promote growth of **aerobic methanotrophic and thiotrophic microbial communities** at the sediment surface [*Succession Stage 1: colonisation by aerobic microbial community – ca. 1–10 y*]

4. Heterotrophic **ampharetid polychaetes** colonise the patch, feeding on abundant microbial production, primarily by aerobic methanotrophic microbes at the sediment surface [17,18,57].

5. Tube building by **high-density ampharetid populations** increases permeability of sediments and thus both upward flux of methane-rich fluids through the patch and downward irrigation of sulphate-rich seawater into the sediments. This increases both aerobic microbial production at the sediment surface and AOM within the sediments [16,57]. [*Succession Stage 2: Colonisation by ampharetid polychaetes – ca. 1–100 y*]

6. Locally elevated supply of sulphide within the patch facilitates colonisation by **Calyptogena clams** (and other vesicomyid taxa), which require high sulphide concentrations in surficial soft sediments. Clams also contribute to irrigation of sulphate-rich seawater into sediments, further enhancing AOM [62–64] [*Succession Stage 3: Colonisation by clams – ca. >50 y*].

- If flow ceases in patch, local die-off of ampharetids and clams follows and disarticulated clam shells accumulate on background sediment.

7. If flow persists, **authigenic carbonate precipitates** form hard substratum particles in the patch, enabling colonisation by **Lamellibrachia tubeworms** (thiotrophic symbionts). [*Succession Stage 4: Carbonate precipitates enable colonisation by lamellibrachid tubeworms – ca. >100 s y*]

8. Continuing build-up of carbonate modifies flow and reduces the extent of sulphide-rich sediments. This causes local reduction of *Calyptogena* populations but favours recruitment of **Bathymodiolin mussels** (requiring hard substrata and methane flux to the water column)

9. Long lived *Lamellibrachia* (100 s y [51]) continue to access deep sulphide below carbonates through their 'root', stimulating continued sulphide and carbonate generation though AOM by actively introducing sulphate to deep sediments [66–68].

10. **Carbonate build-up continues** and eventually caps the original seep site, causing flow to be displaced to periphery of carbonate platform. Widespread failure of clam populations and more gradual decline of tubeworm populations follows before carbonate rocks are colonised by **non-seep epifauna**; corals, non-methanotrophic sponges, etc. [*Succession Stage 5: Colonisation by non-chemosynthetic epifauna – ca. >1000 s y* [29]].

Vulnerability to human activities

Anthropogenic threats to the seep sites include on-going physical disturbance from bottom trawling, and potential modification of fluid flow patterns resulting from future large-scale extraction of methane hydrates. While hydrate extraction is still in the resource evaluation phase, the trawl frequency data here show that most sites were affected by bottom trawling in the period from 1989 to 2005, with trawl intensity ranging from a single trawl at some Omakere Ridge sites to 150 trawls at LM-3 (Figure 8). Despite the wide variability in trawl intensity between regions and sites, detecting quantifiable effects of trawling on the seep fauna is not straightforward for two reasons. First, any effects of trawling are likely to be overlaid on differences in successional stage (i.e. the age of the seep site) outlined above (Figure 11). Thus, while South Tower, Takahe, and Hihi have similar trawl histories, any visible effects of trawling are likely to differ between the low-lying authigenic carbonate habitats at South Tower by comparison with the open muddy sediments at Takahe, or the massive late-stage carbonate platforms at Hihi. Second, the heterogeneous physical structures within carbonate concretions are likely to create localised refugia in which benthic fauna escape trawl impacts. An example of this is the thriving but highly localised (ca. 25 m diameter) chemoautotrophic community observed at LM-3, despite this site having the highest recorded frequency of trawling.

There are, however, qualitative differences between regions that might be related to differences in trawling history. Most obviously, very few live chemoautotrophic taxa were recorded at sites in the Rock Garden region by comparison with any of the other regions: apart from sparse *Lamellibrachia* sp. tubeworms at Rock Garden Knoll and Faure Site, and the single patch of seep fauna at LM-3, only disarticulated clam shells were recorded. It is also clear that *Stelletta* n. sp. sponges were seen, often in high densities, only at sites in the Omakere Ridge region, where trawl intensity was lowest. Also apparent from close examination of still images in particular, are differences in the growth forms of lamellibrachid tubeworms. At most sites, *Lamellibrachia* tubes emerged from beneath carbonate blocks and ledges and characteristically grew along overhangs or in the spaces between blocks [8]. At the least-trawled sites, Southwest Moa, Bear's Paw, and Kea-Kaka, however, tubeworms grew directly from the upper surfaces of carbonate blocks (Figure 2 D), often in association with *Stelletta* n. sp. sponges. Finally, although not a seep-associated taxon, matrix-forming scleractinian corals were seen as intact 'thickets' only at Moa in the Omakere Ridge region (Figure 2 F), whereas on similar rock substrata in the heavily trawled Rock Garden and Uruti Ridge regions, extensive areas of broken fragments of coral were recorded (Figure 6).

Based on their physical form and what is known of their life histories, we can construct a ranking of the principal chemoautotrophic taxa in terms of their likely vulnerability to physical disturbance. *Calyptogena* sp. clams exploit transient sulphide patches, suggesting that their populations may be resilient to intermittent disturbance. However, if, as proposed above, the establishment of dense ampharetid populations is important for generating high-sulphide patches, it is conceivable that the viability of clam populations might be influenced by the resilience of ampharetids to disturbance, as well as that of the clams themselves. Clams potentially reach reproductive maturity in less than five years from first settlement [52] but Dale et al. [57] estimate that the ampharetid patches they studied at North Tower to be in the order of 70 years old. Thus, if the ampharetids were to be particularly susceptible to disturbance, whether from direct contact or from smothering by resuspended sediments, this could

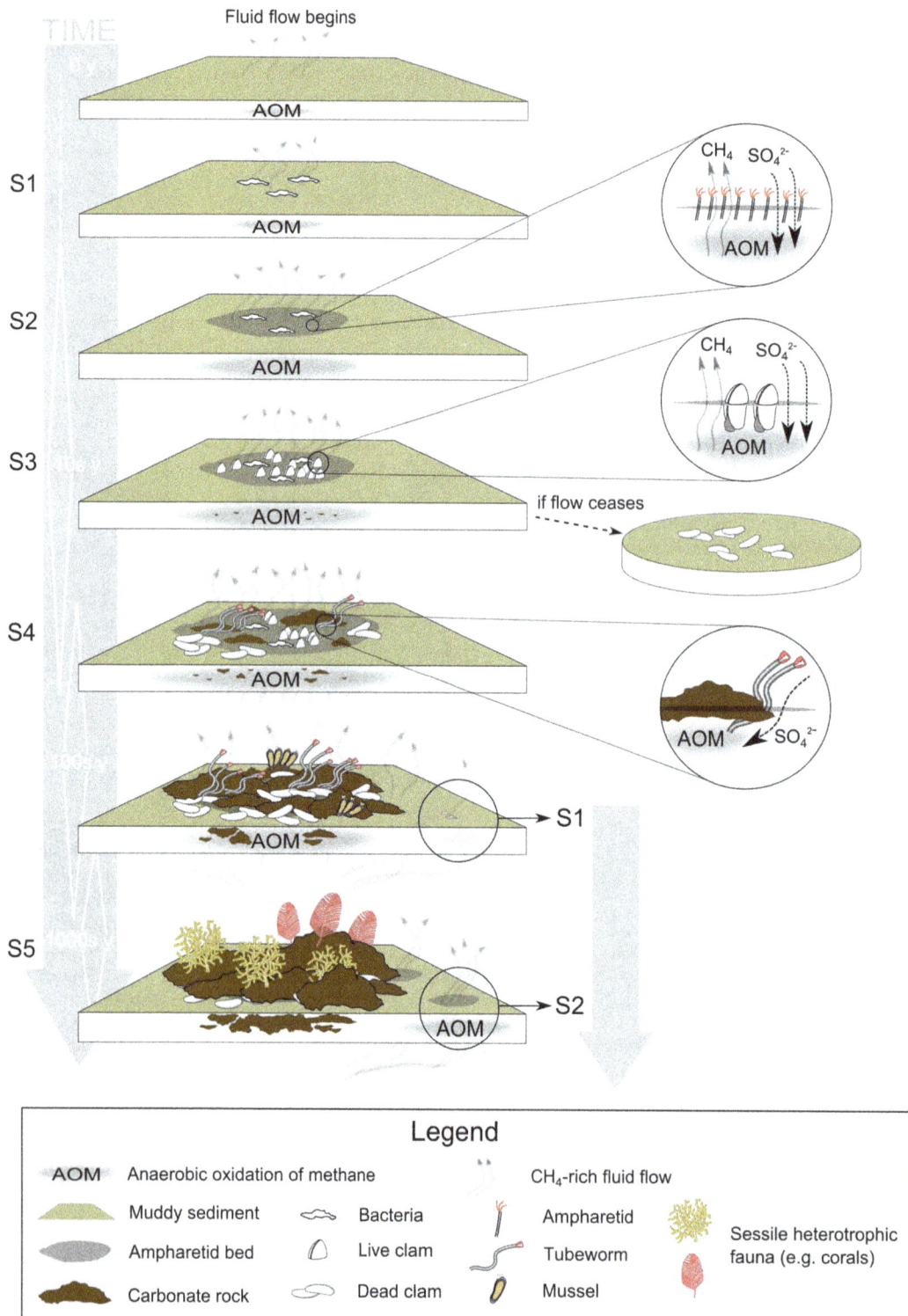

Figure 10. Hypothetical succession sequence at Hikurangi Margin cold seep sites. Labels 'S1', 'S2', etc., indicate Successional Stages as described in the discussion text.

have consequences for clam populations over longer time scales through reduction in the availability of sulphide-rich habitat.

Lamellibrachid tubeworms apparently gain some protection from physical impacts by virtue of their habitat within the authigenic carbonates of the seep. However, if the high population

densities and distinctive growth form of tubeworms at Southwest Moa and other Omakere Ridge sites are representative of undisturbed populations, it seems likely that populations at all other sites have been affected by trawling. The chitinous material of their tubes is very tough, requiring considerable force to cut or

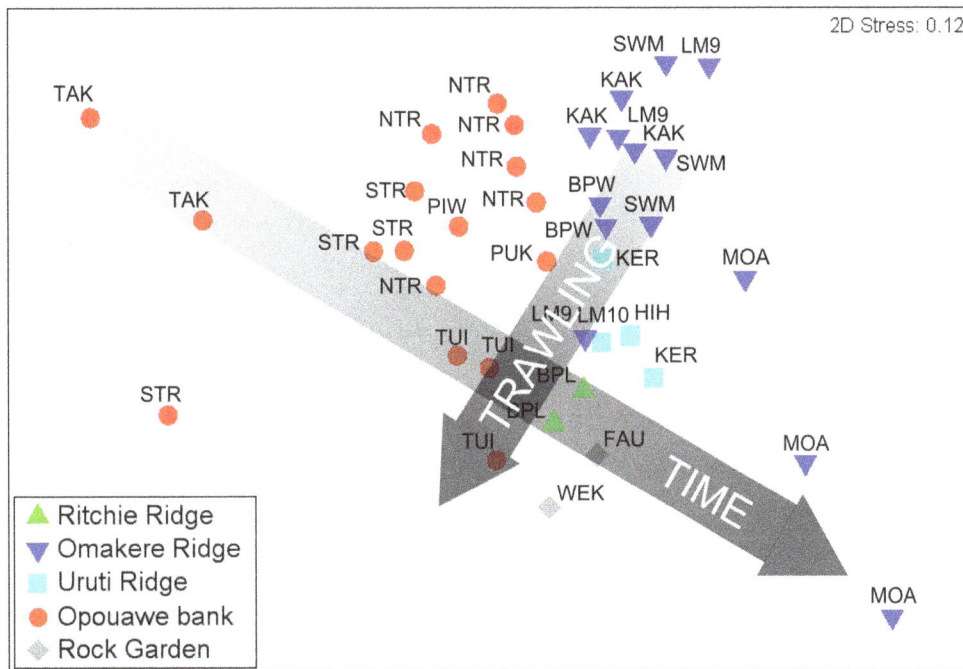

Figure 11. Interaction between seep site age and disturbance from trawling. NMDS ordination of seep site communities (see Figure 7) showing conceptual illustration of the relative influences of seep site age and trawling on observed seep characteristics.

break in recovered specimens (authors' personal observations and [32]), yet splintered and broken tubes were recorded in several transects. While physical impact is the most likely cause of such breakage, no well-defined correlation with trawl intensity was apparent in the present data, other than the observation that tubeworm populations were sparse, highly localised, or absent at sites where trawling was most intense. Observations of live tubeworms in moderate densities at Uruti Ridge sites where there was obvious evidence of trawl damage in the form of coral fragments, trawl marks, and abandoned trawl gear [8], suggest that they are more resilient to this kind of impact than sessile suspension-feeding taxa such as corals and sponges. While persistence of chemoautotrophic communities at even heavily trawled sites might suggest that some of the seep fauna are resistant to physical disturbance, existing data are not sufficient to determine the extent to which whole communities might have been removed, nor yet the longer-term consequences of on-going disturbances. It is clear, however, that coral-dominated communities characteristic of later successional stages at the seep sites are highly vulnerable to direct physical impacts and have long recovery times.

Techniques for large-scale extraction of seabed gas hydrates have not yet been developed and target sites have yet to be identified with any confidence. However, potential impacts on seep communities can be classified into four categories: (1) direct physical disturbances of the seabed; (2) smothering by re-suspended sediments or tailings, (3) indirect effects associated with modification of fluid flow to the seep sites, and (4) large-scale destabilisation of slope sediments. It is currently considered unlikely that extraction would directly target the anticlinal structures on which many seep communities are found, because of their geological complexity and consequent unpredictability of flow [35], and considering the potentially catastrophic consequences of slope destabilisation, its prevention is likely to be a major factor in extraction planning. If this proves to be the case, then

direct physical impacts and smothering might represent lower levels of risk to seep communities than the consequences of flow modification.

Any reduction in fluid flow to sites of established seep communities would affect all successional stages proposed here, with consequent declines in the populations of all chemoautotrophic taxa. For any future large-scale resource extraction, therefore, the potential ecological consequences would need to be assessed by consideration of the specific geologic setting of the hydrate reserves, the proximity of extraction sites to seep sites with live chemoautotrophic communities, the spatial scale over which flow regimes would be affected, and thus how flow at neighbouring seep sites would be affected. A strategy to manage impacts would require consideration of the life-history characteristics of the principal taxa and the time scales associated with each of the successional stages proposed here, which range from years to millennia. In particular, it will be important to develop understanding of spatial and temporal scales of connectivity among populations, and evaluate susceptibility to the different categories of disturbance.

Acknowledgments

We thank the officers, crew, and science teams of R/V Tangaroa, voyage TAN0616 'New Zeeps', and R/V Sonne, voyages SO191 'New Vents' and SO214 'NEMESYS', particularly Jens Greinert and Jörg Bialas as voyage leaders of SO191-2 and SO214-2, respectively. Thanks also to Brent Wood and Arne Pallentin at NIWA for database and GIS support, and to Erik Cordes and an anonymous reviewer for perceptive reviews that helped refine the manuscript.

Author Contributions

Conceived and designed the experiments: DAB AAR ART AB LAL CRS. Performed the experiments: DAB AAR ART AB LAL CRS. Analyzed the data: DAB. Contributed reagents/materials/analysis tools: DAB. Wrote the paper: DAB AAR ART AB LAL CRS.

References

1. Levin LA, Mendoza GF, Gonzalez JP, Thurber AR, Cordes EE (2010) Diversity of bathyal macrofauna on the northeastern Pacific margin: the influence of methane seeps and oxygen minimum zones. Marine Ecology-an Evolutionary Perspective 31: 94–110.

2. Olu-Le Roy K, Sibuet M, Fiala-Medioni A, Gofas S, Salas C, et al. (2004) Cold seep communities in the deep eastern Mediterranean Sea: composition, symbiosis and spatial distribution on mud volcanoes. Deep-Sea Research Part I-Oceanographic Research Papers 51: 1915–1936.

3. Sellanes J, Quiroga E, Neira C (2008) Megafauna community structure and trophic relationships at the recently discovered Concepcion Methane Seep Area, Chile, ∼36°S. ICES Journal of Marine Science 65: 1102–1111.

4. Levin LA (2005) Ecology of cold seep sediments: interactions of fauna with flow, chemistry and microbes. Oceanography and marine Biology: an Annual Review 43: 1–46.

5. German CR, Ramirez-Llodra E, Baker MC, Tyler PA, and the ChEss Scientific Steering Committee (2011) Deep-Water Chemosynthetic Ecosystem Research during the Census of Marine Life Decade and Beyond: A Proposed Deep-Ocean Road Map. PLoS ONE 6(8): e23259.

6. Domack E, Ishman S, Leventer A, Sylva S, Willmott V, et al. (2005) A chemotrophic ecosystem found beneath Antarctic ice shelf. EOS, Transactions American Geophysical Union 86: 269–278.

7. Sibuet M, Olu K (1998) Biogeography, biodiversity and fluid dependence of deep-sea cold-seep communities at active and passive margins. Deep-Sea Research Part II-Topical Studies in Oceanography 45: 517–567.

8. Baco AR, Rowden AA, Levin LA, Smith CR, Bowden DA (2010) Initial characterization of cold seep faunal communities on the New Zealand Hikurangi margin. Marine Geology 272: 251–259.

9. Ramirez-Llodra E, Tyler PA, Baker MC, Bergstad OA, Clark MR, et al. (2011) Man and the Last Great Wilderness: Human Impact on the Deep Sea. PLoS ONE 6(7): e22588.

10. Van Dover CL, Smith CR, Ardron J, Dunn D, Gjerde K, et al. (2012) Designating networks of chemosynthetic ecosystem reserves in the deep sea. Marine Policy 36: 378–381.

11. Lewis KB, Marshall BA (1996) Seep faunas and other indicators of methane-rich dewatering on New Zealand convergent margins. New Zealand Journal of Geology and Geophysics 39: 181–200.

12. Greinert J, Lewis KB, Bialas J, Pecher IA, Rowden A, et al. (2010) Methane seepage along the Hikurangi Margin, New Zealand: Overview of studies in 2006 and 2007 and new evidence from visual, bathymetric and hydroacoustic investigations. Marine Geology 272: 6–25.

13. Leon R, Somoza L, Medialdea T, Maestro A, Diaz-del-Rio V, et al. (2006) Classification of sea-floor features associated with methane seeps along the Gulf of Cadiz continental margin. Deep-Sea Research Part II-Topical Studies in Oceanography 53: 1464–1481.

14. Jones AT, Greinert J, Bowden DA, Klaucke I, Petersen CJ, et al. (2010) Acoustic and visual characterisation of methane-rich seabed seeps at Omakere Ridge on the Hikurangi Margin, New Zealand. Marine Geology 272: 154–169.

15. Klaucke I, Weinrebe W, Petersen CJ, Bowden D (2010) Temporal variability of gas seeps offshore New Zealand: Multi-frequency geoacoustic imaging of the Wairarapa area, Hikurangi margin. Marine Geology 272: 49–58.

16. Sommer S, Linke P, Pfannkuche O, Niemann H, Treude T (2010) Benthic respiration in a seep habitat dominated by dense beds of ampharetid polychaetes at the Hikurangi Margin (New Zealand). Marine Geology 272: 223–232.

17. Thurber AR, Kröger K, Neira C, Wiklund H, Levin LA (2010) Stable isotope signatures and methane use by New Zealand cold seep benthos. Marine Geology 272: 260–269.

18. Thurber AR, Levin L, Rowden A, Sommer S, Linke P, et al. (In press) Microbes, macrofauna, and methane: a novel seep community fueled by aerobic methanotrophy. Limnology and Oceanography.

19. Henrys SA, Ellis S, Uruski C (2003) Conductive heat flow variations from bottom-simulating reflectors on the Hikurangi margin, New Zealand. Geophysical Research Letters 30(2): 1065.

20. Barnes PM, Lamarche G, Bialas J, Henrys S, Pecher I, et al. (2010) Tectonic and geological framework for gas hydrates and cold seeps on the Hikurangi subduction margin, New Zealand. Marine Geology 272: 26–48.

21. Hill P (2009) Designing a deep-towed camera vehicle using single conductor cable. Sea Technology 50: 49–51.

22. Anderson MJ, Crist TO, Chase JM, Vellend M, Inouye BD, et al. (2011) Navigating the multiple meanings of beta diversity: a roadmap for the practicing ecologist. Ecology Letters 14: 19–28.

23. Clarke KR, Gorley RN (2006) PRIMER v6: User manual. Plymouth UK, PRIMER-e Ltd. 192 p.

24. Baird SJ, Wood BA, Bagley NW (2011) Nature and extent of commercial fishing effort on or near the seafloor within the New Zealand 200 n. mile Exclusive Economic Zone, 1989–90 to 2004–05. New Zealand Aquatic Environment and Biodiversity Report No. 73. 143 p.

25. Navalpakam RS, Pecher IA, Stern T (2012) Weak and segmented bottom simulating reflections on the Hikurangi Margin, New Zealand - Implications for gas hydrate reservoir rocks. Journal of Petroleum Science and Engineering 88–89: 29–40.

26. Crutchley GJ, Gorman AR, Pecher IA, Toulmin S, Henrys SA (2011) Geological controls on focused fluid flow through the gas hydrate stability zone on the southern Hikurangi Margin of New Zealand, evidenced from multi-channel seismic data. Marine and Petroleum Geology 28: 1915–1931.

27. Pecher IA, Henrys SA, Wood WT, Kukowski N, Crutchley GJ, et al. (2010) Focussed fluid flow on the Hikurangi Margin, New Zealand — Evidence from possible local upwarping of the base of gas hydrate stability. Marine Geology 272: 99–113.

28. Plaza-Faverola A, Barnes PM, Pecher I, Henrys S, Mountjoy JJ (2012) Evolution of fluid expulsion and concentrated hydrate zones across the southern Hikurangi subduction margin, New Zealand: an analysis from depth migrated seismic data. Geochemistry Geophysics Geosystems 13: Q08018.

29. Liebetrau V, Eisenhauer A, Linke P (2010) Cold seep carbonates and associated cold-water corals at the Hikurangi Margin, New Zealand: New insights into fluid pathways, growth structures and geochronology. Marine Geology 272: 307–318.

30. Campbell KA, Nelson CS, Alfaro AC, Boyd S, Greinert J, et al. (2010) Geological imprint of methane seepage on the seabed and biota of the convergent Hikurangi Margin, New Zealand: Box core and grab carbonate results. Marine Geology 272: 285–306.

31. Sommer S, Linke P, Pfannkuche O, Bowden D, Greinert J, et al. (2007) High sea bed methane emission rates at Hikurangi margin (New Zealand) associated with extremely dense populations of ampharetid polychaetes. Geochimica Et Cosmochimica Acta 71: A955–A955.

32. Southward EC (1991) Three new species of Pogonophora, including two vestimentiferans, from hydrothermal sites in the Lau Back-arc Basin (Southwest Pacific Ocean). Journal of Natural History 25: 859–881.

33. Cordes EE, Bergquist DC, Predmore BL, Jones C, Deines P, et al. (2006) Alternate unstable states: Convergent paths of succession in hydrocarbon-seep tubeworm-associated communities. Journal of Experimental Marine Biology and Ecology 339: 159–176.

34. Naudts L, Greinert J, Poort J, Belza J, Vangampelaere E, et al. (2010) Active venting sites on the gas-hydrate-bearing Hikurangi Margin, off New Zealand: Diffusive- versus bubble-released methane. Marine Geology 272: 233–250.

35. Boswell R, Collett TS, Frye M, Shedd W, McConnell DR, et al. (2012) Subsurface gas hydrates in the northern Gulf of Mexico. Marine and Petroleum Geology 34: 4–30.

36. Krabbenhoeft A, Bialas J, Klaucke I, Crutchley G, Papenberg C, et al. (2013) Patterns of subsurface fluid-flow at cold seeps: The Hikurangi Margin, offshore New Zealand. Marine and Petroleum Geology 39: 59–73.

37. Fohrmann M, Pecher IA (2012) Analysing sand-dominated channel systems for potential gas-hydrate-reservoirs using an AVO seismic inversion technique on the Southern Hikurangi Margin, New Zealand. Marine and Petroleum Geology 38: 19–34.

38. Pecher I, Henrys S (2003) Potential gas reserves in gas hydrate sweet spots on the Hikurangi Margin, New Zealand. GNS Science Report. Institute of Geological and Nuclear Sciences, Lower Hutt, New Zealand. 32 p.

39. Cordes EE, Carney SL, Hourdez S, Carney RS, Brooks JM, et al. (2007) Cold seeps of the deep Gulf of Mexico: Community structure and biogeographic comparisons to Atlantic equatorial belt seep communities. Deep-Sea Research Part I-Oceanographic Research Papers 54: 637–653.

40. Levin LA, Orphan VJ, Rouse GW, Rathburn AE, Ussler W III, et al. (2012) A hydrothermal seep on the Costa Rica margin: middle ground in a continuum of reducing ecosystems. Proceedings of the Royal Society B-Biological Sciences 279: 2580–2588.

41. Govenar B (2010) Shaping vent and seep communities: habitat provision and modification by foundation species. Topics in Geobiology 33: 403–432.

42. Olu-Le Roy K, Caprais J-C, Fifis A, Fabri M-C, Galeron J, et al. (2007) Cold-seep assemblages on a giant pockmark off West Africa: spatial patterns and environmental control. Marine Ecology 28:115–130.

43. Ritt B, Sarrazin J, Caprais J-C, Noel P, Gauthier O, et al. (2010) First insights into the structure and environmental setting of cold-seep communities in the Marmara Sea. Deep-Sea Research Part I-Oceanographic Research Papers 57:1120–1136.

44. Barry JP, Greene HG, Orange DL, Baxter CH, Robison BH, et al. (1996) Biologic and geologic characteristics of cold seeps in Monterey bay, California. Deep-Sea Research Part I-Oceanographic Research Papers 43: 1739–1762.

45. Bernardino AF, Smith CR (2010) Community structure of infaunal macro-benthos around vestimentiferan thickets at the San Clemente cold seep, NE Pacific. Marine Ecology-an Evolutionary Perspective 31: 608–621.

46. Baco AR, Smith CR (2003) High species richness in deep-sea chemoautotrophic whale skeleton communities. Marine Ecology-Progress Series 260: 109–114.

47. Bergquist DC, Urcuyo IA, Fisher CR (2002) Establishment and persistence of seep vestimentiferan aggregations on the upper Louisiana slope of the Gulf of Mexico. Marine Ecology-Progress Series 241: 89–98.

48. Cordes EE, Bergquist DC, Redding ML, Fisher CR (2007) Patterns of growth in cold-seep vestimenferans including *Seepiophila jonesi*: a second species of long-lived tubeworm. Marine Ecology-an Evolutionary Perspective 28: 160–168.

49. Cordes EE, Bergquist DC, Shea K, Fisher CR (2003) Hydrogen sulphide demand of long-lived vestimentiferan tube worm aggregations modifies the chemical environment at deep-sea hydrocarbon seeps. Ecology Letters 6: 212–219.

50. Freytag JK, Girguis PR, Bergquist DC, Andras JP, Childress JJ, et al. (2001) A paradox resolved: Sulfide acquisition by roots of seep tubeworms sustains net chemoautotrophy. Proceedings of the National Academy of Sciences of the United States of America 98: 13408–13413.

51. Fisher CR, Urcuyo IA, Simpkins MA, Nix E (1997) Life in the slow lane: Growth and longevity of cold-seep vestimentiferans. Marine Ecology-Pubblicazioni Della Stazione Zoologica Di Napoli I 18: 83–94.

52. Barry JP, Kochevar RE, Baxter CH (1997) The influence of pore-water chemistry and physiology on the distribution of vesicomyid clams at cold seeps in Monterey Bay: Implications for patterns of chemosynthetic community organization. Limnology and Oceanography 42: 318–328.

53. Barry JP, Whaling PJ, Kochevar RK (2007) Growth, production, and mortality of the chemosynthetic vesicomyid bivalve, Calyptogena kilmeri from cold seeps off central California. Marine Ecology-an Evolutionary Perspective 28: 169–182.

54. Fujikura K, Amaki K, Barry JP, Fujiwara Y, Furushima Y, et al. (2007) Long-term in situ monitoring of spawning behavior and fecundity in Calyptogena spp. Marine Ecology-Progress Series 333: 185–193.

55. Lisin SE, Hannan EE, Kochevar RE, Harrold C, Barry JP (1997) Temporal variation in gametogenic cycles of vesicomyid clams. Invertebrate Reproduction & Development 31: 307–318.

56. Barry JP, Buck KR, Kochevar RK, Nelson DC, Fujiwara Y, et al. (2002) Methane-based symbiosis in a mussel, Bathymodiolus platifrons, from cold seeps in Sagami Bay, Japan. Invertebrate Biology 121:47–54.

57. Dale AW, Sommer S, Haeckel M, Wallmann K, Linke P, et al. (2010) Pathways and regulation of carbon, sulfur and energy transfer in marine sediments overlying methane gas hydrates on the Opouawe Bank (New Zealand). Geochimica Et Cosmochimica Acta 74: 5763–5784.

58. Boetius A, Ravenschlag K, Schubert CJ, Rickert D, Widdel F, et al. (2000) A marine microbial consortium apparently mediating anaerobic oxidation of methane. Nature 407: 623–626.

59. Dubilier N, Bergin C, Lott C (2008) Symbiotic diversity in marine animals: the art of harnessing chemosynthesis. Nature Reviews Microbiology 6: 725–740.

60. Tyler PA, Young CM (1992) Reproduction in Marine-Invertebrates in Stable Environments - the Deep-Sea Model. Invertebrate Reproduction & Development 22: 185–192.

61. Tyler PA, Young CM (1999) Reproduction and dispersal at vents and cold seeps. Journal of the Marine Biological Association of the United Kingdom 79: 193–208.

62. Wallmann K, Linke P, Suess E, Bohrmann G, Sahling H, et al. (1997) Quantifying fluid flow, solute mixing, and biogeochemical turnover at cold vents of the eastern Aleutian subduction zone. Geochimica Et Cosmochimica Acta 61: 5209–5219.

63. Bertics VJ, Treude T, Ziebis W (2007) Vesicomyid Clams Alter Biogeochemical Processes at Pacific Methane Seeps. American Geophysical Union, Fall Meeting 2007, abstract #B43E-1649. American Geophysical Union.

64. Fischer D, Sahling H, Noethen K, Bohrmann G, Zabel M, et al. (2012) Interaction between hydrocarbon seepage, chemosynthetic communities, and bottom water redox at cold seeps of the Makran accretionary prism: insights from habitat-specific pore water sampling and modeling. Biogeosciences 9:2013–2031.

65. Luff R, Greinert J, Wallmann K, Klaucke I, Suess E (2005) Simulation of long-term feedbacks from authigenic carbonate crust formation at cold vent sites. Chemical Geology 216: 157–174.

66. Cordes EE, Arthur MA, Shea K, Arvidson RS, Fisher CR (2005) Modeling the Mutualistic Interactions between Tubeworms and Microbial Consortia. PLoS Biol 3: e77.

67. Dattagupta S, Arthur MA, Fisher CR (2008) Modification of sediment geochemistry by the hydrocarbon seep tubeworm Lamellibrachia luymesi: A combined empirical and modeling approach. Geochimica Et Cosmochimica Acta 72: 2298–2315.

68. Dattagupta S, Miles LL, Barnabei MS, Fisher CR (2006) The hydrocarbon seep tubeworm Lamellibrachia luymesi primarily eliminates sulfate and hydrogen ions across its roots to conserve energy and ensure sulfide supply. Journal of Experimental Biology 209: 3795–3805.

69. Cordes EE, Bergquist DC, Fisher CR (2009) Macro-ecology of Gulf of Mexico cold seeps. Annual Reviews of Marine Science 1: 143–168

19

Phylogenetic and Functional Diversity of Microbial Communities Associated with Subsurface Sediments of the Sonora Margin, Guaymas Basin

Adrien Vigneron[1,2,3]*, **Perrine Cruaud**[1,2,3], **Erwan G. Roussel**[7], **Patricia Pignet**[1,2,3], **Jean-Claude Caprais**[4], **Nolwenn Callac**[1,2,3,5], **Maria-Cristina Ciobanu**[6], **Anne Godfroy**[1,2,3], **Barry A. Cragg**[7], **John R. Parkes**[7], **Joy D. Van Nostrand**[8], **Zhili He**[8], **Jizhong Zhou**[8,9,10], **Laurent Toffin**[1,2,3]

1 Ifremer, Laboratoire de Microbiologie des Environnements Extrêmes, UMR6197, ZI de la pointe du Diable, Plouzané, France, **2** Université de Bretagne Occidentale, Laboratoire de Microbiologie des Environnements Extrêmes, UMR6197, ZI de la pointe du Diable, Plouzané, France, **3** CNRS, Laboratoire de Microbiologie des Environnements Extrêmes, UMR6197, ZI de la pointe du Diable, Plouzané, France, **4** Ifremer, Laboratoire Etude des Environnements Profonds, UMR6197, ZI de la pointe du Diable, Plouzané, France, **5** Université de Brest, Domaines Océaniques IUEM, UMR6538, Place Nicolas Copernic, Plouzané, France, **6** Ifremer, Géosciences Marines, Laboratoire des Environnements Sédimentaires, ZI de la pointe du Diable, Plouzané, France, **7** School of Earth and Ocean Sciences, Cardiff University, Cardiff, United Kingdom, **8** Institute for Environmental Genomics and Department of Microbiology and Plant Biology, University of Oklahoma, Norman, Oklahoma, United States of America, **9** State Key Joint Laboratory of Environment Simulation and Pollution Control, School of Environment, Tsinghua University, Beijing, China, **10** Earth Science Division, Lawrence Berkeley National Laboratory, Berkeley, California, United States of America

Abstract

Subsurface sediments of the Sonora Margin (Guaymas Basin), located in proximity of active cold seep sites were explored. The taxonomic and functional diversity of bacterial and archaeal communities were investigated from 1 to 10 meters below the seafloor. Microbial community structure and abundance and distribution of dominant populations were assessed using complementary molecular approaches (Ribosomal Intergenic Spacer Analysis, 16S rRNA libraries and quantitative PCR with an extensive primers set) and correlated to comprehensive geochemical data. Moreover the metabolic potentials and functional traits of the microbial community were also identified using the GeoChip functional gene microarray and metabolic rates. The active microbial community structure in the Sonora Margin sediments was related to deep subsurface ecosystems (Marine Benthic Groups B and D, Miscellaneous Crenarchaeotal Group, *Chloroflexi* and Candidate divisions) and remained relatively similar throughout the sediment section, despite defined biogeochemical gradients. However, relative abundances of bacterial and archaeal dominant lineages were significantly correlated with organic carbon quantity and origin. Consistently, metabolic pathways for the degradation and assimilation of this organic carbon as well as genetic potentials for the transformation of detrital organic matters, hydrocarbons and recalcitrant substrates were detected, suggesting that chemoorganotrophic microorganisms may dominate the microbial community of the Sonora Margin subsurface sediments.

Editor: Jack Anthony Gilbert, Argonne National Laboratory, United States of America

Funding: The oceanographic cruise and this study was funded by IFREMER and a IFREMER PhD grant. The funders had no role in study design, data collection and analysis, decision to publish, or preparation of the manuscript.

Competing Interests: The authors have declared that no competing interests exist.

* Email: avignero@gmail.com

Introduction

Deep marine subsurface sediments are one of the most extensive microbial habitats on Earth, covering more than two-thirds of the Earth's surface and reaching maximal thickness of more than 10 km at some locations [1]. Microbial populations are widespread in these sediments as deep as temperature permits [2] and cell numbers vary consistently ranging from 10^{10} to 10^3 cells per cm^3 of sediments according to their proximity from land, sedimentary rates and depth [3]. In general, microbial abundance in subsurface sediments (below 1 mbsf) decreases exponentially with depth, as a probable consequence of the decreasing organic carbon quality and availability [4]. Recent investigations based on

NanoSIMS monitoring [5] or intact ribosomal RNA [6] and membrane lipid detection [6,7] demonstrate that sedimentary microbial communities are active as they can incorporate carbon and nitrogen. However, overall metabolic rates are very slow, with biomass turnovers ranging from years to millennia [8]. Numerous of studies have focused on elucidating the microbial diversity of subsurface sediments [6,9–13]. Specific lineages of *Bacteria* (for e.g. *Chloroflexi*, Candidate division JS1) and *Archaea* (for e.g. Miscellaneous Crenarchaeotal Group (MCG), Marine Benthic Group D (MBGD), South African Goldmine Euryarchaeotal Group (SAGMEG) [14,15], distinct from the surface biospheres (above 1 mbsf), appear to occur consistently in marine subsurface sediments. However identification of the metabolism of these microbial populations remains challenging. Isotopic signatures of membrane lipids suggested that heterotrophic strategies dominated in these ecosystems [6,7]. Metagenomic and metatranscriptomic analyzes of subsurface sediments from the deep biosphere of the Peru Margin revealed metabolisms associated with lipids, carbohydrates and amino acids utilization. However detected genes and transcripts were mainly affiliated to *Firmicutes*, *Actinobacteria*, and *Alpha*- and *Gammaproteobacteria* rather than *Archaea*, *Chloroflexi* and candidate divisions [16,17]. Finally, recent single cell genomic approaches indicated the capacity of peptides degradation for members of MCG and MBGD archaeal lineages [18]. Despite these recent advances the metabolic pathways associated to the dominant microbial communities in subsurface sediments remain unclear.

The cold seeps of the Sonora Margin in the Guaymas Basin (Gulf of California), colonized by visible microbial mats and faunal assemblages, were previously characterized as highly active areas with abundant concentrations of methane and sulfur cycle microorganisms (Anaerobic methanotrophs, sulfate-reducing bacteria) in the shallow sediments (0–20 cmbsf) [19,20]. However, the subsurface microbial communities and processes that occur in the deeper sediments of the Sonora Margin have not yet been explored. The aim of this study was therefore to estimate the phylogenetic and functional biodiversity of the Sonora Margin sediments by comparing the geochemical composition, the microbial taxonomic diversity and abundance, and the Geo-Chip-based metagenome from subsurface sediments sampled in proximity with active cold seeps of the Sonora Margin. We analyzed the archaeal and bacterial diversity, abundance and distribution in correlation with geochemical gradients and elementary composition of the sediments and compared with the Sonora Margin surface cold seep sediments. Furthermore, we identified the metabolic processes and the functional potentials in term of carbon utilization and energy for both bacterial and archaeal communities and present insights into the microorganism adaptability and capacity to use various substrates in marine subsurface sediments.

Materials and Methods

Core sampling and abiotic variables

Sediment samples were collected from Sonora Margin cold seeps in the Guaymas Basin, during the Ifremer "BIG" cruise on the research vessel *L'Atalante* in June 2010. This cruise has benefited from a work permit in Mexican waters by the Mexican Secretariat of Foreign Relations (DAPA/2/281009/3803, October 28th, 2009). Gravity core BCK1 (N 27°35.804, W 111°28.697), 10 meters in length, was recovered from an observed gas depression in methane plume fields, 600 meters distant from visible active cold seeps (WM14 and EWM14 in Vasconcelos area [20]), at 1723 meters water depth. *In situ* temperatures, measured using thermal sensors (THP, Micrel) attached to the core, increased gradually from 3.5°C at the water-sediment interface to 5°C in the bottom of the core (9 mbsf). Immediately after retrieval, BCK1 core was sectioned in 1 meter long sections and transferred into the cold room. The plastic core liner was opened every 50 cm for sub-sampling. Samples for molecular analysis were collected aseptically using cut-off sterile 5 mL syringes, and frozen at −80°C. Sediment samples for activity rate estimations were taken using five cut-off sterile 5 mL syringes per section. These syringes were hermetically and anaerobically sealed with nitrogen in aluminum bags (Grüber-Folien, Germany) and stored at 4°C for processing back to laboratory. Methanogenic activity

measurements from Acetate, Di-methylamines and CO_2 substrates were carried out at Cardiff University, UK, as detailed in Methods S1.

Pore water was obtained by spinning down approximately 10 grams of crude sediment then was fixed as previously described [20]. Sulfate concentrations were determined by ion exchange chromatography as previously described [21]. Hydrogen sulfide and ammonium concentrations were measured by colorimetry [22]. Methane concentrations were quantified using the headspace technique (HSS Dani 86.50) and a gas chromatograph (Perichrom 2100) equipped with a flame-ionization detector [23]. Total organic carbon (TOC) of the sediments were measured by combustion in a LECO CS 125 carbon analyzer, as previously detailed [24]. Quantitative elemental chemical compositions of unfiltered pore waters were measured using Inductively Coupled Plasma-Atomic Emission Spectrophotometry (ICP-AES, Ultima 2, Horiba, JobinYvon), as previously detailed [25]. Effect of eventual particle contaminations was limited by normalization of the elemental concentrations by conservative element (Na) concentrations. The stable-isotope composition of methane was measured by ISOLAB b.v. company (Neerijen, The Netherlands) in the first meter deep section (0.5 mbsf) and in the deepest sediment layer (8.5 mbsf) as previously described [20].

Nucleic acids extraction and amplifications

Total nucleic acids (DNA and RNA) were directly extracted in duplicate from 2.5 grams of sediments [26], then pooled and purified [27]. Total RNA was purified from crude nucleic acids using Nucleospin RNA II Kit (Macherey Nagel, Düren, Germany) prior to RT-PCR. Aliquots of rRNA were reverse transcribed using Quanta qScript kit according to manufacturer's protocol (Quanta Bioscience, Gaithersburg, MD, USA). As control for DNA contamination, no amplification was obtained by PCR on RNA aliquots. All molecular experiments were carried out as previously monitored in surface cold seep sediments of the Sonora Margin [20]. PCR primers and appropriate annealing temperatures are listed in Table S1. Sequencing of 16S rRNA transcripts and their analysis including, taxonomic affiliations and phylogenetic trees were performed as detailed in Methods S1. Automated ribosomal intergenic spacer analysis (ARISA) of the archaeal and bacterial communities and real-time (q)PCR experiments targeting various sedimentary microbial lineages (*Archaea*, ANME-1, ANME-2a, ANME-2c, ANME-3, Methanosarcinales, Methanomicrobiales, Methanococcales, Methanobacteriales, Methanopyrales, MCG, MBGB, MBGD, *Bacteria*, *Chloroflexi*, Candidate division JS1, *Desulfosarcina/Desulfococcus*, *Desulfobulbus*, SEEP SRB2; Table S1) were carried out on purified DNA samples every 50 cm from 1 mbsf to 9 mbsf as presented in Methods S1. Statistical tests were carried out using the software PAST [28]. Nucleic acid sequences are available in the EMBL database under the following accession numbers: HF543837–HF543861 for archaeal, HF545450–HF545524 for bacterial 16S rRNA sequences and HF935025–HF935037 for *mcrA* gene sequences.

GeoChip analysis

The GeoChip 4.0 microarray, containing 83992 oligonucleotide probes and targeting 152414 gene variants in 401 categories for different microbial functional and biogeochemical processes was monitored as previously detailed [29]. Although the GeoChip was initially based on the genome of cultured microorganisms, the new generation of GeoChip has been extensively enriched with metagenome data from various environments and contains now an important number of relevant probes targeting genes from cultured and uncultured microorganisms involved in key biogeochemical cycles. Total purified DNA samples were labeled then hybridized on GeoChip slides. Signal intensities were scanned and spots with signal-to-noise ratios lower than 2 were removed before analyses [29]. The phylogenetic design of the data acquisition enabled confident assignment of metabolic capabilities to bacterial and archaeal phyla [30,31], thus dataset were sorted according to the taxonomic affiliation of the genes (*Bacteria*, *Euryarchaeota* and *Crenarchaeota*). Output was analyzed using the GeoChip 4.0 data analysis pipeline [32] and tested using the statistical software PAST [28]. Relative signal intensity was normalized by the number of the probes for each indicated metabolic pathway. List of targeted genes for each category are provided in Table S2. Visualization of the bacterial and archaeal functional potential was achieved using spider dendrograms, where each arm of the plot corresponded to a metabolic pathway. The raw GeoChip dataset is available at http://ieg.ou.edu/4download/.

Results

Geochemical description

The BCK1 core, from the Sonora Margin sediments, showed the typical geochemical signatures of continental margin sediments (Figure 1a), with a sulfate to methane transition zone (SMTZ) located around 5 mbsf. Sulfate pore water concentrations decreased from 25 mM at the sediment-water interface down to 2 mM at 5.5 mbsf. Hydrogen sulfide concentrations were only detected in the deeper sediment layers with a maximum of 32 mM at 5 mbsf decreasing to around 8 mM at 8 mbsf. Methane pore water concentrations increased with depth reaching 500 µM at the bottom of the sediment core (8.5 mbsf). and were positively correlated with the methanogenesis rates (Pearson correlation coefficient $r = 0.72$, $P = 0.001$; <45 pmol/cm^3/d at 8.5 mbsf) (Figure 1d). Isotopic signature of methane was $-97.3‰$ at the bottom of the core (9 mbsf) and $-82‰$ at 1 mbsf, confirming that most of methane produced was from biogenic origin and indicating that methane oxidation potentially occurred towards the sediment surface. Ammonium concentrations, likely resulting of organic matter degradation, increased with depth until reaching 2.5 mM at 5 mbsf (Figure 1b). Total organic carbon (TOC) content varied between 3.1 and 4.3% (w/w) throughout the sediment with peaks at 3.5, 5 and 7.5 mbsf (Figure 1c). Analysis of the element composition of the pore water highlighted both a manganese reduction zone in the first meter of sediment and specific horizons (3.5, 5–6, 7 and 8 mbsf) with significant enrichment of metallic elements (Fe, Al, Si, Mn, Ti) (Figure S1). These increases of metal concentrations in pore water suggest detrital terrigenous inputs in the sediment layers, as previously detected in the Guaymas Basin [33].

Microbial community structure and composition

Microbial community structure variations with depth were compared from the Sonora Margin cold seep surface sediments using ARISA. The archaeal and bacterial community structures of the BCK1 were significantly different from the surface sediments of both cold seep (WM14 and EWM14 samples [20]) and outside active seepage areas (REF samples [20]) as shown by clustering and ANOSIM (p<0.0008) on ARISA dataset (Figure 2) Dendrogram and Nonmetric Multidimensional Scaling (NMDS) analysis, based on Bray-Curtis similarity measure also indicated that BCK1 samples clustered according to sediment depths (1–4 mbsf, 4.5–6 mbsf and 6.5–9 mbsf). However this observation was not statistically supported by ANOSIM and seemed to rather reflect a difference in signal intensity more than in community composition.

Based on geochemical features, representative sediment depth horizons (1, 4, 5, 7 and 8 mbsf) were selected for the 16S rRNA survey. A total of 565 partial 16S rRNA sequences (303 for Archaea and 262 for Bacteria) were obtained and used as a proxy for active microbial communities [34–36]. Overall, statistical analysis of the microbial community structure of the samples indicated that the microbial community was nearly constant throughout the sediment core (SIMPER average similarities between paired samples above 74.05%).

Archaeal 16S rRNA libraries showed a very limited diversity throughout the sediment core (1-H$_{Simpson}$ = 0.615±0.08; Figure 2, Figure S2, Figure S3), including three uncultivated phylotypes, mainly found in the deep biosphere: the Marine Benthic Groups B and D (MBGB, MBGD) and the Miscellaneous Crenarchaeotal Group (MCG), mainly represented by the MCG-8 and MCG-10 sub-groups [37]. Other groups such as South Africa Gold Mine Euryarchaeotal Group (SAGMEG), Marine Hydrothermal Vent Group (MHVG) and Terrestrial Miscellaneous Euryaechaeotal group (TMEG) were also detected in lower proportions in the deepest sediment layers.

In contrast, the bacterial 16S rRNA libraries indicated a larger diversity (1-H$_{Simpson}$ = 0.712±0.06), dominated by *Chloroflexi* and diverse bacterial candidate divisions including JS1, OP11, OP1, OP8 and OP3 (Figure 2, Figure S2, Figure S4). The *Chloroflexi* lineage included different sub-groups and most of the amplified sequences were relatives to *Dehalococcoidetes* or subphylum IV groups. A few *Deltaproteobacteria*, usually related to sulfate-reducers and hydrocarbon degraders in cold seep sediments were detected in 4, 5 and 7 meters depth sediment horizons.

Microbial 16S rRNA gene abundance and distributions

Depth distributions and relative abundance of microorganisms were analyzed every 50 cm by real-time PCR (Figure 3). 16S rRNA gene abundance of *Bacteria* was around 10 fold higher than *Archaea* throughout the sediment core, and decreased with depth from 4×10^9 16S rRNA gene copies per gram of sediment in the top of the core to 2.8×10^8 copies at the bottom. Bacterial relative abundance showed elevated concentrations in particular at 5, 7 and 8 meters below the seafloor with 1.95×10^9, 1.45×10^9 and 1.1×10^9 16S rRNA gene copies g^{-1} respectively. As sequences affiliated to *Chloroflexi* and candidate division JS1 dominated bacterial 16S rRNA gene libraries, the 16S rRNA genes of these groups were specifically quantified. *Chloroflexi* 16S rRNA gene abundance was estimated by subtracting JS1 16S rRNA gene copy numbers from quantifications with JS1 and *Chloroflexi* groups specific primers [38]. *Chloroflexi* 16S rRNA gene abundance appeared to mirror the bacterial distribution profile (Pearson correlation coefficient $r = 0.914$, $P<0.0001$) and strongly dominated the bacterial community throughout the sediment core. In contrast, JS1 16S rRNA gene copy numbers increased with depth until reaching maximum values between 2.5 and 5 mbsf with 4.62×10^8 copies g^{-1}. No cold seep sulfate-reducing bacteria (*Desulfosarcina/Desulfococcus* and *Desulfobulbus* groups) were detected.

Total archaeal 16S rRNA gene copy numbers, represented 4–10% of the total number of 16S rRNA gene and decreased with depth, from 1.9×10^8 16S rRNA gene copies g^{-1} at 1 mbsf to 2.67×10^7 16S rRNA gene copies at the bottom of the sediment core. However, specific horizons (1 mbsf, 5 mbsf, 7 and 8 mbsf) showed peaks of elevated archaeal 16S rRNA gene concentrations with 4.8×10^8, 1×10^8, 1.1×10^8 and 6.8×10^7 16S rRNA gene copies respectively. Within the *Archaea*, uncultivated groups MBGD, MBGB and MCG were detected throughout the sediment core. Their distributions were correlated with the

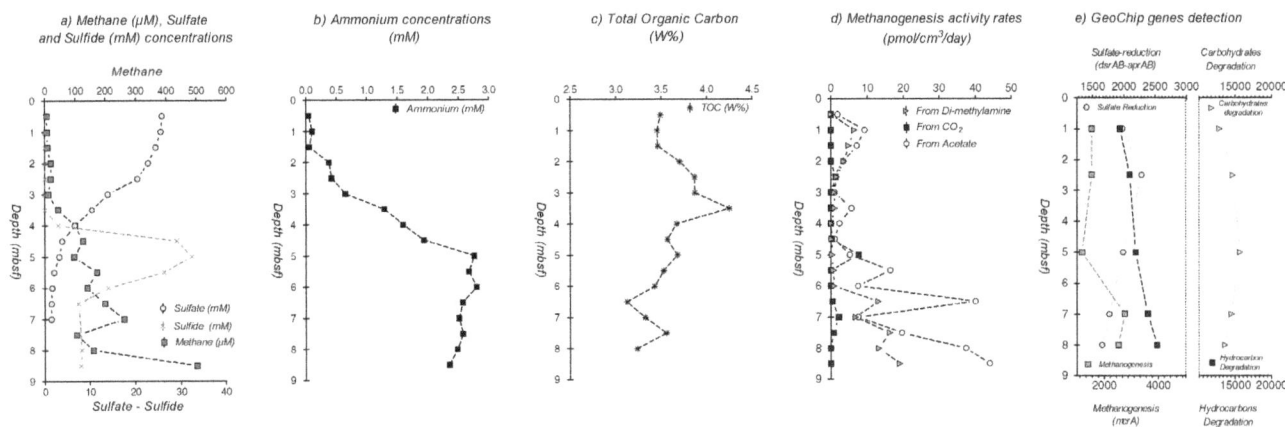

Figure 1. Geochemical depth profiles, putative methanogenesis activity rates and GeoChip genes detection of the sediment core BCK1. 1a) Dissolved methane (grey square, µM), sulfate (white circle, mM) and sulfide (grey cross, mM) concentrations in pore waters. 1b) Dissolved ammonium concentrations (mM) in pore waters. 1c) Total organic carbon (TOC) content in the sediments (% w/w). 1d) Methanogenesis activity rates from acetate (white circle), bicarbonate (black square) and di-methylamine (grey triangle) in the sediments (pmol/cm³/day). 1e) Relative signal intensity of the GeoChip microarray for sulfate-reduction (circle), methanogenesis (grey square), carbohydrates degradation (triangle) and hydrocarbon degradation (black square) pathways, normalized by the number of the probes for each indicated metabolic pathway.

archaeal distribution (Pearson correlation coefficient $r = 0.98$, $P < 0.0001$) and no specific niche repartition was detected along the sulfate and methane concentration gradients. Assuming the same 16S rRNA copy number for each microbial lineage, MCG were fivefold less abundant than marine benthic groups except at 7 mbsf with 4.8×10^7 16S rRNA gene copies g^{-1}. Consistently with 16S rRNA library results, ANME lineages were below the detection limit ($<10^4$ 16S rRNA gene copies g^{-1}) and methanogens were only represented by Methanosarcinales at 1 mbsf with 2.4×10^6 16S rRNA gene copies g^{-1}.

Functional gene diversity and GeoChip array

In order to investigate the ecophysiology of the microbial community associated to subsurface Sonora Margin sediments, an array targeting functional genes was used for sediments collected at selected depths (1, 2.5, 5, 7 and 8 mbsf). The microarray results indicated a small but significant variation between the metabolic potential of microbial communities from each sediment horizon (ANOVA: F = 5.64, P = 0.002). Similarity percentages (SIMPER) and clustering analyses using Bray-Curtis similarity measure showed that the microbial communities associated with the 2.5 and 5 mbsf sediment horizons and the two deeper sediment horizons (7 and 8 mbsf) shared the greatest number of functional genes (93.3% and 91.92% similarity respectively), and that divergence between these metabolic potentials increased with sediment depth. These analyses indicated that this divergence was mainly due to the highest presence, in deepest sediment layer communities, of genes involved in hydrocarbon degradation (13% of variation) and in the upper sediment layers the predominance of genes involved in cellulose degradation (6.79% of variation, Figure 4). Using the taxonomic nature of the GeoChip probes [31,32], putative metabolic functions were sorted according to specific taxonomic ranks: *Archaea* (3% of the total prokaryotic signal) or *Bacteria* (97%) super kingdoms and *Euryarchaeota* or *Crenarchaeota* phyla. *Crenarchaeota* phylum was recently revised to include only thermophilic lineages, excluding lineages such as MCG and MBGB [39]. However, GeoChip array was designed on the former phylogeny, thus the crenarchaeotal metabolic

pathways detected in this study are likely to include MCG and MBGB lineages.

Carbon metabolism

A large variety of bacterial genes for carbon utilization were identified (Figure 4). Genes coding for the RuBisCo, the propionyl-CoA/acetyl-CoA carboxylase (*ppc*), the ATP citrate lyase (*aclB*) and the carbon-monoxide dehydrogenase (CODH) were detected throughout the sediment core, indicating an autotrophic carbon fixation potential for both bacterial and archaeal lineages. Genes involved in heterotrophic metabolic pathways were also detected, indicating an important potential to transform a large variety of organic compounds. Bacterial genes associated with metabolic pathways for carbohydrates degradation (starch, cellulose, hemicellulose, chitin; lignin and pectin degradation), notably with extracellular enzyme genes, were detected in slightly higher proportion in the surface sediments. Hydrocarbon degradation pathway genes such as *chnA*, involved in ethylphenol and ethylbenzene catabolism, the *tut* operon, involved in toluene degradation and *alk* genes in the alkane degradation pathway [40] were also detected in increasing proportion with depth. The ability to degrade chlorinated, aromatic, polycyclic and xenobiotic compounds were also detected for bacteria, particularly with genes involved in the superpathway of aromatic compound degradation *via* 2-oxopent-4enoate and in the metacleavage of aromatic compounds [41]. Finally, the bacterial potential to use methylated amines was also identified throughout the sediment core. Archaeal metabolic genes for carbon utilization involved in carbohydrates and complex organic matter degradation as well as autotrophic metabolisms associated with *Euryarchaeota* and *Crenarchaeota*-related lineages were also detected. Finally, *mcrA* euryarchaeotal genes, involved in both methane production and anaerobic oxidation [42] were detected in increasing proportion with depth consistently with methane concentrations (Pearson correlation coefficient $r = 0.832$, $P = 0.08$; Figure 1e).

Sulfate and Nitrogen metabolisms

The elevated ammonium concentrations measured in the sediments suggested that nitrogen cycle might be significant in

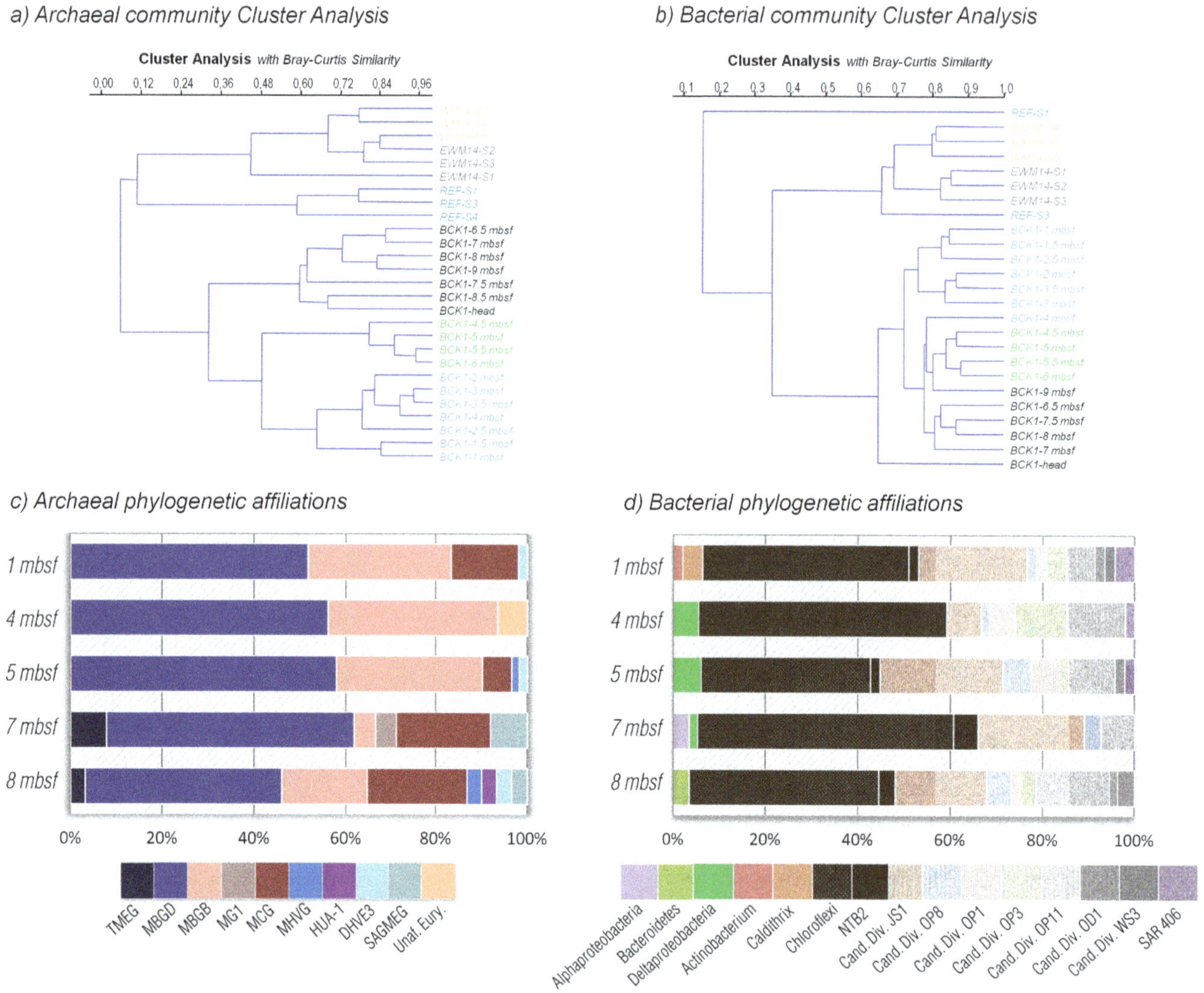

Figure 2. Microbial diversity. Clustering analyses using unweighted pair-group average (UPGMA) and Bray-Curtis Similarity measure of the a) archaeal and b) bacterial community structures visualizing the ARISA dataset. Depth distribution of the c) archaeal and d) bacterial phylogenetic affiliations of the 16S rRNA-derived sequences at 1, 4, 5, 7 and 8 mbsf sediment layers of BCK1. WM14 (White Microbial mat), EWM14 (Edge of White Microbial mat) and REF (reference outside active seepage area) samples were previously analyzed with the same material and method in Vigneron et al 2013 and corresponded to archaeal community structure of the surface sediments of the Sonora Margin. TMEG, Terrestrial Miscellaneous Euryarchaeotal Group; MBGD/B, Marine Benthic Group D/B; MG I, Marine Group I; MCG, Miscellaneous Crenarchaeotic Group; MHVG, Marine Hydrothermal Vent Group; Hua1, Huasco archaeal group 1; DHVE3, Deep-Sea Hydrothermal Vent Euryarchaeotal Group 3; SAGMEG, South Africa Gold Mine Euryarchaeotal Group.

the Sonora Margin sediments. Analyses of the functional gene array detected essential genes involved in the major pathways of the nitrogen cycle (Figure 5). Genes suggesting metabolic potentials for nitrogen fixation and mineralization (Glutamate dehydrogenase and urea amidohydrolase genes), allowing nitrogen input to the microbial ecosystem, were observed in both bacterial and euryarchaeotal lineages, while nitrification genes were detected in *Bacteria* and *Crenarchaeota*. Denitrification potential was identified in *Bacteria* and in higher proportion in *Archaea*. Hydrazine oxidoreductase genes involved in the anaerobic oxidation of ammonium (anammox) were also detected throughout the sediment core and in higher proportion (1.5 times) at 5 mbsf. Finally, genes involved in sulfate-reduction (*dsrAB*, *aprAB*) were identified throughout the sediments and in higher intensity at 1, 2.5 and 5 mbsf sediment horizons, which coincided with the sulfate-rich sediment layers (Figure 1).

Discussion

Microbial community structure

In this study, we document the taxonomic and functional diversity of the microbial community associated with subsurface sediments from a site adjacent (600 m) to cold seep sediment sites of the Sonora Margin [19,20]. Although identical molecular methods were used in both studies, the microbial diversity associated with the subsurface sediments (0.5–9 mbsf) was different from the surface cold seeps (0–0.2 mbsf) of the Sonora Margin. For example, anaerobic methanotrophs and associated sulfate-reducing bacteria, observed in high concentrations in the cold seep surface sediments [19,20] were not detected in subsurface sediments despite presence of a sulfate and methane transition zone. In contrast, the subsurface bacterial community was strongly dominated by members of *Chloroflexi* and candidate division

a) Bacteria

b) Archaea

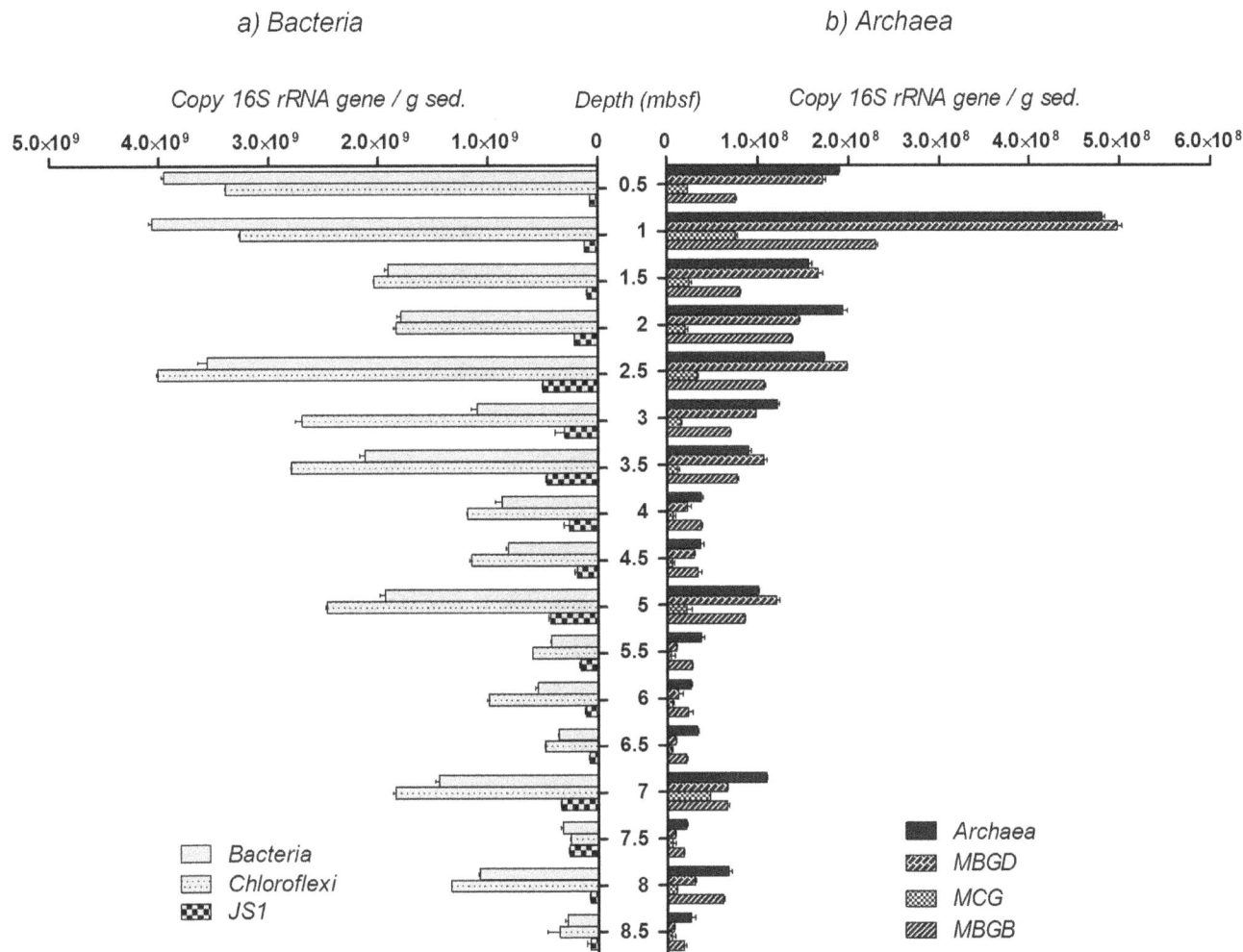

Figure 3. Q-PCR estimations. Q-PCR estimation of 16S rRNA gene copy numbers per gram of sediment for a) total *Bacteria* and bacterial groups of *Chloroflexi*, candidate division JS1 and b) total *Archaea* and archaeal groups of Marine Benthic Group B (MBGB), D (MBGD), Miscellaneous Crenarchaeotal Group (MCG), from BCK1 sediment core. Methanosarcinales were only detected at 1 mbsf with 2.4×10⁶ 16S rRNA gene copies g⁻¹ but were not represented in the figure. ANaerobic MEthanotrophs (ANME), *Desulfosarcina/Desulfococcus* (DSS), *Desulfobulbus* (DBB) and other methanogens orders were not detected in analyzed samples.

phyla (JS1, OP8, etc.), and the major archaeal lineages detected were MCG, MBGB (also known as DSAG [43]) and MBGD. All these microbial populations have been frequently encountered in continental margin sediments and in the deep subsurface marine biosphere [9,10,13], but only in minor proportion in highly active ecosystems (hydrothermal vent, cold seeps) [20,44] and in low carbon environments (open ocean sediments) [45]. Interestingly, no significant variation of the microbial community structure, excepted for the candidate division JS1, was detected throughout the sediment core, despite the presence of marked geochemical gradients (sulfate, methane). These results suggest that dominant microbial lineages were probably not directly involved in these biogeochemical cycles, as previously proposed for archaeal lineages [6,37]. Overall, estimated cell abundance decreased with depth as commonly observed in marine sediments [3,4]. In the Sonora Margin, elevated amounts of organic matter, derived from both marine production and continental inputs, sedimented in the seafloor with an estimated rate of 2 mm/y [46]. The accumulation of these sedimented particles led to an elevated sedimentary TOC content (3.5~4%). Distance to land, geochemical gradients and organic carbon quality and abundance can control the microbial

community structure and abundance in marine sediments [3,7,14], thus the high cellular abundance in the Sonora Margin sediments could be a consequence of the high concentrations of organic carbon. Elevated Q-PCR-based cell abundance estimations in the first meters of sediment could be due to higher concentrations of several electron acceptors (oxygen, nitrate, manganese and sulfate). Furthermore, significant correlations were found between TOC percentage and total *Bacteria*, *Chloroflexi*, candidate division JS1 and MBGD cell abundance estimations below 1.5 mbsf (Pearson correlation coefficients $r = 0.58$, 0.66, 0.75 and 0.66 respectively; $P<0.04$), which are consistent with reports of correlation between TOC and subsurface microbial biomass [7,47]. Likewise, fluctuations below 3 mbsf of all microbial lineage cell abundances, appeared to be positively correlated with the local elementary composition of the sediments (Fe, Ti and Al, Pearson correlation coefficients $r>0.67$, $P<0.04$; Table S3). These results clearly indicate that in subsurface margin sediments microbial communities are influenced directly or indirectly by the geochemical composition of the sediments and suggest that the microbial abundance in margin ecosystems could be enhanced by the continental detrital inputs rather than by

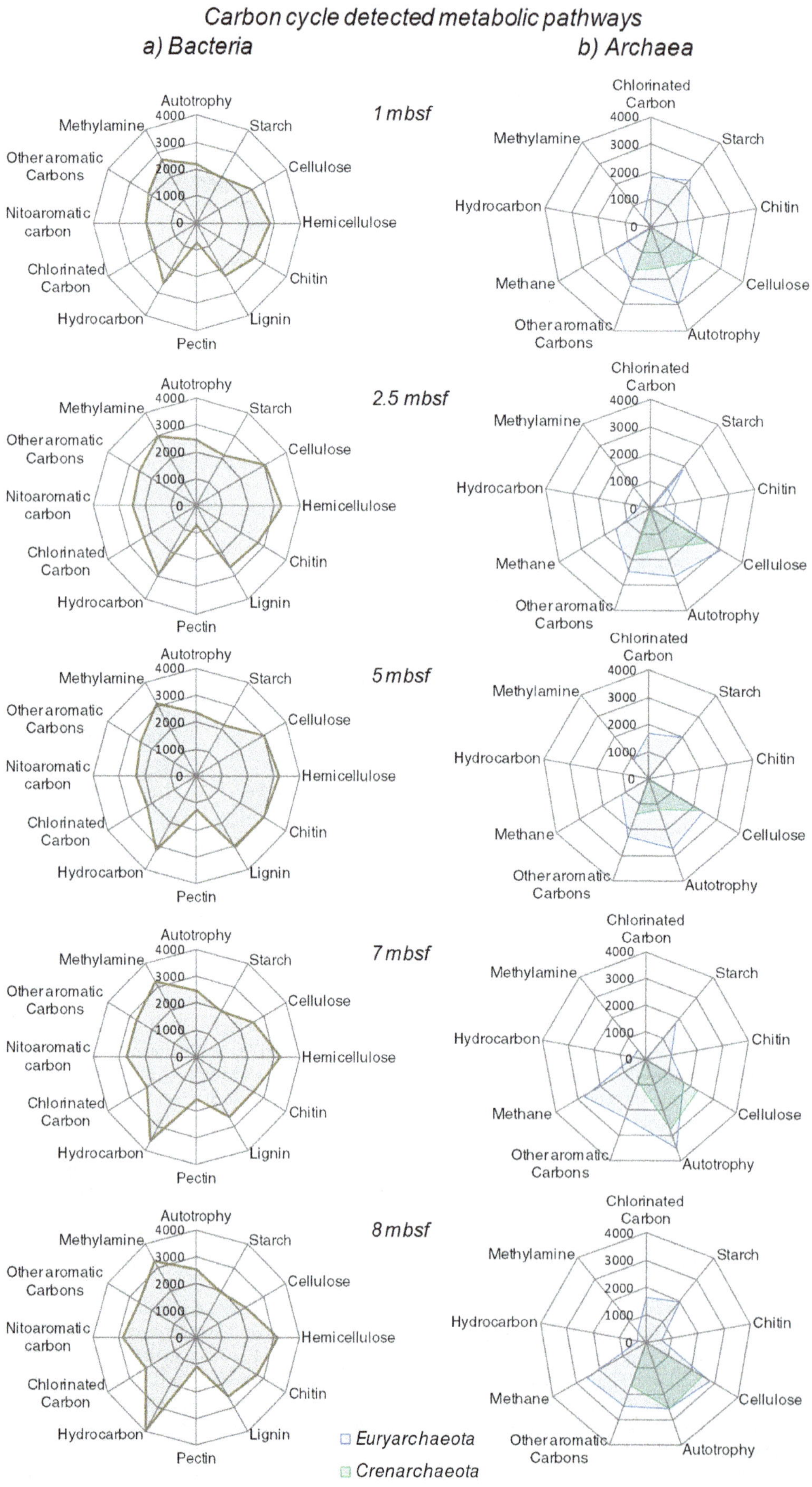

Carbon cycle detected metabolic pathways
a) Bacteria
b) Archaea
1 mbsf
2.5 mbsf
5 mbsf
7 mbsf
8 mbsf
Autotrophy
Starch
Cellulose
Hemicellulose
Chitin
Lignin
Pectin
Hydrocarbon
Chlorinated Carbon
Nitoaromatic carbon
Other aromatic Carbons
Methylamine
Methane
4000
3000
2000
1000
0
Euryarchaeota
Crenarchaeota

Figure 4. Carbon-cycling methabolic pathways detected by GeoChip. Carbon-cycling metabolic pathways identified for a) *Bacteria* and b) Archaeal *Euryarchaeota* (Blue) and *Crenarchaeota*-related (Green) lineages at different depths for BCK1 sediment core. Relative signal intensity was normalized by the number of the probes for each indicated metabolic pathway. List of targeted genes for each category are provided in Table S2.

oceanic production, as indicated the correlations with terrigenous-derived metallic elements [48,49]. This result is congruent with recent model calculations in subsurface sediments, indicating that buried organic carbon is sufficient to fuel microbial communities over turnover of millions of years [50].

Organic matter degradation

Based on single cell genomics, it was recently proposed that archaeal MCG and MBGD lineages could degrade detrital organic matter [18]. Moreover, genes and transcripts, involved in anaerobic metabolism of amino acids, carbohydrates and lipids have been previously detected in the deep subsurface biosphere [16,17]. However, it remains unclear how the microbial community is organized to degrade the detrital inputs and which microbial processes are involved. Although the GeoChip cannot be considered to be a comprehensive array with respect to marine sediment environments, it does contain an important number of relevant probes targeting genes involved in key biogeochemical

cycles and represents an interesting approach to analyze the genomic potential in environments. The microbial metabolic potential analyzed using the GeoChip showed that the majority of the genes detected were related to various bacterial metabolic pathways for the transformation and the anaerobic degradation of simple and complex organic matter (Figure 1e). The high ammonium concentrations in these sediments could therefore be a consequence of the degradation of large amounts of organic matter by microbial communities associated to the Sonora Margin subsurface sediments. Genes associated with several metabolic pathways including extracellular and intracellular enzymes involved in the degradation and assimilation of decaying wood were detected, supporting the importance of subsurface microbial communities degrading organic matter such as plants and starch. For example, genes for transformation of lignin and complex organic aromatic substrates were also identified, notably involved in the superpathway of the aromatic compound cleavage, indicating that even the more recalcitrant wood particles could

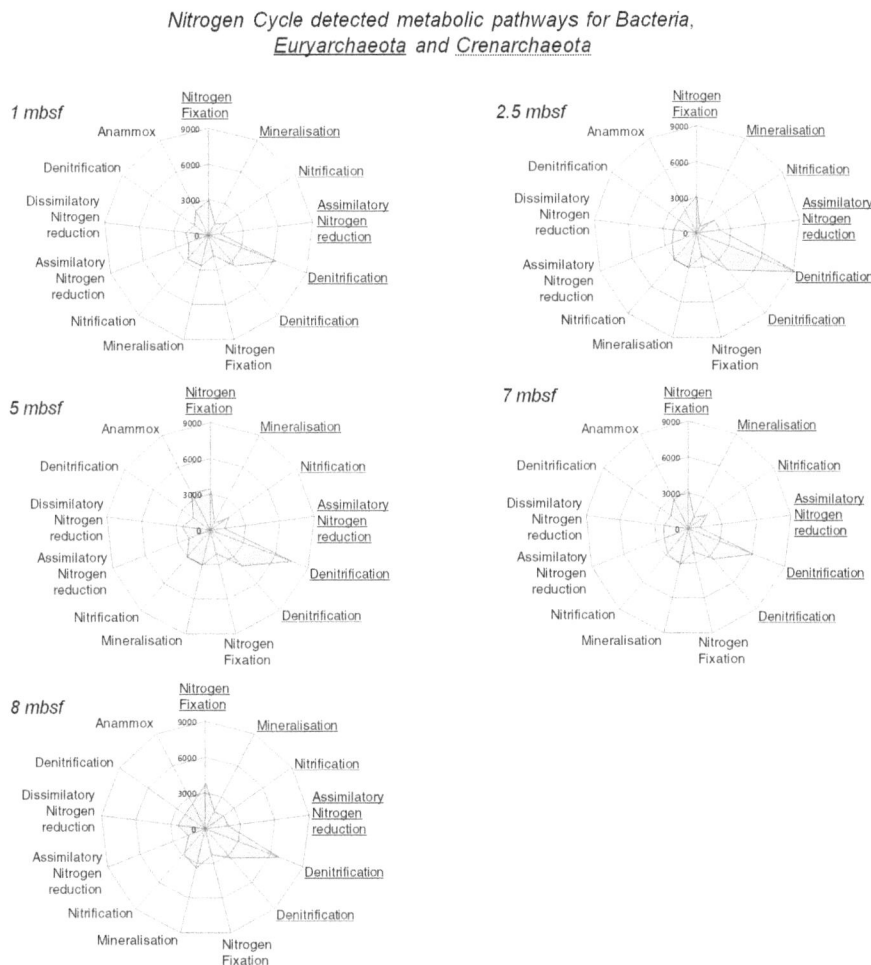

Nitrogen Cycle detected metabolic pathways for Bacteria,
Euryarchaeota and Crenarchaeota

Figure 5. Nitrogen-cycling metabolic pathways identified at different depths for BCK1 sediment cores. Bacterial metabolic pathways are not underlined while *Euryarchaeota* and *Crenarchaeota*-related pathways are underlined with solid and dotted line respectively. Relative signal intensity was normalized by the number of the probes for each indicated metabolic pathway. List of targeted genes for each category are provided in Table S2.

potentially be degraded by the bacterial community in the Sonora Margin (Figure 4a). This wood-based degradation metabolism appeared to be predominant in the upper sediment layers while hydrocarbon catabolism predominated the deeper sediment horizons. The Guaymas Basin sediments are well known to harbor various C_1 to C_8 hydrocarbon compounds such as ethane, butane, pentane and other alkanes [51]. Thus the bacterial community may be able to degrade this upward migrating organic carbon source as well as sedimented particles.

Other genes implicated in metabolic pathways for carbon assimilation have also been identified, indicating that different strategies for carbon assimilation occur amongst the different bacterial lineages (Figure 4a). For example, potential for degradation of chlorinated compounds was present, which is congruent with previous detection of dehalogenase enzymes and dehalogenation activities in similar deep biosphere sediments dominated by *Chloroflexi* lineages [52]. Degradation of chlorinated compounds derived from decaying marine phytoplankton pigments [53], suggests that in addition to terrestrial input the Sonora Margin bacterial community, (e.g. *Chloroflexi* members) could catabolize marine production and phytoplankton [54]. This metabolic specialization, which is energetically more favorable than sulfate reduction, may also explain the overall abundance of *Chloroflexi* representatives in marine sediments [45]. In addition, part of the bacterial community could also decompose decaying macrofauna with metabolic pathways involved in chitin and methylamine degradation. Finally, genetic potential for autotrophic metabolism was identified in both *Bacteria* and *Archaea* domains, suggesting that carbon dioxide could be either assimilated by specific microbial groups or that some subsurface microorganisms might be facultative heterotrophs, as previously suggested [5].

In contrast to bacterial lineages, the detected metabolic potential of *Archaea* appeared to be less diverse, maybe due to the more limited number of genes targeted by the GeoChip. Even if we could not exclude that our representation of the archaeal metabolic potential may be biased by unknown or non-targeted archaeal genes that escape to the microarray detection, various archaeal functional genes were identified. Crenarchaeotal-related lineages, likely including MCG and MBGB phyla, appeared to have the metabolic potential for complex organic carbon degradation (cellulose and aromatic polymers; Figure 4b). This result is supported by single cell MCG genomes [18] and distribution [37] suggesting heterotrophic metabolisms, possibly linked to aromatic compounds degradation [55]. Likewise, euryarchaeotal lineages, dominated by MBGD (95% based on Q-PCR estimations), appeared to have mainly the potential to degrade wood detrital polymers like starch, cellulose and aromatic compounds (Figure 4b). Hence, MBGD members could be anaerobic and heterotrophic degraders of complex organic matter, as previously suggested [18]. Resulting peptides from enzymatic degradations could be further assimilated by MBGD cells *via* peptidases and oligopeptide transporters, recently detected in their genome [18].

Methane and Sulfate cycles

Interestingly, the low GeoChip signal intensity for the *mcrA* gene, a gene coding for an enzyme involved in production and anaerobic oxidation of methane [42] was correlated with methane concentrations and methanogenesis rates measured in the sediments (Figure 1). However, Q-PCR quantification and *mcrA* gene clone libraries (data not shown) only detected putative methane cycling *Archaea* related to *Methanococcoides* in sediments at 1 mbsf. Detection of these methanogens degrading noncompetitive substrates, such as methylated amines [56] is consistent

with the presence of methanogenesis from dimethylamine and the detection of euryarchaeotal genes involved in methylamine degradation (Figure 4b). In deeper sediments with low methanogenesis rates (10–100 fold lower than in cold seeps [57]), relative abundances of known methanogens were probably below the PCR and Q-PCR detection limits (<1000 16S rRNA gene copy per gram of sediment) or escape amplification due to primer deficiencies [13]. As suggested by the changing $\delta13\text{-}CH_4$ signature, anaerobic methanotrophs could also be present in extremely low abundance or with altered key genes that would escape molecular detection [58]. These methanotrophs might be coupled directly or indirectly with sulfate-reducing *Deltaproteobacteria*, detected between 4 and 7 mbsf by 16S rRNA libraries and *dsrAB AprAB* GeoChip probes and thereby, lead to the formation of the SMTZ in these sediments (Figure 1).

Nitrogen cycle

Key bacterial metabolic genes involved in the nitrogen cycle were also detected with the microarray approach in the Sonora Margin sediments (Figure 5). In addition to nitrogen fixation, denitrification and anammox by bacterial communities, *Euryarchaeota* showed genetic potential for nitrogen fixation. Nitrogen assimilation is an important metabolic process for deep subsurface sediment microbial communities [5] and various members of the *Euryarchaeota* such as methanogenic lineages [59,60], ANME-2 [61] and ANME-1 [62] were previously found to anaerobically fix nitrogen. The detection of euryarchaeotal nitrogen fixation genes in our results suggested that members of MBGD, representing 95% of the *Euryarchaeota* could also be diazotrophic *Archaea*. Nitrification (ammonium oxidation) genes (*amoA*) were identified as a potential metabolism in crenarchaeotal-related lineages. Although ammonium, a potential electron donor, is abundant in the Sonora Margin sediments, probably due to organic matter microbial degradation, the presence of such oxygenase enzymes in this anoxic environment remains enigmatic [47,63]. It was therefore suggested that ammonium oxidation could be performed using an alternative electron acceptor [47] or that *amo* genes in anoxic environments could have an alternative function [64]. Consistently with the detection of *nar* transcripts in deep marine sediments [16], archaeal and bacterial denitrification genes were present throughout the sediment core, which could contribute to the elevated ammonium concentrations. Anaerobic ammonium oxidation was previously suggested for the nitrate origin in the deepest sediments as it could potentially be produced as a by-product of the process [16]. This would be supported by the detection of the *hzo* genes by the GeoChip probes, as well as the previously report of anammox process in the Sonora Margin sediments [65].

Conclusion

This study clearly indicated that Sonora Margin sub-surface sediment microbial communities, probably controlled by terrigeneous inputs, are composed of deep biosphere-related microorganisms, distinct of the Sonora Margin surface cold seep communities. Consistently, genetic potentials for the catabolism of complex organic matters (decaying wood, macrofauna, phytoplankton and hydrocarbon) were identified, suggesting that various heterotrophic strategies occur amongst sedimentary microbial communities. Further specific measurements of rates of degradation these different substrates could confirm these results and lead to a better understanding of the biochemical processes driving subseafloor microbial communities.

Supporting Information

Figure S1 Geochemical depth profiles of: total iron, aluminum, potassium, manganese, total sulfur, silica and titanium concentrations in the unfiltered pore waters of the BCK1 core. The blue shade represent important changes in elemental composition profiles.

Figure S2 Rarefaction curves for A) archaeal and B) Bacterial 16S rRNA gene libraries.

Figure S3 Maximum Likelihood phylogenetic tree of the archaeal 16S cDNA sequences amplified from sections 1, 4, 5, 7 and 8 mbsf (labeled S1, S4, S5, S7 and S8 respectively) of the BCK1 sediment core. Phylogenetic tree was performed using RAxML 7.2.8. and GTRCAT model approximation with 1000 replicates. Only bootstrap values up to 70% are shown. Only one representative sequence (>97% identical) per sediment horizon is shown. Number in brackets shown the number of clones analyzed from RNA clone libraries. MBG-D/B, Marine Benthic Group D/B; TMEG, Terrestrial Miscellaneous Euryarcheotal Group; MCG, Miscellaneous Crenarchaeotal Group; MHVG, Marine Hydrothermal Vent Group; SAGMEG, South Africa Gold Mine Euryarchaeotal Group.

Figure S4 Maximum Likelihood phylogenetic tree of the bacterial 16S cDNA sequences amplified from sections 1, 4, 5, 7 and 8 mbsf (labeled S1, S4, S5, S7 and S8 respectively) of the BCK1 sediment core. Phylogenetic tree was performed using RAxML 7.2.8. and GTRCAT model approximation with 1000 replicates. Only bootstrap values up to 70% are shown. Only one representative sequence (>97% identical) per sediment horizon is

shown. Number in brackets shown the number of clones analyzed from RNA clone libraries.

Table S1 Primer sets and annealing temperatures used for real-time PCR of 16S rRNA gene.

Table S2 Details of GeoChip-targeted genes, related proteins and processes corresponding to each identified metabolic pathways.

Table S3 Correlation statistical tests and associated P values for microbial lineages and elementary composition of the sediment pore-waters.

Methods S1 Detailed methods for methanogenesis activity measurements, gene library constructions and phylogenetic affiliations, Q-PCR and ARISA experiments.

Acknowledgments

We would like to thank the crew of *L'Atalante* for their help on core preparation, François Harmegnies for temperature gradient recovery and Olivier Rouxel for geochemical interpretation.

Author Contributions

Conceived and designed the experiments: AV LT JP JZ. Performed the experiments: AV PC EGR PP JCC NC MCC BAC JVN ZH. Analyzed the data: AV PC EGR BAC JVN ZH. Contributed reagents/materials/analysis tools: NC MCC. Contributed to the writing of the manuscript: AV PC EGR JP JZ LT AG.

References

1. Divins DL (2003) Total Sediment Tchickness of the World's Oceans & Marginal Seas. NOAA National Geophysical Data Center, Boulder CO.
2. Roussel EG, Bonavita MA, Querellou J, Cragg BA, Webster G, et al. (2008) Extending the sub-sea-floor biosphere. Science 320: 1046.
3. Kallmeyer J, Pockalny R, Adhikari RR, Smith DC, D'Hondt S (2012) Global distribution of microbial abundance and biomass in subseafloor sediment. Proceedings of the National Academy of Sciences.
4. Parkes RJ, Cragg BA, Wellsbury P (2000) Recent studies on bacterial populations and processes in subseafloor sediments: A review. Hydrogeology Journal 8: 11–28.
5. Morono Y, Terada T, Nishizawa M, Ito M, Hillion F, et al. (2011) Carbon and nitrogen assimilation in deep subseafloor microbial cells. Proceedings of the National Academy of Sciences 108: 18295–18300.
6. Biddle JF, Lipp JS, Lever MA, Lloyd KG, Sorensen KB, et al. (2006) Heterotrophic Archaea dominate sedimentary subsurface ecosystems off Peru. Proc Natl Acad Sci U S A 103: 3846–3851.
7. Lipp JS, Morono Y, Inagaki F, Hinrichs KU (2008) Significant contribution of Archaea to extant biomass in marine subsurface sediments. Nature 454: 991–994.
8. Jørgensen BB (2011) Deep subseafloor microbial cells on physiological standby. Proceedings of the National Academy of Sciences 108: 18193–18194.
9. Inagaki F, Suzuki M, Takai K, Oida H, Sakamoto T, et al. (2003) Microbial communities associated with geological horizons in coastal subseafloor sediments from the Sea of Okhotsk. Applied and Environmental Microbiology 69: 7224–7235.
10. Inagaki F, Nunoura T, Nakagawa S, Teske A, Lever M, et al. (2006) Biogeographical distribution and diversity of microbes in methane hydrate-bearing deep marine sediments, on the Pacific Ocean Margin. Proc Natl Acad Sci U S A 103: 2815–2820.
11. D'Hondt S, Jørgensen BB, Miller DJ, Batzke A, Blake R, et al. (2004) Distributions of Microbial Activities in Deep Subseafloor Sediments. Science 306: 2216–2221.
12. Webster G, Parkes RJ, Fry JC, Weightman AJ (2004) Widespread occurrence of a novel division of bacteria identified by 16S rRNA gene sequences originally found in deep marine Sediments. Appl Environ Microbiol 70: 5708–5713.
13. Newberry CJ, Webster G, Cragg BA, Parkes RJ, Weightman AJ, et al. (2004) Diversity of prokaryotes and methanogenesis in deep subsurface sediments from

the Nankai Trough, Ocean Drilling Program Leg 190. Environmental Microbiology 6: 274–287.
14. Orcutt BN, Sylvan JB, Knab NJ, Edwards KJ (2011) Microbial ecology of the dark ocean above, at, and below the seafloor. Microbiol Mol Biol Rev 75: 361–422.
15. Teske A, Sorensen KB (2008) Uncultured archaea in deep marine subsurface sediments: have we caught them all? Isme Journal 2: 3–18.
16. Orsi WD, Edgcomb VP, Christman GD, Biddle JF (2013) Gene expression in the deep biosphere. Nature 499: 205–208.
17. Biddle JF, Fitz-Gibbon S, Schuster SC, Brenchley JE, House CH (2008) Metagenomic signatures of the Peru Margin subseafloor biosphere show a genetically distinct environment. Proceedings of the National Academy of Sciences 105: 10583–10588.
18. Lloyd KG, Schreiber L, Petersen DG, Kjeldsen KU, Lever MA, et al. (2013) Predominant archaea in marine sediments degrade detrital proteins. Nature.
19. Vigneron A, Cruaud P, Pignet P, Caprais J-C, Gayet N, et al. (2014) Bacterial communities and syntrophic associations involved in anaerobic oxidation of methane process of the Sonora Margin cold seeps, Guaymas Basin. Environ Microbiol: n/a–n/a.
20. Vigneron A, Cruaud P, Pignet P, Caprais J-C, Cambon-Bonavita M-A, et al. (2013) Archaeal and anaerobic methane oxidizer communities in the Sonora Margin cold seeps, Guaymas Basin (Gulf of California). ISME J.
21. Lazar CS, L'Haridon S, Pignet P, Toffin L (2011) Archaeal populations in hypersaline sediments underlying orange microbial mats in the Napoli mud volcano. Appl Environ Microbiol 77: 3120–3131.
22. Fonselius S, Dyrssen D, Yhlen B (2007) Determination of hydrogen sulphide. Methods of Seawater Analysis: Wiley-VCH Verlag GmbH. pp. 91–100.
23. Sarradin P-M, Caprais J-C (1996) Analysis of dissolved gases by headspace sampling gas chromatography with column and detector switching. Preliminary results. Analytical Communications 33.
24. Ciobanu MC, Rabineau M, Droz L, Révillon S, Ghiglione JF, et al. (2012) Paleoenvironmental imprint on subseafloor microbial communities in Western Mediterranean Sea Quaternary sediments. Biogeosciences Discuss 9: 253–310.
25. Callac N, Rommevaux-Jestin C, Rouxel O, Lesongeur F, Liorzou C, et al. (2013) Microbial colonization of basaltic glasses in hydrothermal organic-rich sediments at Guaymas Basin. Frontiers in Microbiology 4.

26. Zhou J, Bruns MA, Tiedje JM (1996) DNA recovery from soils of diverse composition. Appl Environ Microbiol 62: 316–322.

27. Lazar CS, Dinasquet J, Pignet P, Prieur D, Toffin L (2010) Active archaeal communities at cold seep sediments populated by Siboglinidae tubeworms from the Storegga Slide. Microb Ecol 60: 516–527.

28. Hammer Ø, DA TH, PD R (2001) PAST: Paleontological Statistics Software Package for Education and Data Analysis. Palaeontologia Electronica 4.

29. Lu ZM, He ZL, Parisi VA, Kang S, Deng Y, et al. (2012) GeoChip-Based Analysis of Microbial Functional Gene Diversity in a Landfill Leachate-Contaminated Aquifer. Environmental Science & Technology 46: 5824–5833.

30. Zhou J (2009) GeoChip: A high throughput genomics technology for characterizing microbial functional community structure. Phytopathology 99: S164–S164.

31. Chan Y, Van Nostrand JD, Zhou J, Pointing SB, Farrell RL (2013) Functional ecology of an Antarctic Dry Valley. Proceedings of the National Academy of Sciences 110: 8990–8995.

32. He ZL, Deng Y, Van Nostrand JD, Tu QC, Xu MY, et al. (2010) GeoChip 3.0 as a high-throughput tool for analyzing microbial community composition, structure and functional activity. Isme Journal 4: 1167–1179.

33. Cheshire H, Thurow J, Nederbragt AJ (2005) Late Quaternary climate change record from two long sediment cores from Guaymas Basin, Gulf of California. Journal of Quaternary Science 20: 457–469.

34. Kemp PF, Lee S, LaRoche J (1993) Estimating the Growth Rate of Slowly Growing Marine Bacteria from RNA Content. Applied and Environmental Microbiology 59: 2594–2601.

35. Kerkhof L, Ward BB (1993) Comparison of Nucleic Acid Hybridization and Fluorometry for Measurement of the Relationship between RNA/DNA Ratio and Growth Rate in a Marine Bacterium. Applied and Environmental Microbiology 59: 1303–1309.

36. Danovaro R, Dell'anno A, Pusceddu A, Fabiano M (1999) Nucleic acid concentrations (DNA, RNA) in the continental and deep-sea sediments of the eastern Mediterranean: relationships with seasonally varying organic inputs and bacterial dynamics. Deep Sea Research Part I: Oceanographic Research Papers 46: 1077–1094.

37. Kubo K, Lloyd KG, J FB, Amann R, Teske A, et al. (2012) Archaea of the Miscellaneous Crenarchaeotal Group are abundant, diverse and widespread in marine sediments. ISME J.

38. Blazejak A, Schippers A (2010) High abundance of JS-1-and Chloroflexi-related Bacteria in deeply buried marine sediments revealed by quantitative, real-time PCR. FEMS Microbiol Ecol 72: 198–207.

39. Guy L, Ettema TJG (2011) The archaeal TACK superphylum and the origin of eukaryotes. Trends Microbiol 19: 580–587.

40. Carmona M, Zamarro MT, Blázquez B, Durante-Rodríguez G, Juárez JF, et al. (2009) Anaerobic catabolism of aromatic compounds: a genetic and genomic view. Microbiology and Molecular Biology Reviews 73: 71–133.

41. Arensdorf JJ, Focht DD (1995) A meta cleavage pathway for 4-chlorobenzoate, an intermediate in the metabolism of 4-chlorobiphenyl by Pseudomonas cepacia P166. Appl Environ Microbiol 61: 443–447.

42. Knittel K, Boetius A (2009) Anaerobic Oxidation of Methane: Progress with an Unknown Process. Annual Review of Microbiology 63: 311–334.

43. Vetriani C, Tran HV, Kerkhof LJ (2003) Fingerprinting microbial assemblages from the oxic/anoxic chemocline of the Black Sea. Appl Environ Microbiol 69: 6481–6488.

44. Lloyd KG, Albert DB, Biddle JF, Chanton JP, Pizarro O, et al. (2010) Spatial structure and activity of sedimentary microbial communities underlying a Beggiatoa spp. mat in a Gulf of Mexico hydrocarbon seep. PLoS One 5: e8738.

45. Durbin AM, Teske A (2011) Microbial diversity and stratification of South Pacific abyssal marine sediments. Environ Microbiol 13: 3219–3234.

46. Simoneit BRT, Lonsdale PF, Edmond JM, Shanks WC (1990) Deep-Water Hydrocarbon Seeps in Guaymas Basin, Gulf of California. Applied Geochemistry 5: 41–49.

47. Jorgensen SL, Hannisdal B, Lanzen A, Baumberger T, Flesland K, et al. (2012) Correlating microbial community profiles with geochemical data in highly stratified sediments from the Arctic Mid-Ocean Ridge. Proc Natl Acad Sci U S A 109: E2846–2855.

48. Govin A, Holzwarth U, Heslop D, Ford Keeling L, Zabel M, et al. (2012) Distribution of major elements in Atlantic surface sediments (36°N–49°S): Imprint of terrigenous input and continental weathering. Geochemistry, Geophysics, Geosystems 13: Q01013.

49. Nath BN, Rao VP, Becker KP (1989) Geochemical evidence of terrigenous influence in deep-sea sediments up to 8°S in the Central Indian Basin. Marine Geology 87: 301–313.

50. Lomstein BA, Langerhuus AT, D'Hondt S, Jorgensen BB, Spivack AJ (2012) Endospore abundance, microbial growth and necromass turnover in deep sub-seafloor sediment. Nature 484: 101–104.

51. Simoneit BRT, Mazurek MA, Brenner S, Crisp PT, Kaplan IR (1979) Organic geochemistry of recent sediments from Guaymas Basin, Gulf of California. Deep Sea Research Part A Oceanographic Research Papers 26: 879–891.

52. Futagami T, Morono Y, Terada T, Kaksonen AH, Inagaki F (2009) Dehalogenation Activities and Distribution of Reductive Dehalogenase Homologous Genes in Marine Subsurface Sediments. Appl Environ Microbiol 75: 6905–6909.

53. Roy R (2010) Short-term variability in halocarbons in relation to phytoplankton pigments in coastal waters of the central eastern Arabian Sea. Estuarine, Coastal and Shelf Science 88: 311–321.

54. Löffler FE, Yan J, Ritalahti KM, Adrian L, Edwards EA, et al. (2012) Dehalococcoides mccartyi gen. nov., sp. nov., obligate organohalide-respiring anaerobic bacteria, relevant to halogen cycling and bioremediation, belong to a novel bacterial class, Dehalococcoidetes classis nov., within the phylum Chloroflexi. International Journal of Systematic and Evolutionary Microbiology.

55. Meng J, Xu J, Qin D, He Y, Xiao X, et al. (2013) Genetic and functional properties of uncultivated MCG archaea assessed by metagenome and gene expression analyses. ISME J.

56. Sowers KR, Ferry JG (1983) Isolation and Characterization of a Methylotrophic Marine Methanogen, Methanococcoides methylutens gen. nov., sp. nov. Appl Environ Microbiol 45: 684–690.

57. Parkes RJ, Cragg BA, Banning N, Brock F, Webster G, et al. (2007) Biogeochemistry and biodiversity of methane cycling in subsurface marine sediments (Skagerrak, Denmark). Environ Microbiol 9: 1146–1161.

58. Parkes RJ, Webster G, Cragg BA, Weightman AJ, Newberry CJ, et al. (2005) Deep sub-seafloor prokaryotes stimulated at interfaces over geological time. Nature 436: 390–394.

59. Leigh JA (2000) Nitrogen fixation in methanogens: the archaeal perspective. Curr Issues Mol Biol 2: 125–131.

60. Raymond J, Siefert JL, Staples CR, Blankenship RE (2004) The natural history of nitrogen fixation. Molecular Biology and Evolution 21: 541–554.

61. Dekas AE, Poretsky RS, Orphan VJ (2009) Deep-sea archaea fix and share nitrogen in methane-consuming microbial consortia. Science 326: 422–426.

62. Meyerdierks A, Kube M, Kostadinov I, Teeling H, Glockner FO, et al. (2010) Metagenome and mRNA expression analyses of anaerobic methanotrophic archaea of the ANME-1 group. Environ Microbiol 12: 422–439.

63. Roussel EG, Sauvadet A-L, Chaduteau C, Fouquet Y, Charlou J-L, et al. (2009) Archaeal communities associated with shallow to deep subseafloor sediments of the New Caledonia Basin. Environ Microbiol 11: 2446–2462.

64. Mussmann M, Brito I, Pitcher A, Sinninghe Damste JS, Hatzenpichler R, et al. (2011) Thaumarchaeotes abundant in refinery nitrifying sludges express amoA but are not obligate autotrophic ammonia oxidizers. Proc Natl Acad Sci U S A 108: 16771–16776.

65. Russ L, Kartal B, Op Den Camp HJM, Sollai M, Le Bruchec J, et al. (2013) Presence and diversity of anammox bacteria in cold hydrocarbon-rich seeps and hydrothermal vent sediments of the Guaymas Basin. Frontiers in Microbiology 4.

20

Differences in Life-Histories Refute Ecological Equivalence of Cryptic Species and Provide Clues to the Origin of Bathyal *Halomonhystera* (Nematoda)

Jelle Van Campenhout[1,2]*, **Sofie Derycke**[1,2], **Tom Moens**[1,2], **Ann Vanreusel**[1]

1 Biology Department, Research Group Marine Biology, Ghent University, Ghent, Belgium, **2** Center for Molecular Phylogenetics and Evolution (CeMoFe), Ghent University, Ghent, Belgium

Abstract

The discovery of morphologically very similar but genetically distinct species complicates a proper understanding of the link between biodiversity and ecosystem functioning. Cryptic species have been frequently observed to co-occur and are thus expected to be ecological equivalent. The marine nematode *Halomonhystera disjuncta* contains five cryptic species (GD1-5) that co-occur in the Westerschelde estuary. In this study, we investigated the effect of three abiotic factors (salinity, temperature and sulphide) on life-history traits of three cryptic *H. disjuncta* species (GD1-3). Our results show that temperature had the most profound influence on all life-cycle parameters compared to a smaller effect of salinity. Life-history traits of closely related cryptic species were differentially affected by temperature, salinity and presence of sulphides which shows that cryptic *H. disjuncta* species are not ecologically equivalent. Our results further revealed that GD1 had the highest tolerance to a combination of sulphides, high salinities and low temperatures. The close phylogenetic position of GD1 to *Halomonhystera hermesi*, the dominant species in sulphidic sediments of the Håkon Mosby mud volcano (Barent Sea, 1280 m depth), indicates that both species share a recent common ancestor. Differential life-history responses to environmental changes among cryptic species may have crucial consequences for our perception on ecosystem functioning and coexistence of cryptic species.

Editor: Diego Fontaneto, Consiglio Nazionale delle Ricerche (CNR), Italy

Funding: This work was supported by GOA funds from Ghent University and by the 7th FP MIDAS (Grant agreement No. 603418). The funders had no role in study design, data collection and analysis, decision to publish, or preparation of the manuscript.

Competing Interests: The authors have declared that no competing interests exist.

* Email: Jelle.Vancampenhout@ugent.be

Introduction

DNA sequencing of multiple independently evolving gene regions has revealed that distinct taxonomic units, previously classified as single species due to a similar morphology, can be distinguished based on nuclear and mitochondrial sequence markers [1,2]. Such cryptic diversity is almost evenly distributed among major metazoan taxa and biogeographical regions [2]. However, it has been suggested that cryptic diversity is more common in marine environments because many marine species rely on chemical cues for mate recognition [3–5].

The co-occurrence of species with highly similar morphologies has been frequently observed [6–8] and indicates that they are adapted to ecologically similar environments. Similarities in body size and life histories of cryptic species may form important equalizing mechanisms which can minimize average fitness differences between species and, consequently, slow down competitive exclusion [9]. However, these equalizing mechanisms are unlikely to lead to stable coexistence [9,10]. Stable coexistence of species can only exist if both equalizing- and stabilizing mechanisms, such as fitness differences [9,11], niche differentiation and/or density-dependent life history adjustments [12,13], are present [14]. Even then, the sufficiency of both mechanisms highly

depends on the species composition and the strength of each individual mechanism [9]. To date, ecological characterization of cryptic invertebrate species has focused on nutritional [15] and habitat preferences [16,17]. However, susceptibility to predation [17,18], response to abiotic factors [19] and species-specific symbioses [20] are additional examples of stabilizing mechanisms which may also contribute to stable coexistence of cryptic species. Whether and to what extent this implies that cryptic species are ecologically equivalent remains unknown.

In marine sediments, nematodes are usually the most abundant and diverse Metazoa [21,22]. Cryptic diversity has been reported in marine nematodes belonging to different orders [23–25]. These cryptic species can exhibit differential dispersal capacities [26] and differential resource use, and environmental factors such as salinity and temperature may affect the outcome of their competitive interactions when they co-occur [27]. However, at this point, we lack any information on if and how life-history traits of cryptic nematode species differ in response to changes in abiotic factors. Such information is, however, important to understand mechanisms driving the coexistence cryptic species, which in turn is crucial to correctly understand the biodiversity and ecosystem functioning relationship [28].

In this study, we focus on the marine nematode *Halomonhystera disjuncta* (previously named *Geomonhystera disjuncta*), a representative of a genus which has a widespread geographical distribution and which has been reported both in shallow-water [23,29–31] and in deep-sea environments [32,33]. In the Westerschelde estuary (located in the Southwestern part of The Netherlands), five cryptic species (GD1-5) of *H. disjuncta* have been reported based on nuclear and mitochondrial sequence data; these five species exhibit subtle morphometric differences [23,34]. *Halomonhystera disjuncta* has most often been isolated from macroalgal holdfasts and wrack deposits, but has also been frequently observed in the sediment, successfully exploiting organically enriched substrata [21,35,36]. In the Westerschelde estuary, the five cryptic species (GD1-5) show sympatric distributions [23]. In view of their mainly intertidal occurrence, these species are subject to strong physical, chemical and biological gradients. Ecological and physiological responses of *H. disjuncta* from the intertidal zone to changes in salinity, temperature, food quality, food density and heavy metals have been investigated [21,37–42]. However, it is unknown whether the observed broad tolerance to e.g. temperature and heavy metals is the same for different cryptic species.

Halomonhystera is also the dominant nematode genus in the sulphide rich bacterial mats of the Nyegga pockmark (Nordic Norwegian margin) [32] and the Håkon Mosby mud volcano (HMMV, Barent Sea slope) [33], situated at depths of 730 m and 1280 m, respectively. At both locations, fluids escape from the deep-sea floor at lower temperatures and flow rates compared to mid-oceanic ridges, and are, therefore, called cold seeps. The *Halomonhystera* from both environments were originally identified as *H. disjuncta*, but the *Halomonhystera* from the HMMV has morphological and genetic differences compared to the shallow-water *H. disjuncta* species GD1-5, and has been described as a new species *Halomonhystera hermesi* [43]. Phylogenetic analysis revealed three intertidal clades (GD1/GD4, GD3, GD2/GD5) and one deep-sea clade *Halomonhystera hermesi* [44] and showed that *H. hermesi* is more closely related to GD1 and GD4 than to the other species. A deep-sea invasion from shallow-water regions has hence been hypothesized [44]. Nordic seep colonization from intertidal regions implies that early colonizers had to adapt to low temperatures, higher salinities and the presence of high concentrations of sulphides. If cryptic species of *H. disjuncta* differ in their tolerance to one or more of these environmental differences, then we could expect that the species with the highest degree of tolerance to bathyal cold seep conditions is phylogenetically closest related to *H. hermesi*.

Here we investigate the effect of three abiotic factors (salinity, temperature and sulphide) on life-history traits of three cryptic *H. disjuncta* species (GD1, 2 and 3). Each of the three abiotic factors varied from control conditions (salinity of 25 psu, 16°C and no sulphide) to environmental conditions representative of the HMMV (salinity of 34–35 psu, 2°C and hydrogen sulphide concentrations of ca. 1 mM) [45,46] in a fully crossed factorial design with three salinity levels (25, 29.5 and 34 psu), three temperatures (16, 10 and 4°C) and two sulphide treatments (no sulphide vs. 1 mM). Because these three cryptic species co-occur in the Westerschelde estuary [23], are phenotypically almost identical [34], and were isolated from the same species of macroalgae (*Fucus vesiculosus*), they are expected to show a high degree of ecological equivalence with respect to abiotic environmental factors. In contrast, if differences in life history traits would be observed, then we would expect GD1 to better tolerate bathyal cold seep conditions than GD2 and GD3, because it is phylogenetically more closely related to *H. hermesi* compared to

GD2-3. This prediction follows the rationale of the phylogenetic niche conservatism theory (PNC). PNC is the tendency of lineages to retain their niche-related traits through speciation events and over long evolutionary periods [47,48]. Consequently, phylogenetically closely related species are expected to be ecologically similar [49].

Materials and Methods

Nematode cultures

Halomonhystera disjuncta species GD1 and GD3 were retrieved from decaying *Fucus vesiculosus* from the Paulina tidal flat in the Westerschelde estuary (The Netherlands, 51° 20′ 56.79″ N, 3° 43′ 29.56″ E), while GD2 was isolated from the same decaying macroalgae species from a different location (Kruispolderhaven, 51° 21′ 34.35″ N, 4° 5′ 52.02″ E) in the same estuary. All species were collected in April 2012. Permission for the field work was issued by the Provincie Zeeland, the Netherlands (Directie Ruimte, Milieu en Water). Pieces of algae were inoculated on Petri dishes filled with 0.8% nutrient:bacto agar (ratio of 1/7) prepared in artificial seawater [50] with a salinity of 25 psu and placed at a constant temperature of 16°C. These conditions were chosen based on the temperature and salinity values observed in the field locations from where the species were isolated and are perfectly suited for cultivation of *H. disjuncta* species. Monospecific cultures were established by transferring a single gravid female to a new Petri dish (5.5 cm inner diameter) containing the same agar medium. By doing so, we reduced population variation, which was necessary to ensure monospecific status of the cultures. This, however, implies that differences observed here between cryptic species may partly reflect individual and/or population level differences, for instance resulting from local adaptation, rather than consistent differences between species. The monospecific cultures were maintained at the same temperature (16°C) with frozen-and-thawed *Escherichia coli* K12 as a food source. Species identity and monospecificity of the cultures was confirmed by PCR and sequencing the ITS gene of ten nematodes per culture. Primers, thermo-cycling conditions and GenBank accession numbers can be found in [23]. Nematodes for the experiment were harvested from the stock cultures.

Experimental design

To test the effect of three abiotic factors (salinity, temperature and sulphide) between cryptic species, a fully crossed design with four factors: salinity (25, 29.5 and 34 psu), temperature (16, 10 and 4°C), species (GD1, GD2 and GD3) and sulphide (present or absent) was set up. For each treatment x species combination, four replicates were prepared, resulting in 216 microcosms. The experiment was conducted on experimental microcosms which consisted of Petri dishes (3.5 cm i.d.) containing 1.5 ml 0.8% nutrient:bacto agar (1/7). Because stock cultures are maintained at a salinity of 25 psu and 16°C, these conditions served as a control. The sulphidic media were prepared from the same agar medium as above, but to which we added sodium thiosulphate ($Na_2S_2O_3$) and sodium sulphide (Na_2S) in a final concentration of 1 μM for both reagents. Lastly, to each microcosm 350 μl *E. coli* K12 (ca. 7.6×10^8 cells/ml) was added as a food source, which is sufficient for the duration of the experiment.

Each replicate microcosm received 10 nematodes. We randomly picked out six female and four male nematodes, in accordance with the sex ratio in the stock cultures. Nematodes were considered adult when all parts of the reproductive system were clearly visible. Only nematodes which had just reached adulthood were randomly selected. Immediately after addition of the nematodes to the Petri dishes, we checked them carefully for specimens which died, were injured or immobilized during manipulation. These were replaced by new specimens of the same sex. The Petri dishes were then closed with parafilm, maintained in temperature-controlled incubators without light, and the experiment was started.

Data collection and analysis

The amounts of eggs, juveniles and adults were counted daily. In addition, we observed mortality of inoculated nematodes. Young F1-adults are easily distinguishable from F0-adults, up to a few days after they reached adulthood, because of their smaller body size. Therefore, the experiment was stopped before F1-adults started to deposit eggs and/or when we were no longer able to distinguish new adults from the inoculated nematodes. This was after ca. 13, 18 and 38 days at 16°C, 10°C and 4°C, respectively. Hence, our results encompassed a single nematode generation.

Four response variables were used for statistical analysis: (1) minimum time until first egg(s) deposition (MEGD) is the minimum time (in days) from the inoculation of the nematodes until we observed the first egg(s); (2) the minimum time for embryonic development (MEMD) was calculated by subtracting MEGD from the time at which we observed the first juvenile; (3) the minimum time for the development of juveniles into adults (MJD) was calculated by subtracting the time at which the first juvenile was observed from the time at which the first F1 adult was detected; (4) the minimum generation time (MGT) was recorded as the time from the start of the experiment (parental generation) until the first new adult (filial generation) was observed.

In addition, fertility was approximated by summing the counts of eggs, juveniles and new adults at the end of the incubation. Fertility was expressed per parental female by dividing total offspring (eggs, juveniles and new adults) of a microcosm, by the number of inoculated females. Furthermore, mortality of the inoculated nematodes was monitored. We, therefore, created an additional variable, the minimum adult lifespan (MALS), which corresponds to the day at which we observed the death of at least one inoculated adult nematode. It must be noted that the experiment for some treatments was stopped before any mortality of inoculated nematodes was observed. Small standard deviations and visual inspection of our data revealed that MALS was not importantly influenced by stochasticity, i.e. the incidental death of an inoculated individual in random replicates and treatments.

Because our data did not meet the assumptions for parametric variance analysis, even after transformation, a Permutational Based Multivariate Analysis of Variance [51], on the basis of Euclidean distances with 9999 permutations, was used. In comparison to ANOVA/MANOVA, both assuming normal distributions and, implicitly, Euclidean distance, PERMANOVA works with any distance measure that is appropriate to the data, and uses permutations to make it distribution free [51]. Each analysis was performed using a fully crossed PERMANOVA design with four fixed factors (salinity, temperature, sulphide presence and species). All analyses were performed within PRIMER v6 with PERMANOVA+ add-on software [51]. Because of the relatively limited level of replication (4 or 3, the latter in few cases where we observed agar dehydration as a result of an imperfect Petri dish closure), a Monte Carlo test was performed. The components of variation, estimated by PERMANOVA, were used to attribute the amount of variation to different factors and interaction terms [52]. A high component value corresponds with a high relative importance of the respective term in the model [51]. Pairwise comparisons were performed on the full dataset and p-values can be found in the supplementary

information. If the four-way interaction was significant, we were unable to reliably interpret two-way and three-way interaction terms. In these cases we selected subsets of our data for which only two or three factors varied. However, we tested these two- and three-way interaction effects for all levels and combinations of the factor(s) that were kept constant. PERMDISP [53] was executed to test the homogeneity of multivariate dispersions in order to discriminate between real location or factor effects and effects explained by differences in dispersion for the significant factors.

Figures were made with Graphpad prism 5.

Results

Minimum generation time (MGT)

MGT is the sum of the minimum time until first egg(s) deposition (MEGD), the minimum time for embryonic development (MEMD) and the minimum time for the development of juveniles into adults (MJD). A significant four-way interaction effect on MGT was found (PERMANOVA; p = 0.0001, Table 1) which was not caused by differences in multivariate dispersion ($F_{43, 109} = 0.408$, p = 0.9994). The estimation of the components of variation showed that temperature was the most important factor influencing MGT (Table 2).

MGT increased with decreasing temperature (T) for all species (Sp; Table 3). When species are compared in presence and absence of sulphide (S), GD2 generally had the shortest MGT at 16°C and at 4°C. Except at cold seep conditions (at a temperature of 4°C, a salinity of 34 psu and in the presence of sulphide), GD1 had a significantly shorter MGT than GD2 (p = 0.0156, Table S1). A significant shorter MGT for GD1, in comparison to GD2-3, was also observed at 10°C (all p-values <0.0018, Table S1). The difference in MGT between species in relation to temperature was underlined by a significant Sp x T interaction (all p-values = 0.0001, Table 1) independent of salinity and sulphide addition.

Increasing the salinity (Sal) resulted in an increase of MGT. The response to higher salinities is more pronounced for GD3 (2–10 days) than for GD1-2 (0–6 days; Table 3), in the absence of sulphide. This observation was supported by a significant Sp x Sal interaction at all temperatures (all p-values <0.0468, Table 1).

The effect of adding sulphide to the cultures resulted in the death of all inoculated GD3 nematodes during the first two minutes and no life-history traits could, therefore, be determined. The addition of sulphide resulted in a significant increase in MGT of 1–10 days for GD1 and GD2 at 10°C and 4°C (all p-values < 0.0138, Table S2) at all salinity levels (Table 3).

Minimum time until first egg(s) deposition (MEGD)

GD3 was the only species showing an ovoviviparous reproduction strategy throughout the whole experiment. Therefore, we did not include GD3 in the MEGD dataset. For the two other species, abiotic factors significantly affected MEGD (four-way PERMANOVA; p = 0.0065, Table 1). Homogeneity of multivariate dispersions was observed ($F_{35, 89} = 0.736$, p = 0.8453). The estimation of the components of variation revealed that temperature is the most import factor influencing MEGD (Table 2).

A strong significant increase in MEGD from 10°C to 4°C (all p-values <0.0003, Table S3) was observed for both species (Figure 1). This increase was observed at all salinity levels and in presence/absence of sulphide. Increasing salinity usually resulted in a significant increase in MEGD for both species (p-values in Table S4) and was most pronounced at 4°C for both species, supported by a significant T x Sal interaction term (all p-values < 0.0026, Table 1).

Table 1. P-values of a PERMANOVA test on all possible interaction.

Stable Factor 1	Stable Factor 2	Interaction tested	MGT	MEGD	MEMD	MJD	MALS	OS
No S	GD1	T x Sal	0.0003*	0.0026*	0.0219*	0.9134	0.0007*	0.6495
No S	GD2	T x Sal	0.0006*	0.0001*	0.0071*	0.5990	0.0345*	0.0001*
No S	GD3	T x Sal	0.0001*	/	/	0.0180*	0.1420	0.0001*
S	GD1	T x Sal	0.0001*	0.0001*	0.9742	0.1448	0.0001*	0.9725
S	GD2	T x Sal	0.0001*	0.0001*	0.0162*	0.0001*	0.0034*	0.1032
No S	16°C	Sp x Sal	0.0468*	0.6952	0.1942	0.0909	0.0067*	0.0079*
No S	10°C	Sp x Sal	0.0001*	0.0455*	0.0113*	0.0004*	0.0700	0.0001*
No S	4°C	Sp x Sal	0.0001*	0.0017*	0.0108*	0.0645	0.2617	0.0489*
S	16°C	Sp x Sal	0.0074*	/	0.8087	0.0129*	0.4497	0.3492
S	10°C	Sp x Sal	0.1938	0.1040	0.5424	0.0861	0.4497	0.4837
S	4°C	Sp x Sal	0.0001*	0.5787	0.0420*	0.0409*	0.4390	0.1006
No S	25 psu	Sp x T	0.0001*	0.0003*	0.001*	0.0002*	0.0001*	0.0001*
No S	29.5 psu	Sp x T	0.0001*	0.0694	0.0001*	0.0002*	0.0001*	0.0001*
No S	34 psu	Sp x T	0.0001*	0.0024*	0.0001*	0.0001*	0.0002*	0.0001*
S	25 psu	Sp x T	0.0001*	0.2228	0.0001*	0.0001*	0.0002*	0.0001*
S	29.5 psu	Sp x T	0.0001*	0.1190	0.0001*	0.0001*	0.0001*	0.0001*
S	34 psu	Sp x T	0.0001*	0.6028	0.0003*	0.0003*	0.0001*	0.0001*
16°C	25 psu	S x Sp	/	/	0.1121	0.6600	/	0.0797
16°C	29.5 psu	S x Sp	0.0593	0.0383*	0.0237*	0.0327*	0.0946	0.4270
16°C	34 psu	S x Sp	0.0707	0.0155*	0.0009*	0.0090*	0.0006*	0.0009*
10°C	25 psu	S x Sp	0.1231	0.0201*	0.4999	0.0049*	0.2640	0.0001*
10°C	29.5 psu	S x Sp	0.7980	0.8596	0.1066	0.0624	0.8394	0.0001*
10°C	34 psu	S x Sp	1	0.2908	0.3875	0.2098	1	0.0001*
4°C	25 psu	S x Sp	1	0.0182*	0.0257*	0.6736	0.3305	0.7430
4°C	29.5 psu	S x Sp	0.3984	0.3390	0.4784	0.7818	0.0462*	0.6679
4°C	34 psu	S x Sp	0.0001*	/	0.0023*	0.0079*	0.2772	0.0013*
GD1	16°C	S x Sal	0.6674	0.4896	0.4436	0.8028	0.0834	0.3377
GD1	10°C	S x Sal	0.7077	0.2439	0.2491	0.9511	0.2970	0.5969
GD1	4°C	S x Sal	0.0093*	0.0128*	0.4725	0.2240	0.0088	0.9318
GD2	16°C	S x Sal	0.0091*	0.6131	0.7348	0.0201*	0.0001	0.0539
GD2	10°C	S x Sal	0.5374	0.3693	0.5552	0.1469	0.8020	0.0001*
GD2	4°C	S x Sal	0.0001*	0.0184*	0.0001*	0.0005*	0.4469	0.0186*
GD1	25 psu	S x T	0.0007*	0.0002*	0.1599	0.7149	0.8802	0.0003*
GD1	29.5 psu	S x T	0.0001*	0.0002*	0.6982	0.0574	0.0038*	0.005*
GD1	34 psu	S x T	0.0001*	0.0001*	0.0311*	0.0895	0.0975	0.0001*

Table 1. Cont.

Stable Factor 1	Stable Factor 2	Interaction tested	MGT	MEGD	MEMD	MJD	MALS	OS
GD2	25 psu	S x T	0.0003*	0.0001*	0.1905	0.0154*	0.8404	0.0001*
GD2	29.5 psu	S x T	0.0001*	0.0003*	0.0266*	0.7216	0.1859	0.0007*
GD2	34 psu	S x T	0.0001*	0.0001*	0.0005*	0.0005*	0.0004*	0.0150*
GD1		S x T x Sal	0.2698	0.0017*	0.2798	0.8041	0.1377	0.8175
GD2		S x T x Sal	0.0001*	0.0184*	0.0004*	0.0001*	0.0628	0.0001*
25 psu		S x Sp x T	0.4376	0.0002*	0.0732	0.1945	0.8810	0.0001*
29.5 psu		S x Sp x T	0.6984	0.4486	0.1112	0.1344	0.0276*	0.0002*
34 psu		S x Sp x T	0.0001*	0.0196*	0.0001*	0.2219	0.0012*	0.0001*
No S		Sp x T x Sal	0.0001*	0.0008*	0.0002*	0.0823	0.1520	0.0001*
S		Sp x T x Sal	0.0001*	0.1830	0.1730	0.0029*	0.4053	0.2825
16°C		S x Sp x Sal	0.3157	0.6495	0.4232	0.1867	0.0037*	0.0151*
10°C		S x Sp x Sal	0.4492	0.1358	0.2546	0.2737	0.6765	0.0001*
4°C		S x Sp x Sal	0.0001*	0.0326*	0.001*	0.0133*	0.0795	0.0528
		S x Sp x T x Sal	0.0001*	0.0065*	0.0002*	0.0433*	0.3177	0.0001*

A Permutational Based Multivariate Analysis of Variance (Euclidean distances, 9999 permutations), was used to statistically test two-way, three-way and four-way interaction terms (column 3). Each interaction component (two-way and three-way) was tested for all possible combinations of the factor(s) that were kept constant (column 1–2). Significant p-values are indicated with an asterisk.

Abbreviations: No S, absence of sulphide; S, presence sulphide; T, temperature (16, 10 and 4°C); Sal, salinity (25, 29.5 and 34 psu); Sp, cryptic *Halomonhystera disjuncta* species GD1–3 (previously named *Geomonhystera disjuncta*); MGT, minimum generation time; MEGD, minimum time until first egg(s) deposition; MEMD, minimum time for embryonic development; MJD, minimum time for the development of juveniles into adults; MALS, minimum adult life span; OS, offspring per female at the end of the experiment; /, no test available.

Table 2. Estimates of components of variation results of the four-way PERMANOVA on single factors and interaction terms for all analyzed variables.

Single factor and/or interaction terms	MGT	MEGD	MEMD	MJD	MALS	OS
S	2.893	0.526	−0.004	0.870	0.637	183.480
Sp	7.217	0.002	0.311	0.748	3.776	452.880
T	169.230	25.561	10.500	14.894	124.740	93.748
Sal	9.277	1.634	0.106	0.455	0.230	71.800
T x Sal	0.223	0.005	0.046	0.371	0.080	30.517
Sp x Sal	3.163	1.484	0.039	0.197	0.201	25.923
Sp x T	0.384	0.010	0.050	0.101	0.144	3.972
S x Sp	5.498	0.039	2.793	1.826	12.950	160.550
S x Sal	1.744	0.020	0.016	0.214	0.090	13.748
S x T	1.931	0.768	0.005	0.061	0.482	12.609
S x T x Sal	0.139	0.126	0.213	−0.031	0.474	142.200
S x Sp x T	0.210	−0.007	0.026	0.072	0.176	7.579
Sp x T x Sal	0.790	0.090	0.021	0.189	0.082	13.487
S x Sp x Sal	0.232	0.070	0.073	0.151	0.077	20.431
S x Sp x T x Sal	0.696	0.130	0.550	0.197	0.024	44.797
V(Res)	0.269	0.157	0.356	0.432	0.451	20.371

The components of variation was estimated by **PERMANOVA** (Euclidean distances, 9999 permutations) for MGT, minimum generation time; MEGD, minimum time until first egg(s) deposition; MEMD, minimum time for embryonic development; MJD, minimum time for the development of juveniles into adults; MALS, minimum adult life span; OS, offspring per female at the end of the experiment. High values correspond to a high relative importance of the respective factor.
Abbreviations: S, absence/presence of sulphide; T, temperature (16, 10 and 4°C); Sp, cryptic *Halomonhystera disjuncta* species GD1–3 (previously named *Geomonhystera disjuncta*); Sal, salinity (25, 29.5 and 34 psu).

The effects of sulphide were heavily dependent on temperature for both species (significant S x T interaction term, all p-values < 0.0003 at all salinity levels, Table 1). Addition of sulphide resulted in a significant decrease (all p-values <0.0031, Table S2) in MEGD of GD1 at 16°C, (Figure 1A, 1C) but not of GD2 (Figure 1B, 1C). The observed difference between species was corroborated by significant S x Sp interaction term 16°C at all salinity levels (p-values <0.0383, Table 1). At 10°C, adding sulphide resulted in a significant increase in MEGD of GD1 (all p-values <0.0429, table S2) but not for GD2. At the lowest

temperature (4°C), adding sulphide resulted in an increase in MEGD for both species (all p-values <0.0134, Table S2).

Minimum time for embryonic development (MEMD)

Due to GD3's ovoviviparous reproduction strategy we were unable to estimate MEMD for this species. Similar to the MEGD, a significant four-way interaction (p = 0.0002, Table 1) and homogeneity of multivariate dispersions were found ($F_{1,123} = 0.042$, p = 0.8381). The estimates of components of variation revealed that temperature and the interaction term Sp x T were the most important effects (Table 2).

Table 3. Minimum generation time of GD1-3 under changing environmental conditions.

Temperature	Salinity	GD1– No S	GD2 - No S	GD3 - No S	GD1 - S	GD2 - S
16°C	25 psu	9.3±0.6	7.3±0.6	10.3±0.6	9.0±0.0	7.0±0.0
	29.5 psu	9.3±0.6	8.0±0.0	12.0±0.0	9.5±0.6	9.0±0.0
	34 psu	11.25±0.5	9.3±0.5	/	11.3±0.5	10.3±0.4
10°C	25 psu	10.3±0.6	13.7±0.6	13.5±0.6	11.7±0.6	16.0±0.0
	29.5 psu	10.5±0.6	15.7±0.6	19.3±0.6	12.3±0.6	17.7±0.5
	34 psu	12.8±0.5	17.3±0.5	24.3±0.5	14.5±0.6	19.0±0.0
4°C	25 psu	28.3±0.6	25.5±0.6	33.0±0.8	31.3±0.6	28.5±0.5
	29.5 psu	29.7±0.6	28.3±0.5	38.3±0.5	34.3±0.5	33.3±0.5
	34 psu	34.3±0.5	30.3±0.6	43.3±0.6	39.3±0.5	40.5±0.5

Minimum generation time (mean ± standard deviation) of *H. disjuncta* cryptic species (GD1-3) in the absence (No S) and presence of sulphide (S) at different temperatures (16, 10 and 4°C) and salinity (25, 29.5 and 34 psu). No results were obtained for GD3 in the presence of sulphide because these nematodes died within the two first minutes after adding Su. The minimum generation time was depicted from the day we observed the first adult.
Abbreviations: No S, absence of sulphide; S, presence sulphide; GD1-3, cryptic *Halomonhystera disjuncta* species 1–3 (previously named *Geomonhystera disjuncta*).

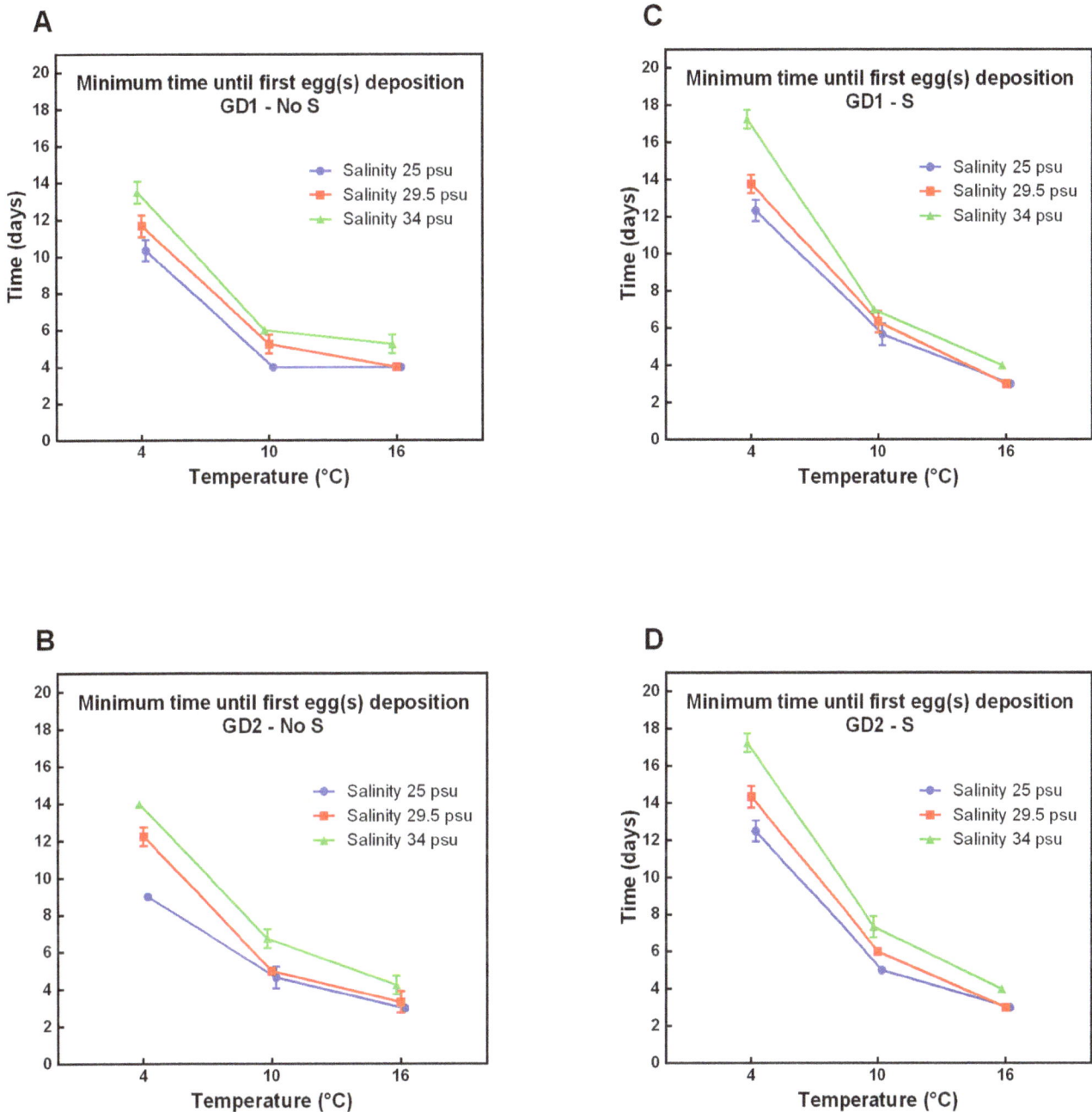

Figure 1. Minimum time until first egg(s) deposition of GD1 and GD2 under changing environmental conditions. Minimum time until first egg(s) deposition in relation to temperature and salinity of cryptic *Halomonhystera disjuncta* species: GD1 (A, C) and GD2 (B, D), in the absence (No S; A, B) and presence (S; C, D) of sulphide. The minimum egg deposition time corresponds to the day after the start of the experiment on which the first egg was observed within the experimental microcosms. Data shown are mean values ±1stdev of 3 or 4 replicates per treatment.

Comparable with MEGD, lowering the temperature increased MEMD (Figure 2A–D). Both GD1 and GD2 had a stable MEMD between 16°C and 10°C, while a significant increase was observed at 4°C (all p-values <0.0124, Table S3). This increase was much more pronounced for GD1 than for GD2, underlined by a significant Sp x T interaction effect (all p-values <0.001, Table 1) independent of salinity and sulphide.

Increasing the salinity had no significant effect on MEMD for GD1, but had an opposite effect on MEMD for GD2 at 4°C between different sulphide treatments (Figure 2B, 2D). This observation is supported by a significant S x Sp x Sal interaction

component at 4°C (p-value = 0.001, Table 1). The addition of sulphide had no further consistent or opposing effects on MEMD. Interestingly, at bathyal cold seep conditions, GD2 had a significantly shorter MEMD than GD1 (p<0.0029, Table S1).

Minimum time for the development of juveniles into adults (MJD)

MJD showed a significant four-way interaction effect (p = 0.0433, Table 1) and homogeneity of multivariate dispersions was found ($F_{43,109} = 0.870$, p = 0.6917). According to the estimates

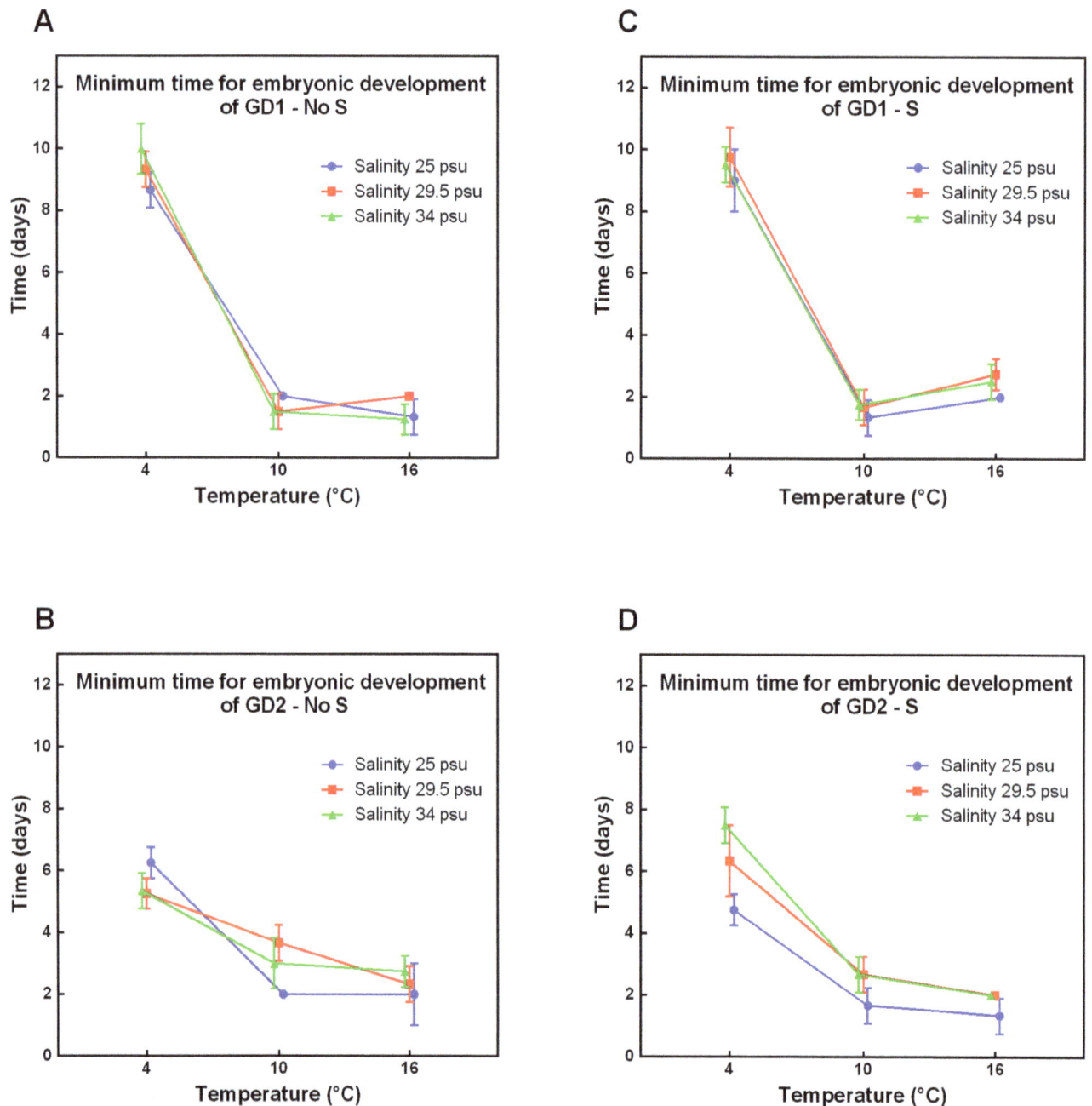

Figure 2. Minimum time for embryonic development of GD1 and GD2 under changing environmental conditions. Minimum time for embryonic development in relation to temperature and salinity of cryptic *Halomonhystera disjuncta* species: GD1 (A, C) and GD2 (B, D), in the absence (No S; A, B) and presence (S; C, D) of sulphide. The minimum time for embryonic development was calculated by subtracting the day we observed the first egg from the day we observed the first juvenile within the experimental microcosms. Data shown are mean values ±1stdev of 3 or 4 replicates per treatment.

of components of variation, temperature affects MJD the most (Table 2).

Lowering temperature resulted in a significant increase in MJD (Figure 3A–E) for both GD2 and GD3 independent of other factors (all p-values <0.0189, Table S3). GD1 also showed a significant increase in MJD from 10°C and 16°C to 4°C (all p-values <0.0024, Table S3; Figure 3A, 3B) but not between 16°C to 10°C (all p-values >0.1092, Table S3). This differential effect of temperature changes on species was confirmed by a significant Sp x T interaction effects at all salinity and sulphide levels (all p-values

<0.0003, Table 1). Interestingly, GD3 did not produce offspring at 16°C and a salinity of 34 psu. Even though female nematodes had eggs inside the uterus, which developed into juveniles, they were never deposited.

Our data also shows that at 4°C the effect of salinity interacted with the addition of sulphide and was more pronounced for GD2 than for GD1 (significant S x Sp x Sal interaction term, p = 0.0133, Table 1). In addition, at bathyal cold seep conditions GD1 had a significantly shorter MJD compared to GD2 (p = 0.0009, Table S1).

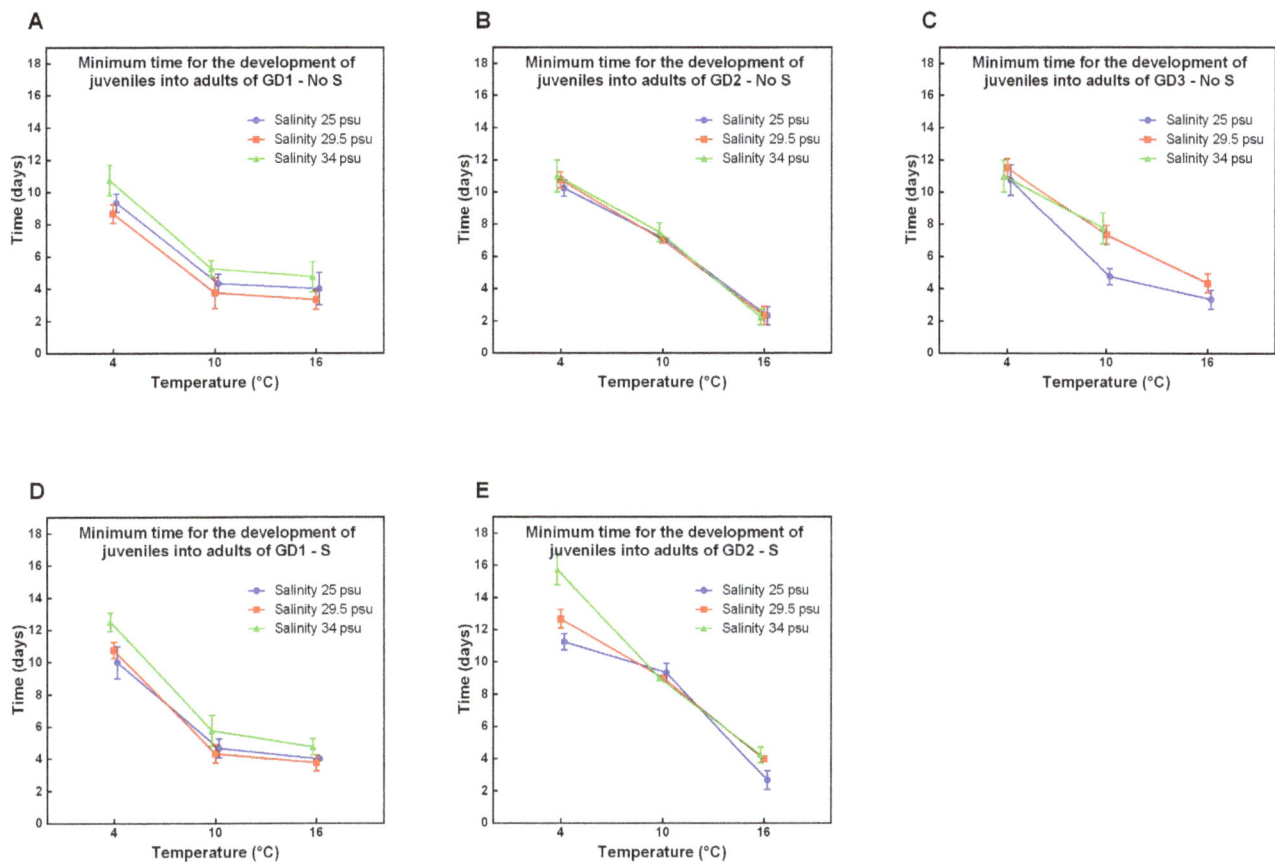

Figure 3. Minimum time for the development of juveniles into adults of GD1-3 under changing environmental conditions. Minimum time for the development of juveniles into adults in relation to temperature and salinity of cryptic *Halomonhystera disjuncta* species: GD1 (A, D), GD2 (B, E) and GD3 (C), in the absence (No S; A–C) and presence (S; D–E) of sulphide. The minimum time for embryonic development was calculated by subtracting the day we observe the first juvenile from the day we observed the first adult within the experimental microcosms. Data shown are mean values ±1stdev of 3 or 4 replicates per treatment. No results were obtained for GD3 in the presence of sulphide because inoculated nematodes died within the first 1–2 minutes after adding Su.

Minimum adult life span (MALS)

The four-way interaction of MALS was not significant (p = 0.3177, Table 1) and homogeneity of multivariate dispersions ($F_{41, 99}$ = 1.085, p = 0.3638) was observed. Some data points are missing because in some treatments we stopped the experiment before any mortality of inoculated nematodes occurred.

A decrease in temperature generally resulted in a significant increase in MALS for all species (Figure 4). However, no significant increase was observed for GD1 from 16°C to 10°C (all p-values >0.1167, Table S3), in contrast with GD2 and GD3 (all p-values <0.0010, Table S3). At 4°C, in the absence of sulphide, GD2 and GD3 had the shortest and largest MALS, respectively. This difference between species was supported by a significant Sp x T interaction effect at all salinity and sulphide levels (all p-values <0.0002, Table 1).

Addition of sulphide resulted in the death of all inoculated GD3 nematodes within minutes. For both other species the addition of sulphide commonly decreased MALS. At 4°C and at bathyal cold seep conditions, GD1 had a higher MALS than GD2 (all p-values <0.0023, Table S1, Figure 4D–E). Sulphide-induced mortality decreased with decreasing temperature: in the presence of sulphide and at a temperature of 16°C, ca. 99% of both inoculated and new adults of GD1 and GD2 died after one generation. However, this high mortality rate was decreased to ca. 50% at 10°C and was negligible at 4°C.

Offspring per female (OS)

Variances were homogeneous ($F_{44, 112}$ = 0.526, p = 0.9931) and a significant four-way interaction effect was found (p = 0.0001, Table 1). Estimates of components of variation revealed that most factors and interactions had an effect on the OS (Table 2) with temperature being the most important one. Consequently, sub-datasets were analyzed to better understand effects of single factors and their two- and three-way interactions.

All cryptic species had the highest OS at 16°C in the absence of sulphide and at a salinity of 25 psu (except for GD2, Figure 5), and the lowest OS at 4°C in the presence of sulphide and at a salinity of 34 psu. A reduction in OS was observed for all species with decreasing temperature, increasing salinity and addition of sulphide. However, the declination in OS for GD2 was clearly more influenced by temperature than for GD1 and GD3 as was supported by a significant Sp x T interaction component (all p-values = 0.0001, Table 1). Variation in OS, caused by temperature and salinity, was much lower in GD3 in the absence of sulphide (p-value $_{Sp x T x Sal}$ = 0.0001, Table 1), compared to the two other species. However, no offspring was observed at 16°C and a salinity of 34 psu. Generally, adding sulphide resulted in a reduction in OS for both GD1 and GD2. Furthermore, GD1 produced significantly more OS than GD2 at bathyal cold seep conditions (p = 0.006, Table S1).

A

B

C

D

E

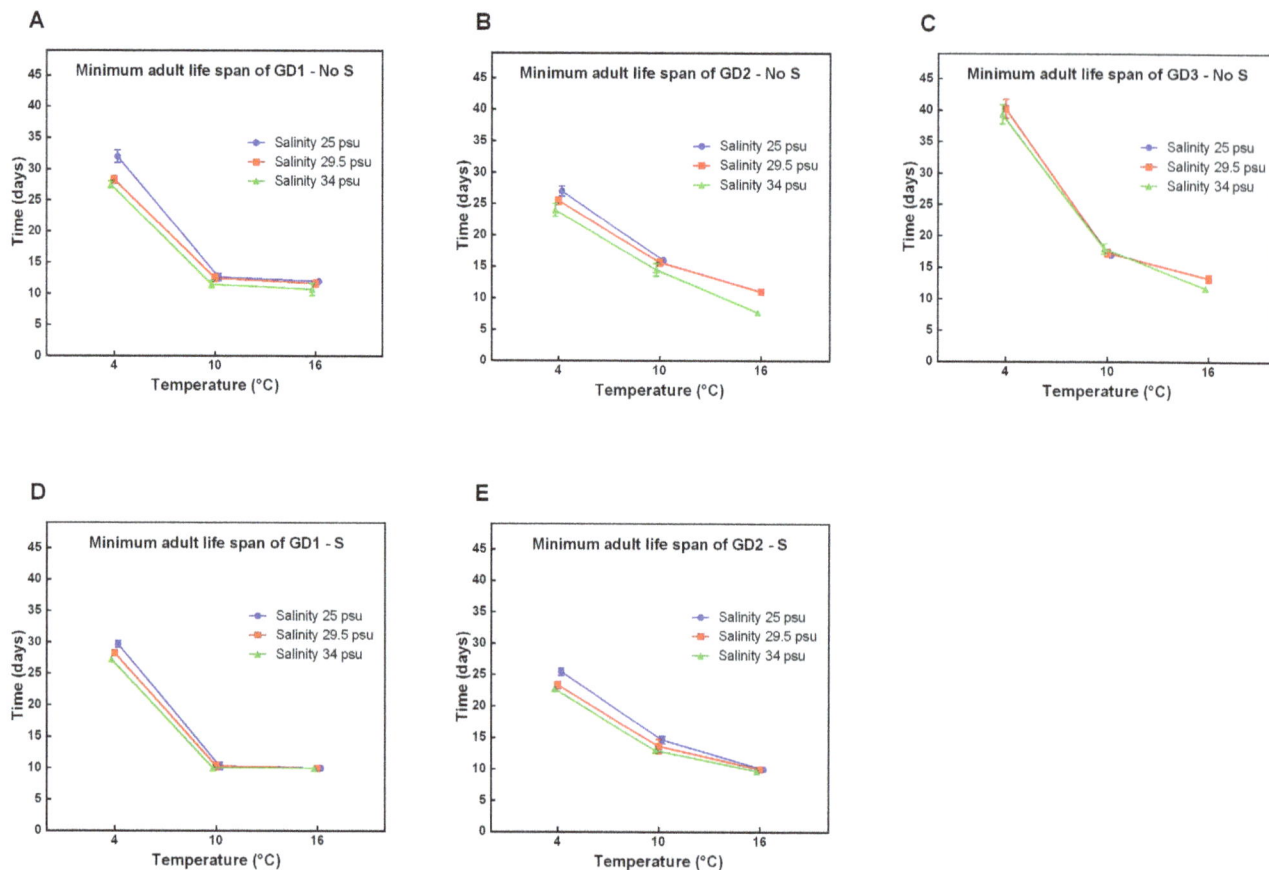

Figure 4. Minimum adult life span under changing environmental conditions. Minimum adult life span in relation to temperature and salinity of cryptic *Halomonhystera disjuncta* species: GD1 (A, D), GD2 (B, E) and GD3 (C), in the absence (No S; A–C) and presence (S; D–E) of sulphide. The minimum adult life span corresponds to the day after the start of the experiment on which the first inoculated nematode died. Data shown are mean values ±1stdev of 3 or 4 replicates per treatment. No results were obtained for GD3 in the presence of sulphide because inoculated nematodes died within the first 1–2 minutes after adding Su. Some data points are missing because the experiment was stopped before the death of any of the inoculated nematodes.

Discussion

The success of nematodes in marine environments relates to their ability to adapt to changing environmental conditions [54,55]. Because the three cryptic *H. disjuncta* species commonly occur on decaying macroalgae in the high intertidal, they are regularly exposed to short- and long-term fluctuations in abiotic factors such as salinity and temperature. Moreover, they can be exposed to hydrogen sulphide produced by sulphate reducing bacteria thriving on rotting macroalgae [56]. In summary, the three species were expected to show wide tolerances to the abiotic factors tested in this study. In our experiments, temperature had the strongest impact on life-history traits of all cryptic species. The minimum generation time and adult life span increased with decreasing temperature, which is most likely caused by lower metabolic rates at lower temperatures [57]. Faster generation times at higher temperatures shorten the vulnerable period of the embryo, thus preventing embryonic deformities and/or arrest [54], and allow species to rapidly increase in numbers during favorable episodes, a common feature of opportunistic species [58]. *Halomonhystera disjuncta* is indeed an opportunistic species capable of quickly colonizing suitable patches such as new algal deposits [35]. Salinity, on the other hand, had a relatively minor impact on the life-history traits studied here. The salinity and temperature ranges tested in our study reflect fluctuations in their

natural habitat and indicate a strong capacity for adaptation to natural fluctuations, which is consistent with previous studies on other marine nematodes from very similar intertidal habitats [59,60].

Cryptic *H. disjuncta* species are not ecologically equivalent

Co-occurrence of cryptic species raises questions regarding their ecological similarity and the mechanisms influencing their coexistence [15,19]. The three cryptic species studied here have a very similar morphology, have sympatric occurrences and all thrive on the same habitat, i.e. decaying macroalgae in the Westerschelde estuary. We may, therefore, expect that they have evolved comparable adaptations to similar environmental conditions (equalizing mechanism). The general response to temperature and salinity, i.e. an increase in time for life-history traits and a decrease in number of offspring with lower temperatures and higher salinities, was indeed similar in all cryptic species and points towards ecological equivalence. However, our results also uncovered specific life-history responses to different abiotic factors. Life-history traits of GD1 were more stable in the temperature interval between 10°C and 16°C than those of GD2. Interestingly, GD3 did not survive in the presence of sulphides, whereas GD1 and GD2 not only survived but also reproduced and developed normal

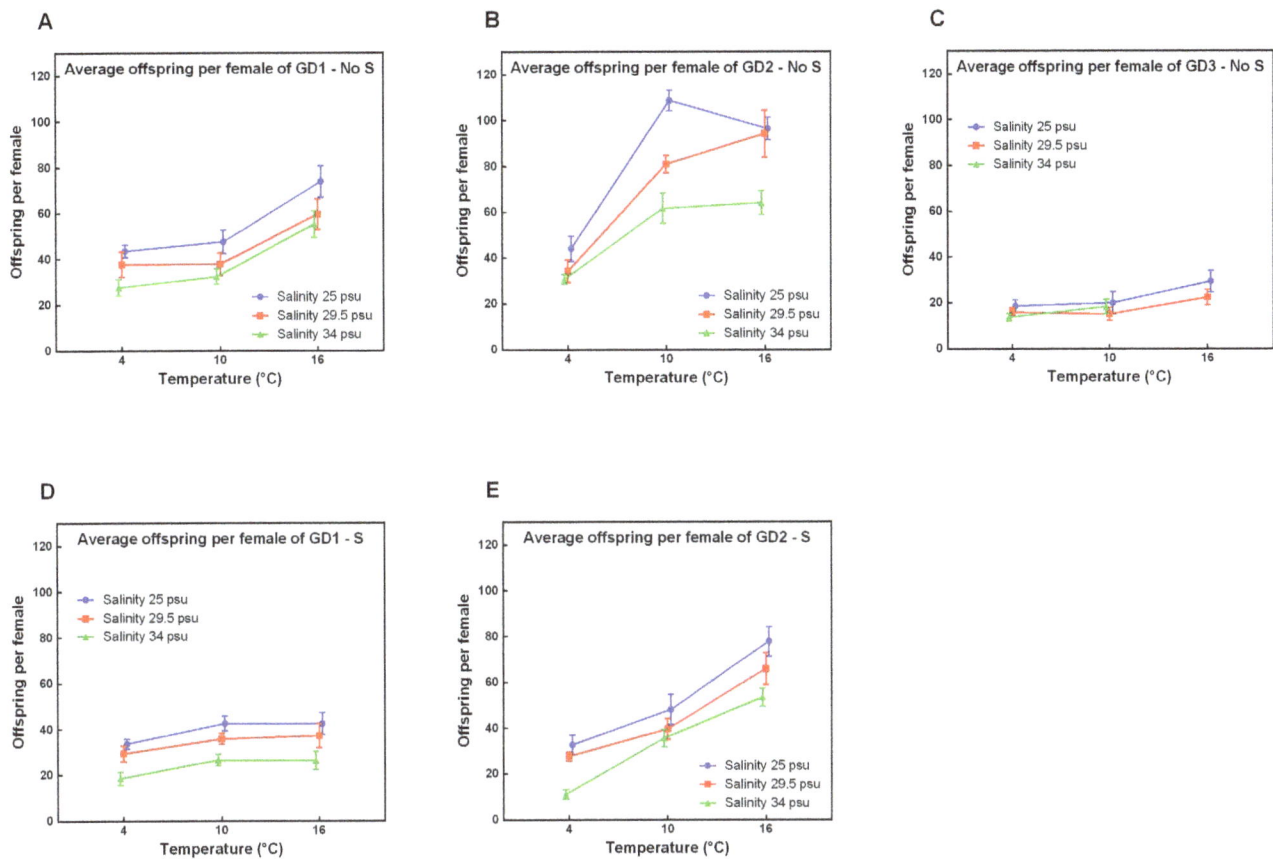

Figure 5. Offspring per female of GD1-3 under changing environmental conditions. Offspring per female in relation to temperature and salinity of cryptic *Halomonhystera disjuncta* species: GD1 (A, D), GD2 (B, E) and GD3 (C), in the absence (No S; A–C) and presence (S; D–E) of sulphide. The offspring corresponds to the amount of eggs, juveniles and new adults at the end of the experiment, averaged across six females. Data shown are mean values ±1stdev of 3 or 4 replicates per treatment. No results were obtained for GD3 in the presence of sulphide because inoculated nematodes died within the first 1–2 minutes after adding sulphide.

offspring for at least one generation. These results reveal that cryptic species are not completely ecologically equivalent. Our results might be partly biased because a single gravid female was used to generate the monospecific cultures. Nevertheless, the observed differential effects of abiotic factors on life-history traits, may provide important stabilizing mechanisms facilitating the coexistence of cryptic species [9].

In addition to its sensitivity to sulphides, the most obvious difference between GD3 and the other two cryptic species was its ovoviviparous reproduction strategy. Intrauterine egg hatching (ovoviviparity) allows a more secure survival and early development of offspring and can be an adaptation strategy in toxic environmental conditions [61]. The low number of GD3 offspring compared to GD1 and GD2 suggests a trade-off between the reproductive strategy and fecundity. This is further supported by the observation that the fecundity of GD3 was less affected by changes in salinity and temperature than that of GD1 and GD2. In addition, the minimum adult life span was always the highest for GD3 at all temperatures. The comparatively smaller impact of temperature on GD3 could also in part explain its occurrence throughout the year and its high relative abundance in winter in the Westerschelde estuary [23]. In contrast to GD3, GD2 was often absent in winter but had the highest relative abundance in spring [23]. Our experimental data revealed that GD2 had the fastest minimum generation time at 16°C and the highest number of offspring at 10°C and 16°C. In spite of this, its number of

offspring was more sensitive to temperature changes. These observations may explain the high relative abundance of GD2 in spring and its disappearance in winter.

Differential responses of nematodes to abiotic factors can also affect interspecific interactions [27,62] and dispersal capacities [26], which could in turn facilitate coexistence. Recently, significant differences in resource use among sympatric cryptic nematode species have been demonstrated, and other potentially discriminating biotic/abiotic factors, such as differential susceptibility to predation and competitive interaction with other taxa, should be taken into account to fully grasp the coexistence and spatial/temporal distribution patterns of these cryptic species because the small morphological differences between GD1-3 [34] may not be sufficient to predict ecological segregation [63].

GD1 and the Nordic seep nematode *Halomonhystera hermesi* share a common ancestor

It is often suggested that current deep-sea biodiversity has largely resulted from recurrent invasions from bathyal and abyssal depths followed by speciation [64]. The existence of morphologically and/or genetically close relatives from intertidal and deep-sea environments [65] tends to support this contention. In an evolutionary context, understanding the adaptations which have allowed deep-sea colonization is an important prerequisite. GD1 is phylogenetically most closely related to the Nordic seep nematode

Halomonhystera hermesi [44] and, therefore, shares a most recent common ancestor with *H. hermesi*.

In the reduced environment of the Håkon Mosby mud volcano (HMMV), organisms encounter stress levels which selectively favor sulphide tolerant species [66]. In view of the 100% mortality of GD3 in the sulphide treatments, it is very unlikely that this species successfully colonized the HMMV. It must, however, be mentioned that GD3 did survive under lower concentrations of sodium thiosulphate and sodium sulphide (up to 0.25 µM) in a preliminary test. Such a mild exposure could increase resistance to subsequent higher doses (hormesis) of the same or different stressors [67]. Nevertheless, GD3 was unable to cope with high concentrations of sulphides. GD1 and GD2, by contrast, survive and produce offspring at the sulphide concentrations tested here. Such a high sulphide tolerance may be an advantage for colonizing reduced environments [66]. Species living in sulphidic environments often have ecto- or endosymbionts which could either serve as a food source and/or play a role in sulphur detoxification [68,69]. In addition, sulphur inclusions could temporarily reduce the toxic effect of H_2S [70]. However, no signs of bacterial symbionts in *H. hermesi* were detected [45], and we did not observe sulphur inclusions in GD1-2. Interestingly, however, low temperature had a positive effect on the survival of GD1 and GD2 (but not GD3) under sulphide exposure. Similarly, low temperatures reduced the mortality rate of *H. disjuncta* subjected to toxic chromium concentrations [40]. Hence, the low temperatures in the deep sea may have been an important factor enabling *Halomonhystera* to cope with otherwise toxic sulphide concentrations.

Our results further reveal that GD1 had a higher number of offspring at bathyal cold seep conditions (salinity of 34 psu – temperature of 4°C – presence of sulphides) than GD2. In addition, GD1 had the fastest minimum generation time and a longer minimum life span than GD2 at these conditions. This implies that GD1 is more resistant to a combination of sulphides, low temperature and higher salinities, and is in agreement with the close phylogenetic relationship between GD1 and *H. hermesi*, supporting the idea that it could have successfully colonized Nordic seeps.

Conclusion

We have shown that life-histories of cryptic were differentially affected by changes in temperature, salinity and presence of sulphides. These observed differences imply that closely related cryptic *H. disjuncta* species are not necessarily ecologically equivalent and that abiotic factors can strongly impact their life-history traits, which in turn can affect their coexistence. Our results further indicate that *Halomonhystera hermesi* and GD1 share a common ancestor as the latter appears to be more resistant to bathyal seep conditions compared to other cryptic species. The observed limited ecological equivalence among cryptic species may have important repercussions for our understanding of the link between biodiversity and ecosystem functioning [71].

Supporting Information

Table S1 P-values of pairwise comparisons between species with the same treatment. PERMANOVA (Euclidean distances, 9999 permutations) was used for the pairwise comparisons. The table part between double lines contains the variable factor species (GD1-3) on which we performed pairwise comparisons for the variables in column 2. Row 1–2 and column 1 contain fixed factor levels for the respective column and row. Significant p-values are indicated with an asterisk.

Table S2 P-values of pairwise comparisons between treatments which differ in presence or absence of sulphide. PERMANOVA (Euclidean distances, 9999 permutations) was used for the pairwise comparisons. The table part between double lines contains the variable factor sulphide (sulphide present = S, sulphide absent = No S) on which we performed pairwise comparisons for the variables in column 2. Significant p-values are indicated with an asterisk.

Table S3 P-values of pairwise comparisons between treatments which differ in one level of the factor temperature. PERMANOVA (Euclidean distances, 9999 permutations) was used for the pairwise comparisons. The table part between double lines contains the variable factor temperature (16, 10 and 4°C) on which we performed pairwise comparisons for the variables in column 2. Significant p-values are indicated with an asterisk.

Table S4 P-values of pairwise comparisons between treatments which differ in one level of the factor salinity. PERMANOVA (Euclidean distances, 9999 permutations) was used for the pairwise comparisons. The table part between double lines contains the variable factor salinity (25, 29.5 and 34 psu) on which we performed pairwise comparisons for the variables in column 2. Significant p-values are indicated with an asterisk.

Acknowledgments

The data analysis of this research has benefitted from a statistical consult with Ghent University FIRE (Fostering Innovative Research based on Evidence). The authors would like to thank Marti Anderson (Department of Statistics, University of Auckland, New Zealand) and Jan Vanaverbeke (Research Group Marine Biology, Biology Department, Ghent University, Belgium) for statistical guidance in the use of PERMANOVA. S.D. acknowledges a postdoctoral fellowship from the Flemish Fund for Scientific Research (F.W.O).

Author Contributions

Conceived and designed the experiments: JVC SD TM AV. Performed the experiments: JVC. Analyzed the data: JVC SD TM AV. Contributed reagents/materials/analysis tools: JVC SD TM AV. Contributed to the writing of the manuscript: JVC SD TM AV.

References

1. Bickford D, Lohman DJ, Sodhi NS, Ng PKL, Meier R, et al. (2007) Cryptic species as a window on diversity and conservation. Trends Ecol Evol 22: 148–155.
2. Pfenninger M, Schwenk K (2007) Cryptic animal species are homogeneously distributed among taxa and biogeographical regions. Bmc Evol Biol 7: 121.
3. Palumbi SR (1994) Genetic-Divergence, Reproductive Isolation, and Marine Speciation. Annu Rev Ecol Syst 25: 547–572.
4. Lonsdale DJ, Frey MA, Snell TW (1998) The role of chemical signals is copepod reproduction. J Marine Syst 15: 1–12.

5. Stanhope MJ, Connelly MM, Hartwick B (1992) Evolution of a Crustacean Chemical Communication Channel - Behavioral and Ecological Genetic-Evidence for a Habitat-Modified, Race-Specific Pheromone. J Chem Ecol 18: 1871–1887.

6. Knowlton N (2000) Molecular genetic analyses of species boundaries in the sea. Hydrobiologia 420: 73–90.

7. Derycke S, Remerie T, Backeljau T, Vierstraete A, Vanfleteren J, et al. (2008) Phylogeography of the Rhabditis (Pellioditis) marina species complex: evidence for long-distance dispersal, and for range expansions and restricted gene flow in the northeast Atlantic. Mol Ecol 17: 3306–3322.

8. Gomez A, Serra M, Carvalho GR, Lunt DH (2002) Speciation in ancient cryptic species complexes: Evidence from the molecular phylogeny of Brachionus plicatilis (Rotifera). Evolution 56: 1431–1444.

9. Chesson P (2000) Mechanisms of maintenance of species diversity. Annu Rev Ecol Syst 31: 343–366.

10. Hubbell SP (2006) Neutral theory and the evolution of ecological equivalence. Ecology 87: 1387–1398.

11. Adler PB, HilleRisLambers J, Levine JM (2007) A niche for neutrality. Ecol Lett 10: 95–104.

12. Zhang DY, Hanski I (1998) Sexual reproduction and stable coexistence of identical competitors. Journal of Theoretical Biology 193: 465–473.

13. Montero-Pau J, Serra M (2011) Life-Cycle Switching and Coexistence of Species with No Niche Differentiation. PLoS ONE 6: 1399–1410.

14. Leibold MA, McPeek MA (2006) Coexistence of the niche and neutral perspectives in community ecology. Ecology 87: 1399–1410.

15. Gabaldon C, Montero-Pau J, Serra M, Carmona MJ (2013) Morphological Similarity and Ecological Overlap in Two Rotifer Species. PLoS ONE 8: e57087.

16. Ortells R, Gomez A, Serra M (2003) Coexistence of cryptic rotifer species: ecological and genetic characterisation of Brachionus plicatilis. Freshwater Biol 48: 2194–2202.

17. Wellborn GA, Cothran RD (2007) Niche diversity in crustacean cryptic species: complementarity in spatial distribution and predation risk. Oecologia 154: 175–183.

18. Cothran RD, Henderson KA, Schmidenberg D, Relyea RA (2013) Phenotypically similar but ecologically distinct: differences in competitive ability and predation risk among amphipods. Oikos 122: 1429–1440.

19. Montero-Pau J, Ramos-Rodriguez E, Serra M, Gomez A (2011) Long-Term Coexistence of Rotifer Cryptic Species. PLoS ONE 6: e21530.

20. Cunning R, Glynn PW, Baker AC (2013) Flexible associations between Pocillopora corals and Symbiodinium limit utility of symbiosis ecology in defining species. Coral Reefs 32: 795–801.

21. Heip C, Vincx M, Vranken G (1985) The Ecology of Marine Nematodes. Oceanogr Mar Biol 23: 399–489.

22. Lambshead PJD, Boucher G (2003) Marine nematode deep-sea biodiversity - hyperdiverse or hype? J Biogeogr 30: 475–485.

23. Derycke S, Backeljau T, Vlaeminck C, Vierstraete A, Vanfleteren J, et al. (2007) Spatiotemporal analysis of population genetic structure in Geomonhystera disjuncta (Nematoda, Monhysteridae) reveals high levels of molecular diversity. Marine Biology 151: 1799–1812.

24. Derycke S, Remerie T, Vierstraete A, Backeljau T, Vanfleteren J, et al. (2005) Mitochondrial DNA variation and cryptic speciation within the free-living marine nematode Pellioditis marina. Mar Ecol-Prog Ser 300: 91–103.

25. De Oliveira DAS, Decraemer W, Holovachov O, Burr J, De Ley IT, et al. (2012) An integrative approach to characterize cryptic species in the Thoracostoma trachygaster Hope, 1967 complex (Nematoda: Leptosomatidae). Zool J Linn Soc-Lond 164: 18–35.

26. De Meester N, Derycke S, Moens T (2012) Differences in time until dispersal between cryptic species of a marine nematode species complex. Plos ONE 7: e42674.

27. De Meester N, Derycke S, Bonte D, Moens T (2011) Salinity effects on the coexistence of cryptic species: a case study on marine nematodes. Marine Biology 158: 2717–2726.

28. Bik HM, Porazinska DL, Creer S, Caporaso JG, Knight R, et al. (2012) Sequencing our way towards understanding global eukaryotic biodiversity. Trends Ecol Evol 27: 233–243.

29. Mokievsky VO, Filippova KA, Chesunov AV (2005) Nematode fauna associated with detached kelp accumulations in the subtidal zone of the White Sea. Oceanology+45: 689–697.

30. Vranken G, Herman PMJ, Heip C (1988) Studies of the Life-History and Energetics of Marine and Brackish-Water Nematodes: 1. Demography of Monhystera disjuncta at Different Temperature and Feeding Conditions. Oecologia 77: 296–301.

31. Trotter D, Webster JM (1983) Distribution and Abundance of Marine Nematodes on the Kelp Macrocystis integrifolia. Marine Biology 78: 39–43.

32. Portnova D, Haflidason H, Todt C (2010) Nematode species distribution patterns at the Nyegga pockmarks. 14th International Meiofauna Conference (FourtMCo) Ghent, Belgium, pp 174.

33. Van Gaever S, Moodley L, de Beer D, Vanreusel A (2006) Meiobenthos at the Arctic Hakon Mosby Mud Volcano, with a parental-caring nematode thriving in sulphide-rich sediments. Mar Ecol-Prog Ser 321: 143–155.

34. Fonseca G, Derycke S, Moens T (2008) Integrative taxonomy in two free-living nematode species complexes. Biol J Linn Soc 94: 737–753.

35. Derycke S, Van Vynckt R, Vanoverbeke J, Vincx M, Moens T (2007) Colonization patterns of Nematoda on decomposing algae in the estuarine environment: Community assembly and genetic structure of the dominant species Pellioditis marina. Limnol Oceanogr 52: 992–1001.

36. Moens T (1999) Feeding ecology of free-living estuarine nematodes: an experimental approach. Gent: Universiteit Gent. 302 p.

37. Gerlach SA, Schrage M (1971) Life Cycles in Marine Meiobenthos - Experiments at Various Temperatures with Monhystera disjuncta and Theristus pertenuis (Nematoda). Marine Biology 9: 274–280.

38. Vranken G (1987) An autecological study of free-living marine nematodes. Academia Analecta 49: 71–97.

39. Vranken G, Tire C, Heip C (1988) The Toxicity of Paired Metal Mixtures to the Nematode Monhystera disjuncta (Bastian, 1865). Mar Environ Res 26: 161–179.

40. Vranken G, Tire C, Heip C (1989) Effect of Temperature and Food on Hexavalent Chromium Toxicity to the Marine Nematode Monhystera disjuncta. Mar Environ Res 27: 127–136.

41. Vranken G, Vanderhaeghen R, Van Brussel D, Heip C, Hermans D (1984) The toxicity of mercury on the free-living marine nematode Monhystera disjuncta Bastian, 1865. Coomans, A and Heip, CHR (Ed) (1984), Coordinated Research Actions Oceanography: progress report 1983 Ecological and ecotoxicological studies of benthos of the Southern Bight of the North Sea. Brussel: Programmatie van het Wetenschapsbeleid. pp. 186–201.

42. Vranken G, Vanderhaeghen R, Heip C (1985) Toxicity of Cadmium to Free-Living Marine and Brackish Water Nematodes (Monhystera microphthalma, Monhystera disjuncta, Pellioditis marina). Dis Aquat Organ 1: 49–58.

43. Tchesunov AV, Portnova DA, Van Campenhout J Description of two free-living nematode species of Halomonhystera disjuncta complex (Nematoda: Monhysterida) from two peculiar habitats in the sea. Helgoland Mar Res. DOI :10.1007/s10152-014-0416-1.

44. Van Campenhout J, Derycke S, Tchesunov A, Portnova D, Vanreusel A (2013) The Halomonhystera disjuncta population is homogeneous across the Håkon Mosby mud volcano (Barents Sea) but is genetically differentiated from its shallow-water relatives. Journal of Zoological Systematics and Evolutionary Research 52: 203–216.

45. Van Gaever S, Moodley L, Pasotti F, Houtekamer M, Middelburg JJ, et al. (2009) Trophic specialisation of metazoan meiofauna at the Håkon Mosby Mud Volcano: fatty acid biomarker isotope evidence. Marine Biology 156: 1289–1296.

46. Sauter EJ, Muyakshin SI, Charlou JL, Schluter M, Boetius A, et al. (2006) Methane discharge from a deep-sea submarine mud volcano into the upper water column by gas hydrate-coated methane bubbles. Earth Planet Sc Lett 243: 354–365.

47. Ackerly DD (2003) Community assembly, niche conservatism, and adaptive evolution in changing environments. Int J Plant Sci 164: S165–S184.

48. Wiens JJ (2004) Speciation and ecology revisited: Phylogenetic niche conservatism and the origin of species. Evolution 58: 193–197.

49. Harvey PH, Pagel MD (1991) The comparative method in evolutionary biology. Oxford UK: Oxford University Press. 248p.

50. Moens T, Vincx M (1998) On the cultivation of free-living marine and estuarine nematodes. Helgolander Meeresun 52: 115–139.

51. Anderson M, Gorley RN, Clarke KR (2008) PERMANOVA+ for PRIMER: Guide to Software and Statistical Methods: PRIMER-E. Plymouth.

52. Schmidt AL, Wysmyk JKC, Craig SE, Lotze HK (2012) Regional-scale effects of eutrophication on ecosystem structure and services of seagrass beds. Limnol Oceanogr 57: 1389–1402.

53. Anderson M (2004) PERMDISP: a FORTRAN computer program for permutation analysis of multivariate dispersions (for any two-factor ANOVA design) using permutation tests. Department of Statistics, University of Auckland, New Zealand.

54. Tahseen Q (2012) Nematodes in aquatic environments: adaptations and survival strategies. Biodiversity Journal 3: 13–40.

55. Levins R (1968) Evolution in Changing Environments: Some Theoretical Explorations: Princeton University Press. 120p.

56. Bottcher ME, Hespenheide B, Llobet-Brossa E, Beardsley C, Larsen O, et al. (2000) The biogeochemistry, stable isotope geochemistry, and microbial community structure of a temperate intertidal mudflat: an integrated study. Cont Shelf Res 20: 1749–1769.

57. Ferris H, Lau S, Venette R (1995) Population Energetics of Bacterial-Feeding Nematodes - Respiration and Metabolic Rates Based on Co2 Production. Soil Biol Biochem 27: 319–330.

58. Levinton JS (1970) The paleoecological significance of apportunistic species. Lethaia 3: 69–78.

59. Moens T, Vincx M (2000) Temperature and salinity constraints on the life cycle of two brackish-water nematode species. J Exp Mar Biol Ecol 243: 115–135.

60. Moens T, Vincx M (2000) Temperature, salinity and food thresholds in two brackish-water bacterivorous nematode species: assessing niches from food absorption and respiration experiments. J Exp Mar Biol Ecol 243: 137–154.

61. Boffe A (1985) Vergelijkend toxicologisch onderzoek naar sublethale effecten bij de mariene nematode Monhystera disjuncta (Bastian 1865). University Ghent: pp. 70.

62. Lowe CD, Kemp SJ, Diaz-Avalos C, Montagnes DJS (2007) How does salinity tolerance influence the distributions of Brachionus plicatilis sibling species? Marine Biology 150: 377–386.

63. Nicholls B, Racey PA (2006) Habitat selection as a mechanism of resource partitioning in two cryptic bat species *Pipistrellus pipistrellus* and *Pipistrellus pygmaeus*. Ecography 29: 697–708.

64. Smith KE, Thatje S (2012) The Secret to Successful Deep-Sea Invasion: Does Low Temperature Hold the Key? PLoS ONE 7: e51219.

65. Bik HM, Thomas WK, Lunt DH, Lambshead PJD (2010) Low endemism, continued deep-shallow interchanges, and evidence for cosmopolitan distributions in free-living marine nematodes (order Enoplida). Bmc Evol Biol 10: 389.

66. Bernardino AF, Levin LA, Thurber AR, Smith CR (2012) Comparative Composition, Diversity and Trophic Ecology of Sediment Macrofauna at Vents, Seeps and Organic Falls. PLoS ONE 7: e33515.

67. Zhao YL, Wang DY (2012) Formation and regulation of adaptive response in nematode *Caenorhabditis elegans*. Oxid Med Cell Longev 2012: 564093.

68. Ott J, Bright M, Bulgheresi S (2004) Symbioses between marine nematodes and sulfur-oxidizing chemoautotrophic bacteria. Symbiosis 36: 103–126.

69. Levin LA (2005) Ecology of cold seep sediments: Interactions of fauna with flow, chemistry and microbes. Oceanography and Marine Biology - an Annual Review, Vol 43 43: 1–46.

70. Thiermann F, Vismann B, Giere O (2000) Sulphide tolerance of the marine nematode *Oncholaimus campylocercoides* - a result of internal sulphur formation. Mar Ecol Prog Ser 193: 251–259.

71. Westram AM, Jokela J, Keller I (2013) Hidden Biodiversity in an Ecologically Important Freshwater Amphipod: Differences in Genetic Structure between Two Cryptic Species. PLoS ONE 8: e69576.

21

High Connectivity of Animal Populations in Deep-Sea Hydrothermal Vent Fields in the Central Indian Ridge Relevant to Its Geological Setting

Girish Beedessee[1¤]**, Hiromi Watanabe**[2]*****, Tomomi Ogura**[2,3]**, Suguru Nemoto**[4]**, Takuya Yahagi**[5]**,
Satoshi Nakagawa**[6]**, Kentaro Nakamura**[7]**, Ken Takai**[2,7]**, Meera Koonjul**[8]**, Daniel E. P. Marie**[1]

1 Mauritius Oceanography Institute, Quatre-Bornes, Mauritius, 2 Institute of Biogeosciences, Japan Agency for Marine-Earth Science and Technology, Yokosuka, Kanagawa, Japan, 3 Graduate School of Marine Science and Technoloy, Tokyo University of Marine Science and Technology, Minato, Tokyo, Japan, 4 Enoshima Aquarium, Fujisawa, Kanagawa, Japan, 5 Atmosphere and Ocean Research Institute, the University of Tokyo, Kashiwa, Chiba, Japan, 6 Faculty of Fisheries Sciences, Hokkaido University, Hakodate, Hokkaido, Japan, 7 Precambrian Ecosystem Laboratory, Japan Agency for Marine-Earth Science and Technology, Yokosuka, Kanagawa, Japan, 8 Albion Fisheries Research Centre, Ministry of Fisheries, Petite Rivière, Mauritius

Abstract

Dispersal ability plays a key role in the maintenance of species in spatially and temporally discrete niches of deep-sea hydrothermal vent environments. On the basis of population genetic analyses in the eastern Pacific vent fields, dispersal of animals in the mid-oceanic ridge systems generally appears to be constrained by geographical barriers such as trenches, transform faults, and microplates. Four hydrothermal vent fields (the Kairei and Edmond fields near the Rodriguez Triple Junction, and the Dodo and Solitaire fields in the Central Indian Ridge) have been discovered in the mid-oceanic ridge system of the Indian Ocean. In the present study, we monitored the dispersal of four representative animals, *Austinograea rodriguezensis*, *Rimicaris kairei*, *Alviniconcha* and the scaly-foot gastropods, among these vent fields by using indirect methods, i.e., phylogenetic and population genetic analyses. For all four investigated species, we estimated potentially high connectivity, i.e., no genetic difference among the populations present in vent fields located several thousands of kilometers apart; however, the direction of migration appeared to differ among the species, probably because of different dispersal strategies. Comparison of the intermediate-spreading Central Indian Ridge with the fast-spreading East Pacific Rise and slow-spreading Mid-Atlantic Ridge revealed the presence of relatively high connectivity in the intermediate- and slow-spreading ridge systems. We propose that geological background, such as spreading rate which determines distance among vent fields, is related to the larval dispersal and population establishment of vent-endemic animal species, and may play an important role in controlling connectivity among populations within a biogeographical province.

Editor: John F. Valentine, Dauphin Island Sea Lab, United States of America

Funding: Part of this study was supported by Kakenhi (TAIGA project). GB wishes to acknowledge the support of the Western Indian Ocean Marine Science Association (WIOMSA) for a MARG grant and the InterRidge/ISA Endowment Fund. The funders had no role in study design, data collection and analysis, decision to publish, or preparation of the manuscript.

Competing Interests: One of the authors is an employee of a commercial company (Enoshima Aquarium). All the other authors are graduate students, researchers or faculty with financial support from public organizations, and the authors declare that no other competing interests exist.

* E-mail: hwatanabe@jamstec.go.jp

¤ Current address: Department of Chemistry and Biomolecular Sciences, Macquarie University, Sydney, New South Wales, Australia

Introduction

More than 30 years have passed since the discovery of the first hydrothermal vents on the Galapagos Rift in the eastern Pacific Ocean [1]. Numerous vents continue to be found and catalogued along global mid-ocean ridge systems, back-arc spreading centers, and off-axis submarine volcanoes [2]. The vents are usually associated with dense assemblages of organisms, which are patchily distributed on the deep-sea floor [3,4]. These communities are typically separated by tens to hundreds of kilometers along an actively spreading ridge, and by even greater distances between ridge segments. The existence of such communities highlights the significant contribution to geologically produced energy sources to chemosynthetic biomass production, and also the remarkable adaptability of life in deep-sea hydrothermal vent ecosystems [5].

Considering the vast geographical distribution of the hydrothermal vent sites [6], the animals inhabiting them are likely to possess exceptional colonization abilities, including high rates of dispersal, growth, and reproduction [7,8,9]; further, dispersal probably occurs mainly via the planktonic larval or juvenile stages [10]. Dispersal and subsequent reproduction results in a decrease in genetic differences among populations and an increase in the genetic variability within populations [11]. Negatively buoyant larvae mainly move with bottom currents in line with the axial valley of the ridge system [12,13,14,15], while positively buoyant larvae can disperse more than several tens of meters above the seafloor in hydrothermal plumes [16]. Colonization is not a simple process, because dispersal is affected by several factors, particularly plate tectonics, which have contributed to the current biogeographical patterns of faunal diversity in vents around the globe

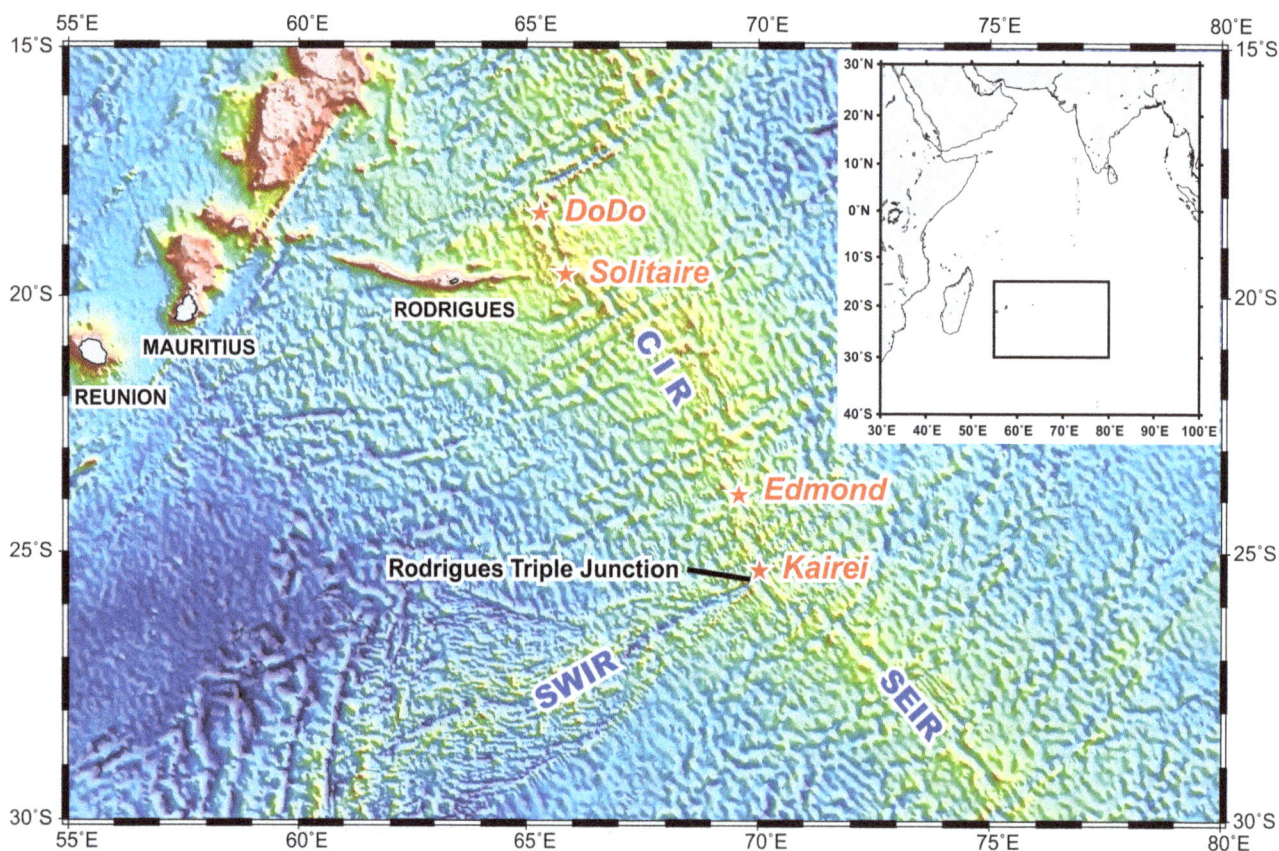

Figure 1. Location of the collection sites.

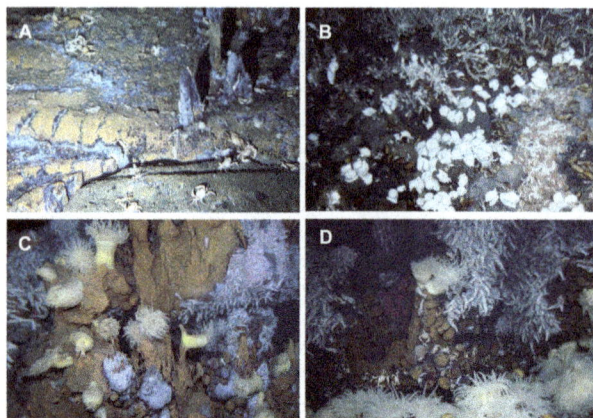

Figure 2. Photographs of the collection sites. A, Dodo site; B, Solitaire site; C, Edmond site; and D, Kairei site.

[17,18]. Dispersers may never reach a suitable habitat or may suffer from high mortality during transition. On the other hand, dispersal may allow them to explore new favorable habitat [19].

The dispersal, isolation, and speciation of vent endemic species constitute one of the key questions of deep-sea biology, particularly in deep-sea hydrothermal vents [9]. Several field-based studies have been conducted in different geographical provinces [17,18,20]. In the present study, we aimed to address these phenomena, with respect to hydrothermal systems, the Kairei and Edmond vent fields [21,22], and two recently discovered vent fields, namely, the Dodo and Solitaire fields [23], located in segments 16 and 15, respectively, of the Central Indian Ridge (CIR) (Figures 1 and 2). Biogeographical features appear to be influenced by the timing and spacing of hydrothermal activities in a spreading axis [5]. The factors generally considered when elucidating the relationships of biogeography are spreading rate and spacing. Previous studies have indicated the presence of one active vent site per 100–350 km in the Mid-Atlantic Ridge (MAR) and one active vent site per 5 km in the East Pacific Rise (EPR) [5]. However, few investigations of the CIR have been conducted, and the spacing of vent fields in this axis remains unclear. The CIR has an intermediate spreading rate (50–60 mm/year), while the EPR and MAR have fast (180 mm/year) and slow (25 mm/year) spreading rates, respectively [5,20]. These differences in spreading rates are expected to influence the biogeographical pattern, e.g., the distance between vent fields represents a greater biogeographical barrier in a slow-spreading axis than in a fast-spreading axis [5]. However, the results of recent analyses on gene flow contradict this hypothesis. On the basis of genetic analyses of deep-sea vent species in the EPR, some dispersal barriers such as faults, fracture zones, and intercalation of microplate and topographic depressions have been proposed [9]. Further, in the hydrothermal vent fields of the MAR, no genetic difference has been detected among populations of *Rimicaris exoculata* [24] or other animals [25]. Elucidation of the biogeographical patterns of the Indian Ocean hydrothermal vents will provide an insight into the relationships between geological background such as spreading rate, and larval dispersal or connectivity of vent animals.

Table 1. Number of specimens used in the present study.

Site	No. of specimens used			
	A. rodriguezensis	*R. kairei*	*Alviniconcha* sp.	scaly-foot gastropod [23]
Dodo	10 (ND)	6 (ND)	-	-
Solitaire	20 (18)	40 (18)	17 (12)	ND (14)
Edmond	ND (ND)	373 (18)	ND (19)	-
Kairei	20 (18)	383 (18)	15 (13)	ND (19)

ND: no data in the present study, -: not observed. Numbers in parentheses are those used for population genetic analyses.

In the present study, we investigated genetic diversity and population differentiation of four dominant species in the four hydrothermal vent fields along the CIR to discuss larval dispersal ability indirectly. In addition, we measured the size of the animals constituting each population, because this reflects many aspects of the life-history trait, and also variations among categories such as

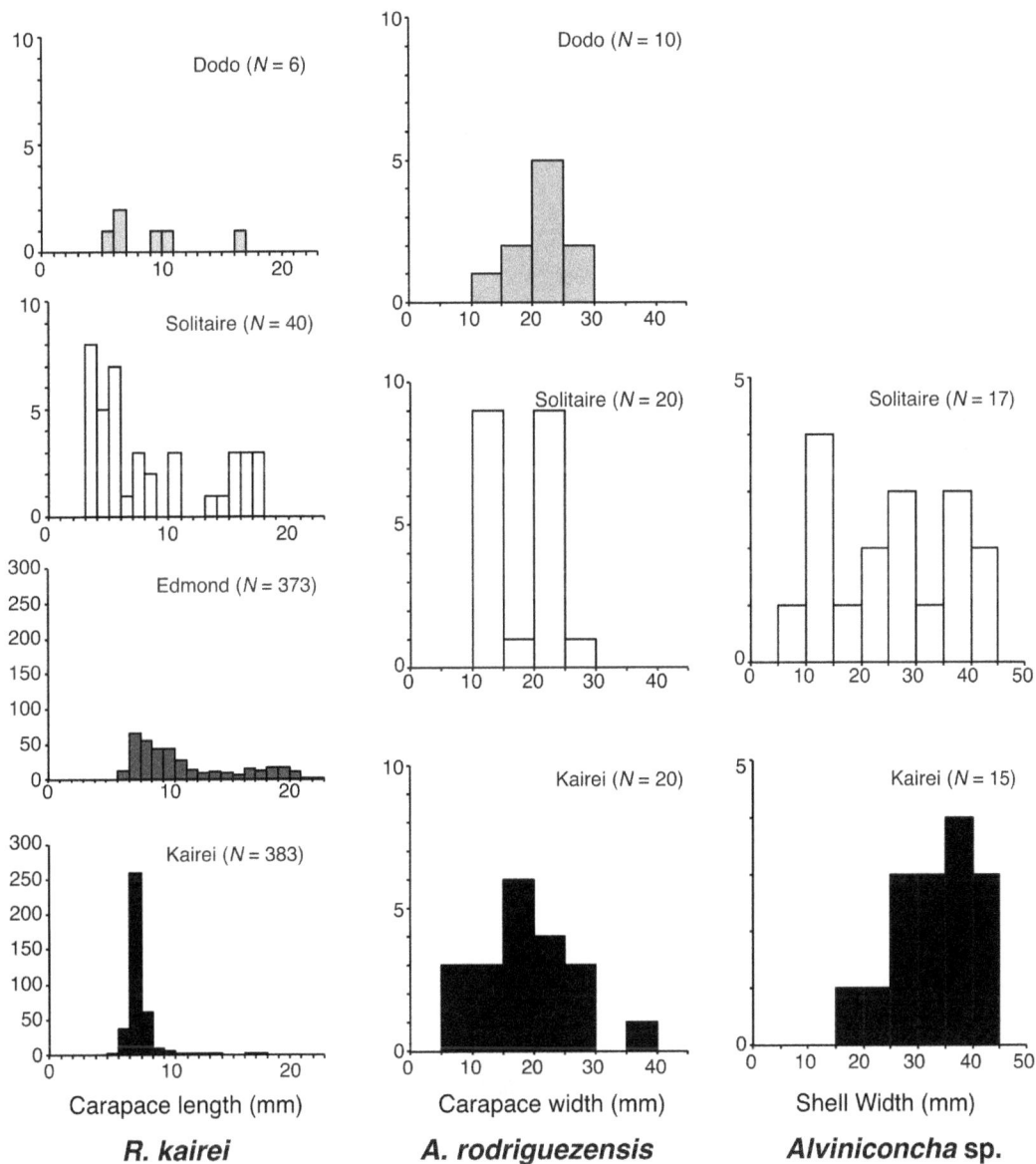

Figure 3. Size histograms of the four representative species in the CIR. A, *Austinograea rodriguezensis*; B, *Rimicaris kairei*; and C, *Alviniconcha* sp. type 3.

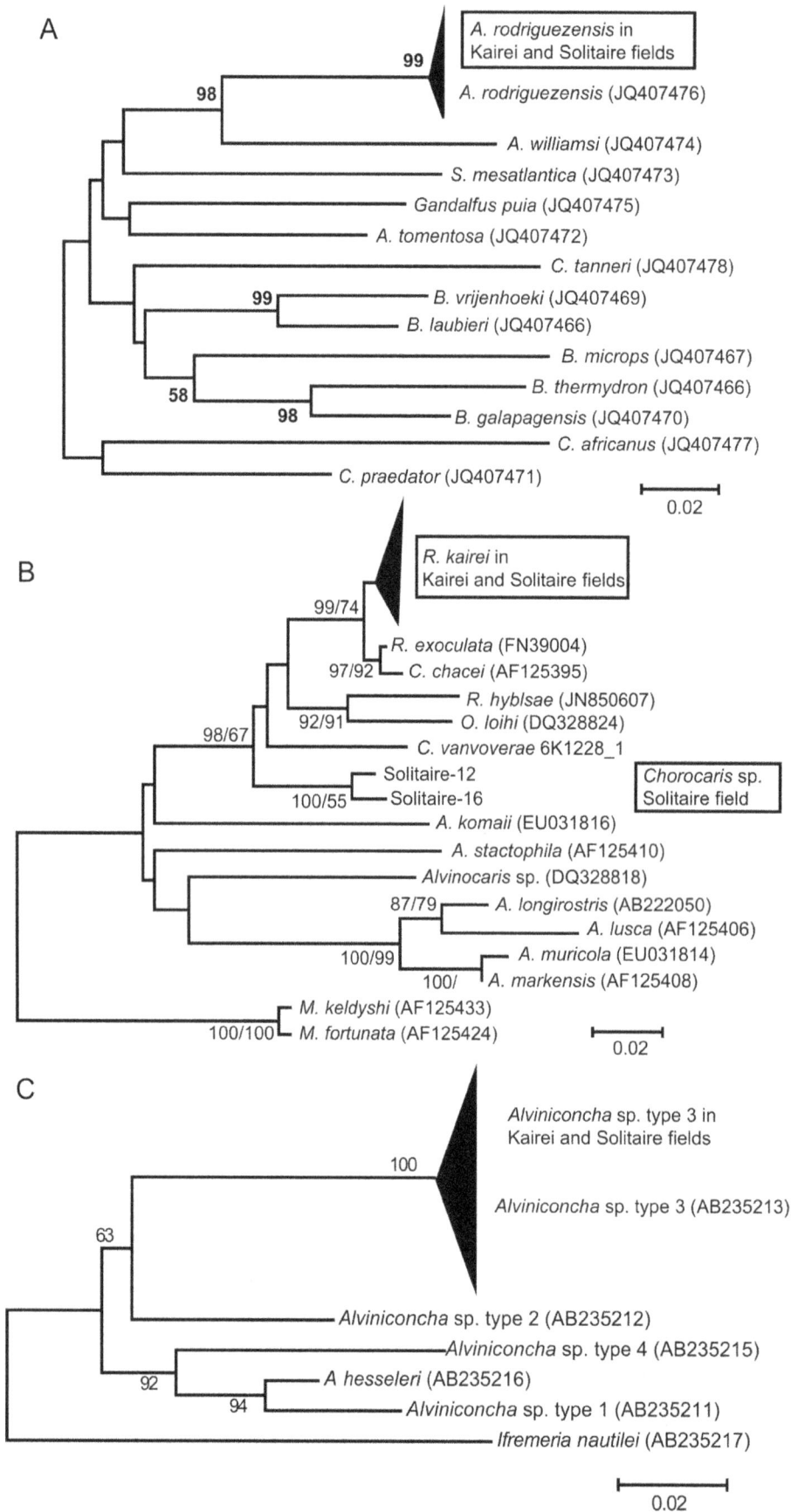

Figure 4. Molecular phylogenetic trees of the four representative species in the CIR. A, *A. rodriguezensis*; B, *R. kairei*; and C, *Alviniconcha* sp. type 3.

A. *A. rodriguezensis*

B. *R. kairei*

C. *Alviniconcha* sp. type 3

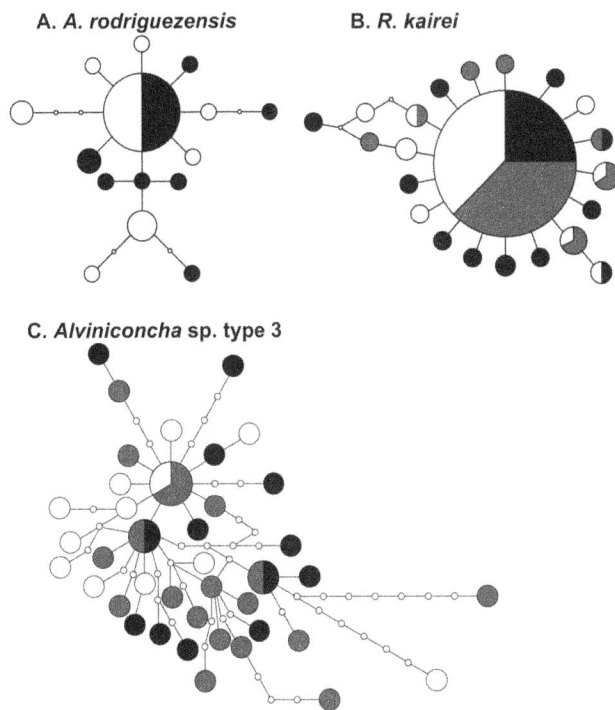

Figure 5. Haplotype networks of the four representative species in the CIR. A, *A. rodriguezensis*; B, *R. kairei*; and C: *Alviniconcha* sp. type 3. Black, population in the Kairei field; gray, population in the Edmond site; and white, population in the Solitaire field.

taxa and population [26]. In order to elucidate the potential effects of geological setting and geographical distance on dispersal, connectivity of the four vent species were compared to those observed in EPR and MAR, and discuss potential dispersal barriers encountered by animal populations associated with deep-sea hydrothermal vents.

Materials and Methods

Sample collection

The samples were collected in the four deep-sea hydrothermal vent fields in the CIR (Dodo, Solitaire, Edmond and Kairei fields). Among them, the Dodo and Solitaire fields are located in the Exclusive Economic Zone of Mauritius, and we conducted these researches with the permission of the Mauritius Prime Minister's Office, during the YK09-13 cruise (October to November 2009), by using the Human Observing Vehicle (HOV) *Shinkai6500* and R/V *Yokosuka* of the Japan Agency for Marine-Earth Science and Technology (JAMSTEC). We collected four representative animal species of the CIR vent fields, *Austinograea rodriguezensis* Tsuchida &

Hashimoto 2002, *Rimicaris kairei* Watabe & Hashimoto 2002, *Alviniconcha* sp. type 3 (refer to [27]), and the scaly-foot gastropod from the hydrothermal vent fields shown in Figure 1. The geological background, geochemical characteristics, and species composition of the associated fauna were reported previously [27]. Immediately after recovery of HOV Shinkai6500, the specimens were stored in either 10% seawater-buffered formalin or 99.5% ethanol, or at $-80°C$ (Table 1).

Size distribution

With the exception of *Alviniconcha* sp. type 3, all of the collected specimens were used in size distribution analysis (Accidentally, only some of the *Alviniconcha* sp. type 3 specimens were measured.). The sizes of all individuals of the representative species in the four hydrothermal vent sites were measured. We examined differences in the average sizes of each local population by using non-parametric tests (Mann-Whitney U-test or Kruskal-Wallis test, depending on the number of populations), based on the following measurements: carapace width (CW) for *A. rodriguezensis*; carapace length (CL) for *R. kairei*; and shell width (SW) for *Alviniconcha* sp. type 3. Measurements for the scaly-foot gastropod were reported previously [23].

DNA extraction, amplification, and sequencing

Genomic DNA was extracted from ethanol-fixed and frozen samples by using the DNeasy Tissue Extraction Kit (QIAGEN, Hilden, Germany). The extracted DNA of molluscs was treated with GeneReleaser (BioVenture, Marfreesboro, USA) before Polymerase Chain Reaction (PCR). The target DNA sequence of the present study was a mitochondrial gene, cytochrome oxidase *c* subunit 1 (COI). For *R. kairei*, a 535-bp fragment of COI was amplified by using the primers LCO1490 and HCO2198 [28]. For *A. rodriguezensis*, an 810-bp fragment was amplified by using the primers LCO1490 and COI-6 [29]. For, *Alviniconcha* sp. type 3, a 491-bp fragment was amplified by using the primers COI-B [30] and COI-6. The reactions were performed in 20-μL reaction mixtures containing genomic DNA, *ExTaq* buffer, 0.3 mM of dNTP mix, 1 mM of each primer, and 0.75 units of ExTaq DNA polymerase (TaKaRa Bio, Ohtsu, Japan). The cycling parameters were 94°C for 2 min, 30 cycles of 94°C for 30 s, 50°C for 30 s, 72°C for 1 min, and a final elongation step at 72°C for 2 min. The PCR products were purified by using ExoSAP-IT® (USB®, Affymax, Santa Clara, USA), and sequenced bidirectionally by using the same primers as those described above for PCR, and the BigDye® Terminator Cycle Sequencing Kit Version 3.1 (Applied Biosystems®, Life Technologies Corporation, Carlsbad, USA). The products were purified by using a Gel Filtration Cartridge (Edge BioSystems, Gaithersburg, USA) or BigDye XTerminator® Kit (Applied Biosystems®) treatment, before sequencing analyses by using ABI 3130 or ABI 3730 DNA sequencers (Applied Biosystems®).

Table 2. Haplotype diversities of the populations of the representative species of Indian Ocean hydrothermal vent fields.

	A. rodriguezensis	*R. kairei*	*Alviniconcha* sp.	scaly-foot gastropod
Solitaire	0.8718±0.0670	0.7516±0.1031	1.0000±0.0340	0.9170±0.0419
Edmond		0.7582±0.1056	0.9942±0.0193	
Kairei	0.9103±0.0683	0.8954±0.0653	1.0000±0.0302	0.8975±0.0443

Table 3. Nucleotide diversities of the populations of the representative species of Indian Ocean hydrothermal vent fields.

	A. rodriguezensis	R. kairei	Alviniconcha sp.	scaly-foot gastropod
Solitaire	0.003138±0.002026	0.002162±0.001636	0.004568±0.002700	0.005602±0.003341
Edmond		0.002211±0.001663	0.005140±0.002894	
Kairei	0.002817±0.001856	0.003091±0.002132	0.005889±0.003362	0.005236±0.003107

Molecular data analysis

The obtained forward and reverse sequences were assembled into contigs by using the program ATGC (Genetyx® Version 6, Genetyx, Tokyo, Japan), and aligned by eye. All of the detected haplotypes were used for phylogenetic analyses in neighbor-joining (NJ) and maximum likelihood (ML) algorithms. The phylogeny of each animal was clarified by using the software package MEGA 5 [31]. The molecular phylogenetic tree was reconstructed with previously published data in DNA databanks (NCBI, DDBJ, and EMBL; accession numbers are shown in the phylogenetic trees). Additionally, all of the obtained sequences were used for population-level analyses. The parsimonious haplotype network of each species was estimated by using the software package TCS 1.21 [32]. For each population, we estimated genetic diversity indices (H: gene diversity, π: nucleotide diversity, and mismatch distribution), and conducted statistical tests for genetic structure, by using Arlequin ver. 3.11 [33]. In mismatch distribution analysis, we applied the goodness-of-fit test to compare the observed mismatch distribution, with the mismatch distribution obtained by using the sudden expansion model. For those species with more than three populations (R. kairei and Alviniconcha sp. type 3), we used analysis of molecular variance (AMOVA) to examine the genetic structure between northern segment populations (Dodo and Solitaire fields), and populations near the triple junction (Edmond and Kairei fields). The relative number of migrants per generation was estimated by using MIGRATE ver. 3.2.1 [34]. A part of the population-level analyses included the previously reported sequences for the scaly-foot gastropod [23].

Results

Size distribution

The obtained size distributions of the four representative species are summarized in Figure 3. The results for the scaly-foot gastropod were reported previously [23].

A. rodriguezensis was observed in all four vent fields of the Indian Ocean. The Dodo population had the largest CW (25.19±5.00 mm; average ± standard deviation), while the Solitaire population had the smallest CW (20.99±4.79 mm). The Kairei population showed an intermediate CW (22.90±7.61 mm), but only a single peak in size distribution

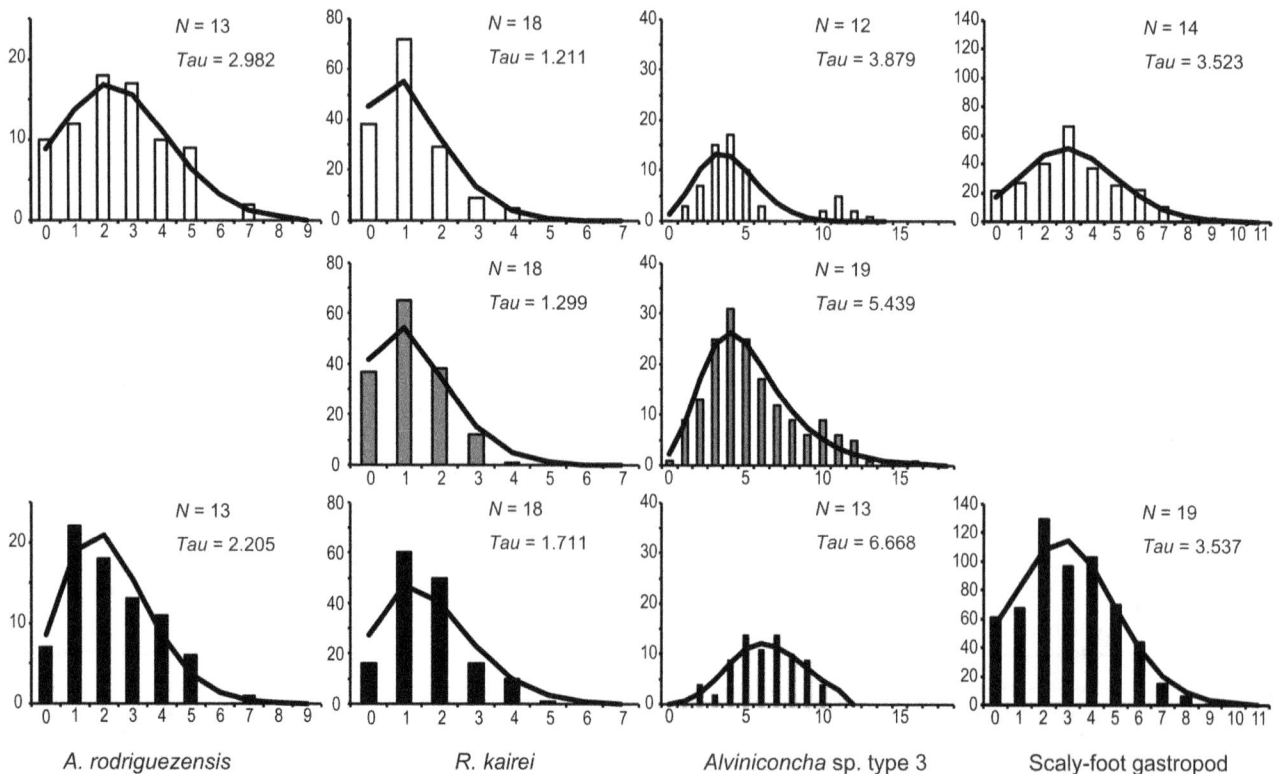

Figure 6. Mismatch distribution of the four representative species in the CIR. A, A. rodriguezensis; B, R. kairei; C, Alviniconcha sp. type 3; and D, scaly-foot gastropod. Black, population in the Kairei field; gray, population in the Edmond site; and white, population in the Solitaire field.

Table 4. Results of genetic structure analyses on
A. rodriguezensis.

	Kairei	Solitaire
Kairei		NS
Solitaire	0.00407	

Lower left: pairwise F_{ST}, upper right: Wright's Exact test. NS: not significant in 5% level.

Table 6. Results of genetic structure analyses on *Alviniconcha* sp. type 3.

	Kairei	Edmond	Solitaire
Kairei		NS	NS
Edmond	0.0044		NS
Solitaire	0.01394	0.06738**	

Lower left: pairwise F_{ST}, upper right: Wright's Exact test. NS: not significant in 5% level.

(Figure 3A). The Kruskal-Wallis test revealed no significant difference among the CW values of local *A. rodriguezensis* populations ($H = 3.848$, $P = 0.146$).

R. kairei was dominant in the vent fields near the Rodriguez Triple Junction (Edmond and Kairei fields), but sparse in the northern segments (Dodo and Solitaire fields; Figure 2). Therefore, fewer individuals were collected from the Dodo and Solitaire fields (6 and 40, respectively) than from the Edmond and Kairei fields (373 and 383, respectively). The Edmond population had the largest CL (11.61 ± 4.43 mm), while the Kairei population had the smallest CL (7.70 ± 1.39 mm; Figure 3B). A single prominent peak in size distribution was present in the Kairei population; however, other populations showed multiple peaks or gently sloping size distributions (CL: Dodo population, 10.10 ± 4.12 mm; Solitaire population, 9.41 ± 5.06 mm). The Kruskal-Wallis test revealed a significant difference among the CL values of local *R. kairei* populations ($H = 260.3$, $P < 0.001$).

Alviniconcha sp. type 3 was not observed in the Dodo field but was present in the remaining three fields. The Kairei population had a larger SW (37.89 ± 7.48 mm) than did the Solitaire population (29.82 ± 11.66 mm; Figure 3C). A single prominent peak in size distribution was present in the Kairei population; however, the Solitaire population showed multiple peaks. Mann-Whitney's *U*-test revealed no significant difference between the SW values of the Kairei and Solitaire populations ($U = 78.50$, $P = 0.0651$).

Molecular phylogenetic analyses

In total 167 COI sequences (78 haplotypic sequences were registered as AB817031–AB813148 in DDBJ, and also in NCBI and EMBL; 33 sequences for the scaly-foot gastropod were reported in [23]) were obtained. All of the individuals collected and analyzed comprised a single lineage in each species. The results of phylogenetic analyses revealed that the haplotype lineages of *A. rodriguezensis* and *Alviniconcha* sp. type 3 were specific to the CIR, with high bootstrapping values (99 and 100; Figure 4A, C). On the other hand, the lineage of *R. kairei* in the CIR was supported by a bootstrapping value of <50 (Figure 4B).

Genetic diversity and population structure analyses

We reconstructed and estimated haplotype networks (Figure 5) and genetic diversity indices (Tables 2 and 3), including mismatch distribution (Figure 6), for the four representative animals from the CIR vent fields. In addition, we calculated pairwise F_{ST} and Wright's exact tests for each population of the four species (Tables 4, 5, 6 and 7), and conducted AMOVA for *R. kairei* and *Alviniconcha* sp. type 3 (Tables 8 and 9).

A. rodriguezensis showed the simple network and low genetic diversity among the four investigated species. Genetic diversity was higher in the Solitaire population than in the Kairei population. Pairwise F_{ST} and Wright's exact tests indicated that the populations in the two CIR vent fields were not divided genetically (Table 4). The goodness-of-fit test revealed no significant difference between the observed and model frequencies (Solitaire population, $P = 0.86$; Kairei population, $P = 0.55$); further, sudden population expansion was not rejected. Migration analyses showed that the relative number of migrants per generation was about twice of those from the Solitaire population to the Kairei population than from the Kairei population to the Solitaire population (Figure 7A).

The haplotype network for *R. kairei* was estimated as a simple star-like network (Figure 5). Each of the estimated genetic diversity indices was the lowest among the four investigated species (Tables 2 and 3). Mismatch distribution of each population showed a single peak that did not differ significantly from the model frequency (goodness-of-fit test: Solitaire population, $P = 0.20$; Edmond population, $P = 0.41$; Kairei population, $P = 0.14$); further, sudden population expansion was not rejected. Pairwise F_{ST} and Wright's exact tests revealed no significant difference among the examined populations (Table 5), even between the northern population (Solitaire) and the near-triple junction populations (Edmond and Kairei) (Table 8). The estimated number of migrants per generation was lower for all migrations directed toward the Kairei population (284 from the Edmond population; 621 from the Solitaire population) than for any other migration direction (>4085; Figure 7).

Table 5. Results of genetic structure analyses on *R. kairei*.

	Kairei	Edmond	Solitaire
Kairei		NS	NS
Edmond	−0.00774		NS
Solitaire	0.00591	−0.00814	

Lower left: pairwise F_{ST}, upper right: Wright's Exact test. NS: not significant in 5% level.

Table 7. Results of genetic structure analyses on scaly-foot gastropod.

	Kairei	Solitaire
Kairei		NS
Solitaire	−0.00957	

Lower left: pairwise F_{ST}, upper right: Wright's Exact test. NS: not significant in 5% level.

Table 8. Results of AMOVA on *R. kairei*.

Source of Variation	d.f.	Sum of squares	Variance components	Percentage of variation
Among groups	1	0.648	0.00154	0.23
Among populations within groups	1	0.611	−0.00303	−0.46
Within populations	51	33.944	0.66558	100.22

In comparison with the other investigated species, *Alviniconcha* sp. type 3 showed a more highly diversified haplotype network, i.e., a larger number of haplotypes constituting a more complicated network (Figure 5). The genetic diversity indices were high in the Kairei population and low in the Solitaire population (Tables 2 and 3, and Figure 6). The goodness-of-fit test revealed no significant difference between the observed and model frequencies (Solitaire population, $P=0.25$; Edmond population, $P=0.78$; Kairei population, $P=0.65$); however, mismatch peaks were observed in mismatch numbers 10 and 11 for the Edmond and Solitaire populations, respectively. With the exception of the pairwise F_{ST} test between the Solitaire and Edmond populations (Table 6), pairwise F_{ST} and Wright's exact test revealed no significant difference among the examined populations. AMOVA between the northern population and the near-triple junction populations revealed no significant differences (Table 9). The relative number of migrants per generation was notably high for migrations from the Edmond population (1500 to the Solitaire population; 1710 to the Kairei population; Figure 7) than for any other migration direction (<500).

The genetic diversity indices for the scaly-foot gastropod were relatively high and comparable with those of *Alviniconcha* sp. type 3 (Tables 2 and 3). The goodness-of-fit test showed a single peak that did not differ significantly between the observed and model frequencies (Solitaire population, $P=0.49$; Kairei population, $P=0.76$; Figure 6); further, sudden expansion was not rejected. Pairwise F_{ST} and Wright's exact tests revealed no genetic difference between the two populations (Table 7). The estimated number of migrants per generation was higher from the Kairei population to the Solitaire population (1907) than from the Solitaire population to the Kairei population (365; Figure 7).

Discussion

The results of our present study indicate for the first time to examine population differentiation in the four representative species (*A. rodriguezensis*, *R. kairei*, *Alviniconcha* sp. type 3 and scaly-foot gastropod) in the four hydrothermal vent communities (Dodo, Solitaire, Edmond, and Kairei) along the CIR.

The vent crab *A. rodriguezensis* is distributed in all 4 hydrothermal vents at CIR and appears to be an important predator in the area, while not dominating the chemosynthetic communities in the

region [35]. The size distribution pattern of *A. rodriguezensis* in the Solitaire population showed two peaks, rather than a single peak as observed for the Dodo and Kairei populations (Figure 3). The result indicates that two cohorts may exist in the Solitaire population, while the other two populations only consisted of a single cohort. The different number of cohort of the Solitaire population may reflect the difference in the colonization process in the Solitaire field, for example, those resulted from different reproductive period or growth rate. Differences in environmental cues were previously reported to cause variations in reproductive periodicity among hydrothermal vent crabs belonging to the genus *Bythograea* [36]. Reproductive features are known sometimes to differ among populations of the same species [26]. The environmental characteristics of the four hydrothermal vents differ in many aspects, as indicated by the physical and chemical characteristics of high-temperature end member (-like) hydrothermal fluids (Table 10). The results of our present study may indicate differing reproductive periods among populations of a single species in CIR. To increase the reliability of our results, the number of individuals should be increased. Our present genetic analyses did not include individuals from the Dodo and Edmond populations; however, we detected no significant difference in genetic diversity between the Solitaire and Kairei populations (Tables 4–7). This finding is consistent with the results of previous investigation of *A. rodriguezensis* populations in the Kairei and Edmond fields, which showed that the genetic diversity differed by only 0.2% in mitochondrial cytochrome *b* [22]; taken together, these findings suggest the existence of gene flow among the populations in CIR vent fields. In the present study, the results of mismatch distribution and other genetic diversity indices indicate that the Solitaire population has a higher genetic diversity than does the Kairei population. The high genetic diversity and broad size range of the *A. rodriguezensis* population in the Solitaire field suggest that this field represents the potential source population for the four investigated populations (Figure 7). Larvae of *Bythograea thermydron* have developed eyes and show phototactic behavior [37]. Therefore larvae of *A. rodriguezensis*, which probably use the surface current as a means of dispersal, are able to disperse among the four hydrothermal vent fields, which are separated by several transform faults and non-transform offsets.

The vent shrimp, *R. kairei*, was the dominant species in all four hydrothermal vent fields along the CIR. The size distribution

Table 9. Results of AMOVA on *Alviniconcha* sp. type 3.

Source of Variation	d.f.	Sum of squares	Variance components	Percentage of variation
Among groups	1	4.806	0.10181	3.54
Among populations within groups	1	3.074	0.02074	0.72
Within populations	41	112.915	2.75403	95.74

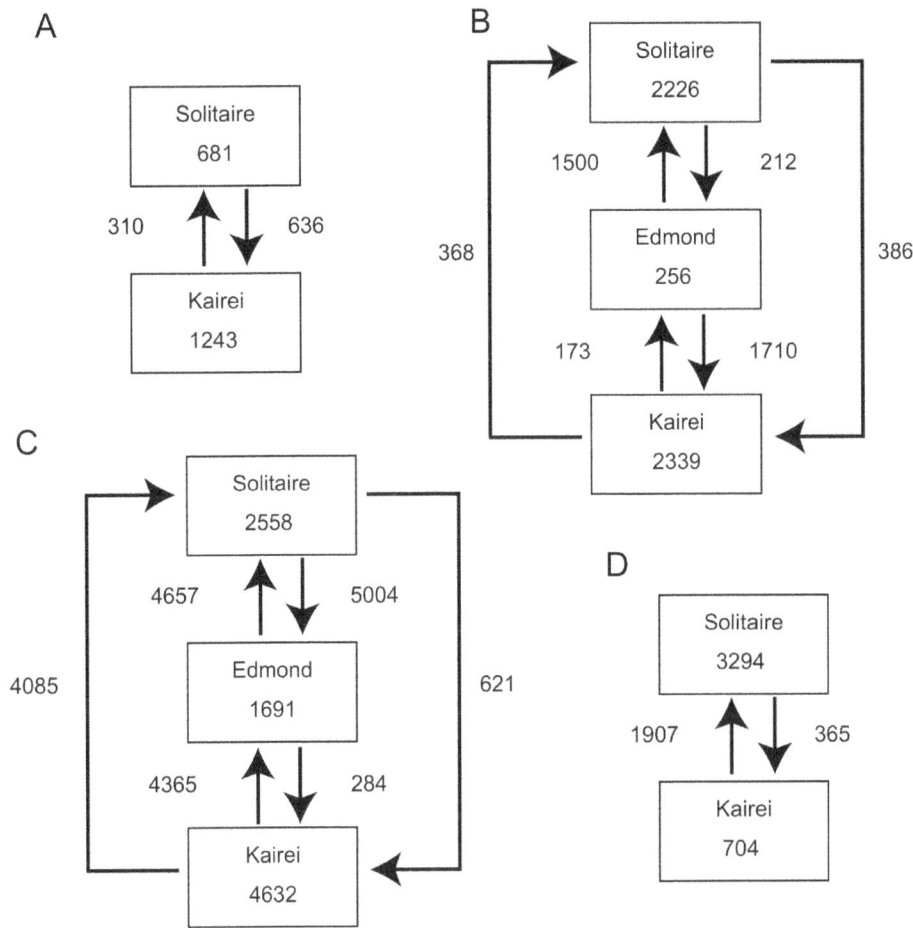

Figure 7. Schematic images of the results of MIGRATE analyses. A, *A. rodriguezensis*; B, *R. kairei*; C, *Alviniconcha* sp. type 3; and D, scaly-foot gastropod.

pattern differed among the populations, the Kairei population showed a single prominent peak, whereas the other populations showed broad and multiple peaks (Figure 3). Similar to *A. rodriguezensis*, differences in size distribution among the *R. kairei* populations may reflect differences in reproductive features. The Edmond population showed the broadest size distribution (Figure 3), and therefore probably had the largest number of individuals. The results of phylogenetic analyses revealed no clear endemicity of this shrimp in the Indian Ocean; in other words, *R. kairei* belongs to the same lineage as *R. exoculata* in the hydrothermal vent fields of the MAR. However, *R. kairei* differs morphologically from *R. exoculata* in the MAR [38]. The results of AMOVA analysis further revealed no significant genetic differences between the four examined *R. kairei* populations (Tables 5 and 8). This finding is consistent with that of a previous study, which showed no genetic differences between the Kairei and Edmond populations of swarming shrimps [22]. Vent shrimps in the family Alvinocarididae are believed to have long larval dispersal periods, during which they consume photosynthetic-based nutrition in the euphotic zone [39,40]. Further, some larvae of vent shrimps have potential to survive for long periods at a distance of up to 100 km from the vent sites [41,42]. On the other hand, the results of migration analyses estimated a biased number of migrants per generation among three populations, i.e., prominent exportation from the Edmond population to the Solitaire and Kairei populations (Figure 7). Among the four vent

field populations, *R. kairei* was less abundant in the two northern vent populations (Dodo and Solitaire) than in the two vent populations near the Rodriguez Triple Junction (Kairei and Edmond) (Figures 2 and 3). Taken together, our results suggest that the Edmond field is the largest, and provides a larval supply for the other vent field populations along the CIR.

The large hairy gastropod, *Alviniconcha* sp. type 3, was present in the Solitaire, Edmond and Kairei fields. The size distribution pattern differed between the two fields; a single peak was present in the Kairei population and the broad-ranged population in the Solitaire field (Figure 3), implying that the population conditions were influenced by the different environments of the two vent fields (Table 10). In accordance with a previous study of Kairei field samples [25], the results of phylogenetic analyses revealed the specificity of *Alviniconcha* sp. type 3 to the CIR region. Further, in accordance with a previous study of the Kairei and Edmond populations based on the mitochondrial 16SrRNA sequence [22], we revealed no significant genetic differences among the three examined populations (Tables 6 and 9). The results of migration analyses indicated the presence of biased migration, i.e., migration was higher from the Edmond field than from the other two vent fields (Figure 7). The egg of *Alviniconcha* sp. type 3 shows neutral buoyancy under atmospheric observation (Watanabe et al. unpublished data). Planktotrophic larval development is inferred from the protoconch morphology of *Alviniconcha* sp. type 3 [43], and enables potentially wide geographical dispersal and distribu-

Table 10. Characteristics of hydrothermal fluid in the four vent fields in CIR.

Site	Temperature (°C)	pH	H_2 (mmol/L)	CH_4 (mmol/kg)	CO_2 (mmol/kg)	Fe (mmol/kg)	Chlorinity
Dodo [23]	356	3.2	>2	~0.02	~4		brine-rich
Solitaire [23]	296	4.8	0.46	~0.05	~8		vapor-rich
Edmond [22]	382	~3	0.2	0.4		~14	brine-rich
Kairei [22]	365	~3	8.5	0.2		~5	brine-rich

tion. Gene flow of *Alviniconcha* gastropods was discussed previously [29], and was reported to differ between lineages, i.e., gene flow between the Manus Basin and the North Fiji Basin was attributed to *Alviniconcha* sp. type 1 and *Alviniconcha* sp. type 2; in contrast, gene flow of *Alviniconcha* sp. type 4 was limited to the Lau Basin. It appears that dispersal of this species is not constrained by horizontal distance, but by dispersal barriers that probably differ among species. Taken together, our results suggest that the Edmond population represents the potential source population for *Alviniconcha* sp. type 3 in the four vent fields, and that transform faults do not act as dispersal barriers for this species.

The scaly-foot gastropod was present in the Solitaire field and at a single chimney in the Kairei field. The average shell width in the Solitaire population was slightly smaller than that in the Kairei population [23]. The results of statistical analyses revealed no genetic differentiation between the two populations (Table 7), inferring potential connectivity between the two vent fields; however, the egg of the scaly-foot gastropod shows negative buoyancy under atmospheric pressure (Watanabe et al. unpublished data). Topological depressions were previously shown to act as dispersal barriers for polychaete species with negatively buoyant eggs [44]. The results of migration analyses indicated that the relative number of migrants per generation was higher from the Kairei field to the Solitaire field than from the Solitaire field to the Kairei field (Figure 7). Taken together, our results suggest that the Kairei population represents the potential source population for the two populations in the CIR. Recently, an additional scaly-foot assemblage was discovered in the Southwestern Indian Ridge [45]. Further studies, including investigations of this newly discovered population, will provide an insight into the dispersal ability and evolution of this unusual gastropod.

Potential dispersal barriers of vent-endemic animal species have mainly been reported for populations in the fast-spreading EPR, along which topological depression and interception of microplates appear to act as barriers for gene flow [9]. On the other hand, very little genetic differentiation has been reported for population in the slow-spreading MAR. The results of our present investigation of CIR vent-endemic animal species revealed almost no genetic differentiation among the four vent populations. Taken together, these findings indicate that the existence of relatively high connectivity among populations in slow- and intermediate- (<60 mm/year) spreading ridge systems. High variability in physical and chemical features of hydrothermal activities and

fluids has been demonstrated for slow- and ultraslow-spreading ridge systems, such as the MAR and Mid-Cayman Rise [46]; this variability provides diversification of habitats and niches for animals associated with vent environments. Habitat diversity strongly affects the adaptation and connectivity of animals with various dispersal strategies. On the other hand, in arc-backarc systems of the western Pacific, in which vent-endemic animal communities are believed to form a single biogeographical province [18,20], connectivity among vent populations is more complex than that occurring in mid-oceanic ridge systems; this complexity arises because arc-backarc systems are generally surrounded by island arcs, which prohibit dispersal of thermally intolerant larvae [47,48].

In summary, the results of our present study provide an overview of the dispersal and population conditions of representative vent-associated fauna in four deep-sea hydrothermal vent fields along the CIR. Few previous investigations of the Indian Ocean Ridge system have been conducted; thus, our present findings not only provide new insights into larval dispersal and population establishment, but also help to clarify previously reported geochemical and biogeographical diversification [22,23]. Additional biogeographical and ecophysiological studies of vent animals in the known and newly discovered vent fields of the CIR, and in other geographical and geological settings such as the East Scotia Ridge and Mid-Cayman Rise [46,49], will further contribute to our understanding of larval dispersal and population establishment in global deep-sea chemosynthetic faunal communities.

Acknowledgments

We thank the crews of R/V Yokosuka and HOV Shinkai6500 for sampling and on-board analyses, Professor Shigeaki Kojima and Ms. Seiko Honma for help with laboratory analysis, and the late Professor Kensaku Tamaki for organizing this collaborative study. We grateful to the editor and two anonymous reviewers for their helpful advice on this study.

Author Contributions

Conceived and designed the experiments: HW KT SN S. Nakagawa KN MK DEPM. Performed the experiments: GB HW TO SN S. Nemoto TY. Analyzed the data: GB HW TO TY. Contributed reagents/materials/analysis tools: GB HW KT SN S. Nakagawa KN MK DEPM. Wrote the paper: GB HW.

References

1. Lonsdale P (1977) Clustering of suspension-feeding macrobenthos near abyssal hydrothermal vents at oceanic spreading centers. Deep Sea Res 24: 857–863.
2. Tunnicliffe V, Juniper SK, Sibuet M (2003) Reducing environments of the deep-sea floor. In: Tyler PA, editor. Ecosystems of the world (ecosystems of the deep Ocean). Amsterdam: Elsevier. pp. 81–110.
3. Van Dover CL (1990) Biogeography of hydrothermal vent communities along seafloor spreading centers. Trends Ecol Evol 5: 242–246.

4. Hessler RR, Lonsdale PF (1991) Biogeography of Mariana Trough hydrothermal vent communities. Deep Sea Res Part I Oceanogr Res Pap 38: 185–199.
5. Van Dover CL (2000) The ecology of deep-sea hydrothermal vents. Princeton: Princeton University Press. 352 p.
6. Ramirez-Llodra E, Shank TM, German CR (2007) Biodiversity and biogeography of hydrothermal vent species. Oceanography 20: 30–41.
7. Lutz RA, Shank TM, Fornari DJ, Haymon RM, Lilley MD, et al. (1994) Rapid growth at deep-sea vents. Nature 371: 663–664.

8. Tyler PA, Young CM (1999) Reproduction and dispersal at vents and cold seeps. J Mar Biol Assoc UK 79: 193–208.

9. Vrijenhoek RC (2010) Genetic diversity and connectivity of deep-sea hydrothermal vent metapopulations. Mol Ecol 19: 4391–4411.

10. Lutz RA, Jablonski D, Turner RD (1984) Larval development and dispersal at deep-sea hydrothermal vents. Science 226: 1451–1454.

11. Billiard S, Lenormand T (2005) Evolution of migration under kin selection and local adaptation. Evolution 59: 13–23.

12. Kim SL, Mullineaux LS (1998) Distribution and near-bottom transport of larvae and other plankton at hydrothermal vents. Deep Sea Res Part II Top Stud Oceanogr 45: 423–440.

13. Pradillon F, Shillito B, Young CM, Gaill F (2001) Deep-sea ecology: Developmental arrest in vent worm embryos. Nature 413: 698–699.

14. Thomson RE, Mihaly SF, Rabinovich AB, McDuff RE, Veirs SR, et al. (2003) Constrained circulation at Endeavour ridge facilitates colonization by vent larvae. Nature 424: 545–549.

15. Mullineaux L, Mills SW, Sweetman AK, Beaudreau AH, Metaxas A, et al. (2005) Vertical, lateral and temporal structure in larval distributions at hydrothermal vents. Mar Ecol Prog Ser 293: 1–16.

16. Kim SL, Mullineaux LS, Helfrich KR (1994) Larval dispersal via entrainment into hydrothermal vent plumes. J Geophys Res: Oceans 99: 12655–12665.

17. Tunnicliffe V, Fowler MRC (1996) Influence of sea-floor spreading on the global hydrothermal vent fauna. Nature 379: 531–533.

18. Bachraty C, Legendre P, Desbruyéres D (2009) Biogeographic relationships among deep-sea hydrothermal vent faunas at global scale. Deep Sea Res Part I Oceanogr Res Pap 56: 1371–1378.

19. Olivieri I, Michalakis Y, Gouyon PH (1995) Metapopulation genetics and the evolution of dispersal. Am Nat 146: 202–228.

20. Van Dover CL, German CR, Speer KG, Parson LM, Vrijenhoek RC (2002) Evolution and biogeography of deep-sea vent and seep invertebrates. Science 295: 1253–1257.

21. Hashimoto J, Ohta S, Gamo T, Chiba H, Yamaguchi T, et al. (2001) First hydrothermal vent communities from the Indian Ocean discovered. Zoolog Sci 18: 717–721.

22. Van Dover CL, Humphris SE, Fornari D, Cavanaugh CM, Collier R, et al. (2001) Biogeography and ecological setting of Indian Ocean hydrothermal vents. Science 294: 818–823.

23. Nakamura K, Watanabe H, Miyazaki J, Takai K, Kawagucci S, et al. (2012) Discovery of new hydrothermal activity and chemosynthetic fauna on the Central Indian Ridge at 18–20°S. PLoS One 7: e32965.

24. Teixeira S, Serrão EA, Arnaud-Haond S (2012) Panmixia in a fragmented and unstable environment: The hydrothermal shrimp Rimicaris exoculata disperses extensively along the Mid-Atlantic Ridge. PLoS One 7: e38521.

25. German CR, Ramirez-Llodra E, Baker MC, Tyler PA, ChEss Scientific Steering Committee (2011) Deep-water chemosynthetic ecosystem research during the Census of Marine Life decade and beyond: a proposed deep-ocean road map. PLoS One 6: e23259.

26. Begon M, Harper JL, Townsend CR (1996) Ecology. Oxford: Blackwell, pp. 1068.

27. Suzuki Y, Kopp RE, Kogure T, Suga A, Takai K, et al. (2006) Sclerite formation in the hydrothermal-vent gastropod' control of iron sulfide biomineralization by the animal. Earth Planet Sci Lett 242: 39–50.

28. Folmer O, Black MB, Hoeh W, Lutz R, Vrijenhoek R (1994) DNA primers for amplification of mitochondrial cytochrome c oxidase subunit I from diverse metazoan invertebrates. Mol Mar Biol Biotechnol 3: 294–299.

29. Kojima S, Segawa R, Fijiwara Y, Fujikura K, Ohta S, et al. (2001) Phylogeny of hydrothermal-vent-endemic gastropods Alviniconcha spp. from the western Pacific revealed by mitochondrial DNA sequences. Biol Bull 200: 298–304.

30. Hasegawa T, Yamaguchi T, Kojima S, Ohta S (1996) Phylogenetic analysis among three species of the genus Tetraclita (Cirripedia: Balanomorpha) by

31. Tamura K, Peterson D, Peterson N, Stecher G, Nei M, et al. (2011) MEGA5: Molecular evolutionary genetics analysis using maximum likelihood, evolutionary distance, and maximum parsimony methods. Mol Biol Evol 28: 2731–2739.

32. Clement M, Posada D, Crandall KA (2000) TCS: a computer program to estimate gene genealogies. Mol Ecol 9: 1657–1660.

33. Excoffier L, Laval G, Schneider S (2005) Arlequin ver. 3.0: An integrated software package for population genetics data analysis. Evol Bioinform Online 1: 47–50.

34. Beerli P (2009) How to use migrate or why are Markov chain Monte Carlo programs difficult to use? In: Bertorelle G, Bruford MW, Haffe HC, Rizzoli A, Vernesi C, editors. Population genetics for animal conservation. Cambridge: Cambridge University Press. pp. 42–79.

35. Van Dover CL (2002) Trophic relationships among invertebrates at the Kairei hydrpthermal vent field (Central Indian Ridge). Mar Biol 141: 761–772.

36. Hilario A, Vilar S, Cunha MR, Tyler PA (2009) Reproductive aspects of two bythograeid crab species from hydrothermal vents in the Pacific-Antarctic Ridge. Mar Ecol Prog Ser 378: 153–160.

37. Epifanio CE, Perovich G, Dittel AI, Cary SC (1999) Development and behavior of megalopa larvae and juveniles of the hydrothermal vent crab Bythograea thermydron. Mar Ecol Prog Ser 185: 147–154.

38. Watabe H, Hashimoto J (2002) A new species of the genus Rimicaris (Alvinocarididae: Caridea: Decapoda) from the active hydrothermal vent field, Kairei Field on the Central Indian Ridge, the Indian Ocean. Zoolog Sci 19: 1167–1174.

39. Gebruk AV, Pimenov NV, Savvichev AS (1993) Feeding specialization of bresiliid shrimps in the TAG site hydrothermal community. Mar Ecol Prog Ser 98: 237–246.

40. Koyama S, Nagahama T, Ootsu N, Takayama T, Horii M, et al. (2005) Survival of deep-sea shrimp (Alvinocaris sp.) during decompression and larval hatching at atmospheric pressure. Mar Biotechnol 7: 272–278.

41. Pond DW, Gebruk A, Southward EC, Southward AJ, Fallick AE, et al. (2000) Unusual fatty acid composition of storage lipids in the bresilioid shrimp Rimicaris exoculata couples the photic zone with MAR hydrothermal vent sites. Mar Ecol Prog Ser 198: 171–179.

42. Herring PJ, Dixon DR (1998) Extensive deep-sea dispersal of postlarval shrimp from a hydrothermal vent. Deep Sea Res Part I Oceanogr Res Pap 45: 2105–2118.

43. Warén A, Bouchet P (1993) New records, species, genera, and a new family of gastropods from hydrothermal vents and hydrocarbon seeps. Zool Scr 22: 1–90.

44. Hurtado LA, Lutz RA, Vrijenhoek RC (2004) Distinct patterns of genetic differentiation among annelids of eastern Pacific hydrothermal vents. Mol Ecol 13: 2603–2615.

45. Tao C, Lin J, Guo S, Chen YJ, Wu G, et al. (2012) First active hydrothermal vents on an ultraslow-spreading center: Southwest Indian Ridge. Geology 40: 47–50.

46. German CR, Bowen A, Coleman ML, Honig DL, Huber JA, et al. (2010) Diverse styles of submarine venting on the ultraslow spreading Mid-Cayman Rise. Proc Natl Acad Sci U S A 107: 14020–14025.

47. Watanabe H, Tsuchida S, Fujikura K, Yamamoto H, Inagaki F, et al. (2005) Population history associated with hydrothermal vent activity inferred from genetic structure of neoverrucid barnacles around Japan. Mar Ecol Prog Ser 288: 233–240.

48. Watanabe H, Fujikura K, Kojima S, Miyazaki J, Fujiwara Y (2010) Japan: Vents and seeps in close proximity. In: Kiel, S, editor. The vent and seep biota: Aspects from microbes to ecosystems. Heidelberg: Springer. pp. 379–402.

49. Rogers AD, Tyler PA, Connelly DP, Copley JT, James R, et al. (2012) The discovery of new deep-sea hydrothermal vent communities in the Southern Ocean and implications for biogeography. PLoS Biol 10: e1001234.

nucleotide sequences of a mitochondrial gene. Benthos Res 51: 33–39 (in Japanese with English abstract).

Permissions

All chapters in this book were first published in PLOS ONE, by The Public Library of Science; hereby published with permission under the Creative Commons Attribution License or equivalent. Every chapter published in this book has been scrutinized by our experts. Their significance has been extensively debated. The topics covered herein carry significant findings which will fuel the growth of the discipline. They may even be implemented as practical applications or may be referred to as a beginning point for another development.

The contributors of this book come from diverse backgrounds, making this book a truly international effort. This book will bring forth new frontiers with its revolutionizing research information and detailed analysis of the nascent developments around the world.

We would like to thank all the contributing authors for lending their expertise to make the book truly unique. They have played a crucial role in the development of this book. Without their invaluable contributions this book wouldn't have been possible. They have made vital efforts to compile up to date information on the varied aspects of this subject to make this book a valuable addition to the collection of many professionals and students.

This book was conceptualized with the vision of imparting up-to-date information and advanced data in this field. To ensure the same, a matchless editorial board was set up. Every individual on the board went through rigorous rounds of assessment to prove their worth. After which they invested a large part of their time researching and compiling the most relevant data for our readers.

The editorial board has been involved in producing this book since its inception. They have spent rigorous hours researching and exploring the diverse topics which have resulted in the successful publishing of this book. They have passed on their knowledge of decades through this book. To expedite this challenging task, the publisher supported the team at every step. A small team of assistant editors was also appointed to further simplify the editing procedure and attain best results for the readers.

Apart from the editorial board, the designing team has also invested a significant amount of their time in understanding the subject and creating the most relevant covers. They scrutinized every image to scout for the most suitable representation of the subject and create an appropriate cover for the book.

The publishing team has been an ardent support to the editorial, designing and production team. Their endless efforts to recruit the best for this project, has resulted in the accomplishment of this book. They are a veteran in the field of academics and their pool of knowledge is as vast as their experience in printing. Their expertise and guidance has proved useful at every step. Their uncompromising quality standards have made this book an exceptional effort. Their encouragement from time to time has been an inspiration for everyone.

The publisher and the editorial board hope that this book will prove to be a valuable piece of knowledge for researchers, students, practitioners and scholars across the globe.

List of Contributors

Laurie L. Baker, Ian D. Jonsen, Sara J. Iverson and Damian C. Lidgard
Department of Biology, Dalhousie University, Halifax, Nova Scotia, Canada

Joanna E. Mills Flemming
Department of Mathematics and Statistics, Dalhousie University, Halifax, Nova Scotia, Canada

William D. Bowen
Population Ecology Division, Bedford Institute of Oceanography, Dartmouth, Nova Scotia, Canada

Dale M. Webber
VEMCO Ltd., Halifax, Nova Scotia, Canada

Xiao-hong Chen and Long Cheng
Wuhan Center of China Geological Survey, Wuhan, Hubei, P. R. China

Ryosuke Motani
Department of Earth and Planetary Sciences, University of California Davis, Davis, California, United States of America

Da-yong Jiang
Laboratory of Orogenic Belt and Crustal Evolution, Ministry of Education, Department of Geology and Geological Museum, Peking University, Beijing, P.R. China

Olivier Rieppel
Center of Integrative Research, The Field Museum, Chicago, Illinois, United States of America

Ewan Hunter, Derek Eaton, Christie Stewart, Andrew Lawler and Michael T. Smith
Centre for Environment, Fisheries and Aquaculture Science, Lowestoft Laboratory, Lowestoft, Suffolk, United Kingdom

Daphne Cuvelier, Pierre-Marie Sarradin and Jozée Sarrazin
Institut Carnot Ifremer EDROME, Centre de Bretagne, REM/EEP, Laboratoire Environnement Profond, Plouzané, France

Pierre Legendre
Département de Sciences Biologiques, Université de Montré al, succursale Centre-ville, Montréal, Québec, Canada

Agathe Laes
Institut Carnot Ifremer EDROME, Centre de Bretagne, REM/RDT, Laboratoire Détection, Capteurs et Mesures, Plouzané, France

Helge Niemann
Department of Environmental Sciences, University of Basel, Basel, Switzerland
Max Planck Institute for Marine Microbiology, Bremen, Germany

Katrin Knittel and Gaute Larvik
Max Planck Institute for Marine Microbiology, Bremen, Germany

Antje Boetius
Max Planck Institute for Marine Microbiology, Bremen, Germany
Alfred Wegener Institute for Marine and Polar Research, Bremerhaven, Germany

Enoma Omoregie
Max Planck Institute for Marine Microbiology, Bremen, Germany
Centro de Astrobiología (CSIC/INTA), Instituto Nacional de Técnica Aeroespacial Torrejón de Ardoz, Madrid, Spain

Ulrike Schacht
Sonderforschungsbereich 574, University of Kiel, Kiel, Germany

Peter Linke, Warner Brückmann and Klaus Wallmann
Sonderforschungsbereich 574, University of Kiel, Kiel, Germany
Helmholtz Centre for Ocean Research Kiel, GEOMAR, Kiel, Germany

Gregor Rehder
Sonderforschungsbereich 574, University of Kiel, Kiel, Germany
Leibniz Institute for Baltic Sea Research Warnemu¨nde (IOW), Rostock, Germany

Enrique MacPherson
Centro de Estudios Avanzados de Blanes (CEAB-CSIC), Blanes, Spain

David Hilton and Kevin Brown
Scripps Institution of Oceanography, University of California, San Diego, United States of America

Andrew David Thaler, Sophie Plouviez, Alixandra Jacobson, Emily A. Boyle, Thomas F. Schultz and Cindy Lee Van Dover
Marine Laboratory, Nicholas School of the Environment, Duke University, Beaufort, North Carolina, United States of America

William Saleu
Nautilus Minerals, Port Moresby, NCD, Papua New Guinea

Freddie Alei
Environmental Science and Geography Division, School of Natural and Physical Sciences, University of Papua New Guinea, Port Moresby, Papua New Guinea

Jens Carlsson
School of Biology & Environmental Science, University College Dublin, Dublin, Ireland

Ellen C. Garland
National Marine Mammal Laboratory, Alaska Fisheries Science Center, National Marine Fisheries Service, National Oceanic and Atmospheric Administration, Seattle, Washington, United States of America

Jason Gedamke
Ocean Acoustics Program, Office of Science and Technology, National Marine Fisheries Service, National Oceanic and Atmospheric Administration, Silver Spring, Maryland, United States of America

Melinda L. Rekdahl and Michael J. Noad
Cetacean Ecology and Acoustics Lab, School of Veterinary Science, University of Queensland, Gatton, Queensland, Australia

Claire Garrigue
Opération Cétacés, Noumea, New Caledonia

Nick Gales
Australian Marine Mammal Centre, Australian Antarctic Division, Kingston, Tasmania, Australia

Aurélie Tasiemski, Céline Boidin-Wichlacz and Virginie Cuvillier-Hot
Université de Lille1-CNRS UMR8198, Laboratoire GEPV, Ecoimmunology of Marine Annelids (EMA), Villeneuve d'Ascq, France

Sascha Jung, Oliver Hecht and Joachim Grötzinger
Institute of Biochemistry, Christian- Albrechts-Universität, Kiel, Germany

Didier Jollivet
Université Pierre et Marie Curie-CNRS UMR7144, Laboratoire AD2M, Adaptation et Biologie des Invertébrés en Conditions Extrêmes (ABICE), Station Biologique, Roscoff, France

Florence Pradillon
IFREMER, Centre de Brest, REM/EEP/LEP, Plouzané, France

Costantino Vetriani
Department of Biochemistry and Microbiology and Institute of Marine and Coastal Sciences, Rutgers University, New Brunswick, New Jersey, United States of America

Frank D. Sönnichsen
Otto Diels Institute for Organic Chemistry, Christian-Albrechts- Universität, Kiel, Germany

Christoph Gelhaus and Matthias Leippe
Institute of Zoology, Zoophysiology, Christian-Albrechts-Universität, Kiel, Germany

Chien-Wen Hung and Andreas Tholey
Division of Systematic Proteome Research, Institute for Experimental Medicine, Christian-Albrechts-Universität, Kiel, Germany

Françoise Gaill
Université Pierre et Marie Curie-Muséum National d'Histoires Naturelles CNRS BOREA IRD, Paris, France

Bautisse Postaire, J. Henrich Bruggemann and Hélène Magalon
Laboratoire d'ECOlogie MARine, Université de la Réunion, FRE3560 INEE-CNRS, Saint Denis, La Réunion, France Labex CORAIL, Perpignan, France

Baptiste Faure
Laboratoire d'ECOlogie MARine, Université de la Réunion, FRE3560 INEE-CNRS, Saint Denis, La Réunion, France, Biotope, Service Recherche et Développement, Mèze, France

Dominique Le Guen, Franc¸ois H. Lallier and Didier Jollivet
Université Pierre et Marie Curie-Paris 6, Laboratoire Adaptation et Diversité en Milieu Marin, Station Biologique de Roscoff, Roscoff, France
CNRS UMR 7144, Station Biologique de Roscoff, Roscoff, France

Sophie Plouviez
Université Pierre et Marie Curie-Paris 6, Laboratoire Adaptation et Diversité en Milieu Marin, Station Biologique de Roscoff, Roscoff, France

CNRS UMR 7144, Station Biologique de Roscoff, Roscoff, France
Division of Marine Science and Conservation, Nicholas School of the Environment, Duke University, Beaufort, North Carolina, United States of America

Baptiste Faure
Université Pierre et Marie Curie-Paris 6, Laboratoire Adaptation et Diversité en Milieu Marin, Station Biologique de Roscoff, Roscoff, France
CNRS UMR 7144, Station Biologique de Roscoff, Roscoff, France
Université Montpellier 2, Montpellier, France
CNRS UMR 5554, Institut des Sciences de l'Evolution, Station Méditerranéenne de l'Environnement Littoral, Sète, France

Nicolas Bierne
Université Montpellier 2, Montpellier, France
CNRS UMR 5554, Institut des Sciences de l'Evolution, Station Méditerranéenne de l'Environnement Littoral, Sète, France

Christopher K. Pham, Eduardo Isidro, Telmo Morato and José Nuno Gomes-Pereira
Center of the Institute of Marine Research (IMAR) and Department of Oceanography and Fisheries, University of the Azores, Horta, Portugal
Laboratory of Robotics and Systems in Engineering and Science (LARSyS), Lisbon, Portugal

Joan B. Company
Institut de Ciències del Mar (ICM-CSIC), Barcelona, Spain,

Eva Ramirez-Llodra
Institut de Ciències del Mar (ICM-CSIC), Barcelona, Spain
Norwegian Institute for Water Research (NIVA), Marine Biology section, Oslo, Norway

Claudia H. S. Alt and Paul A. Tyler
Ocean and Earth Science, University of Southampton, National Oceanography Centre, Southampton, United Kingdom

Teresa Amaro
Norwegian Institute for Water Research, Bergen, Norway

Melanie Bergmann
Alfred-Wegener-Institut, Helmholtz-Zentrum für Polar- und Meeresforschung, Bremerhaven, Germany

Miquel Canals, Galderic Lastras and Xavier Tubau
GRC Geociències Marines, Departament d9Estratigrafia, Paleontologia i Geociències Marines, Facultat de Geologia, Universitat de Barcelona, Campus de Pedralbes, Barcelona, Spain

Jaime Davies and Kerry L. Howell
Marine Biology & Ecology Research Centre, Marine Institute, Plymouth University, Plymouth, United Kingdom

Gerard Duineveld
Netherlands Institute for Sea Research (NIOZ), Texel, The Netherlands

François Galgani
Institut Français de Recherche pour l9Exploitation de la Mer (IFREMER), Bastia, France

Veerle A. I. Huvenne and Daniel O. B. Jones
National Oceanography Centre, University of Southampton Waterfront Campus, Southampton, United Kingdom

Autun Purser
OceanLab, Jacobs University Bremen, Bremen, Germany

Heather Stewart
British Geological Survey, Murchison House, Edinburgh, United Kingdom

Inês Tojeira
Portuguese Task Group for the Extension of the Continental Shelf (EMEPC), Paço de Arcos, Portugal

David Van Rooij
Renard Centre of Marine Geology (RCMG), Department of Geology and Soil Science, Ghent University, Gent, Belgium

William D. K. Reid, Christopher J. Sweeting and Nicholas V. C. Polunin
School of Marine Science and Technology, Newcastle University, Newcastle upon Tyne, United Kingdom

Ben D. Wigham
Dove Marine Laboratory, School of Marine Science and Technology, Newcastle University, Cullercoats, United Kingdom

Katrin Zwirglmaier and Katrin Linse
British Antarctic Survey, Natural Environment Research Council, High Cross, Madingley Road, Cambridge, United Kingdom

Jeffrey A. Hawkes
Ocean and Earth Science, University of Southampton, National Oceanography Centre Southampton, Southampton, United Kingdom

Rona A. R. McGill
Natural Environment Research Council Life Sciences Mass Spectrometry Facility, Scottish Universities Environmental Research Centre, East Kilbride, United Kingdom

Rika E. Anderson and John A. Baross
School of Oceanography and Astrobiology Program, University of Washington, Seattle, Washington, United States of America

Mitchell L. Sogin
Josephine Bay Paul Center, Marine Biological Laboratory, Woods Hole, Massachusetts, United States of America

Michael E. Burns
Biological Sciences, University of Alberta, Edmonton, Alberta, Canada

Matthew J. Vavrek
Pipestone Creek Dinosaur Initiative, Clairmont, Alberta, Canada

Candice St. Germain
Department of Biology, University of Victoria, Victoria, British Columbia, Canada

Verena Tunnicliffe
Department of Biology, University of Victoria, Victoria, British Columbia, Canada
School of Earth & Ocean Sciences, University of Victoria, Victoria, British Columbia, Canada

Ana Hilário
Departamento de Biologia and Centro de Estudos do Ambiente e do Mar, Universidade de Aveiro, Campus de Santiago, Aveiro, Portugal

Naraporn Somboonna
Department of Microbiology, Faculty of Science, Chulalongkorn University, Bangkok, Thailand

Alisa Wilantho, Duangjai Sangsrakru, Sithichoke Tangphatsornruang and Sissades Tongsima
Genome Institute, National Center for Genetic Engineering and Biotechnology, Pathumthani, Thailand

Kruawun Jankaew
Department of Geology, Faculty of Science, Chulalongkorn University, Bangkok, Thailand

Anunchai Assawamakin
Department of Pharmacology, Faculty of Pharmacy, Mahidol University, Bangkok, Thailand

Anastasia A. Lunina and Alexandr L. Vereshchaka
Laboratory of structure and dynamics of plankton communities, P.P. Shirshov Institute of Oceanology of Russian Academy of Sciences, Moscow, Russia

David A. Bowden and Ashley A. Rowden
Coasts and Oceans Centre, National Institute of Water and Atmospheric Research, Wellington, New Zealand

Andrew R. Thurber
College of Earth, Ocean, and Atmospheric Sciences, Oregon State University, Corvallis, Oregon, United States of America

Amy R. Baco
Department of Earth, Ocean and Atmospheric Sciences, Florida State University, Tallahassee, Florida, United States of America

Lisa A. Levin
Center for Marine Biodiversity and Conservation, Integrative Oceanography Division, Scripps Institution of Oceanography, La Jolla, California, United States of America

Craig R. Smith
Department of Oceanography, School of Ocean and Earth Science and Technology, University of Hawaii at Manoa, Honolulu, Hawaii, United States of America

Adrien Vigneron, Perrine Cruaud, Patricia Pignet, Laurent Toffin and Anne Godfroy
Ifremer, Laboratoire de Microbiologie des Environnements Extrêmes, UMR6197, ZI de la pointe du Diable, Plouzané, France
Université de Bretagne Occidentale, Laboratoire de Microbiologie des Environnements Extrêmes, UMR6197, ZI de la pointe du Diable, Plouzané , France
CNRS, Laboratoire de Microbiologie des Environnements Extrêmes, UMR6197, ZI de la pointe du Diable, Plouzané, France

Nolwenn Callac
Ifremer, Laboratoire de Microbiologie des Environnements Extrêmes, UMR6197, ZI de la pointe du Diable, Plouzané, France
Université de Bretagne Occidentale, Laboratoire de Microbiologie des Environnements Extrêmes, UMR6197, ZI de la pointe du Diable, Plouzané , France
CNRS, Laboratoire de Microbiologie des Environnements Extrêmes, UMR6197, ZI de la pointe du Diable, Plouzané, France
Université de Brest, Domaines Océaniques IUEM, UMR6538, Place Nicolas Copernic, Plouzané, France

Jean-Claude Caprais
Ifremer, Laboratoire Etude des Environnements Profonds, UMR6197, ZI de la pointe du Diable, Plouzané, France

Maria-Cristina Ciobanu
Ifremer, Géosciences Marines, Laboratoire des Environnements Sédimentaires, ZI de la pointe du Diable, Plouzané, France

Erwan G. Roussel, Barry A. Cragg and John R. Parkes
School of Earth and Ocean Sciences, Cardiff University, Cardiff, United Kingdom

Joy D. Van Nostrand and Zhili He
Institute for Environmental Genomics and Department of Microbiology and Plant Biology, University of Oklahoma, Norman, Oklahoma, United States of America

Jizhong Zhou
Institute for Environmental Genomics and Department of Microbiology and Plant Biology, University of Oklahoma, Norman, Oklahoma, United States of America
State Key Joint Laboratory of Environment Simulation and Pollution Control, School of Environment, Tsinghua University, Beijing, China
Earth Science Division, Lawrence Berkeley National Laboratory, Berkeley, California, United States of America

Ann Vanreusel
Biology Department, Research Group Marine Biology, Ghent University, Ghent, Belgium,

Jelle Van Campenhout, Sofie Derycke and Tom Moens
Biology Department, Research Group Marine Biology, Ghent University, Ghent, Belgium,
Center for Molecular Phylogenetics and Evolution (CeMoFe), Ghent University, Ghent, Belgium

Girish Beedessee and Daniel E. P. Marie
Mauritius Oceanography Institute, Quatre-Bornes, Mauritius

Hiromi Watanabe
Institute of Biogeosciences, Japan Agency for Marine-Earth Science and Technology, Yokosuka, Kanagawa, Japan

Tomomi Ogura
Institute of Biogeosciences, Japan Agency for Marine-Earth Science and Technology, Yokosuka, Kanagawa, Japan
Graduate School of Marine Science and Technoloy, Tokyo University of Marine Science and Technology, Minato, Tokyo, Japan

Suguru Nemoto
Enoshima Aquarium, Fujisawa, Kanagawa, Japan

Takuya Yahagi
Atmosphere and Ocean Research Institute, the University of Tokyo, Kashiwa, Chiba, Japan

Satoshi Nakagawa
Faculty of Fisheries Sciences, Hokkaido University, Hakodate, Hokkaido, Japan

Kentaro Nakamura
Precambrian Ecosystem Laboratory, Japan Agency for Marine-Earth Science and Technology, Yokosuka, Kanagawa, Japan

Ken Takai
Institute of Biogeosciences, Japan Agency for Marine-Earth Science and Technology, Yokosuka, Kanagawa, Japan
Precambrian Ecosystem Laboratory, Japan Agency for Marine-Earth Science and Technology, Yokosuka, Kanagawa, Japan

Meera Koonjul
Albion Fisheries Research Centre, Ministry of Fisheries, Petite Riviére, Mauritius

Index